Design, Manufacturing and Properties of Refractory Materials

Design, Manufacturing and Properties of Refractory Materials

Editors

Ilona Jastrzębska
Jacek Szczerba

Basel • Beijing • Wuhan • Barcelona • Belgrade • Novi Sad • Cluj • Manchester

Editors
Ilona Jastrzębska
AGH University of Krakow
Krakow
Poland

Jacek Szczerba
AGH University of Krakow
Krakow
Poland

Editorial Office
MDPI
St. Alban-Anlage 66
4052 Basel, Switzerland

This is a reprint of articles from the Special Issue published online in the open access journal *Materials* (ISSN 1996-1944) (available at: https://www.mdpi.com/journal/materials/special_issues/refractory_materials).

For citation purposes, cite each article independently as indicated on the article page online and as indicated below:

Lastname, A.A.; Lastname, B.B. Article Title. *Journal Name* **Year**, *Volume Number*, Page Range.

ISBN 978-3-7258-1089-5 (Hbk)
ISBN 978-3-7258-1090-1 (PDF)
doi.org/10.3390/books978-3-7258-1090-1

Cover image courtesy of Ilona Jastrzębska

Contents

About the Editors

Ilona Jastrzębska

Since 2019, Dr. Ilona Jastrzębska has worked as an Assistant Professor at the AGH University of Krakow, Poland. She defended her Ph.D. thesis entitled *"Hercynite solid solutions—synthesis, properties and applications"* in 2017 with distinction. Within this thesis, she construed the arc plasma furnace for synthesizing high-temperature compounds. She has a postgraduate degree in *Business English* and *Management of EU funds*. Her research interests focus on refractory materials. Currently, she leads the project "LIDERXII" (NCBR, PL), in which, together with her research group, she develops a new generation of refractory materials without Cr_2O_3 for the copper industry. She is the author or co-author of over 50 research articles, including 27 publications in JCR journals, 2 patents, and 2 European patent applications, and she has presented her research at over 60 international conferences. She has also won several prizes, including "The Gustav Eirich Award" for the best international Ph.D. dissertation in the field of refractory materials (ECREF, 2018), "Scientist of the Future" award (2022) in the category "Research of the Future", and "Female Scientist Who Will Change the World" award (2023) for innovative research work, as well as being awarded a grant from the European Ceramic Society (JECS Trust, 2017). She serves as an academic teacher, educating students in materials engineering, refractory materials science and the physics and chemistry of colloids; a supervisor of Doctoral, Bachelor's and Master's students; and a Masterclass lecturer for Ph.D. students at the Max-von Laue Institute of Advanced Ceramic Material Properties. She has participated in numerous scientific and technological internships, including at Seul National University, South Korea; Rostock University, Germany; and in the companies Vesuvius and ArcelorMittal; and she has implemented Canadian technical standards in the framework of CETA for the five largest Polish companies from the steel sector.

Jacek Szczerba

Prof. Jacek Szczerba is a Professor of technical engineering sciences. His tenure at the AGH University of Krakow from 2002 to 2023, where he served as the Head of the Refractories Group in the Department of Ceramics and Refractories at the Faculty of Materials Science and Ceramics, is a testament to his unwavering commitment. Before this, he held significant roles, such as the Head of the Laboratory of Refractories (1988–1996) and the Head of the Process Engineering Department (up to 2002) at the Institute of Mineral Building Materials Poland, where he started work in 1973. His primary research focus is refractory materials, especially sintered and fused refractory semi-products, the recycling and alternative resources of refractory raw materials, novel hydraulic binder systems, microstructure and thermomechanical properties of refractories. Prof. Szczerba has written over 320 publications and over 30 patents. He is a Fellow of the European Ceramic Society; a Science Fellow of the Max-von-Laue Institute of Advanced Ceramic Material Properties Studies, University of Koblenz Landau; a Member of the Gustav Eirich International Award Committee, ECREF Germany; a Member of the MSCA ITN-ETN ATHOR project; Vice President of the Association of Refractory Materials Manufacturers in Poland; Vice President of the SITPMB, Krakow, Poland (1998–2014); and Vice President of the SITPMB Main Board (2007–2014). He has received awards many times for his scientific and educational activities, including a Medal from the National Education Commission; a First Prize award from the Refractory Materials journal and the Association of Engineers and Technicians of the Steel Industry in Poland, in 1994; a Medal and Badges from the Polish Engineering Associations; and a Polish Ceramic Society Award.

Preface

Refractory materials are crucial for our industrial and civil development.

The refractory industry is, however, sensitive to global changes and economic instabilities. The COVID-19 pandemic and war instability have acted as the litmus paper for the refractories market and development in recent years. On the other hand, growing legislation requirements and approaches toward using more environmentally neutral technology in the production of refractory materials are stimulating continuous runs to improve these materials' properties and extend their lifetimes, ensuring more stable processes and longer campaigns of high-temperature vessels.

Sharing knowledge on refractories is the only way we can upgrade the quality of refractories. Therefore, the authors contributing to this reprint by sharing their state-of-the-art research are greatly acknowledged.

This Special Issue reprint covers 18 research papers, including 1 review. The topics focus on the most crucial aspects of refractory development: advanced methods for developing refractories, including Artificial Intelligence and image analysis, and other cardinal aspects, such as corrosion and fracture mechanics, binders, and new designs for environmentally friendly refractories.

Ilona Jastrzębska and Jacek Szczerba

Editors

Editorial

Design, Manufacturing and Properties of Refractory Materials

Ilona Jastrzębska * and Jacek Szczerba

Faculty of Materials Science and Ceramics, AGH University of Krakow, al. A. Mickiewicza 30, 30-059 Krakow, Poland
* Correspondence: ijastrz@agh.edu.pl

Citation: Jastrzębska, I.; Szczerba, J. Design, Manufacturing and Properties of Refractory Materials. *Materials* **2024**, *17*, 1673. https://doi.org/10.3390/ma17071673

Received: 20 March 2024
Accepted: 25 March 2024
Published: 5 April 2024

With pleasure, we present this Special Issue of *Materials*, titled "Design, Manufacturing and Properties of Refractory Materials". Refractory materials are both economically and socially strategic materials as they enable the production of other crucial products, including steel, non-ferrous metals, cement clinker, lime, glass and many others. They are designed to operate at high temperatures, temperature gradients and under severe chemical and mechanical loadings. Therefore, extensive research of their properties is critical in the development of new types of longer-lifetime and pro-ecologic materials as well as in enhancing the properties of existing ones. At present, artificial intelligence contributes progressively to making refractory production and application more sustainable, which is one of the scopes of this issue. This Special Issue consists of 18 papers, including 1 review article on the application of machine learning in the refractories field, which is the first review work on this modern topic so far. This issue covers the following research aspects in refractories: artificial intelligence and computer-aided methods, simulation of refractories properties, new refractory materials, castables and binders, corrosion, and, finally, it is closed with environmental aspects in refractories. We hope this issue will bring new insights into the development of refractories, allowing them to serve us better and longer.

1. Artificial Intelligence and Computer-Aided Methods

Sado et al. [1] presented a review article on the application of machine learning (ML) technology in the investigation of MgO–C refractories from the perspective of their key properties prediction. Their work also presented different ML algorithms currently used in materials engineering. The authors critically assessed works on the up-to-date application of ML in the refractories field, which is currently developing rapidly. Challenges in the development of reliable models were discussed.

Zelik et al. [2] used Bayesian modeling for the prediction of MgO–C refractory unit wear rate in the slag-spout zone of a 350 t-oxygen converter that completed 2063 heats, applying data of lining thickness collected via laser measurements. A 420 µm usage of refractory thickness per one heat of convertor was determined using this type of modeling.

By using neural networks—an artificial intelligence tool—Jančar et al. [3] ordered parameters which positively and negatively impact the refractory lining in a steel ladle. He showed that the time of empty ladle most adversely affects the a ladle' campaign due to thermal shocks of refractory lining. Other parameters which were observed to shorten the lifetime of the lining are electrical energy consumption, number of vacuum heats and temperature of steel tapped to the ladle. On the other hand, Ar consumption and time of full ladle were found to positively influence the lifetime of refractory lining.

Stec et al. [4] used the innovative technique of corrosion testing combining hot metal penetration with subsequent X-ray computed tomography to investigate the microporous carbon refractory lining dedicated to blast furnace hearth walls. The results showed the volume percentage of permeable pores and metal inclusions of 2.5% and 3.2%, respectively, with their respective maximum sizes of 410 µm and 843 µm. This shows that during 1 h-penetration at 1500°C crude iron dissolved the carbon skeleton of the refractory lining.

Jastrzębska and Piwowarczyk [5] showed how computer image analysis can be utilized to extract more valuable data from SEM images via traditional stereology-based methods vs. automated method developed in the work for spinel refractory material. The automated image analysis allows for fast retrieval of the data, like quantity of phases at SEM image, which in some cases—like partly amorphous materials—is complementary to the XRD method.

2. Simulation of Refractory Properties

Grigoriev et al. [6] presented computer modeling of 96% SiO_2 refractories (for coke ovens, heat exchangers of blast furnaces, and glass-making furnaces) using the method of homogeneously deformable discrete elements to simulate the uniaxial compressive and tensile failure in a wide range of quasi-static and dynamic loading rates. Also, crack patterns were presented for real and model materials. They determined the real and virtual contribution of cracks within grains, matrix and along the interface. This is important as brittleness at failure correlates with an increase in the relative crack length along the grain/matrix interface. For composites, which are refractory materials, the strength of the interface between the matrix and reinforcing constituents (large grains, fibers) strongly correlates with the material's strength, but an extremely strong interface leads to brittle failure.

3. New Refractory Materials

Two articles presented by Zienert et al. [7] and Storti et al. [8] show a novel $Nb-Al_2O_3$ composite aggregate which can be used for the production of refractories with enhanced shock resistance. Cold crushing strength measured at 1300 °C revealed their semi-plastic behavior. Also, these novel composites can be candidates for new high-temperature heating elements. This area seems to have plenty of room for further research.

4. Castables and Binders

αAl_2O_3 hollow spheres with sizes 5–100 μm were introduced by Stonys et al. [9] to bauxite refractory castables to check their influence on castable thermal shock resistance. In their study, additions of 2.5% and 5% of hollow Al_2O_3 were found to improve the thermal shock resistance of castables, which withstood 30 thermal cycles, whereas a 10% addition caused their deterioration due to clustering, poor bonding of spheres with the surrounding matrix and formed irregular-shape voids. Thermal shock resistance was tested by heating them at 950 °C and water quenching, and this was assessed by determining their loss in ultrasonic pulse velocity.

Tang et al. [10] investigated the influence of $CaTiO_3$, $MgTiO_3$ and nano-TiO_2 additions on the physical and mechanical properties of alumina-magnesia castables in order to increase castable densification, thus, inhibiting the effects of CA_6 and spinel expansion. They found that $MgTiO_3$ is unstable in all tested materials, while 2 wt.% $CaTiO_3$ results in the best castable properties. The latter one, after being subjected to 1450 °C, received linear shrinkage of -0.5%, apparent porosity of 12%, bulk density of 3.2 g/cm^3, CMOR of 45 MPa, and Young's modulus of 194 GPa, which is significantly better in comparison to the castables with nano-TiO_2 and $MgTiO_3$ additions.

Zemanek and Nevřinová [11] tested the performance of castables with the use of silica sol and compared them to castables with calcium aluminate cement. They found that sol-bonded castables perform successfully compared to traditional calcium aluminate-bonded castables in terms of permanent linear change, compactness and corrosion resistance.

Wang et al. [12] tested the influence of 5 μm-andalusite addition on thermal shock resistance of Al_2O_3–SiC–C castables dedicated for the runners of blast furnaces. They found noticeably improved oxidation resistance and residual CMOR after 10 thermal cycles compared to castables without the addition. Such improvement was related to an enhanced mullitization of andalusite, which started at 1250 °C and finished at 1450 °C. Thus, the formation of a liquid phase (due to andalusite decomposition) was enabled earlier, which densified the microstructure via the crystallization of secondary mullite. The maximum 5% addition of micro-andalusite resulted in over 40% higher residual CMOR after 10 thermal cycles at 950 °C and water quenching, and 2% lower open porosity (20%) when compared to the reference. With respect to the previous work of the authors, the oxidation resistance of castables with a 5% addition of micro-andalusite was over 30% better than a 19% addition of andalusite 1–3 mm.

Inorganic chemical binders for refractory materials were characterized by Hopp et al. [13] from the perspective of structure–property relations. The presented work excellently shows the comparative instrumental methods that can be applied for the complete characterization of binder action mechanisms, including XRD, FTIR, Raman spectroscopy, NMR and DMA (dynamic mechanical analysis). In particular, phosphate and silicate binders were deeply analyzed.

5. Corrosion

Ovćaćíková et al. [14] investigated alkali corrosion, which commonly accompanies aluminosilicate refractories working in boilers for wood biomass combustion. The ashes generated during such biomass combustion are rich in SiO_2, CaO, Al_2O_3, Fe_2O_3 and alkalis like Na_2O and K_2O. Moreover, they tend to slag, thus, forming a sticky layer of ash particles on the surface of the refractory lining. After a corrosion test at 1200 °C for 2 h, SiO_2–Al_2O_3 refractories with a SiC addition were found to better protect the refractory lining from the alkali corrosion process compared to traditional aluminosilicate refractories.

Darban et al. [15] revealed the corrosion of an alumina-spinel refractory against secondary steel slag via a coating corrosion test at 1350 °C and 1450 °C. They summarized the study by showing the passive corrosion mechanism of the refractory, with the formation of CA, CA_2 and CA_6 layers around the Al_2O_3 core and C_2AS in the matrix.

Ludwig et al. [16] investigated the corrosion of Cr_2O_3-bearing refractory raw materials against PbO-rich copper slags via hot-stage microscopy. They found that among the four tested raw materials—namely, two magnesia-chromite co-clinkers, Pakistani chrome ore and fused spinel—the latter was characterized by the lowest chemical resistance expressed by the lowest melting and highest final shrinkage (8%). Forsterite, Mg_2SiO_4, was the main corrosion product for all the tested materials.

6. Environmental Aspects in Refractories

Interesting from the environmental point of view are the results shown by Xu et al. [17], who investigated alumina castables bonded with calcium aluminate cement, in which Cr was introduced to composition in the form of pre-synthesized solid solution $(Al_{1-x}Cr_x)_2O_3$. Chromium (III) — previously bounded in this solid solution—is therefore much less prone to oxidize and form Cr(VI) compounds, and this effect is especially enhanced when more CA_6 is present in the system. This smart approach can greatly help to reduce the negative impact of the utilization of Cr_2O_3 in refractory materials.

Nguyen and Sokolář [18] presented the concept of utilizing fly ash as a source of Al_2O_3 to create in situ spinel in new forsterite refractories. The formation of spinel helps to decrease the global, large expansion of forsterite-based materials. They revealed that refractoriness under load was at least 1500 °C (0.5% deformation), HMOR above 20 MPa, and corrosion resistance (expressed as penetration depth) was up to 2 mm for Fe, 3 mm for the clinker, 1 mm for Al and 0.01 mm for Cu. Thus, these materials show the perspective to be adapted in the Cu industry. Therefore, a further investigation of coarse-grain material is needed.

Acknowledgments: The author (I.J.) acknowledge the support from LIDER/14/0086/L-12/20/NCBR/2021, The National Centre for Research and Development, and from the Research project "Excellence initiative—research university" for the AGH University Grant ID9053.

Conflicts of Interest: The authors declare no conflicts of interest.

References

1. Sado, S.; Jastrzębska, I.; Zelik, W.; Szczerba, J. Current State of Application of Machine Learning for Investigation of MgO-C Refractories: A Review. *Materials* **2023**, *16*, 7396. [CrossRef] [PubMed]
2. Zelik, W.; Sado, S.; Lech, R. The Wear Rate Forecast of MgO-C Materials Type MC95/10 in the Slag Spout Zone of an Oxygen Converter in Terms of the Bayesian Estimation. *Materials* **2022**, *15*, 3065. [CrossRef] [PubMed]
3. Jančar, D.; Machů, M.; Velička, M.; Tvardek, P.; Kocián, L.; Vlček, J. Use of Neural Networks for Lifetime Analysis of Teeming Ladles. *Materials* **2022**, *15*, 8234. [CrossRef] [PubMed]
4. Stec, J.; Tarasiuk, J.; Wroński, S.; Kubica, P.; Tomala, J.; Filipek, R. Investigation of Molten Metal Infiltration into Micropore Carbon Refractory Materials Using X-ray Computed Tomography. *Materials* **2021**, *14*, 3148. [CrossRef] [PubMed]
5. Jastrzębska, I.; Piwowarczyk, A. Traditional vs. Automated Computer Image Analysis—A Comparative Assessment of Use for Analysis of Digital SEM Images of High-Temperature Ceramic Material. *Materials* **2023**, *16*, 812. [CrossRef] [PubMed]
6. Grigoriev, A.S.; Zabolotskiy, A.V.; Shilko, E.V.; Dmitriev, A.I.; Andreev, K. Analysis of the Quasi-Static and Dynamic Fracture of the Silica Refractory Using the Mesoscale Discrete Element Modelling. *Materials* **2021**, *14*, 7376. [CrossRef] [PubMed]
7. Zienert, T.; Endler, D.; Hubálková, J.; Günay, G.; Weidner, A.; Biermann, H.; Kraft, B.; Wagner, S.; Aneziris, C.G. Synthesis of Niobium-Alumina Composite Aggregates and Their Application in Coarse-Grained Refractory Ceramic-Metal Castables. *Materials* **2021**, *14*, 6453. [CrossRef] [PubMed]
8. Storti, E.; Neumann, M.; Zienert, T.; Hubálková, J.; Aneziris, C.G. Metal-Ceramic Beads Based on Niobium and Alumina Produced by Alginate Gelation. *Materials* **2021**, *14*, 5483. [CrossRef] [PubMed]
9. Stonys, R.; Malaiškienė, J.; Škamat, J.; Antonovič, V. Effect of Hollow Corundum Microspheres Additive on Physical and Mechanical Properties and Thermal Shock Resistance Behavior of Bauxite Based Refractory Castable. *Materials* **2021**, *14*, 4736. [CrossRef] [PubMed]
10. Tang, H.; Zhou, Y.; Yuan, W. Investigating the Action Mechanism of Titanium in Alumina–Magnesia Castables by Adding Different Ti-Bearing Compounds. *Materials* **2022**, *15*, 793. [CrossRef] [PubMed]
11. Zemánek, D.; Nevřivová, L. Development and Testing of Castables with Low Content of Calcium Oxide. *Materials* **2022**, *15*, 5918. [CrossRef] [PubMed]
12. Wang, X.; Wang, S.; Mu, Y.; Zhao, R.; Wang, Q.; Parr, C.; Ye, G. Enhancing the Oxidation Resistance of Al2O3-SiC-C Castables via Introducing Micronized Andalusite. *Materials* **2021**, *14*, 4775. [CrossRef] [PubMed]
13. Hopp, V.; Alavi, A.M.; Hahn, D.; Quirmbach, P. Structure–Property Functions of Inorganic Chemical Binders for Refractories. *Materials* **2021**, *14*, 4636. [CrossRef] [PubMed]
14. Ovčačíková, H.; Velička, M.; Vlček, J.; Topinková, M.; Klárová, M.; Burda, J. Corrosive Effect of Wood Ash Produced by Biomass Combustion on Refractory Materials in a Binary Al–Si System. *Materials* **2022**, *15*, 5796. [CrossRef] [PubMed]
15. Darban, S.; Reynaert, C.; Ludwig, M.; Prorok, R.; Jastrzębska, I.; Szczerba, J. Corrosion of Alumina-Spinel Refractory by Secondary Metallurgical Slag Using Coating Corrosion Test. *Materials* **2022**, *15*, 3425. [CrossRef] [PubMed]
16. Ludwig, M.; Śnieżek, E.; Jastrzębska, I.; Prorok, R.; Li, Y.; Liao, N.; Nath, M.; Vlček, J.; Szczerba, J. Corrosion Resistance of MgO and Cr2O3-Based Refractory Raw Materials to PbO-Rich Cu Slag Determined by Hot-Stage Microscopy and Pellet Corrosion Test. *Materials* **2022**, *15*, 725. [CrossRef] [PubMed]
17. Xu, T.; Xu, Y.; Liao, N.; Li, Y.; Nath, M. High-Temperature Chemical Stability of Cr(III) Oxide Refractories in the Presence of Calcium Aluminate Cement. *Materials* **2021**, *14*, 6590. [CrossRef] [PubMed]
18. Nguyen, M.; Sokolář, R. Corrosion Resistance of Novel Fly Ash-Based Forsterite-Spinel Refractory Ceramics. *Materials* **2022**, *15*, 1363. [CrossRef] [PubMed]

Review

Current State of Application of Machine Learning for Investigation of MgO-C Refractories: A Review

Sebastian Sado [1,2,*], Ilona Jastrzębska [2,*], Wiesław Zelik [1] and Jacek Szczerba [2,*]

[1] Zaklady Magnezytowe "ROPCZYCE" S.A., Research and Development Centre of Ceramic Materials, ul. Przemysłowa 1, 39-100 Ropczyce, Poland; wieslaw.zelik@ropczyce.com.pl

[2] Faculty of Materials Science and Ceramics, AGH University of Kraków, al. A. Mickiewicza 30, 30-059 Kraków, Poland

* Correspondence: sebastian.sado@ropczyce.com.pl (S.S.); ijastrz@agh.edu.pl (I.J.); jszczerb@agh.edu.pl (J.S.)

Abstract: Nowadays, digitalization and automation in both industrial and research activities are driving forces of innovations. In recent years, machine learning (ML) techniques have been widely applied in these areas. A paramount direction in the application of ML models is the prediction of the material service time in heating devices. The results of ML algorithms are easy to interpret and can significantly shorten the time required for research and decision-making, substituting the trial-and-error approach and allowing for more sustainable processes. This work presents the state of the art in the application of machine learning for the investigation of MgO-C refractories, which are materials mainly consumed by the steel industry. Firstly, ML algorithms are presented, with an emphasis on the most commonly used ones in refractories engineering. Then, we reveal the application of ML in laboratory and industrial-scale investigations of MgO-C refractories. The first group reveals the implementation of ML techniques in the prediction of the most critical properties of MgO-C, including oxidation resistance, optimization of the C content, corrosion resistance, and thermomechanical properties. For the second group, ML was shown to be mostly utilized for the prediction of the service time of refractories. The work is summarized by indicating the opportunities and limitations of ML in the refractories engineering field. Above all, reliable models require an appropriate amount of high-quality data, which is the greatest current challenge and a call to the industry for data sharing, which will be reimbursed over the longer lifetimes of devices.

Keywords: machine learning; MgO-C; refractory; steel; artificial neural networks; ANN

Citation: Sado, S.; Jastrzębska, I.; Zelik, W.; Szczerba, J. Current State of Application of Machine Learning for Investigation of MgO-C Refractories: A Review. *Materials* **2023**, *16*, 7396. https://doi.org/10.3390/ma16237396

Academic Editor: Murali Mohan Cheepu

Received: 20 October 2023
Revised: 19 November 2023
Accepted: 25 November 2023
Published: 28 November 2023

1. Introduction

Magnesia–carbon refractories (MgO-C) belong to the most significant type of refractories for steel and iron industry devices. They thermally protect basic oxygen furnaces (BOFs), steel ladles, and electric arc furnaces (EAFs), and they are used in the production of special products, like purging shapes and taphole sleeves [1]. The wear of MgO-C refractories is caused mainly by the attack of metallurgical slag, the oxidation of C by oxygen or other oxidizing compounds, and the interaction with CO/CO_2, which occur at temperatures of 1600–1750 °C [1]. Also, the thermomechanical impact, associated with thermal shocks and the turbulent flow of hot metal, significantly influences the MgO-C refractory service time [1]. The typical service time of the MgO-C lining in BOFs varies from around 2000 up to 10,000 heats or more, depending on the maintenance conditions [2–4]. In steel ladles, the differentiation in the service time is substantial, as the ladle campaign might be finished after 123–183 heats [5], 70–85 heats [6], or even after only 8–20 heats [7]. The refractory lining service time in EAFs is also highly differentiated. The typical EAF lining service time is 500–1000 heats. But, even in one steel plant, it can vary from approximately 500 heats up to 1200 heats [8]. Refractory wear generates high maintenance costs. The high costs derive not only from the purchase and replacement of new refractory products but also from work stoppage and urgent repairs. The recent significant progress in R&D activities has influenced the extended service times of heating devices (e.g., the service times of refractories in steel ladles increased from 128 to 157 heats via the optimization of the lining materials and service conditions) [9]. Another example is increasing the VOD (vacuum oxygen decarburization) ladle service time from 8.5 heats (2017) to 20 heats (2021) via the addition of $ZrSiO_4$, which enhanced the mechanical and thermomechanical properties of MgO-C bricks [7].

Although MgO-C refractories have been used since 1950 in steel and refining plants [10], their service times are still being extended owing to progressing research efforts. One of the main research directions in MgO-C improvement is the application of various metallic and non-metallic additions (e.g., Al, Mg, Si, SiC, Al-Mg, Fe) [11–15], as well as the development of new ones (e.g., c-ZrN nanopowder, Ti_3AlC_2, Ti_3SiC_2, Cr_3C_2C, spinel micro-powder, YAG nanopowder, and other oxide composites) [16–25]. Their addition to MgO-C improves decarburization resistance as well as hot properties, like hot strength and thermal shock resistance. The corrosion resistance of MgO-C has also been broadly investigated, with multiple techniques used, like the induction furnace test [26–28], finger test [29,30], sessile-drop technique [31–33], cup test [34–37], single hot thermocouple technique [38], and observations of in situ changes using a high-stage microscope [39]. Recently, new calcium–magnesium–aluminate raw materials have been developed, which promote the formation of a protective layer at the hot face of MgO-C bricks during operation [40–42]. With the increased demand for widely understood decarbonization and sustainable development, much effort is also being put into the recycling of MgO-C materials [43–46]. Ludwig et al. [43] obtained satisfactory results for 20% and even 30% additions of recycled MgO-C aggregate in a composition of new MgO-C brick. These areas have great potential for further improvements and research, as around 28 million tons of spent refractories are generated annually [45], while the total worldwide production of refractories is 35–40 million (70% for the steel industry).

However, the commonality in all these experimental studies is that both the experiment and result interpretations are always conducted in a traditional way, with a relatively low quantity of data taken for analysis. For MgO-C refractories, researchers are focused on very detailed investigations of the mechanisms responsible for the particular hot-temperature behavior of MgO-C bricks. Simultaneously, researchers have to face the high quantity of various data types [47]. For this reason, the refractory industry should take the opportunity of the available data and introduce techniques that allow for their better usage.

Recently, more companies have become interested in collecting data and finding relationships with refractories' wear rates to optimize the process and make it more efficient as well as environmentally friendly. The implementation of Industry 4.0 [48] has created a new reality for many companies. This strategy has blurred the difference between the work of people and machines [48]. One of the objectives of Industry 4.0 is to achieve a higher level of digitalization and automation of and improvement in decision-making processes with automated data exchange [49]. An invaluable tool is machine learning (ML), the outstanding performance of which has so far been reported in numerous materials science studies [49]. ML algorithms refer to computational systems that can be trained to perform specific tasks, with no need to implement any explicit programming. Moreover, the quality of the algorithms' performance improves with extended experience [49]. Interest in using ML techniques is constantly growing. The Web of Science database, when searched with the keyword "machine learning" 10 years ago (2013), showed 1908 papers, while, in 2022, 2021, and 2020, it showed 34,934, 30,053, and 22,335 papers, respectively. This 56% increase in the number of publications over the last 2 years and the 18-fold increase over the last 10 years permit the prediction of a forthcoming boom in ML utilization. Furthermore, data in the global datasphere are predicted to reach 175 zettabytes by 2025 (33 zettabytes in 2018) [50].

According to Pilania's work [51], ML algorithms can be applied in various applications in materials science. One of its applications is the development of efficient and surrogate models which map and find relationships between a material's composition, structure, morphology, and processing to select properties or performance criteria. Moreover, the author indicates numerous other fields of machine learning applications, like material characterization and design, designing of experiments, prioritizing of experiments, property prediction, and molecular and atomistic simulations [51].

Taking into account the relatively newly applied ML techniques in the refractories field and their vast innovation potential, this work aims to evaluate the most important published works on the application of various machine learning techniques in the investigation of MgO-C refractories. This review is divided into three main parts. Firstly, in Section 2 we present the most commonly applied ML algorithms and their utilization in different fields. Then, the current state of ML application in laboratory-scale examinations (Section 3) and in industrial-scale tests (Section 4) is revealed. The laboratory-scale works focus on the most critical properties of MgO-C refractories, including oxidation mechanisms, optimization of carbon content, corrosion resistance, and thermomechanical properties. The industrial-scale tests are aimed at the prediction of the service time of refractories in industrial heating devices. Finally, we summarize by indicating the benefits and limitations of ML utilization in research practice (Section 5). This work aims to be a reference for researchers who are searching for new capabilities and techniques to improve R&D activities in the technology of MgO-C refractories.

2. Machine Learning Algorithms—An Overview

Machine learning is a subset of artificial intelligence. Algorithms are dedicated to building computational tools that make decisions without explicit coding. One of the main aims of the application of ML algorithms is taking the historical data and training the algorithms to further use these data in the prediction of specific features. The main advantage of ML algorithms is their powerful performance and speed of data processing compared to hand-coding. ML algorithms have proven their performance and utility in a variety of fields, such as speech recognition, text mining, medicine, data analysis, aeronautics, data analysis, stock market analysis, and many others [52,53]. This wide range of applications is possible due to a variety of existing algorithms which are presented in Figure 1 based on [51–55] (the graph does not exhaust all currently used algorithms).

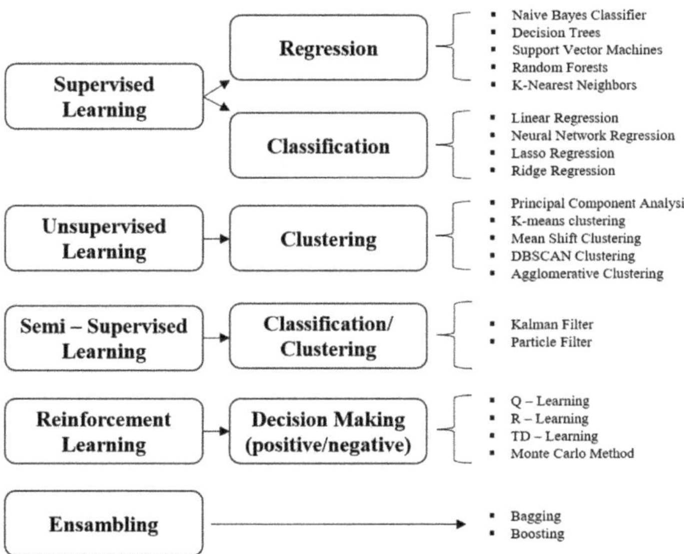

Figure 1. Overview of commonly used ML algorithms.

Sarker [52] has divided ML algorithms into four groups, including supervised learning (algorithms: classification and regression), unsupervised learning (clustering), semi-supervised learning (classification and clustering, based on labelled and unlabeled data) and reinforcement learning (positive and negative).

Jain and Kumar [53] described three groups of ML algorithms, indicating specific ones in each of the groups. The first group is supervised learning with classification (algorithms: naïve Bayes, decision trees, support vector machines, random forest, K-nearest neighbors) and regression (linear regression, neural network regression, lasso regression, ridge regression). The second group is unsupervised learning (principal component analysis, K-means, mean shift clustering, DBSCAN clustering, agglomerative clustering). The third group is reinforcement learning (Q-learning, R- learning, TD-learning and the Monte Carlo method).

According to Sarker [52], the algorithms with the highest popularity index worldwide are assigned to the group of reinforcement learning, but their popularity decreased in 2020. Pugliese et al. [54] showed that, in 2021, the popularity index of reinforced learning was still the highest, while supervised and unsupervised learning popularity indexes were on a similar level. As Pugliese et al. explains in [54], the popularity of reinforcement algorithms (algorithms based on interactions with the environment) reflects their use to solve real-world problems in a variety of fields, such as game theory, control theory, operation analysis, information theory, simulation-based optimization, manufacturing, supply chain logistics, swarm intelligence, aircraft control, robot motion control, laparoscopic surgery, traffic forecasting service, smart cities development, etc. [55].

3. Machine Learning in Investigation of MgO-C Materials

3.1. Application of ML in Laboratory-Scale Examinations

3.1.1. Oxidation Mechanism of MgO-C Refractories

Oxidation of carbon in MgO-C refractories, especially below 1400 °C, is one of the main problems in the application of these materials [56]. The decarburized part of the refractory is loose and porous; thus, the slag and hot metal can easily penetrate the matrix. The mechanism of oxidation is widely examined using traditional techniques [57–62]. The decarburization resistance of MgO-C refractories is mostly affected by graphite content and the overall compactness of the brick.

Artificial neural networks (ANN), which represent one of the supervised ML techniques (belonging to Supervised Learning—Regression, Figure 1), were used by Nemati et al. [63] to predict the oxidation behavior of MgO-C materials. The authors tested several MgO-C materials with different carbon contents. ANN was used to predict the activation energy of oxidation, effective diffusion coefficient, and diffusion activation energy of oxidation. An input variable was the weight loss of MgO-C materials at different temperatures depending on graphite content from 4.5% to 17%. The model was developed using a standard feed-forward backpropagation network with one hidden layer. Oxidation of carbon in MgO-C refractories was found to be driven mainly by diffusion. The ANN model was also utilized to predict the effective diffusion coefficient at different temperatures. The obtained results were of good quality of fit, expressed by the determination coefficient R^2 in the range 0.986–0.999. Finally, the three-layer back propagation ANN model was used to predict the oxidation kinetics of MgO-C specimens based on their weight loss at different temperatures. The authors developed reliable models with excellent fit between experimental and calculated data. The oxygen diffusion was reported as responsible for carbon oxidation in MgO-C refractories.

The oxygen diffusion mechanism in MgO-C composites was also investigated by A. Nemati et al. [64] with the use of the ANN approach. The authors used a standard feed-forward backpropagation network with one hidden layer. For training purposes, the Bayesian regularization algorithm was used (Levenberg–Marquardt modified back-propagation algorithm). Training was conducted with the use of different numbers of neurons in the hidden layer to find the optimal architecture. The external dataset from other authors' experiments was utilized in this study. Similar to previous work [63], the input variables were the carbon content in the materials, oxidation temperature, and weight loss of the MgO-C specimens. It was assumed that three mechanisms control the oxidation rate: chemical adsorption, diffusion, and chemical reaction. Calculations were performed for a wide range of MgO-C materials, with graphite content varying from 5% to 30%. The authors developed models which enabled prediction of the effective diffusion coefficient for selected materials with the R^2 coefficient in the range 0.986–0.999, depending on the carbon content and temperature of the test. If only one diffusion mechanism occurred, the low-temperature diffusion activation energy of oxidation was predicted to be in the range 21.2–35.0 kJ·mole^{-1}, depending on the carbon content. The high-temperature diffusion activation energy was predicted to be in the range 42.1–109.6 kJ·mole^{-1} depending on the carbon content. If three diffusion mechanisms occurred, the low-temperature diffusion activation energy of oxidation was also predicted to be 16.5–25.7 kJ·mole^{-1} depending on the carbon content. The high-temperature diffusion activation energy was predicted to be 31.3–219.7 kJ·mole^{-1} depending on the carbon content. Authors also confirmed that predicted data are comparable with experimental data obtained by other authors in [59,60,65] The increased temperature resulted in an activation energy drop due to the increased oxygen diffusion rate. It was also confirmed that oxygen diffusion through the pores is the most significant factor controlling the oxidation intensity of MgO-C material.

In described works [63,64], the authors conducted advanced calculations to predict the oxidation kinetic parameters of MgO-C samples depending on the carbon content and test temperatures. However, it is necessary to extend this research and find if it is possible to apply ML techniques for predicting the oxidation behavior of MgO-C bricks including other parameters like the compactness of the bricks, as far more factors affect the decarburization resistance of MgO-C bricks. Also, a greater number of samples should be used to obtain more reliable ML results.

3.1.2. Optimization of Carbon Content in MgO-C Refractories

Graphite is a main source of carbon in MgO-C refractories, which, due to its low thermal expansion coefficient and poor wettability by slag, provides high slag corrosion resistance and good thermal shock resistance, respectively. However, too much carbon in the MgO-C composition leads to heat loss in industrial devices during operation due to the increased conductivity of MgO-C materials [66]. Greater amounts of carbon in the MgO-C composition decrease the hot strength and oxidation resistance of MgO-C refractories [66]. Therefore, the optimization of graphite content in MgO-C bricks is crucial.

Mazloom et al. [67] used artificial neural networks to optimize graphite content in MgO-C refractories. The work aimed to find the optimal ratio of graphite to resin to provide the highest possible compressive strength and minimize the apparent porosity of the materials. Overall, 25 formulations of MgO-C refractories were selected, which varied in the amount of resin (1.0–3.0%) and graphite (7.5–17.5%). In total, 100 specimens were prepared (four specimens for each of 25 formulations) for experimental testing. According to the obtained results, it was found that replacing magnesia powder with graphite leads to a decrease in compressive strength of up to 10% of the graphite content. If the graphite content was 12.5%, the compressive strength increased, but further increasing the amount of graphite to 15% and 17% caused a decrease in compressive strength. The experimental results showed that an increase in the content of synthetic resin was always associated with an increase in cold compressive strength. The compactness of specimens with fixed resin content (determined by open porosity measurements) decreased with increased graphite content to 15% in MgO-C, while, above 15% graphite, only a slight increase in apparent porosity was observed. The larger amount of synthetic resin was considered to reinforce the effect of increased graphite content. For ML model development, backpropagation of training error and a three-layer network for training were used. Two variables were selected as input data, namely, resin and graphite content, while the output variables were ultimate compressive strength and apparent porosity of 100 specimens prepared experimentally. Approximately 250 cycles of training were conducted with the use of different numbers of neurons to find the best model. Applying ANN, it was reported that 13.5% graphite and 3.0% synthetic resin in formulation provide the highest ultimate compressive strength with the lowest apparent porosity. The model was validated experimentally on seven specimens, based on the ANN-proposed formulation. The ultimate cold compressive strength predicted by ANN was 365.16 MPa, while the experimental value was 376.47 MPa, which means a 1.3% error. ANN predicted an apparent porosity of 7.08% while the experimental was 7.11%, which gives a 0.35% error. The obtained results are shown in Table 1. As the authors stated, a reliable and accurate model is feasible to develop using ANN to predict the MgO-C material's properties.

Table 1. Experimental results for the compressive strength and apparent porosity of samples based on the optimum formulations from the ANN model [67].

Sample	Compressive Strength [MPa]	Apparent Porosity [%]
F1	381.20	-
F2	375.91	-
F3	371.25	-
F4	377.54	-
F5	-	7.05
F6	-	7.18
F7	-	7.09
Average experimental value	376.47	7.11
Predicted value (ANN)	365.16	7.08
Error, % *	1.30	0.35

* difference between experimental and predicted values.

In [67], the authors determined the optimal amount of graphite and resin (13.5% graphite, 3% resin) to provide the demanded corrosion resistance with no loss in mechanical behavior. The authors show the prediction accuracy for selected specimens together with their open porosity and compressive strength. The accuracy of prediction is high (prediction error 1.30%), and the results of prediction seem to almost ideally fit the data. Therefore, it is good practice to describe the procedure applied to avoid overfitting (algorithmic learning by heart). Apparent porosity and the compressive strength of the samples were measured only on two specimens (out of each 25 formulations); thus, the results of measurements may not be reliable and need to be extended. Also, it would be recommended to investigate the influence of different raw material compositions together with different carbon and resin contents when modelling the basic properties of specimens. After all, the developed model is a useful tool to model and design MgO-C bricks with desirable properties.

3.1.3. Corrosion Resistance of MgO-C Refractories

The corrosion resistance of refractory materials is widely studied due to its exceptional importance for the service life of heating devices. Corrosion of MgO-C refractories mostly limits the duration of a campaign of devices, which increases maintenance costs for the end users of refractories.

Optimization of MgO-C refractories' composition for improved corrosion resistance was studied with unsupervised learning techniques using clustering algorithms [68]. A total of 20 different variants of MgO-C materials were prepared based on four different main raw materials. From each of the variants, eight industrially produced MgO-C bricks were selected for further examinations. Basic physicochemical properties (apparent porosity and bulk density, apparent porosity and bulk density after coking, decarburization resistance at 900 °C and 1100 °C, chemical composition, and graphite content) were experimentally measured. Principal component analysis (PCA) and the K-medoids algorithm were applied to develop a model which clusters the MgO-C materials into groups of comparable properties. PCA analysis showed that it is possible to use two variables, instead of eight, to characterize the prepared MgO-C materials. A new variable, PC1, was obtained, which explained approximately 81% of the variability in the dataset and referred to the basic properties of MgO-C materials. The second variable, PC2, explained about 12.3% of the data variability and referred to the values of pressure used for shaping the materials. In the K-medoids algorithm with PAM (partitioning around medoids), PC1 and PC2 were used as input variables. The algorithm was able to distinguish nine groups with materials of considerably comparable properties. It was assumed that materials assigned to the same clusters by the PAM algorithm have comparable corrosion resistance. Experimental tests of corrosion resistance were conducted with the use of an induction furnace to verify the obtained ML results. The algorithm indicated that a material consisting of fused magnesia of standard quality (shaped at 120 MPa) should perform similarly to a material consisting of 65% sintered and 12% fused magnesia of the highest quality (shaped at 180 MPa). Moreover, the algorithm suggested that materials composed of fused magnesia of standard quality and 27% sintered magnesia (shaped at 180 MPa) should perform similarly at high temperatures to test materials containing of 65% sintered magnesia and 12% of the highest-quality fused magnesia. With statistical tests (Wilcoxon test) applied to the measured wear rates after corrosion tests, it was confirmed that the described material variants were located in the same cluster indicated by the PAM algorithm and performed similarly after being exposed to slag attack at high temperature. Therefore, the algorithms properly indicated materials of comparable corrosion resistance. The conducted examinations, coupled with computer calculations, show the directions and possibilities to substitute fused raw materials with sintered ones with no loss of corrosion resistance.

The described research [68] reveals an extremely important issue in terms of sustainable development, as the production of fused magnesia demands 15 times more energy than the production of sintered aggregates [69]. Even if the corrosion of MgO-C refractories is widely described in the literature [27–32,36–39,70–72], scarce information can be found regarding the comparison of high-temperature properties of MgO-C materials based on different magnesia raw materials. Using ML techniques, it was possible to group MgO-C compositions with different ratios of sintered to fused magnesia characterized by comparable corrosion resistance. The results were validated at a semi-industrial scale by conducting corrosion tests in an induction furnace at 1720 °C. This could contribute to extending the corrosion test to a wider temperature range of 1600–1750 °C, as impurities in magnesia raw materials affect MgO-C materials' performance. Above all, it would be beneficial to test the designed materials in industrial conditions to assess the MgO-C materials' real performance, e.g., in steel ladles which are typically characterized by the shortest service time of refractory lining.

The slag corrosion resistance of MgO-C refractories was also examined by Akkurt [73] with the use of artificial neural networks. The work aimed to predict the wear rate of MgO-C refractories for steel ladles based on the results of laboratory corrosion tests. The data were collected from a series of corrosion finger tests (without rotating the samples). The architecture of the designed ANN was as follows: three layers of the feed-forward type and six neurons in the input layer. The input variables were the percentage content of CO in the atmosphere, the time of brick exposure for slag attacks, the temperature of a test, and the CaO/SiO_2 ratio of the slag. The measured surface of the lost area in the cross-section of tested specimens was taken as the output data. In the testing stage, the average testing error was reported at 14.2% with $R^2 = 0.92$. A detailed comparison of predicted and experimental data is presented in Table 2. Surface plots showing the relationship between input variables and the percentage of lost area during corrosion were generated as a complementation of the results. It was shown that an increase in temperature lead to an increase in refractory wear. An insignificant interaction was observed between temperature and the time of lining exposure for corrosive factors. Some values of prediction error in Table 2 exceed 15%. This phenomenon is probably associated with the relatively low number of experimental measurements done by the authors; however, the results of ANN performance are consistent with the current state of knowledge concerning the MgO-C corrosion mechanism.

Table 2. Results of model testing, based on [73].

% Area Loss—Measured	% Area Loss—Predicted	Difference	% Error (Absolute)
10.57	12.43	−1.86	17.6
10.85	14.57	−3.72	34.3
14.65	14.85	−0.20	1.4
18.99	18.94	0.05	0.3
19.20	18.05	1.15	6.0
32.34	24.80	7.54	23.3
15.67	18.27	−2.60	16.6
Average	-	-	14.2

The conducted calculations are important to model process parameters in steel plants (e.g., % CO, temperature, heat time, slag basicity) and to provide appropriate corrosion resistance for refractories. It is worth noting that only seven observations (measured area loss) were used for model testing. Probably, the low number of measurements (both in the training procedure and testing) is the reason why the percent prediction error is high for some observations (23.4%, 34.3%). The presented research included only a few factors influencing the refractories' corrosion resistance. If the model is allocated for direct use in steel plants, more factors should be included, e.g., number of heats/day, slag chemistry, tapping temperatures, secondary treatment temperatures, types of additives used for refining, etc. Moreover, a greater number of samples should be tested to provide higher accuracy of the model [74].

3.1.4. Thermomechanical Properties of MgO-C Refractories

MgO-C refractories are exposed to extreme thermal, mechanical, and chemical stresses during operation in the steel plant. The highest thermal stresses occur during the preheating stage of the heating device and the tapping of the hot molten steel into the ladle, where the refractory lining suffers mostly from a high temperature gradient (from about 300 °C at the shell to 1600–1700 °C at the lining). For some applications, MgO-C materials have to withstand additional mechanical stresses, e.g., due to rocking of the BOF vessel during preheating [75,76].

An advanced investigation of the thermomechanical behavior of different lining concepts in steel ladles was conducted by Hou et al. [77]. Artificial neural networks were used, among others, to predict the thermal and thermomechanical responses of refractory lining during operation. Overall, 160 different configurations of lining were investigated. In this research, the finite element (FE) method was used to obtain the input data for the ANN architecture design. The FE calculation included preheating of the refractory bricks' hot face in the ladle to 1200 °C for 20 h and direct exposure for tapping temperatures up to 1600 °C. A 95 min refining process was assumed. For the experiment and calculation, 10 different variables were used, assuming various steel shell lining thicknesses (insulation, permanent, and working linings), different thermal conductivities, and different Young's modulus values for the bricks. Three-layer backpropagation ANN was used for prediction. Hyperbolic tangent sigmoid was selected as an activation function. Three tests were used to establish the optimal ANN architecture. At the first test, all 160 samples were selected for the training, where gradient descent with the adaptive learning rate backpropagation (GDX) algorithm was selected. In the second test, the data set was divided into three groups (96, 126, and 160 samples) to find the minimum sample size for the study. In the third test, eight different algorithms were used to find the most favorable one for the steel ladle. Model assessment was conducted with the use of various errors: RE_MAX (maximum relative error), MRE (mean relative error), RRMSE (relative root-mean-squared error), and coefficient of determination B. Out of eight algorithms, two were selected (CFG—conjugate gradient backpropagation with Fletcher–Reeves updating and BR—Bayesian regularization backpropagation) as the most suitable for calculations. The ANN was then built to compare the performance of the selected algorithms in the prediction of the end temperature (the temperature at the cold end of the steel shell), maximum tensile stress, and maximum compressive stress. The comparison results are shown in Table 3. Low values of RE_MAX and MRE and high values of B are desirable. For the maximum tensile strength and maximum compressive strength, the BR model performed more efficiently than the CFG model (for BR: higher values of the coefficient of determination B, lower values of MRE, and a lower value of RE_MAX for tensile strength). Based on the obtained results, the BP-ANN model with BR was utilized for final calculations.

Table 3. Thermomechanical response prediction with the use of ANN based on CGF and BR, based on [77].

	End Temperature [°C]		Maximum Tensile Stress [MPa]		Maximum Compressive Stress [MPa]	
Used algorithm	CFG	BR	CFG	BR	CFG	BR
RE_MAX [%]	7.15	7.15	16.62	12.43	3.12	4.09
MRE [%]	1.02	1.76	2.43	2.37	0.93	0.78
B	0.9967	0.9908	0.9279	0.9348	0.9963	0.9966

RE_MAX, MRE, and B—coefficients evaluating the error between the results of the two used algorithms, CFG and BR.

The optimal ANN architecture was found for seven nodes in the hidden layer and Bayesian regularization with 160 samples for training. Two insulation lining concepts (linings 1 and 2 according to Table 4) were compared with the use of optimized ANN architecture. For this lining concept, predicted (ANN) and simulated (FE modelling) values of selected properties (steel shell temperature, maximum tensile stress, and maximum compressive stress) were shown in Table 5.

Table 4. Refractory lining concepts selected for prediction, based on [77].

	Thickness [mm]	Thermal Conductivity [$W \cdot m^{-1} K^{-1}$]	Young's Modulus [GPa]	Thermal Expansion Coefficient [$10^{-6} K^{-1}$]
Working lining	155.0	9	40	12
Permanent lining	52.5	2.2	45	5
Insulation (lining concept 1)	37.5	0.5	3	6
Insulation (lining concept 2)	37.5	0.38	4	5.6
Steel shell	30	50	210	12.0

Table 5. Comparison of simulated and predicted values of two proposed optimal lining concepts from FE modelling and as predicted by BP-ANN, based on [77].

	Steel Shell Temperature [°C]		Maximum Tensile Stress [MPa]		Maximum Compressive Stress [MPa]	
	Modelling (FE)	Predicted (BP-ANN)	Modelling (FE)	Predicted (BP-ANN)	Modelling (FE)	Predicted (BP-ANN)
Lining concept 1	280	276	1495	1433	512	517
Lining concept 2	259	259	1539	1576	517	515

The results presented in Table 5 confirm that ANN performed outstandingly. The predicted values of selected properties (steel shell temperature, maximum tensile stress, maximum compressive stress) were close to the FE-simulated ones. The temperature difference between the predicted value and the value obtained through FE modelling for lining concept 1 was only 4 °C. Furthermore, for lining concept 2, the model predicted the same temperature of a steel shell as modelled through FE, 259 °C. The predicted maximum tensile stress for lining concept 1 was 1433 MPa, while it was 1495 MPa for FE modelling, which is a 4.1% error. For lining concept 2, the predicted maximum tensile stress was equal to 1576 MPa, while for the FE model it was 1539 MP, which is a 2.4% error. As for compressive stress for lining concepts 1 and 2, the differences between the predicted and FE-modelled maxima were 5 MPa and 2 MPa, respectively. The presented model was also reported as promising for material recipe improvements and steel production optimization.

The variation in the study [77] was analyzed in [78] to optimize the number of nodes used in the hidden layer of ANN. The Taguchi method was used to find the minimum numbers of input variables. Variation/response complexity was found to be crucial for establishing well-performed ANN architecture. The developed methodology and models were used to investigate higher numbers of lining concepts (192 linings) in the case of thermomechanical response in steel ladles [79].

The conducted calculations [77] have significant practical meaning in the case of assessing the thermomechanical behavior of ladle linings during operation. Based on 160 different lining concepts, two of them were selected, and their performance was adequately predicted using the ANN algorithm. The tapping temperature of molten steel was assumed to be 1600 °C, but it would be interesting for scientists and engineers to see the behavior of lining concepts up to 1700 °C. Even though the presented model has an indisputable influence on reducing the time, materials, and cost-consuming labor for trial on site, industrial verification should be done to investigate the designed lining performance under real conditions. Then, the models could be successfully used in industrial practice to design linings with appropriate thermal and thermomechanical properties. The presented work may not only affect the lining design but may also indicate directions for MgO-C materials or safety lining composition development. Furthermore, thermal and thermomechanical linings' response may make it possible to avoid one of the most common failures of ladles associated with MgO-C thermal behavior (vertical cracks) [80].

4. Application of ML in Industrial-Scale Examinations

From the industrial point of view, the most important thing is to provide the longest possible service time for refractories in heating devices, which allows for the optimization of the cost-to-service time ratio. The service time of refractories is affected by several factors, including metallurgical conditions, refractory brick quality, the maintenance schedule of devices, etc. The service time is difficult to assess and predict. However, it seems to have become more feasible with the implementation of computational technologies in the development of refractories.

Borges et al. [81] applied self-organizing maps (SOM), which represent one of the unsupervised algorithms, to identify the main factor influencing the wear rate of MgO-C materials at the slag lines of the steel ladles. Around 6700 data points collected from the industrial database were analyzed. The authors compared the results of the traditional statistical approach with the SOM results. SOMs consisted of seven neurons vertically and six neurons horizontally for the selected properties. Approximately 23 metallurgical parameters were investigated. The SOM maps showed the relations between ladle service time and hot metal treatment with CaSi, Ar bubbling without CaSi, Ar bubbling with CaSi, steel permanence time, steel temperature after tapping, steel weight, and type of product (thick plates, hot strips, and boards for sale). At each step of the analysis, the results were verified with the use of typical regression and correlation analysis. Based on the SOM results, the authors indicated the numerous reasons responsible for the premature or intense wear of MgO-C materials in steel ladles, including the number of chemical additions (like nepheline and CaSi), the interaction between the desulphurization route and the intensity of ladle furnace use, and the extended contact time of the refractories with slag. The ML algorithm results were found to agree with traditional statistics calculations. The important fact in the described work [81] is that the authors verified the calculation results vs. traditional statistics and *post-mortem* results for selected MgO-C bricks. SOM maps are considered a useful tool, not only to indicate the parameters affecting ladle service time, but, thanks to the applied technique, direct recommendations may be allocated for steelmakers to improve the production process and extend the ladles' performance. Considering industrial practice, it would be valuable to investigate not only the metallurgical factors but also the type of ladle maintenance (e.g., using gunning mixes or other protection techniques).

Another industrial study concerning ladle treatment was conducted by Jančar et al. [82]. The authors used selected metallurgical parameters as an input variable to build an artificial neural network for the prediction of ladle service time depending on input values. The output variable was the number of castings. For the calculations, data associated with the secondary treatment of a 230 t ladle in the steel plant Liberty Ostrava was used. 106 ladle campaigns were analyzed. At the first stage of analysis, a statistical evaluation of metallurgical parameters was performed. Parameters which insignificantly influenced data set variability were discarded from further analysis. For building the ANN, seven parameters were selected as input variables, namely, empty ladle time, full ladle period, tapping temperatures, steel temperature after tapping, electricity consumption, number of heats with vacuum treatment, and Ar consumption. A high-quality model was obtained during the network training, with a coefficient of determination R^2 of 0.8927 for analyzing predicted vs. actual values of model output. During testing of the model, the coefficient of determination for actual and predicted ladle service times was $R^2 = 0.66$. The developed model was used to simulate the ladle service time for a wide range of values of selected properties (time of full ladle, Ar consumption, tapping temperature, and electrical energy consumption). The authors showed that if the tapping temperature increases the ladle service time decreases ($R^2 = 0.9719$), and, similarly, if electrical energy consumption increases during secondary treatment the service time of the ladle decreases ($R^2 = 0.9860$). According to the presented studies, the authors show that, if the time of the full ladle (the overall time when metal is present in the ladle) increases, the ladle service time increases ($R^2 = 0.9249$); similarly, if Ar consumption increases, the ladle service time increases ($R^2 = 0.9945$). Furthermore, the authors extracted the selected variables' importance and their influence on ladle service time using a developed criterion. Among the seven selected variables, the most negative impacts on ladle service time were from the time with an empty ladle (assumed importance coefficient of -21.08%), electrical energy consumption (-20.43%), and the number of heats with vacuum treatment (-15.71%). The most positive influences on ladle service time were associated with a higher amount of consumed Ar ($+15.20\%$) and a longer time with the ladle filled with metal ($+9.71\%$). Based on the obtained models [82], the authors expect to achieve increased ladle service times. Nevertheless, the authors will collect more data to expand their research.

The authors of [82] used a large industrial dataset collected from the steel production processes of 106 campaigns to develop a model predicting steel ladle service time. The authors were able to successfully indicate parameters which significantly influence ladle service time (tapping temperatures, energy consumption, time with a full ladle, Ar consumption, etc.). Taking into account overall ANN model performance, the tested model accuracy was not at a high level ($R^2 = 0.66$). A probable reason for this is associated with the types of MgO-C materials installed in the ladle. As the authors indicated [82], 106 ladles were investigated, which were lined with different MgO-C grades. Information about the material's quality should be used in further models as another input variable. Except for that fact, the presented work has significant practical meaning and the potential to be implemented in direct use at steel plants. The developed model could support steelmakers in the optimization of the process to provide safer work and higher performance for MgO-C refractories in steel ladles.

Yemelyanov et al. [83] proposed the use of artificial neural networks to diagnose lining conditions based on refractory lining thermograms. Work [83] showed detailed steps in conducting image recognition using an ANN. A method of preprocessing the thermal images was given by the authors to provide the best possible quality of data for input variables in the ANN. The input parameters were as follows: the mass centers of the thermograms, distance matrixes defining the borders of specific lining zones, and colors spotted on thermograms. The ANN was used in this work to classify the burnout zones of the lining. The training of the network was conducted in two steps. The first step was the typical training of the network with data sampling. In the second step, only experimental data were utilized for training. A total of 480 images of steel ladles and torpedo cars were applied for training. In the second stage, experimental thermograms (620 images) obtained from Alchevsk Iron and Steel Works were examined. The authors tested 22 neural networks to find the optimal architecture. The obtained results enabled them to implement specialized software in the steel plant that inspects the lining conditions.

It is worth emphasizing that the authors used unprecedented real thermograms of the lining collected at the steel plant. In total, 480 standard thermograms and 620 collected experimentally were used, which made it possible to obtain a reliable model with low values of classification error (0.258–0.443). The model performance was satisfactory and was the basis for developing specialist software for lining condition diagnosis. The major advantage of this work is that its results are positively implemented in industry. Moreover, it seems that the model is flexible and can be used successfully in various types of devices, e.g., steel ladles and torpedo ladles.

Zelik et al. [84] showed the application of artificial neural networks to predict the wear rate of MgO-C refractories in the slag spout zone of a basic oxygen furnace. One campaign of BOF was considered in the analysis. Overall, 17 variables, collected automatically at the steel plant, were assigned as input variables, including the chemical composition of hot metal, treatment temperature, the types of additive used in the process, and the type of maintenance operation (gunning and slagging). The residual thickness of the MgO-C bricks in the slag spout zone was taken as the output data. Measurement of the residual thickness of the bricks was conducted with the use of a laser scanner directly at the steel plant. The wear indexes were calculated based on 16 laser measurements of the lining. The values of residual thicknesses were divided into wear classes calculated according to Equation (1):

$$up = t \cdot \frac{w}{10} \tag{1}$$

where up is the upper boundary of the wear class, t is the class number (1...10), and w is the maximum value of the wear index. ANN was used to predict the wear class depending on selected metallurgical parameters. The quality of training was 64.56% and 66.21% with testing performed using the R programming language. Table 6 presents the results of classification using the ANN model. Model performance was verified with the use of Orange 3.21 software. In this case, the classification accuracy reached 63.9%. Evaluation of the variables' importance was done with the use of the Boosted Trees algorithm. The variables influencing the wear rate of MgO-C refractories most significantly were reported accordingly: the number of gunning operations was the most important, then the MgO content in the slag, the amount of lime added to the metal bath, and hot metal weight.

An extension of this work [84] was shown in [85]. The authors used industrial data on the metallurgical process in BOF to predict the wear rate of MgO-C refractories. A total of 13 variables were selected, including Si and C content in the hot metal, the temperature and weight of the hot metal, oxygen activity in the metal bath, the temperature at the end of the refining, the amount of oxygen used during the upper blow, the amount of calcium added to the metal bath, the amount of MgO-containing additive, and the chemical composition of the slag. Data were inspected and prepared in detail to provide the best possible quality. Exponential smoothing was implemented to remove noise from the data. Several ML models were tested to select the most accurate one for prediction, including multivariate adaptive regression splines (MARS), classification and regression trees (CART), boosted trees, and artificial neural networks (ANN, multilayer perceptron, MLP type). Boosted trees were reported to be the most effective in the prediction of the wear rate of MgO-C refractories. A comparison of the model performance was expressed with the use of different statistical measures: SSE (error sum of squares), MSE (mean squared error), RMSE (root-mean-square error), R^2 (coefficient of determination), MAPE (mean absolute percentage error), and MAE (mean absolute error), as shown in Table 7. This extended analysis made it possible to indicate parameters that significantly influence the service time of BOF. The most important factors were found to be hot metal weight, then the Si concentration in the hot metal, scrap mass, and the oxygen activity in the hot metal.

Table 6. Classification of wear rate class conducted with the use of ANN model, based on [84].

		0	1	2	3	4	5	6	7	8	9	Σ
						Real Wear Class						
	0	226	60	19	0	4	0	1	0	0	0	310
	1	63	128	2	0	4	0	12	0	0	0	209
	2	9	1	6	0	0	0	0	0	0	0	16
Predicted	3	0	0	0	0	0	0	0	0	0	0	0
Wear Class	4	12	12	10	0	0	0	7	0	0	8	49
	5	0	0	0	0	0	0	0	0	0	0	0
	6	0	6	0	0	5	0	5	0	0	0	16
	7	0	0	0	0	0	0	0	0	0	0	0
	8	0	0	0	0	0	0	0	0	0	0	0
	9	0	10	0	0	6	0	2	0	0	7	25
	Σ	310	217	37	0	19	0	27	0	0	15	625

Table 7. Comparison of different measures of fit for model performance, based on [85].

Algorithm	SSE	MSE	RMSE	R^2	R	MAPE	MAE
Training Data Set							
CART	6.811	0.004	0.065	0.559	0.747	24.673%	0.057
MARS	4.195	0.002	0.051	0.716	0.846	17.987%	0.047
Boosted Trees	1.590	0.001	0.031	0.899	0.948	11.086%	0.029
ANN	3.521	0.002	0.047	0.789	0.886	16.012%	0.041
Testing Data Set							
CART	5.445	0.008	0.091	0.429	0.655	27.598%	0.066
MARS	3.329	0.005	0.071	0.649	0.805	21.316%	0.054
Boosted Trees	1.458	0.002	0.047	0.849	0.921	13.439%	0.035
ANN	2.932	0.004	0.066	0.687	0.829	20.233%	0.049

Two works [84,85] describe the application of different ML techniques for prediction of the wear rates of MgO-C materials in basic oxygen converters based on metallurgical parameters collected during hot metal treatment. Authors obtained models of different qualities. Among the used techniques, ANN and the boosted trees algorithm were reported as producing the most accurate results. Even though the works carry practical meaning, the model performance needs to be improved. The directions of model performance improvement are associated with the quality of the data. First of all, in both works [84,85], the residual thickness of MgO-C materials in the slag spout zone was used as an output variable. Unfortunately, due to the specific work of the steel plant, only about 20 laser-scanned results for lining thickness were obtained during the campaign. Such a low amount of data in the campaign, which lasts more than 2000 heats, affects the ML model's quality. Industrial data are often not prepared well enough and contain missing or invalid values (e.g., the amount of hot metal exceeding the device's capacity). Collecting quantitative data on the gunning mixes used for sidewall protection would also improve the quality of the models.

5. Benefits and Limitations of the Application of ML Techniques for the Investigation of MgO-C Refractories

Although the number of publications on the application of machine learning is rapidly growing, with a 56% increase over the last 2 years, it is still very low when it comes to ML application in the refractory industry. Based on reviewed works [63,64,67,68,73,77–79,81–85], the most commonly used ML algorithm, and simultaneously the one giving the most accurate predictions, is an artificial neural network. One exception is given in [85], which shows boosted trees are the best-fitting algorithm. The presented articles prove that ML algorithms are highly useful in examinations and in industrial applications of MgO-C materials. In the research process, the most significant advantage of applying ML algorithms is the reduction of time-consuming and expensive experimental investigations in the corrosion testing of MgO-C refractories [63,64].

ML techniques currently have obvious limitations, as the quality of data collected in the industry is still not satisfactory. Thus, it is necessary and highly recommended to improve the process of data registration, especially data involving steel production processes, to avoid missing data, unreal values, or mistake-generative hand typing. Using data of unsatisfactory quality may lead to inaccurate conclusions.

Another important limitation is related to laboratory experiments and the fact that ML algorithms are trained on data collected from specific, highly advanced examinations. It might be difficult to apply external data to such models and obtain reliable results, especially if one—allegedly insignificant—factor is changed.

Nevertheless, interest in using ML techniques in the refractory industry is growing, as it seems that high digitalization in this area is unavoidable. The possibility of predicting the wear rate of refractories depending on their metallurgical data will be especially encouraging for refractory end-users. They should be conscious of the need to improve data collection in order to develop highly predictive models that will serve in industrial practice and help to make the steel process more sustainable.

6. Conclusions

The current state of knowledge on ML techniques—relatively newly applied in refractories investigation—was reviewed in this work for MgO-C materials, which constitute over 70% of total refractories production. The most commonly used ML algorithm type is currently artificial neural networks. The clustering algorithm is also effectively applied in the optimization of MgO-C materials and the identification of factors influencing vessels' service time in steel production.

Nevertheless, the number of papers on the application of ML techniques is still insufficient considering the rapidly growing interest in and high potential of ML techniques. The limited accessibility of reliable data is one of the reasons, which results from the disclosure politics of steel plants. The end users of MgO-C refractories will be conscious of the benefits gained from building high-quality ML models, which can influence the extension of the service time of refractories, thus making the steel production process more efficient and sustainable.

Concerning experimental research activities on MgO-C refractories, it is always cost-intensive to prepare and analyze the great number of samples demanded for ML implementation. The experimental approach has been changing, and wide implementation of ML in the refractory industry is unavoidable to speed up innovation in the industry in the near future, which stands at the front of a fast-changing and challenging environment.

Author Contributions: Conceptualization, S.S. and I.J.; methodology, S.S.; validation, W.Z. and J.S.; investigation, S.S.; resources, S.S. and I.J.; writing—original draft preparation, S.S.; writing—review and editing, I.J. All authors have read and agreed to the published version of the manuscript.

Funding: This research received no external funding.

Institutional Review Board Statement: Not applicable.

Informed Consent Statement: Not applicable.

Data Availability Statement: Data are contained within the article.

Conflicts of Interest: The authors declare no conflict of interest.

References

1. Routschka, G.; Wuthnow, H. *Handbook of Refractory Materials Design, Properties, Testings*, 4th ed.; Vulkan-Verlag Gmbh: Essen, Germany, 2012; pp. 92–98.
2. Dai, Y.; Li, J.; Yan, W.; Shi, C. Corrosion mechanism and protection of BOF refractory for high silicon hot metal steelmaking process. *J. Mater. Res Technol.* **2020**, *9*, 4292–4308. [CrossRef]
3. Guoguang, Z.; Husken, R.; Cappel, J. Experiance with long BOF campaign life and TBM bottom stirring technology. *Stahl Und Eisen.* **2012**, *132*, 61–78.
4. Husken, R.; Pottie, P.; Guoguang, Z.; Cappel, J. Overcoming the conflict between long BOF refractory service time and efficient bottom stirring: A case study at Meishan Steel in China. In *45⁰ Seminário de Aciaria—Internacional*; Porto Alegre, Brazil, 2014; Volume 45, pp. 724–735. Available online: https://abmproceedings.com.br/en/article/overcoming-the-conflict-between-long-bof-refractory-lifetime-and-efficient-bottom-stirring-a-case-study-at-meishan-steel-in-china (accessed on 5 September 2023). [CrossRef]
5. Folco, L.; Kranjc, A. Steel Ladle Lining management: Comparison between different maintenance technologies to increase performance, reduce refractory consumption and waste disposal of used materials. In Proceedings of the Unified International Conference on Refractories UITECR 2023, Frankfurt, Germany, 26–29 September 2023; pp. 351–354.
6. Otunniyi, I.O.; Theko, Z.V.; Mokoena, B.L.E.; Maramba, B. Major deteminantion of service life in magnesia-graphite slagline refractory lining in secondary steelmaking ladle furnace. In *IOP Conference Series: Materials Science and Engineering, Proceedings of the Conference of the South African Advanced Materials Initiative (CoSAAMI 2019), Riverside Sun, Vanderbijlpark, South Africa, 22–25 October 2019*; IOP Publishing: Bristol, UK, 2023; Volume 655, p. 012003. [CrossRef]
7. Sun, C.H.; Zhu, L.L.; Yan, H.; Zhao, W.; Liu, J.X.; Ren, L.; Zhao, X.T.; Tong, X.S.; Yu, S.W. A novel route to enhance high-temperature mechanical property and thermal schock resistance of low-carbon MgO-C bricks by introducing $ZrSiO_4$. *Pol. J. Iron. Steel Res. Int.* **2023**. [CrossRef]
8. Korostelev, A.A.; S'emshchikov, N.S.; Semin, A.E.; Kotel'nikov, G.I.; Murzin, I.S.; Emel'yanov, V.V.; Kolokolov, E.A.; Belonozhko, S.S. Increase in EAF lining life with use of hot-briquetted iron in charge. *Refract. Ind. Ceram.* **2018**, *59*, 107–114. [CrossRef]
9. Gubta, R.B. Innovation in Steel Ladle Life to 157 Heats at Rourkela Steel Plant through Optimization of Refractory Material & Service Conditions. *Int. J. Eng. Res. Technol.* **2017**, *6*, 767–771. [CrossRef]
10. Ewais, E.M.M. Carbon based refractories. *J. Ceram. Soc. Jpn.* **2004**, *112*, 517–532. [CrossRef]
11. Luz, A.P.; Souza, T.M.; Pagliosa, C.; Brito, M.A.M.; Pandolfelli, V.C. In situ hot elastic modulus evolution of MgO-C refractories containing Al, Si or Al-Mg antioxidants. *Ceram. Int.* **2016**, *42*, 9836–9843. [CrossRef]
12. Xiao, J.; Chen, J.; Wei, Y.; Zhang, Y.; Zhang, S.; Li, N. Oxidation behaviors of MgO-C refractories with different Si/SiC ratio in the 1100–1500 °C range. *Ceram. Int.* **2019**, *45*, 21099–21107. [CrossRef]
13. Gao, S.; Xu, L.; Chen, M.; Wang, N. Effect of Fe addition on the microstructure and oxidation behavior of MgO–C refractory. *Mater. Chem. Phys.* **2019**, *238*, 121935. [CrossRef]
14. Atzenhofer, C.; Harmuth, H. Phase formation in MgO-C refractories with different antioxidants. *J. Eur. Ceram. Soc.* **2021**, *41*, 7330–7338. [CrossRef]
15. Yang, P.; Xiao, G.; Ding, D.; Ren, Y.; Yang, S.; Lv, L.; Hou, X.; Gao, Y. Antioxidant properties of low-carbon magnesia-carbon refractories containing AlB_2–Al–Al_2O_3 composites. *Ceram. Int.* **2022**, *48*, 1375–1381. [CrossRef]
16. Chen, Y.; Ding, J.; Deng, C.; Yu, C. Improved thermal shock stability and oxidation resistance of low-carbon MgO–C refractories with introduction of SiC whiskers. *Ceram. Int.* **2023**, *49*, 26871–26878. [CrossRef]
17. Zhang, T.; Chen, J.; Zhang, Y.; Yan, W.; Zhang, S.; Li, N. Elucidating the role of Ti_3AlC_2 and Ti_3SiC_2 in oxidation mechanisms of MgO–C refractories. *Ceram. Int.* **2023**, *49*, 11257–11265. [CrossRef]
18. Chen, Y.; Deng, C.; Wang, X.; Yu, C.; Ding, J.; Zhu, H. Evolution of c-ZrN nanopowders in low-carbon MgO–C refractories and their properties. *J. Eur. Ceram. Soc.* **2021**, *41*, 963–977. [CrossRef]
19. Chandra, K.S.; Sarkar, D. Oxidation resistance, residual strength, and microstructural evolution in Al_2O_3-MgO–C refractory composites with YAG nanopowder. *J. Eur. Ceram. Soc.* **2021**, *41*, 3782–3797. [CrossRef]
20. Zhong, H.; Han, B.; Wei, J.; Wang, X.; Liu, X.; Miao, Z.; Li, N. The microstructure evolution and performance enhancement of MgO-C refractories by the addition of MA90 spinel micro-powder. *J. Eur. Ceram. Soc.* **2023**, *44*, 523–543. [CrossRef]
21. Ren, X.; Ma, B.; Liu, H.; Wang, Z.; Deng, C.; Liu, G.; Yu, J. Designing low-carbon MgO–Al_2O_3–La_2O_3–C refractories with balanced performance for ladle furnaces. *J. Eur. Ceram. Soc.* **2022**, *42*, 3986–3995. [CrossRef]
22. Chen, Q.; Zhu, T.; Li, Y.; Cheng, Y.; Liao, N.; Pan, L.; Liang, X.; Wang, Q.; Sang, S. Enhanced performance of low-carbon MgO–C refractories with nano-sized ZrO_2–Al_2O_3 composite powder. *Ceram. Int.* **2021**, *47*, 20178–20186. [CrossRef]
23. Li, W.; Deng, C.; Chen, Y.; Wang, X.; Yu, C.; Ding, J.; Zhu, H. Application of Cr_3C_2/C composite powders synthesized via molten-salt method in low-carbon MgO–C refractories. *Ceram. Int.* **2022**, *48*, 15227–15235. [CrossRef]
24. Chen, Y.; Ding, J.; Yu, C.; Lou, X.; Wu, Z.; Deng, C. Application of SiC whiskers synthesized from waste rice husk in low-carbon MgO–C refractories. *J. Phys. Chem. Solids* **2023**, *177*, 111304. [CrossRef]

25. Luo, J.-Y.; Ren, X.-M.; Chong, X.-C.; Ding, D.-H.; Ma, B.-Y.; Xiao, G.-Q.; Yu, J.-K. Recent progress in synthesis of composite powders and their applications in low-carbon refractories. *J. Iron Steel Res. Int.* **2022**, *29*, 1041–1051. [CrossRef]
26. Guo, W.; Zhu, T.; Zhao, X.; Li, Y.; Chen, Q.; Xu, X.; Xu, Y.; Dai, Y.; Yan, W. Improved slag corrosion resistance of MgO–C refractories with calcium magnesium aluminate aggregate and silicon carbide: Corrosion behavior and thermodynamic simulation. *J. Eur. Ceram. Soc.* **2023**, *44*, 496–509. [CrossRef]
27. Han, J.S.; Heo, J.H.; Park, J.H. Interfacial reaction between magnesia refractory and 'FeO'-rich slag: Formation of magnesiowüstite layer. *Ceram. Int.* **2019**, *45*, 10481–10491. [CrossRef]
28. Liu, J.; Sheng, H.; Yang, X.; He, Z.; Hou, X. Research on the Wetting and Corrosion Behavior Between Converter Slag with Different Alkalinity and MgO-C Refractories. *Oxid. Met.* **2022**, *97*, 157–166. [CrossRef]
29. Yehorov, A.; Ma, G.; Volkova, O. Interaction between MgO–C-bricks and ladle slag with a 1:1 CaO/Al$_2$O$_3$ ratio and varying SiO$_2$ content. *Ceram. Int.* **2021**, *47*, 11677–11686. [CrossRef]
30. Guo, M.; Parada, S.; Jones, P.; Boydens, E.; Dyck, J.; Blanpain, B.; Wollants, P. Interaction of Al$_2$O$_3$-rich slag with MgO-C refractories during VOD refining-MgO and spinel layer formation at the slag/refractory interface. *J. Eur. Ceram. Soc.* **2009**, *29*, 1053–1060. [CrossRef]
31. Heo, S.H.; Lee, K.; Chung, Y. Reactive wetting phenomena of MgO-C refractories in contact with CaO-SiO$_2$ slag. *Trans. Nonferrous Met. Soc. (Engl. Ed.)* **2012**, *22*, s870–s875. [CrossRef]
32. Liu, Z.; Yuan, L.; Jin, E.; Yang, X.; Yu, J. Wetting, spreading and corrosion behavior of molten slag on dense MgO and MgO-C refractory. *Ceram. Int.* **2019**, *45*, 718–724. [CrossRef]
33. Lao, Y.; Li, G.; Gao, Y.; Yuan, C. Wetting and corrosion behavior of MgO substrates by CaO–Al$_2$O$_3$–SiO$_2$–(MgO) molten slags. *Ceram. Int.* **2022**, *48*, 14799–14812. [CrossRef]
34. Han, B.; Wang, C.; Yang, Q.; Changmin, K.; Chen, F.; Li, N. Corrosive Interaction between MgO-C Refractories and Vanadium-Recovery Slag. *Interceram-Inter. Ceram. Rev.* **2014**, *63*, 99–103. [CrossRef]
35. Borisenko, O.N.; Semchenko, G.D.; Il'icheva, T.V. Slag resistance of periclase-carbon refractories based on modified phenol fomaldehyde resin. *Refract. Ind. Ceram.* **2010**, *51*, 41–44.
36. Liu, Y.; Wang, Q.; Li, G.; Zhang, J.; Yan, W.; Huang, A. Role of graphite on the corrosion resistance improvement of MgO–C bricks to MnO-rich slag. *Ceram. Int.* **2020**, *46*, 7517–7522. [CrossRef]
37. Benavidez, E.; Brandaleze, E.; Musante, L.; Galliano, P. Corrosion Study of MgO-C Bricks in Contact with a Steelmaking Slag. *Procedia Mater. Sci.* **2015**, *8*, 228–235. [CrossRef]
38. Lee, S.; Chung, Y. The effect of C content in MgO–C on dissolution behavior in CaO–SiO$_2$–Al$_2$O$_3$ slag. *Ceram. Int.* **2022**, *48*, 26984–26991. [CrossRef]
39. Bai, R.; Liu, S.; Mao, F.X.; Zhang, Y.Y.; Yang, X.; He, Z. Wetting and corrosion behavior between magnesia–carbon refractory and converter slags with different MgO contents. *J. Iron Steel Res. Int.* **2022**, *29*, 1073–1079. [CrossRef]
40. Gehre, P.; Preisker, T.; Brachhold, N.; Guhl, S.; Wöhrmeyer, C.; Parr, C.; Aneziris, C.G. Thermodynamic calculation and microscopic examination of liquid phase formation in MgO–C refractories contain calcium magnesium aluminate. *Mater. Chem. Phys.* **2020**, *256*, 123723. [CrossRef]
41. Preisker, T.; Gehre, P.; Schmidt, G.; Aneziris, C.G.; Wöhrmeyer, C.; Parr, C. Kinetics of the formation of protective slag layers on MgO–MgAl$_2$O$_4$–C ladle bricks determined in laboratory. *Ceram. Int.* **2020**, *46*, 452–459. [CrossRef]
42. Wöhrmeyer, C.; Gao, S.; Ping, Z.; Parr, C.; Aneziris, C.G.; Gehre, P. Corrosion Mechanism of MgO–CMA–C Ladle Brick with High Service Life. *Steel Res. Int.* **2020**, *91*, 1900436. [CrossRef]
43. Ludwig, M.; Śnieżek, E.; Jastrzębska, I.; Prorok, R.; Sułkowski, M.; Goławski, C.; Fischer, C.; Wojteczko, K.; Szczerba, J. Recycled magnesia-carbon aggregate as the component of new type of MgO-C refractories. *Constr. Build. Mater.* **2021**, *272*, 121912. [CrossRef]
44. Moritz, K.; Kerber, F.; Dudczig, S.; Schmidt, G.; Schemmel, T.; Schwarz, M.; Jansen, H.; Aneziris, C.G. Recyclate-containing magnesia-carbon refractories—Influence on the non-metallic inclusions in steel. *Open Ceram.* **2023**, *16*, 100450. [CrossRef]
45. Horckmans, L.; Nielsen, P.; Dierckx, P.; Ducastel, A. Recycling of refractory bricks used in basic steelmaking: A review. *Resour. Conserv. Recycl.* **2019**, *140*, 297–304. [CrossRef]
46. Moritz, K.; Brachhold, N.; Küster, F.; Dudczig, S.; Schemmel, T.; Aneziris, C.G. Studies on the use of two different magnesia-carbon recyclates as secondary raw material for MgO–C refractories. *Open Ceram.* **2023**, *15*, 100426. [CrossRef]
47. Steiner, R.; Lammer, G.; Spiel, C.; Jandl, C. Refractories 4.0. *Berg Huettenmaenn. Monatsh.* **2017**, *162*, 514–520. [CrossRef]
48. Ślusarczyk, B. Industry 4.0—Are we ready. *Pol. J. Manag. Stud.* **2018**, *17*, 232–248. [CrossRef]
49. Rajendra, P.; Girisha, A.; Naidu, T.G. Advancement of machine learning in materials science. *Mater. Today Proc.* **2022**, *62*, 5503–5507. [CrossRef]
50. Rydning, D.R.-J.G. *The Digitization of the World from Edge to Core*; International Data Corporation: Framingham, MA, USA, 2018.
51. Pilania, G. Machine learning in materials science: From explainable predictions to autonomous design. *Comput. Mater. Sci.* **2021**, *193*, 110360. [CrossRef]
52. Sarker, I.H. Machine Learning: Algorithms, Real-World Applications and Research Directions. *SN Comput. Sci.* **2021**, *2*, 160. [CrossRef]
53. Jain, N.; Kumar, R. A Review on Machine Learning & It's Algorithms. *Int. J. Soft Comp. Eng.* **2022**, *12*, 1–5. [CrossRef]

54. Pugliese, R.; Regondi, S.; Marini, R. Machine learning-based approach: Global trends, research directions, and regulatory standpoints. *Data Sci. Manag.* **2021**, *4*, 19–29. [CrossRef]
55. Bhat, D.; Muench, S.; Roellig, M. Application of machine learning algorithms in prognostics and health monitoring of electronic systems: A review. *Adv. Electr. Electron. Eng.* **2023**, *4*, 100166. [CrossRef]
56. Nadachowski, F.; Kloska, A. *Refractory Wear Processes*; AGH: Kraków, Poland, 1997.
57. Rongti, L.; Wei, P.; Sano, M. Kinetics and Mechanism of Carbothermic Reduction of Magnesia. *Metall. Mater. Trans. B* **2003**, *34*, 433–437. [CrossRef]
58. Lee, J.; Myung, J.; Chung, Y. Degradation Kinetics of MgO-C Refractory at High Temperature. *Metall. Mater. Trans. B* **2023**, *52*, 1179–1185. [CrossRef]
59. Faghihi-Sani, M.-A.; Yamaguchii, A. Oxidation Kinetics of MgO-C refractory bricks. *Ceram. Int.* **2021**, *28*, 835–839. [CrossRef]
60. Li, X.; Riguard, M.; Palco, S. Oxidation Kinetics of Graphite in Magnesia-Carbon Refractories. *J. Am. Ceram. Soc.* **1995**, *78*, 965–971. [CrossRef]
61. Volkova, O.; Scheller, P.; Lychatz, B. Kinetics and Thermodynamics of Carbon Isothermal and Non-isothermal Oxidation in MgO-C Refractory with different air flow. *Metall. Mater. Trans. B* **2014**, *45*, 1782–1792. [CrossRef]
62. Jansson, S.; Brabie, V.; Jonsson, P. Corrosion Mechanism of Commercial MgO-C Refractories in Contact with Different Gas Atmosheres. *ISIJ Int.* **2008**, *48*, 760–767. [CrossRef]
63. Nemati, Z.A.; Moetakef, P. Investigation of graphite oxidation kinetics in MgO-C composite via artificial neural network approach. *Comput. Mater. Sci.* **2007**, *39*, 723–728. [CrossRef]
64. Nemati, A.; Nemati, E. Oxygen diffusion mechanism in MgO-C composites: An artificial neural network approach. *Model Simul. Mat. Sci. Eng.* **2012**, *20*, 015016. [CrossRef]
65. Sadrnezhaad, S.K.; Mahshid, S.; Hashemi, B.; Nemati, Z.A. Oxidation Mechanism of C in MgO-C Refractory Bricks. *J. Am. Ceram. Soc.* **2006**, *89*, 1308–1316. [CrossRef]
66. Mahato, S.; Pratihar, S.K.; Behera, S.K. Fabrication and properties of MgO-C refractories improved with expanded graphite. *Ceram. Int.* **2020**, *40*, 16535–16542. [CrossRef]
67. Mazloom, M.; Sarpoolaky, H.; Savabieh, H.R. Use of neural networks to optimize graphite content in magnesia-graphite refractories. *Refract. Ind. Ceram.* **2012**, *53*, 193–198. [CrossRef]
68. Sado, S. Method of raw materials selection for production of the MgO-C bricks of comparable properties using PCA and K-medoids. *Int. J. Appl. Ceram. Technol.* **2023**. [CrossRef]
69. An, J.; Xue, X. Life-cycle carbon footprint analysis of magnesia products. *Resour. Conserv. Recycl.* **2017**, *119*, 4–11. [CrossRef]
70. Bahtli, T.; Hopa, D.Y.; Bostanci, V.M.; Ulvan, N.S.; Yasti, S.Y. Corrosion behaviours of MgO-C refractories: Incorporation of graphite or pyrolytic carbon black as a carbon source. *Ceram. Int.* **2018**, *44*, 6780–6785. [CrossRef]
71. Chen, L.; Malfliet, A.; Jones, P.T.; Blanpain, B.; Guo, M. Comparison of the chemical corrosion resistance of magnesia-based refractories by stainless steelmaking slags under vacuum conditions. *Ceram. Int.* **2016**, *42*, 743–751. [CrossRef]
72. Liu, Z.; Yu, J.; Yuan, L. The effect of applied voltage on the corrosion resistance of MgO–C refractories. *J. Mater. Sci.* **2019**, *54*, 265–273. [CrossRef]
73. Akkurt, S. Prediction of the slag corrosion of MgO-C ladle refractories by the use of artificial neural networks. *Key Eng. Mater.* **2004**, *264–268*, 1727–1730. [CrossRef]
74. Alwosheel, A.; Cranenburgh, S.; Chorus, C. Is your dataset big enough? Sample size requirements when using artificial neural network for discrete choice analysis. *J. Choice Model.* **2018**, *28*, 167–182. [CrossRef]
75. Benavidez, E.; Brandaleze, E.; Lagorio, Y.; Gass, S.; Martinez, A. Thermal and mechanical properties of commercial MgO-C bricks. *Matéria (Rio De Janeiro)* **2015**, *20*, 571–579. [CrossRef]
76. Lagorio, Y.; Gass, S.; Benavidez, E.; Martinez, A. Thermomechanical evaluation of MgO-C commercial bricks. *Ceram. Int.* **2022**, *48*, 10105–10112. [CrossRef]
77. Hou, A.; Jin, S.; Harmuth, H.; Gruber, D. Thermal and Thermomechanical Responses Prediction of a Steel Ladle Using a Back-Propagation Artificial Neural Network Combining Multiple Orthogonal Arrays. *Steel Res. Int.* **2019**, *90*, 1900116. [CrossRef]
78. Hou, A.; Jin, S.; Gruber, D.; Harmuth, H. Influence of variation/response space complexity and variable completeness on BP-ANN model establishment: Case study of steel ladle lining. *Appl. Sci.* **2019**, *9*, 2835. [CrossRef]
79. Hou, A.; Jin, S.; Gruber, D.; Harmuth, H. Modelling of a Steel Ladle and Prediction of Its Thermomechanical Behavior by Finite Element Simulation Together with Artificial Neural Network Approaches. 2019. Available online: https://www.researchgate.net/publication/334304593 (accessed on 8 September 2023).
80. Griogiev, A.; Dailchenko, S.; Zabolotsky, A.; Migashkin, A.; Turchin, M.; Khadyev, V. Features of the Fracture of Refractory Linings Depending on the Equipment Size. *Refract. Ind. Ceram.* **2023**, *63*, 585–592. [CrossRef]
81. Borges, R.A.A.; Antoniassi, N.P.; Klotz, L.E.; de Carvalho Carneiro, C.; Silva, G.F.B.L.E. A Statistical and Self-Organizing Maps (SOM) Comparative Study on the Wear and Performance of MgO-C Resin Bonded Refractories Used on the Slag Line of Ladles of a Basic Oxygen Steelmaking Plant. *Metall. Mater. Trans. B* **2022**, *53*, 2852–2866. [CrossRef]
82. Jančar, D.; Machů, M.; Velička, M.; Tvardek, P.; Kocián, L.; Vlček, J. Use of Neural Networks for Service time Analysis of Teeming Ladles. *Materials* **2022**, *15*, 8234. [CrossRef]
83. Yemelyanov, V.A.; Yemelyanova, N.Y.; Nedelkin, A.A.; Zarudnaya, M.V. Neural network to diagnose lining condition. *IOP Conf. Ser. Mater. Sci. Eng.* **2018**, *327*, 022107. [CrossRef]

84. Zelik, W.; Lech, R.; Sado, S.; Labuz, A.; Lasota, A.; Lis, S. Modelling the Wear of MC 98/15 Refractory Material in the Slag Spout Zone of an Oxygen Converter with the Use of Artificial Neural Networks. *J. Ceram. Sci. Technol.* **2020**, *11*, 81–90. [CrossRef]
85. Sado, S.; Zelik, W.; Lech, R. Use of Machine Learning for modelling the wear of MgO-C refractories in Basic Oxygen Furnace. *J. Ceram. Process. Res.* **2022**, *23*, 421–429. [CrossRef]

Article

The Wear Rate Forecast of MgO-C Materials Type MC95/10 in the Slag Spout Zone of an Oxygen Converter in Terms of the Bayesian Estimation

Wiesław Zelik [1], Sebastian Sado [1,2,*] and Ryszard Lech [2]

[1] Research and Development Centre of Ceramics, Zakłady Magnezytowe "ROPCZYCE" S.A., ul. Postępu 15c, 02-676 Warszawa, Poland; wieslaw.zelik@ropczyce.com.pl

[2] Faculty of Materials Science and Ceramics, AGH University of Science and Technology, al. Mickiewicza 30, 30-059 Kraków, Poland; rysz.lech@gmail.com

* Correspondence: sebastian.sado@ropczyce.com.pl; Tel.: +48-601-571-270

Abstract: The ceramic–carbon refractory lining of an oxygen converter is subjected to variable thermochemical stresses, causing a progressive loss of material over time, which is expressed in a decreasing residual thickness of the lining. The forecasting method using Bayesian statistics has become a valuable skill in steel production planning and is one of the main conditions constituting the appropriate organization of steel and refractories production. This paper presents examples of Bayesian modelling of the unit wear rate value of the refractory materials for the zone with the highest wear in the refractory lining of a converter. From the experience gained during long-term operation of a steel-producing oxygen converter, it was found that the value of the unit wear rate of the refractory material in the slag spout zone of the steel-producing oxygen converter is subjected to an *a posteriori* normal distribution, with the following parameters: mean value $\mu = 401.23\ \mu$ heat^{-1}, standard deviation $\sigma = 13.74\ \mu$m heat^{-1}. The forecasted mean value of the unit wear rate of the MC95/10 refractories lined in the slag spout zone of the oxygen converter used for steel production, and which operates in intensive exploitation conditions, was equal to $\mu = 420\ \mu$m heat^{-1}.

Keywords: oxygen converter; refractories; wear forecasting; Bayesian statistics

Citation: Zelik, W.; Sado, S.; Lech, R. The Wear Rate Forecast of MgO-C Materials Type MC95/10 in the Slag Spout Zone of an Oxygen Converter in Terms of the Bayesian Estimation. *Materials* **2022**, *15*, 3065. https://doi.org/10.3390/ma15093065

Academic Editor: Andrea Piccolroaz

Received: 22 March 2022
Accepted: 21 April 2022
Published: 22 April 2022

Publisher's Note: MDPI stays neutral with regard to jurisdictional claims in published maps and institutional affiliations.

1. Introduction

The oxygen converter lined with MgO-C refractory materials is one of the two main units used to prepare ferroalloy for secondary metallurgy processing. It is estimated that the amount of steel produced by the converter process is currently about 50% of heated steel [1].

The high process temperatures, changing chemical composition of the slag, and turbulent flow of the heterophase metallic-ceramic liquid caused by blowing oxygen into it, make the working conditions of the refractories in the oxygen converter very variable. The wear process of refractories is dominated by the cyclic processes of oxidation of the carbon part of MgO-C type refractories, dissolution, and abrasion of the near-surface parts of materials, caused by penetration of the liquid alloy [2,3].

The problem of modeling the wear of the refractory materials under the influence of different metallurgical factors has been discussed by various authors [4–8].

In oxygen converters, intensive oxidation of the carbon, silicon, phosphorus, and manganese contained in the batch results in uneven wear of the refractory lining. For this reason, refractories are lined in the converter in a zoned manner, as shown in Figure 1. In the refractory lining zones with particularly unfavorable effects from corrosive factors, the best quality refractories are used. The greatest wear zone of the refractory lining in the oxygen converter is the so-called slag spout zone. The slag spout zone consists of two areas in the converter (left and right trunnion), lying symmetrically in relation to the taphole and with difficult access for maintenance of the refractory lining.

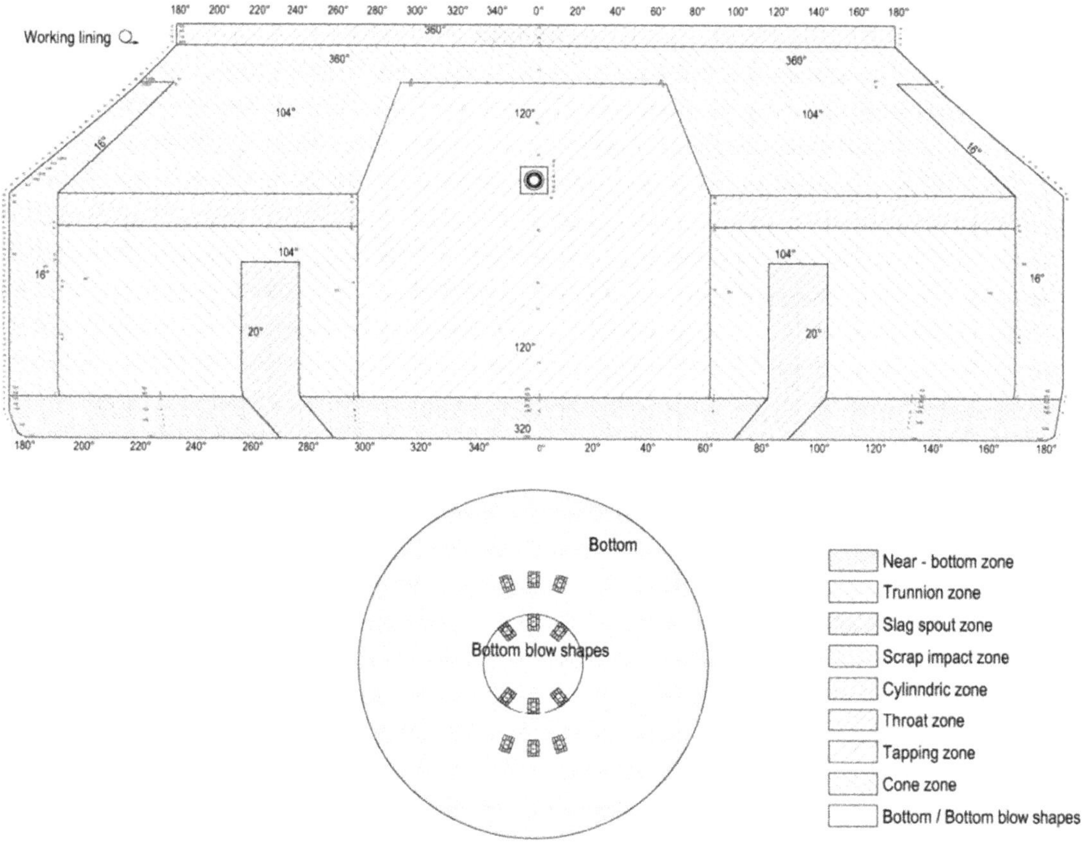

Figure 1. Zone scheme of the working lining with refractory materials.

In the slag spout zone, the thermal and chemical loads are mainly concentrated during the converter campaign. An exception is the initial stage, which usually lasts until about the hundredth heat, where thermomechanical loads caused by thermal stresses generally dominate.

The purpose of the paper is to present the possibility of applying Bayesian modelling in predicting the ranges of reliability of the unit wear rate of the refractory oxygen converter lining in the slag spout zone during the converter campaign. At the same time, the assumed values of the *a priori* distribution of the wear rate value result from the known metallurgical working conditions, to which the MgO-C material of the MC 95/10 type is exposed to. The result of the calculations presented in the article is the *a posteriori* distribution of the wear rate, from which the desired range of reliability of this rate, treated in the calculations as a random variable, is calculated. This paper also gives a procedure for predicting the unit wear rate value of the refractory lining in the slag spout zone of an oxygen converter.

2. Materials and Methods

The subject of the study is unit wear rate of the spout refractory lining of an oxygen converter type TBM (Thyssen bottom metallurgy), with a nominal load capacity of 350 tons. The converter completed 2063 heats and was shut down due to the wear of the spout zone's refractory lining. The converter was operated from August 2018 to January 2019. A high purity MgO-C material bonded with pitch was used in the slag spout zone. The material is classified according to PN-EN ISO 10081-3:2006 (classification of dense shaped refractory products—Part 3: basic products containing from 7 percent to 50 percent residual carbon) as type MC 95/10. Its chemical composition determined by WD XRF technique based on PN-EN ISO 26845:2009 (Chemical analysis of Refractory products by XRF—Fused cast bead method) is presented in Table 1.

Table 1. Material properties of MC95/10.

MgO *	97.7 wt.%	XRF-X'Unique II
CaO *	1.5 wt.%	XRF-X'Unique II
SiO_2 *	0.4 wt.%	XRF-X'Unique II
Fe_2O_3 *	0.3 wt.%	XRF-X'Unique II
Al_2O_3 *	0.1 wt.%	XRF-X'Unique II
Apparent porosity	1.5%	PN—EN 993-1
Bulk density	3.06 g·cm^{-3}	PN—EN 993-1
Cold crushing strength	35 MPa	PN—ISO 10059-1
Total carbon content	14 wt.%	HFF—IR Leco CS300

*—share of the component in the magnesia part (without carbon and antioxidants).

Shaped bricks made of the refractory material were formed on a hydraulic press with a unit pressure of over 1200 bar. The slag spout zone was lined with a combination of 90/80 and 90/20 flat wedges. The dimensions of the wedges are: 900 × 100 × 190/110 mm and 900 × 100 × 160/140 mm. The initial thickness of the working lining in the slag spout zone was 900 mm.

The converter is not equipped with a slag splashing installation. During the campaign, however, the refractory lining of the converter was maintained by gunning and slagging. The chemical compositions of the slags were typical for the converter process. Spectrometric, oxide chemical analysis of samples collected ($n = 1459$) during the campaign showed an average content of MgO, CaO, SiO_2, Fe_2O_3, Al_2O_3 of 6.33 wt.%, 46.85 wt.%, 13.22 wt.%, 15.67 wt.%, 1.26 wt.%, respectively. The average basicity of the slag was about 3.74.

The wear rate of a refractory lining is usually expressed by the material wear rate per heat. The material loss depends on many process factors and on the properties of the lining material in a given zone. The unit wear rate of refractory materials is the basis for assessing the economic efficiency of the used materials, measured by the unit cost of steel production. The result of the lining thickness control, especially in the final stage of the converter operation, is the basis of the decision to end the equipment's operation. On the basis of the measurements made during the campaign, decisions are also made on gunning and/or slagging the working lining of the oxygen converter.

Measurements of the residual thickness of the lining are made with a laser device. The device is placed in front of the empty converter in the slag spout position, so that a concentrated photon beam can scan the inside of the device. Before starting the converter campaign, a baseline measurement of the refractory lining is taken as a reference for the residual thickness measurements of the refractory working lining. The baseline measurement for subsequent measurements of the residual working thickness of the lining is taken at the converter's safety lining. The safety lining is installed behind the working layer and provides refractory protection of the converter against liquid steel leakage when the working lining is already heavily worn. Safety linings are usually made of burnt magnesia materials type M90 or M95 (PN-EN ISO 10081-2:2006 (classification of dense shaped refractory products—Part 2: basic products containing less than 7% residual carbon). The frequency of the residual thickness measurements is dependent on the availability of the device (converter) and the condition of its refractory working lining. The results of the measurements, in the form of wear maps with values of residual thickness in all zones, constitute the basis for assessment of the condition of the refractory lining, qualifying the converter for further operation. The unit wear rate of the refractory materials is calculated from the residual thicknesses measured during the converter's campaign, as the difference of the consecutively measured residual thicknesses of the refractory lining related to the difference in the number of heats, and dividing these measurements according to the formula:

$$w_i = (G_k - G_i)/(i - k) \tag{1}$$

where w_i—unit wear rate, G—laser-measured residual thickness, i and k are the number of successive heats for which measurements have been made, where $i > k$.

In the calculation part of the paper, Bayesian inference was used. This allows the use of *a priori* information (often called initial belief), resulting from knowledge of the research problem and the knowledge of the given phenomenon, in order to obtain *a posteriori* information (often called empirical probability, outcome), obtained using the likelihood function. This methodology is based on Bayes' theorem of conditional probability, known from probability calculus [9–11]:

$$P(A|B) = \frac{P(B|A)P(A)}{P(B)} \tag{2}$$

where $P(A|B)$, $P(B|A)$ are conditional probabilities, and $P(A)$, $P(B)$ are probabilities of events A and B. Bayes' theorem for a continuous probability distribution has the form [12]:

$$P(\mu|dane) = \frac{P(dane|\mu) \cdot P(\mu)}{\int_{-\infty}^{\infty} P(dane|\mu) \cdot P(\mu) d\mu} \tag{3}$$

The derivation of the formulas for the mean value and parameters' precision of the *a posteriori* distribution when the *a priori* distribution is a normal distribution, and the observed data obtained from a normal distribution is given, for example, in [6]. These quantities are expressed by Equations (4) and (5):

$$\mu_{aposteriori} = \frac{\tau_0 \cdot \mu_0 + \tau \cdot \sum_{i=1}^{n} w_i}{\tau_0 + n \cdot \tau} \tag{4}$$

$$\tau_{aposteriori} = \tau_0 + n \cdot \tau \tag{5}$$

where $\tau = 1/\sigma^2$ is the precision, σ—standard deviation, n—number of measurements of analyzed quantity, w_i—consecutive measurement, τ_0—assumed precision of the distribution, indices zero denote hyperparameters of the distribution that result from *a priori* knowledge of the problem. In another notation, without the use of the quantity called precision, formulas for the mean value and variance of *a posteriori* distribution are derived, e.g., in the work [13].

3. Results

The averaged results of the laser measurements of the wear of the oxygen converter's refractory lining in the slag spout zone (left and right spout) are shown in Figure 2.

The thickness of the refractory lining of the converter decreases over time. The points visible in Figure 2, where an increase in the thickness of the lining was found, come from measurements taken after maintenance (repair) of the converter's refractory lining, using auxiliary materials. This means that, in order to protect the refractory lining, a layer of material, in the form of refractory gunning masses and/or processed slag partially saturated with MgO, had been applied to the walls of the converter, which reduces the wear rate of the working lining. In the following calculations, the thickness measurements of the refractory lining for which an increasing residual thickness was observed compared to the previous measurement were not considered. Measurements made at the beginning of the campaign were also not taken into account. The high loss of the refractory lining, expressed by the loss of the refractory material in the slag spout zone, amounted to 182 mm in the initial period of the converter operation, till the 123th heat. As already mentioned, this was connected with thermomechanical loads. Therefore, for further calculations the measured data from the 124th heat and onwards were adopted. The corrected curve of the residual thickness measurements is shown in Figure 3.

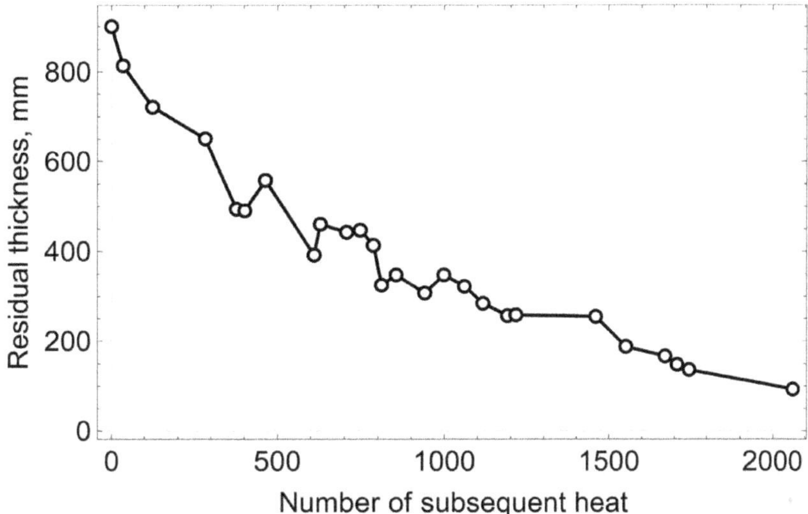

Figure 2. Average residual thickness of the refractory lining of the spout zone.

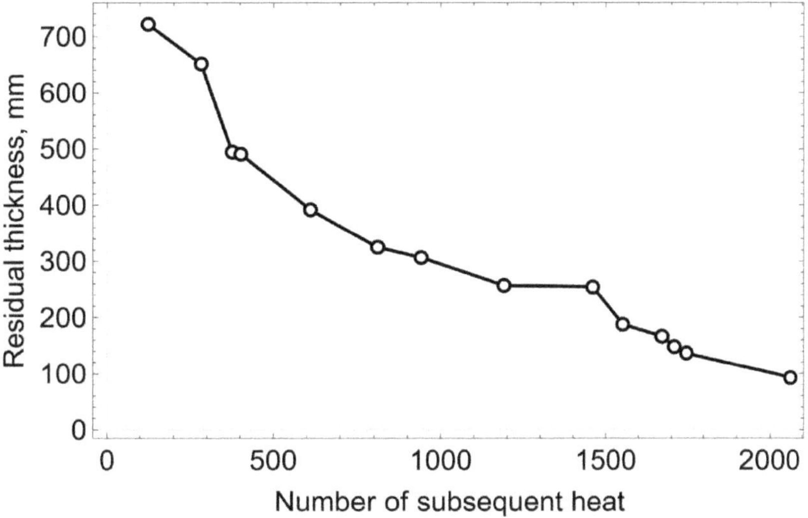

Figure 3. Corrected average residual thickness of the refractory lining of the spout zone.

The wear rate values of the spout zone's refractory lining for the cases considered in the graph shown in Figure 3 are as follows: w_n = 443, 1670, 135, 476, 335, 138, 200, 9, 715, 179, 487, 316, and 137 µm heat^{-1}, where n = 1, ... , 13. From the data analysis of the previous converter campaigns, it is believed that the average wear rate in the campaign will be μ_0 = 300 µm heat^{-1}, with the standard deviation σ_0 = 100 µm heat^{-1}. The *a priori* distribution of the unit wear rate values of the refractory lining expressed by the normal distribution for the assumed assumptions is presented in Figure 4.

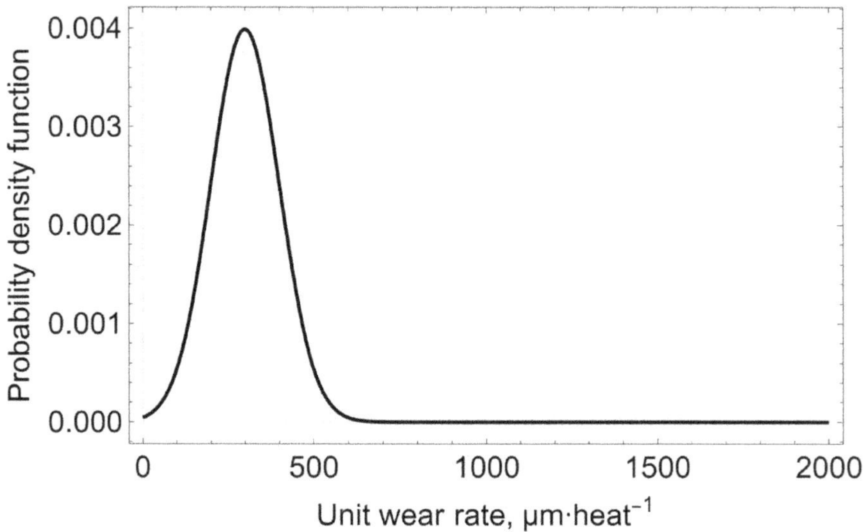

Figure 4. *A priori* normal distribution, expressing the initial belief in the distribution of the unit wear rate with parameters μ_0 = 300 μm heat^{-1}, σ_0 = 100 μm heat^{-1}.

The mean value of the *a posteriori* normal distribution for the conjugate solution, assuming the normality of the *a priori* distribution, is calculated according to Formulas (1) and (4). The precision, which is the inverse of the variance σ^2, is expressed by Formula (5). For a known standard deviation σ = 50 μm heat^{-1} and with the hyperparameters of the *a priori* distribution being μ_0 = 300 μm heat^{-1} and σ_0 = 100 μm heat^{-1}, the values of $\mu_{\text{aposteriori}}$ and $\tau_{\text{aposteriori}}$ calculated from Formulas (4) and (5) for successive measurements are shown in Figures 5 and 6.

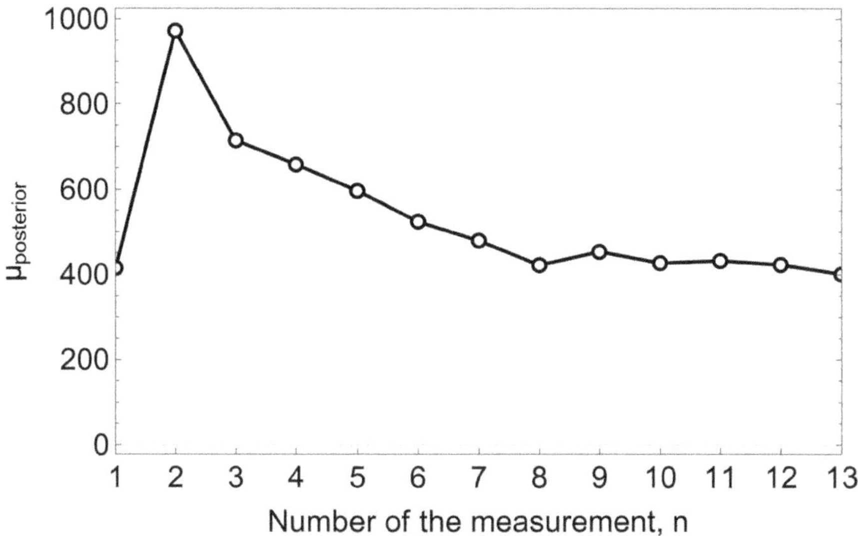

Figure 5. The variation of the mean value of μ μm heat^{-1} of the *a posteriori* distribution of the unit refractory wear rate in the spout zone.

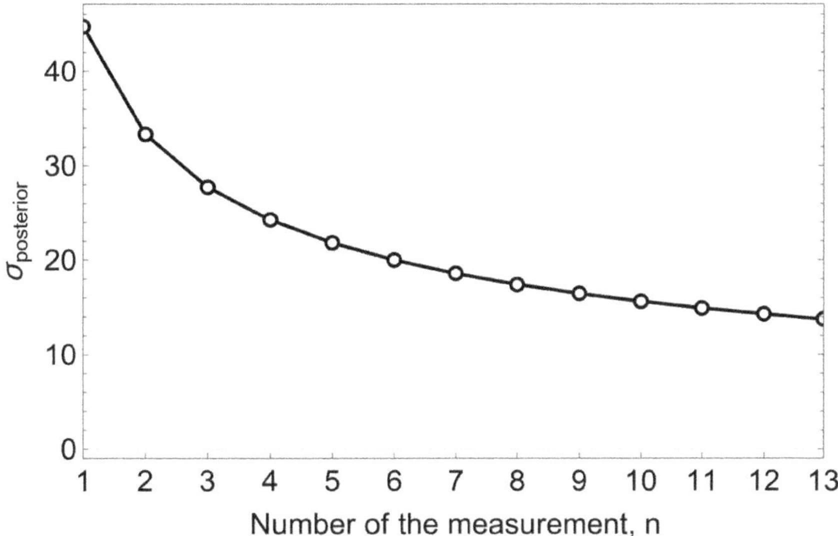

Figure 6. The variation of the standard deviation σ µm heat^{-1} of the *a posteriori* distribution of the unit refractory wear rate in the spout zone.

The evolution of the *a posteriori* distribution of the average refractory wear rate of the spout zone, calculated from Equations (4) and (5) for 13 consecutive rate measurements, is shown in Figure 7.

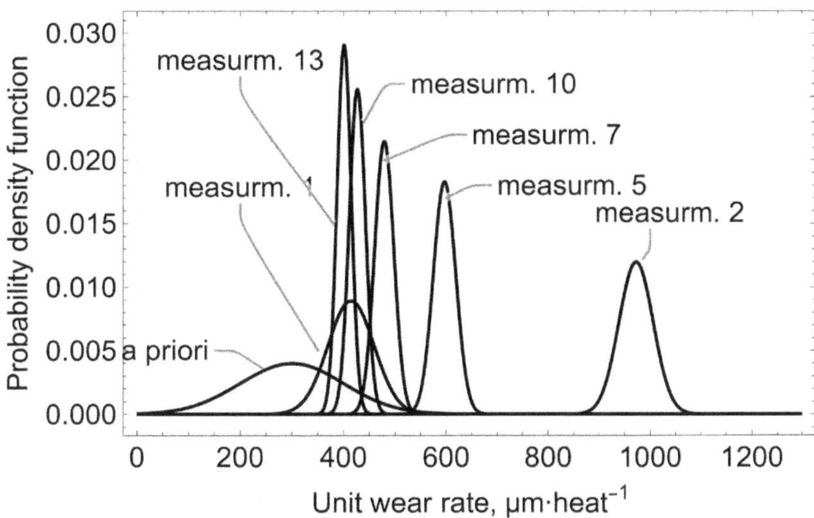

Figure 7. The Bayesian evolution of the mean value of μ µm heat^{-1} unit wear rate of MC 95/10 refractory material in the spout zone, calculated for selected measurements: 1, 2, 5, 7, 10, 13.

The obtained results allow the calculation of credible intervals for the random variable µm heat^{-1}, which are equivalent to the confidence intervals in the frequency approach to statistical inference. Thus, the average unit wear rate of refractories after 13 measurements is $\mu = 401.23$ µm heat^{-1}, standard deviation $\sigma = 13.74$ µm heat^{-1}. The calculated quantiles of the unit wear rate for *a posteriori* distribution are shown in Table 2.

Table 2. Selected quantile values for the unit refractory wear rate in the slag spout zone calculated from the a posteriori distribution.

Quantile	$\mu_{aposteriori}$, $\mu m\ heat^{-1}$
0.01	369.2
0.025	374.2
0.05	378.5
0.25	391.9
0.5	401.1
0.75	410.4
0.95	423.7
0.975	428.1
0.99	433.1

The resulting *a posteriori* distribution for the Bayesian inference can be used as a starting point for further calculations, e.g., to forecast the wear rate. With the new results, a new confidence interval can be calculated, as well as a new mean value and standard deviation of the *a posteriori* distribution.

The calculations above were carried out using conjugated solutions, where the *a priori* distribution and the likelihood function were expressed by normal distributions. The result of the calculation, in the form of an *a posteriori* probability distribution of the unit refractory wear rate value, also belongs to the class of normal probability distributions of random variables.

These calculations were verified using the R software and one of the libraries supporting Bayesian inference. In contrast to the conjugated solution, the use of the Monte Carlo method based on Markov chains involves the generation of random samples from *a posteriori* distributions, together with the approximation of their characteristics. The results obtained using the MCMCpack library and the Monte Carlo algorithm [14–16], sampling in a Markov chains sequence for 10,000 samples, can be obtained using the following R code:

```
library(MCMCpack)
#data
w = c(443, 1670, 135, 476, 335, 138, 200, 9, 715, 179, 487, 316, 137)
#known standard deviation
sigma2 = 50^2
# hyperparameters of the apriori distribution
mu0 = 300
tau20 = 100^2
# function call to calculate the distribution of aposteriori
set.seed(1234)
posterior <- MCnormalnormal(y, sigma2 = sigma2, mu0 = mu0, tau20 = tau20, 10,000)
# summary of results
summary(posterior)
quantile(posterior, c(0.025,0.5,0.975))
mean(posterior)
# apriori and aposteriori distribution diagram
plot(posterior)
grid <- seq(0,1300,0.5)
plot(grid, dnorm(grid, 300, 100), type = "l", col = "red", lwd = 3, ylim = c(0,0.03),
     xlab = "mu", ylab = "density")
lines(density(posterior), col = "blue", lwd = 3)
legend('topright', c("prior", "posterior"), lwd = 3, col = c("red", "blue"))
```

The results obtained using the R code and the MCMCpack library are:

- average wear rate of 401.22 $\mu m\ heat^{-1}$,
- 95% confidence interval: [374.77, 427.86] $\mu m\ heat^{-1}$.

The accuracy of the calculations was also checked using the "MetropolisHastingsSequence" function [17] available in the repository of additional functions of the Wolfram language [18]. The calculations performed with the use of Wolfram Mathematica version 12.31 allowed obtaining the 95% confidence interval of the unit refractory wear rate in the slag spout zone of the converter, being:

$$375.7 \leq \mu_{\text{aposterior}} \leq 428.2 \ \mu\text{m heat}^{-1}. \tag{6}$$

Both ranges are insignificantly different.

The unit wear rate value of the above described refractory lining of the oxygen converter treated as a random variable was experimentally verified after the oxygen converter's campaigns, which were in operation between February 2019 and November 2019. The mean value of the unit wear rate was calculated as $\mu = 420 \ \mu\text{m heat}^{-1}$, and this is within the above calculated confidence ranges.

In order to predict the value of the unit wear rate of the refractory lining in the slag spout zone of the oxygen converter, it was assumed that the hypothetical 14th measurement result of the residual thickness of the residual lining would be $x_p = 650 \ \mu\text{m heat}^{-1}$ then, using Formulas (4) and (5) for i = 14 and n = 14, new mean values of the unit wear rate and its standard deviation were calculated. The new *a posteriori* distribution is shown in Figure 8.

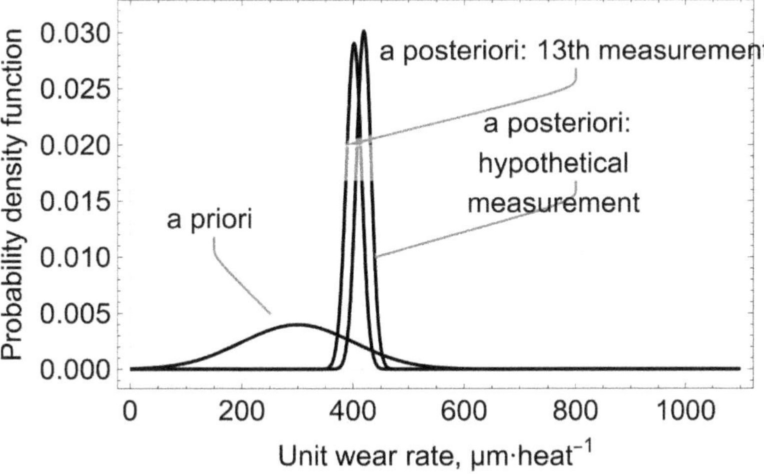

Figure 8. The prediction of the *a posteriori* distribution for the unit wear rate calculated for the hypothetical 14th measurement of $x_p = 650 \ \mu\text{m heat}^{-1}$.

The predicted mean value of the unit wear rate and the standard deviation will be $\mu_p = 418.6 \ \mu\text{m heat}^{-1}$ and $\sigma_p = 13.1 \ \mu\text{m heat}^{-1}$, respectively. The calculated quantiles of the unit wear rate for the predicted *a posteriori* distribution calculated in $\mu\text{m·heat}^{-1}$ are shown in Table 3.

A question that is important for converter users is, what is the probability for the specific wear rate to be, for example, less than $400 \ \mu\text{m heat}^{-1}$. After using a hypothetical 14th measurement of the specific wear rate ($x_p = 650 \ \mu\text{m heat}^{-1}$) in the calculation and determining the form of its *a posteriori* distribution, the probability that the specific wear rate will be less than $400 \ \mu\text{m heat}^{-1}$ is less than 8%, as shown in Figure 9 by the shaded area under the probability density distribution curve.

Table 3. Selected quantile values for the predicted unit refractory wear rate in the slag spout zone for a hypothetical 14th measurement of x_p = 650 µm heat^{-1}, calculated from the a posteriori distribution.

Quantil	$\mu_{aposterior}$, µm heat^{-1}
0.01	387.8
0.025	392.6
0.05	396.8
0.25	409.7
0.5	418.6
0.75	427.5
0.95	440.4
0.975	444.6
0.99	449.4

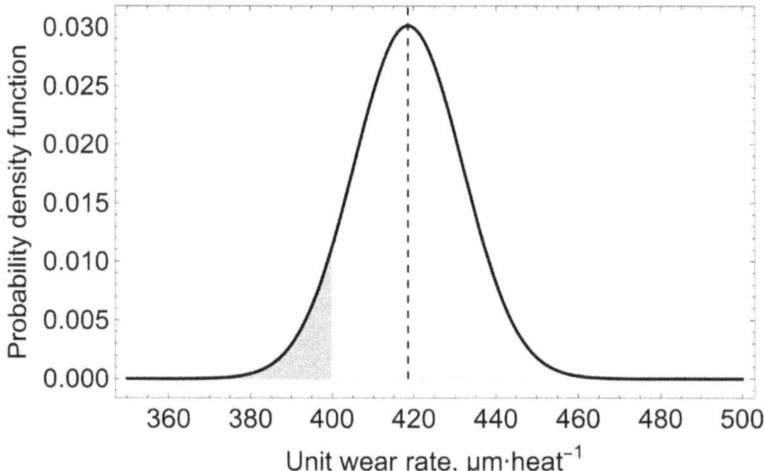

Figure 9. The shaded area shows an 8% probability that the value of the unit refractory wear rate in the oxygen converter slag spout zone will be below 400 µm heat^{-1}.

4. Discussion

The Bayesian evolution of the mean value of μ µm heat^{-1} unit wear rate of the refractory material MC 95/10 in the spout zone, shown in form of graphs in Figures 5–7, illustrates the effect of changing the *a priori* distribution and the likelihood function on changing the *a posteriori* distribution by changing the mean value of the distribution and reducing its variance as the information from the measurement-sampling increases.

Subsequent iterations in the model calculations, carried out by increasing the sample amount (measurements), updated the knowledge of the distribution probability of the value of the unit refractory wear rate in the slag spout zone and its variance.

The results predicting the mean and distribution variance of the unit wear rate value of the steel converter's refractory lining, obtained assuming the value x_p = 650 µm heat^{-1} of the hypothetical 14th measurement, indicate a change in the parameters of the predicted probability distribution to the values of μ_p = 418.6 µm heat^{-1} and σ_p = 13.1 µm heat^{-1}. The results shown in [4–8] cannot be directly compared with the above mentioned results, because they relate to other materials, devices, and research methodologies. The results of the measurements and calculations included in this article extend the range of data concerning the wear of refractory linings in oxygen converters.

The method of searching for an answer to the question of the probability of obtaining a value contained in a given range of values of the unit wear rate of a refractory lining is reduced to finding the size of the area under the curve of the forecast probability density distribution of the unit wear rate of a refractory material.

The choice of Bayesian statistics for forecasting the unit wear rate of MgO-C materials type MC 95/10 in the slag spout zone of an oxygen converter was dictated by the variety and complexity of factors affecting the decision to terminate the converter campaign. Making such a decision is conditioned, both by the laser measurement results of the residual refractory lining thickness of the converter, and by visual assessment of the condition of the converter's refractory lining, which may vary depending on the operating experience of the people making the decision to stop the converter's operation.

5. Conclusions

From the analysis of the carried out experiment and the calculations performed according to the rules of Bayesian statistics, the following conclusions are drawn:

From the experience gained during the long-term operation of the steel-producing oxygen converter, it was found that the value of the unit wear rate of the refractory material in the slag spout zone of the steel-producing oxygen converter is subject to *a posteriori* normal distribution, with the exemplary parameters: mean value $\mu = 401.23$ μ heat^{-1}, standard deviation $\sigma = 13.74$ μm heat^{-1}.

The mean value of the unit wear rate during the oxygen converter's campaigns between February and November 2019 was verified experimentally and was calculated as 420 μm heat^{-1}, which is within the calculated confidence interval $375.7 \leq \mu_{apposterior} \leq 428.2$ μm heat^{-1}.

The calculated results of the forecasted distribution of the unit wear rate value of the refractory material give valuable operational and cost information, which is important for steelmakers, as well as for refractory material manufacturers.

From a prediction point of view, of the value distribution of the refractory unit wear rate, it would be beneficial to distribute the successive measurements of the residual thickness of the refractory lining in the slag spout zone evenly over the course of the converter campaign.

This model allows the creation of "what if" scenarios, which can also be helpful in estimating the cost of a planned converter campaign.

Author Contributions: Data curation, W.Z.; Formal analysis, S.S.; Investigation, W.Z.; Methodology, W.Z. and R.L.; Supervision, R.L.; Validation, R.L.; Writing—original draft, W.Z.; Writing—review & editing, S.S. All authors have read and agreed to the published version of the manuscript.

Funding: This research received no external funding.

Institutional Review Board Statement: Not applicable.

Informed Consent Statement: Not applicable.

Data Availability Statement: Not applicable.

Conflicts of Interest: The authors declare that they have no known competing financial interests or personal relationships that could have appeared to influence the work reported in this paper.

References

1. Lis, T. *Współczesne Metody Otrzymywania Stali*; Wydawnictwo Politechniki Śląskiej: Gliwice, Poland, 2000.
2. Nadachowski, F.; Kloska, A. *Refractory Wear Processes*; AGH: Kraków, Poland, 1997; pp. 43–44, 52–76.
3. Takanaga, S. Wear of Magnesia-Carbon Bricks in BOF. *Taikabutsu Overseas* **1993**, *13*, 8–14.
4. Vollmann, S. Investigation of Rotary Slagging Test by Computational Fluid Dynamics Calculations. *Taikabutsu Overeas* **2010**, *30*, 10–18.
5. Harmuth, H.; Vollmann, S.; Rössler, R. Corrosive wear of refractories in steel ladles—Part I: Fundamentals. In Proceedings of the Unified International Technical Conference of Refractories, UNITECR, Kyoto, Japan, 30 October–2 November 2011.
6. Vollmann, S.; Harmuth, H.; Kronthaler, A. Corrosive Wear of Refractories in Steel Ladles—Part II: Influence of Fluid Flow. In Proceedings of the Unified International Technical Conference of Refractories, UNITECR, Kyoto, Japan, 30 October–2 November 2011.
7. Potschke, J.; Deniet, T. *The Corrosion of Refractory Castables*; Refractories Manual; Interceram, Expert Fachmedien Gmbh: Düsseldorf, Germany, 2005.

8.   Zelik, W.; Lech, R. Forecasting of the Wear of Selected Refractory Material of the MgO-C type in the Slag Zone of a Steel Ladle Using Dimensional Analysis. In Proceedings of the Unified International Technical Conference of Refractories, UNITECR, Yokohama, Japan, 13–16 October 2019; pp. 407–410.

9.   Devore, J.L. *Probability and Statistics for Engineering and the Sciences*, 7th ed.; Thomson Brooks/Cole: Boston, MA, USA, 2008; pp. 72–73.

10.   Fisz, M. *Rachunek Prawdopodobieństwa i Statystyka Matematyczna*; Państwowe Wydawnictwo Naukowe: Warszawa, Poland, 1954; pp. 22–23.

11.   Bolstad, W.M.; Curran, J.M. *Introduction to Bayesian Statistics*, 3rd ed.; John Wiley & Sons, Inc.: Hoboken, NJ, USA, 2017; pp. 68–69, 211–236, 456–457, 461–463.

12.   Therese, M.; Donovan, R.M. *Mickey, Bayesian Statistics for Beginners, A Step by Step Approach*, 1st ed.; Oxford University Press: Oxford, UK, 2018; pp. 198–204, 210, 228–246, 250–260, 379–383.

13.   Grzenda, W. Modelowanie bayesowskie. In *Teoria i Przykłady Zastosowań*; Oficyna Wydawnicza SGH—Szkoła Główna Handlowa w Warszawie: Warszawa, Poland, 2016; pp. 71–90.

14.   Martin, A.D.; Quinn, K.M. *Applied Bayesian Inference in R Using MCMCpack*; R News; The R Foundation: Vienna, Austria, 2006; Volume 6/1, pp. 2–7.

15.   Martin, A.D.; Quinn, K.M.; Park, J.H. MCMCpack: Markov chain Monte Carlo (MCMC) Package. *J. Stat. Softw.* **2011**, *42*. [CrossRef]

16.   Available online: https://cran.r-project.org/web/packages/MCMCpack/index.html (accessed on 5 October 2021).

17.   Available online: https://resources.wolframcloud.com/FunctionRepository/resources/MetropolisHastingsSequence (accessed on 8 October 2021).

18.   Available online: https://wolfram.com/mathematica/?source=nav (accessed on 8 October 2021).

 *materials*

Article

Use of Neural Networks for Lifetime Analysis of Teeming Ladles

Dalibor Jančar [1], Mario Machů [1], Marek Velička [1,*], Petr Tvardek [2], Leoš Kocián [2] and Jozef Vlček [1]

1 Department of Thermal Engineering, Faculty of Materials Science and Technology, Institute of Environmental Technology, VSB-Technical University of Ostrava, 17. Listopadu 2172/15, 708 00 Ostrava, Czech Republic
2 Liberty Ostrava a.s., Vratimovská 689/117, 719 00 Ostrava, Czech Republic
* Correspondence: marek.velicka@vsb.cz; Tel.: +42-05-9732-1538

Abstract: When describing the behaviour and modelling of real systems, which are characterized by considerable complexity, great difficulty, and often the impossibility of their formal mathematical description, and whose operational monitoring and measurement are difficult, conventional analytical–statistical models run into the limits of their use. The application of these models leads to necessary simplifications, which cause insufficient adequacy of the resulting mathematical description. In such cases, it is appropriate for modelling to use the methods brought by a new scientific discipline—artificial intelligence. Artificial intelligence provides very promising tools for describing and controlling complex systems. The method of neural networks was chosen for the analysis of the lifetime of the teeming ladle. Artificial neural networks are mathematical models that approximate non-linear functions of an arbitrary waveform. The advantage of neural networks is their ability to generalize the dependencies between individual quantities by learning the presented patterns. This property of a neural network is referred to as generalization. Their use is suitable for processing complex problems where the dependencies between individual quantities are not exactly known.

Keywords: neural networks; refractory material; ladle; modelling

Citation: Jančar, D.; Machů, M.; Velička, M.; Tvardek, P.; Kocián, L.; Vlček, J. Use of Neural Networks for Lifetime Analysis of Teeming Ladles. *Materials* **2022**, *15*, 8234. https://doi.org/10.3390/ma15228234

Received: 11 October 2022
Accepted: 16 November 2022
Published: 19 November 2022

Publisher's Note: MDPI stays neutral with regard to jurisdictional claims in published maps and institutional affiliations.

1. Introduction

One of the most critical aggregates of a steel plant is the teeming ladle. The lining, made of high-quality, heat-resistant materials, guarantees its long service life. Choosing the appropriate refractory lining requires knowledge of the circulation of the casting ladle during production and knowledge of the individual influences aggressively acting on the refractory lining. As a result of the multiple uses of the ladle, its lining wears out during the ladle campaign. This wear is caused by the erosive and corrosive effect of molten metal and slag. Slowing down the wear process of the lining can be guaranteed not only by choosing the type of refractory materials but also by other measures, especially by a suitable thermal regime of the ladle. Maintaining the temperature of the lining of the casting ladle at high, and, if possible, constant temperatures positively affects the degree of wear of the lining [1]. It is evident that many operating factors affect the lining of the ladle and that there is a mutual connection between them, i.e., that the chosen lifetime solution method must be able to compare a large set of non-linear data [2].

Digitalization of production technologies is also essential in energy intensive industries, such as the steel industry, where emphasis is placed on optimizing the production chain and sustainable production [3]. Artificial intelligence tools [4] are used to estimate the ageing of ladle furnaces, where real data are used for tuning the model and then identifying less critical situations at the end of the ladle's life [5]. Another possibility is to establish a mechanism for predicting the service life of composite ladle structures, which is based on the stress analysis of the steel shell of the ladle and combines conventional fatigue analysis with extended fracture theory. These techniques allow the prediction of the service life of the steel shell by detecting its crack length [6]. A mathematical model of heat transfer through the lining based on technological characteristics and thermal imaging images can be used to diagnose the thickness of the casting ladle in critical areas [7]. Different diagnostics obtain directly measured data using a thermal imaging camera, allowing the determination of the residual thickness of the refractory lining of the metallurgical ladle [8]. For predicting refractory wear of furnaces, linear models are used, which work on the principle of the Kalman filter [9]. For processing a large volume of data intended for diagnostics, the Bayesian network model can be used, which uses the Hadoop software structure and thus achieves a high efficiency of knowledge reasoning [10].

Moreover, special models are proposed for diagnostics and predictions of the current condition of the lining in high-temperature objects in metallurgy [11]. LabVIEW software can also create and monitor the casting ladle system database, which collects data on ladle usage and filling [12]. The use of neural networks has seen a significant upward trend in recent years [13,14]. With the help of neural networks, the need for maintenance can be predicted based on an analysis of the time series of the main parameters of the production equipment [15]. The application of neural networks for device diagnostics is extensive; it is used, for example, to detect critical zones in basins used for the marine industry [16]. Neural networks are also used for the metallographic quality control of metals and for the recognition of their microstructures [17].

The lifetime assessment of casting ladles was carried out using the method of neural networks. A set of input data (operating factors affecting the lifetime of the lining) was compared with an output parameter (number of castings, i.e., the number of cycles on a continuous casting machine, the so-called CMM).

2. Materials and Methods

In general, every computer can be understood as a device that displays a set of input data (e.g., coefficients of a linear system) on a set of other data (e.g., the solution of this system). Neural networks belong to the machine learning methods together with linear regression, decision tree models, support vector machines, probabilistic modelling, or genetic algorithms [18]. If we do not impose further simplifications, it can be stated that even the nervous system of an individual animal is a kind of prototype (every one is different) of a computer. Much mental activity or vegetative activity of an animal can be considered as the implementation of some algorithm (e.g., a more or less meaningful answer to a more or less meaningful question, a reaction to a given environmental stimulus, etc.) [19]. However, there are fundamental differences between these two parallel worlds of living computers and classical electronic computers. Let us try to summarize those that we believe are related to the principle of their operation (see Table 1):

Table 1. Comparison of the basic properties of a classical computer and a neural network.

Classic Computer	Neural System
• the existence of a central processor performing all the work	• the central processor has not yet been discovered in nervous tissues and nothing confirms its existence
• low parallelization, exceptionally many processors ($\leq$1000) communicating with each other	• a very high number of simpler (in terms of information transfer) processors—neurons ($\cong$1012), a very high density of connections between them (min. 10^4 for each neuron)
• exact knowledge of information evaluation and processing (RAM—Random Access Machina, RASP—Random Access Stored Program, Turing machine)	• a very vague idea of the activity of individual elements, almost no idea of the way of communication between elements
• the necessity of detailed algorithmization of calculations, supported by a deep theoretical background (e.g., the necessity of knowledge of linear algebra for solving systems of linear equations)	• it seems that the ability to evaluate different situations is an integral property of nervous systems without the need to understand and be aware of how information is processed
• very high speed of elementary operations computing	• slow information transfer (on the order of milliseconds)
• precisely defined processor architecture	• a large number of local elements connected to each other, with great variability in density and method of connection

It can be expected that an analysis of the structures, which comprise a straightforward model of the nerve centres of animals (based on the currently known findings from the fields of anatomy, biochemistry, and neurology), can be beneficial both for the research of living organisms and vice versa. The development and complementation of new types of parallel computer architectures will make it possible to solve problems for which computers have no effective algorithm [20].

Artificial neural networks find applications in damaged pattern reconstruction, classification, database searching, prediction, approximation, extrapolation, image recognition, and other fields. They can resist both their damage and errors in the input data. Its advantage is a parallel structure allowing for increased speed. They are vital when solving tasks whose data structure contains non-linearities [21].

2.1. Artificial Neuron

An artificial neuron is a simplified mathematical representation of the function of a biological neuron. A formal neuron (see Figure 1) has n generally real inputs $x_1, \ldots, x_n$ corresponding to dendrites. All inputs are evaluated by the respective synaptic weights $w_1, \ldots, w_n$, which are generally also real.

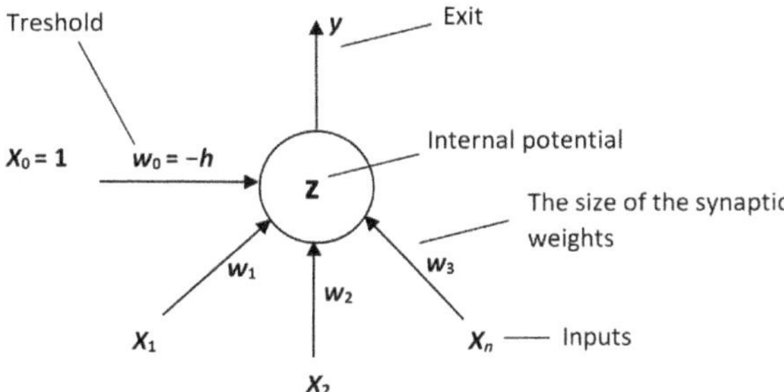

Figure 1. Formal neuron schema.

The weights determine the degree of the throughput of the input signal. The weighted sum of the input values represents the internal potential of the neuron z:

$$z = \sum_{i=1}^{n} w_i \cdot x_i - h \tag{1}$$

where h is the threshold value of the neuron (1), w is the synaptic weight (1), x is the input value (1), z is the internal potential of the neuron (1).

The output (state) of the neuron y modelling the electrical impulse of the axon is given by a generally non-linear transfer function whose argument is the internal potential z. This output is then, at the same time, the input given to other neurons, which is shown in Figure 2.

Among the most important mathematical models consisting of one, a neuron has a continuous perceptron whose potential is defined as a weighted sum of incoming signals. Its transfer function is continuous and differentiable. The most commonly used transfer functions include:

- hyperbolic tangents;
- sharp non-linearity (jump function);
- linear function;
- standard (logistic) sigmoid;
- hyperbolic tangent.

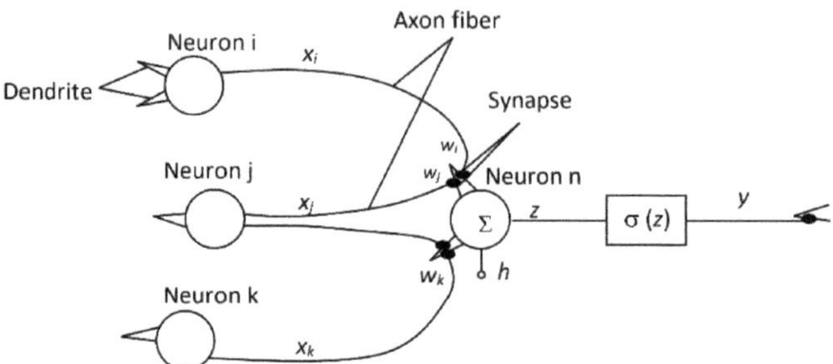

Figure 2. Schema of an artificial neuron inspired by a biological model and how it connects to other neurons.

Historically, the first unit step function (Heaviside step function) was used (Figure 3). It follows from the course of this function and the principle of network operation that, with the exclusive use of this transfer function in the entire network, it is possible to only request a two-state neuron output, which can be disadvantageous for a number of practical applications.

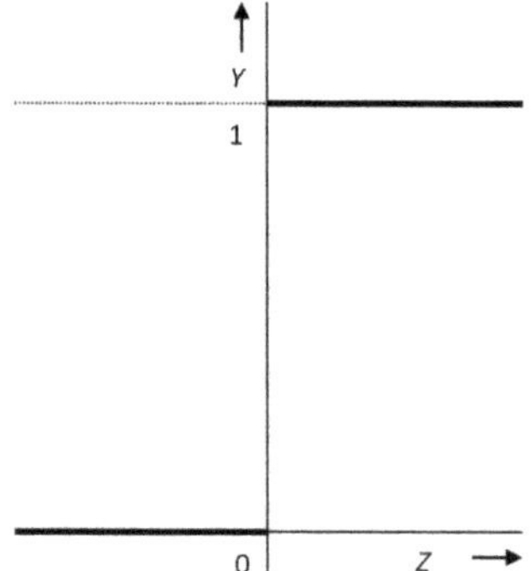

Figure 3. Non-linear Heaviside step function.

Another possible type of transfer function is a linear or piecewise linear function (Figure 4). Research has shown that, when these functions are used, the quality of the network in terms of adaptation speed and generalization ability is lower than that of networks with a non-linear transfer function.

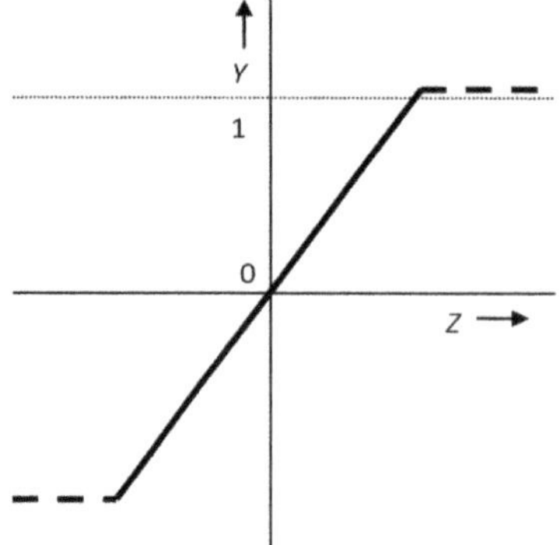

Figure 4. Linear activation function.

The presence of linear transfer functions entails one disadvantage: these functions are not resistant to disturbances, i.e., if an excessively large value comes to the input of the neuron, then the argument is transferred to the output in proportion to the direction of the linear function. However, their use is particularly important for the output layer of the neural network when it is not necessary to transform the required quantities into the interval 0-1, as is often necessary with other transfer functions that move in this interval [22,23].

The third and, at the same time, the most used type of functions are non-linear continuous monotone functions. This group's two most used functions are the sigmoid (Figure 5) and the hyperbolic tangent (Figure 6). Such transfer functions allow the neuron's sensitivity to be significant for small signals. In contrast, for larger signal levels, its sensitivity decreases. They are, therefore, highly resistant to distortions. If an excessively large value arrives at the neuron's input due to the curvature of these functions, values will appear at the output either close to 1 or 0 (−1), depending on the sign of the argument. The steepness of the sigmoid can be taken as a parameter of the network, which contributes significantly to the quality of its activity. Steepness also affects the learning (or adaptation) speed of the network. Therefore, algorithms were developed that, based on the dynamics of the decline of the so-called error function of the network, adjust the values of the steepness of the transfer functions of the entire network or individual neurons [24,25].

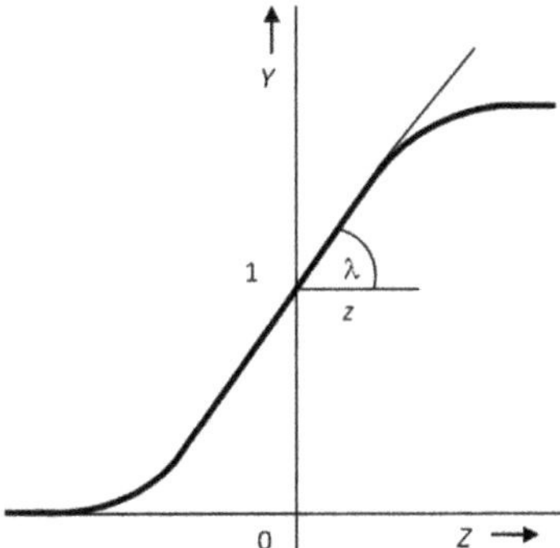

Figure 5. Standard (logistic) sigmoid.

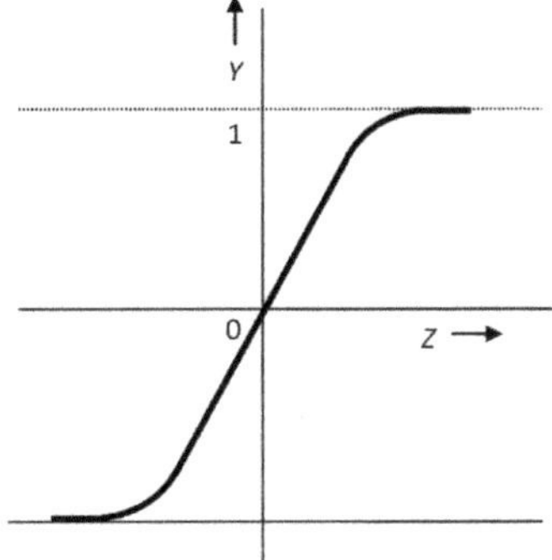

Figure 6. Hyperbolic tangent.

The equation of the standard sigmoid (Figure 5) has the form:

$$y = \sigma(z) = \frac{1}{1 + e^{-\lambda_s z}} \tag{2}$$

where λ_s indicates the slope of the sigmoid and y represents the excitation value.

The tangent equation (Figure 6) has the form:

$$y = tanh(z) \tag{3}$$

where the internal potential z is given by Equation (1).

The following conclusions follow from the above facts:

- the excitation of the neuron varies between 0 and 1, where the value 1 means full excitation of the neuron in contrast to the value 0, which corresponds to the state of inhibition;
- if the internal potential of the neuron approaches the value $+\infty$, then the neuron is fully excited, i.e., $y = 1$;
- on the contrary, if the internal potential of the neuron approaches the value $-\infty$, the neuron is completely inhibited, i.e., $y = 0$.

2.2. Neural Network

Neural networks can have one, two, or three layers, or even thousands of them, in case of deep-learning tasks. More layers increase the time required for learning. There can be multiple outputs from the neural network, or there can be only one output. In multi-layer networks, the first layer is always branching, which means that the neurons in the input layer only distribute the input values to the next layer. Since it is generally a multi-point entry into the network, we are talking about entry or output vectors of information. The required number of neurons in individual layers is variable and depends on the problem being solved. A much more appropriate way to determine the number of neurons is to use a network that itself changes this number according to the evolution of the global error [22]. Each input to the neuron is assigned a so-called weight w. Weight w is a dimensionless number that determines how significant a given input is to the respective neuron (not the network or problem). The learning ability of neural networks lies precisely in the ability to change all the weights in the network according to suitable algorithms—in contrast to biological networks, which are based on the ability to create new connections between neurons.

Physically, both learning abilities are based on different principles, but not logically. In the case of the creation of a new connection (input) in a biological neuron, it is the same as when, in a technical network, the connection between two neurons is initially evaluated with a weight of 0 (and therefore does not exist as an input for the neuron into which it enters) and at the moment when the weight changes to a non-zero number and the given connection becomes visible, i.e., when it is created [26]. Nowadays, there are many types of neural networks, each of which has a predetermined use according to its structure. For most possible applications, the feed-forward, multi-layer neural network is still the most preferred neural network [25].

The most well-known type of feed-forward neural network is the Perceptron. A schematic of a simple perceptron is shown in Figure 7. Essential characteristics:

- simple neuron, or single layer forward propagation network;
- learning: with a teacher;
- neuronal activation function: signum function, or unit jump.

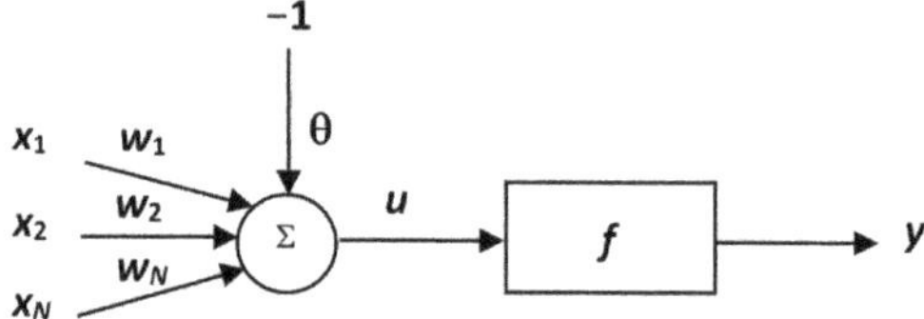

Figure 7. Simple perceptron.

Utilization:

- classification of linearly separable data (see Figure 8).

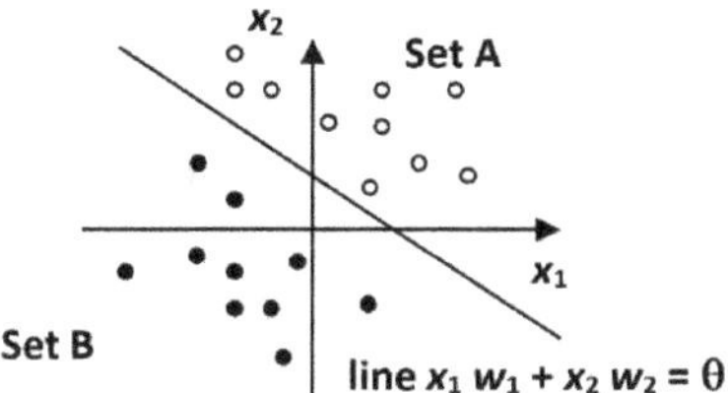

Figure 8. Data classification.

The best-known structure of neural networks used today is the multi-layer perceptron (MLP), which operates with a linear function and, by default, with a non-linear sigmoid activation function [27]. Many examples of MLP use in predictive maintenance, failure analysis, or defect occurrence prediction in industrial processes can be found in the literature, e.g., in works [28–31].

In some cases, a hyperbolic function (tanh) is also used, producing better results. MLP can be used both for quasi-regression tasks and for classification tasks. With different types of features and customizations, it can serve well for both types of tasks. For classification, MLP is used for simple decision-making (single output) and multiple classifications (multi-output). MLP networks are compact with a wide range of applications. MLP training algorithms include Back Propagation, Quick Propagation, Levenberg–Marquardt, Delta-bar-Delta, and several other modifications of training algorithms. Its significant advantage is its very fast solving of tasks. The disadvantage is slow training. An illustration of an MLP network with four layers is shown in Figure 9 [32].

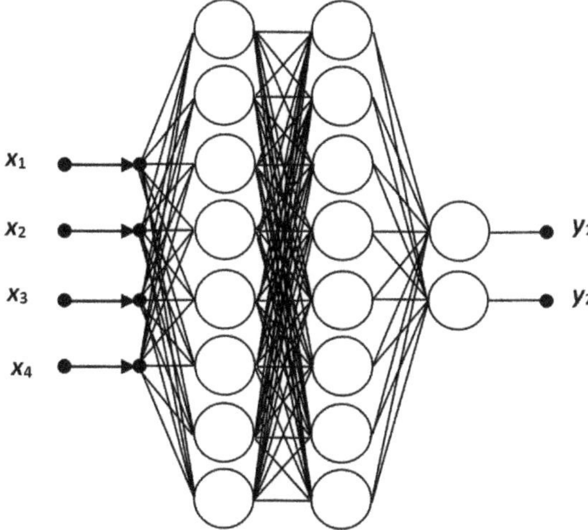

Figure 9. Multi-layer perceptron neural network (MLP).

2.3. Learning Multi-Layer Feed-Forward Neural Networks

Multi-layer, feed-forward neural networks are among the most widely used neural network models. The fundamental element of this type of network is a continuous perceptron. The network consists of at least three layers of neurons: input, output, and at least one inner layer, or hidden layer. There is always a so-called complete connection of neurons between two adjacent layers, i.e., each neuron of a lower layer is connected to all neurons of a higher layer. There are no cycles or mutual connections of neurons from the same layer.

The input to this type of network is a vector of numbers, as many as there are neurons in the input layer, since each input layer neuron has only one input. The neurons in the input layer do not perform any mathematical operations with the input values. The only task of these neurons is to distribute the input signals to the next layer so that each neuron of the first hidden layer has the entire input vector at its inputs. Signal propagation in this type of network is feed-forward propagation.

Learning aims to determine the transformation function using a particular criterion. This error function expresses the general relationship between the input and output variables with a certain approximation. Finding such a relationship means solving a specific problem area. The network learning consists of presenting individual patterns (vectors) from the training set, i.e., the individual elements forming ordered input–output pairs. Each pattern of the training set describes how the input and output layer neurons are excited.

We obtain the network's response by presenting an input vector and known network signal propagation. It will initially be quite different from the output of the specified training set. Learning means adapting the weights so that the network's responses differ as little as possible from the desired outputs in the training set [33].

We can formally define the training set T as a set of elements (patterns) that are defined as ordered pairs in the following way:

$$\begin{aligned}
T &= \left\{ [I_1, O_1], [I_2, O_2] \ldots [I_p, O_p] \right\} \\
I_i &= [i_1, i_2 \ldots i_m] \quad i_j = \langle 0, 1 \rangle \\
O_i &= [o_1, o_2 \ldots o_n] \quad o_j = \langle 0, 1 \rangle
\end{aligned} \tag{4}$$

where p is the number of training patterns, I_i is the vector of excitations of the input layer, which is made up of m neurons, O_i is the vector of excitations of the output layer (required responses to I_i), which is made up of n neurons, i_j, o_j—is the excitation of the jth input neuron and output layers.

3. Results

For our solution, data from the 230-ton casting ladle of Liberty Ostrava a.s. steelwork were used. The ladle lining is shown in Figure 10. The wall and bottom of the working lining is made of AMC fittings with a composition of 84 % Al_2O_3; 8.5% MgO; 7% residual carbon; 2.5% SiO_2. The lining in the area of the MgO-C slag line has a composition of 95% MgO and 10% residual carbon.

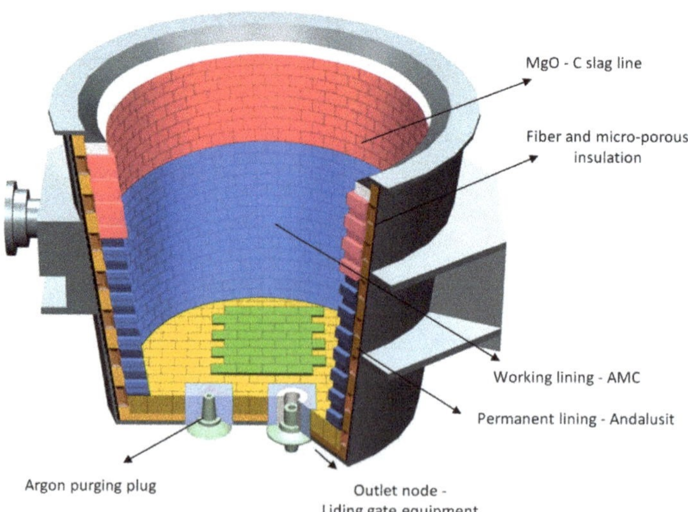

Figure 10. Description of the ladle construction [34].

In order to collect data, i.e., input data, application software was created, in which data from individual database structures created in the control system of the steel mill are processed. Following is a complete data set containing data:

- from the circulation subsystem of the ladles;
- from the so-called furnace side of the steelworks (tandem open hearth furnace, steel processing in a ladle furnace);
- from the so-called displacement side (continuous casting of steel).

The data in this application software is designed to map a ladle campaign from laying to decommissioning. Data from individual ladle cycles are always included within the given campaign. Obtained data analysis and subsequent data cleaning were carried out using the functions of the Visual Basic for Applications (VBA) programming language. The resulting file, containing 106 ladle campaigns, was subsequently used to evaluate the life of the ladle lining. It should be noted here that different steel grades were processed in the campaigns used, and therefore the obtained results are applicable to any steel company.

The problem of the influence of operational factors on the life of the lining of the casting ladle was solved in the first phase using multiple regression analysis in MS Excel. The so-called Student's *t*-test was performed to investigate the statistical significance of the individual parameters on the output parameter (number of castings). Based on this, some parameters with very little significance for the output parameter were discarded from the tested set. In the second phase, the solution was implemented in the Gensym product environment, the so-called NeurOn-Line Studio (Gensym Corporation, Austin, Texas, USA). This is a graphical, object-oriented software product for creating neural network applications. Using NeurOn-Line Studio, it is possible to create dynamic and non-linear models without knowing the structure or analytical description of the model. Data stored in databases, historians or text files, are sufficient to create a model.

3.1. Learning of Operational Factors

Among the many operating factors that were predicted to have the most significant effect on ladle lining life, the following were selected:

- empty ladle time (end of pouring at continuous casting machine (CCM) until tapping)—this is the time from the end of pouring to the next tapping when the ladle is empty. This period usually includes high-temperature heating of the ladle lining;
- full ladle period—this is the period from tapping to the end of casting, when the ladle is full of steel, or the steel level drops during casting;
- tapping temperature—this is the temperature of the steel when tapped from the given melting unit. In our case, it was a tandem furnace;
- steel temperature in the ladle after tapping—since the range of the recommended temperature of steel in the ladle after tapping of the brand of steel produced is set in the technological regulations of the steel plant, this value is not always the same for a certain brand; however, this temperature varies in a certain interval;
- electricity consumption (when heating steel in a ladle furnace)—the effect of heating steel in a ladle furnace on the wear of the ladle lining is represented in our case by the total electricity consumption for melting;
- the number of melts processed in the caisson of the vacuum station—the effect on the lining can be expressed by the number of ladle cycles when a vacuum station is included in the production process;
- argon consumption—for this evaluation, the total argon consumption per melt was used.

3.2. Testing the Training Model

The procedure for creating a model in NeurOn-Line Studio is as follows:

1. The data are imported first. NeurOn-Line Studio uses data in two basic types. The first type is time-based data, so-called Time-Based Data, where a timestamp is attached to each record. The second type is time-independent data, so-called Row-Based Data. Data

import loading data from a text file is similar to loading data from a text file in MS Excel. The imported data can be processed further using the built-in editor and their visualization in graphs;

2. Creation of the so-called preprocessor allows us to select data groups suitable for training;

3. After determining the input variables to the neural networks and the output values to be learned by the model, training or learning of the neural networks follows. The model creator or user can monitor the entire course of training. The training of the network is shown in Figure 11. The left part of the figure represents the course of the least squares deviation between the training results and the actual values, and the right graph shows the correlation between the predicted and the real value of the output.

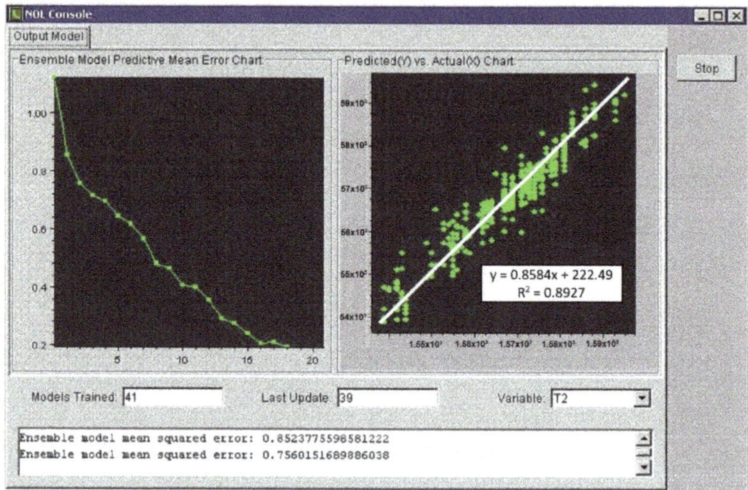

Figure 11. Neural network training progress.

The user is notified of the just-completed training by a message in the middle of the screen. The result is graphically represented by the relationship between the predicted and actual values of the output parameter of the model (Figure 12). Figures 11 and 12 show the training in the NeurOn-Line Studio program. They are not the resulting graphs for the selected periods [35].

The result of the training is a prediction model that can be used using the ActiveX component in any common programming environment under MS Windows or is directly usable in the G2 environment with the NeurOn-Line superstructure. For the model, it is possible to view the basic statistics of the Root Mean Squared Error (RMSE) model and the correlation coefficient immediately after training. The model can then be analyzed using standard statistical methods [36].

3.3. Using the Model for Real Conditions

The actual evaluation of the analysis results was carried out by comparing the set of input data mentioned above and the output parameter (the number of castings). Figure 13 shows the output after training the set, where n and the x-axis are the actual numbers of the castings and the y-axis is the number of simulated (predicted) castings.

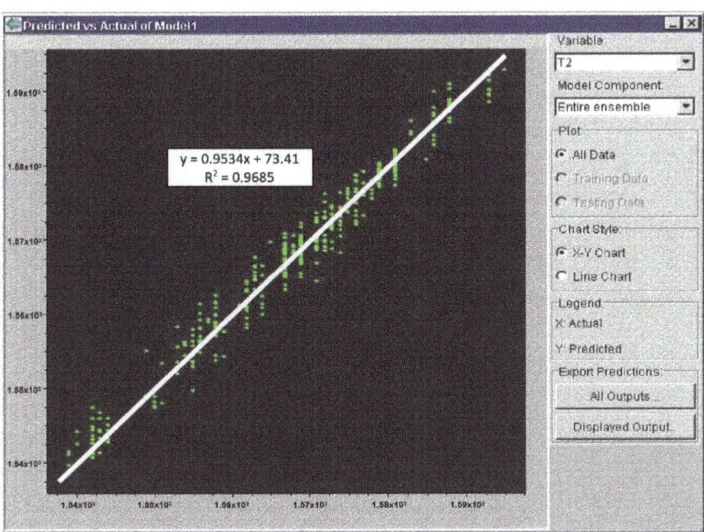

Figure 12. The predicted versus actual value of the model's output parameter.

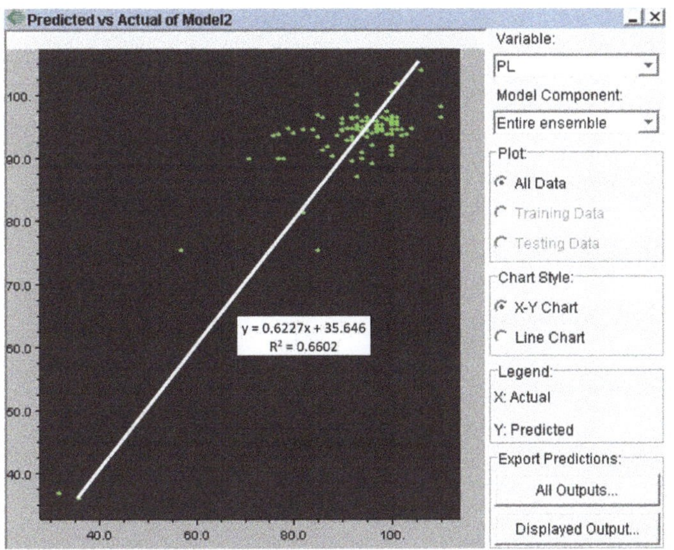

Figure 13. Correlation between calculated (y-axis) and actual (x-axis) ladle lining life.

Therefore, the graph describes to what extent the neural network was able to learn from the given data. The exact correlation coefficient can also be read in the NeurOn-Line Studio environment. In simple terms, it can be said that, if the red line leads from the lower left corner to the upper right corner, it is a very well-trained neural network, i.e., a network with a high correlation coefficient. In this case, the correlation coefficient R^2 was 0.7938.

On the trained data set, simulations of the influence of selected operational factors on lining wear were carried out for both periods while maintaining the interrelationships of other parameters. This analysis was carried out to determine how a change in a particular technological parameter would affect the durability of the ladle lining under the given conditions of steel production and processing, i.e., taking into account the simultaneous action of all factors occurring during ladle cycling. Figure 14 shows the dependencies of four selected parameters affecting lining wear. In the graphs, the x-axis expresses the given parameter and the y-axis the number of castings. At the point where the x-axis crosses the y-axis, on the x-axis, the minimum value of the given parameter is selected either from the input values for the neural network or the minimum possible, which is allowed by the current state of the technological process of steel production in the steel mill.

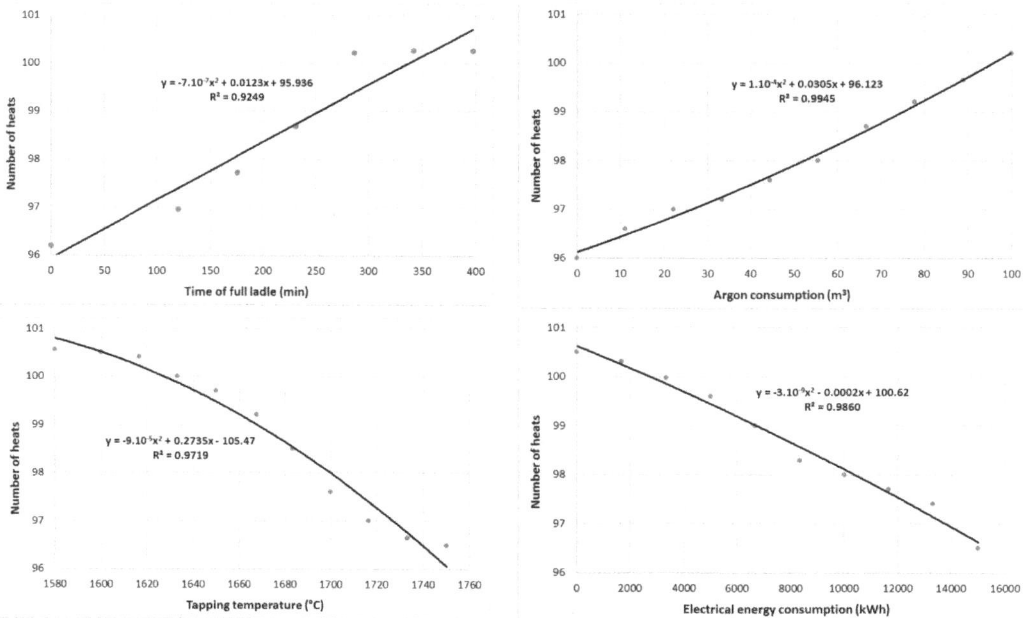

Figure 14. Simulation of the effect of selected operating parameters on lining wear.

Similarly, the maximum value is left as the maximum of the input set of the given parameter, or an extreme state that could occur in a real state was chosen. It is clear from the graphs that, as the time of the empty ladle increases and the amount of argon blown into the ladle increases, so too does the number of castings, which indicates a positive effect of these factors on the life of the lining. The other two factors, i.e., electricity consumption and tap temperature, have the exact opposite effect.

The result of the simulation expressing the effects of all selected operating parameters on the life of the lining is shown in a bar graph in Figure 15.

The bar chart describes the percentage representation of individual parameters, either with positive influence (plus sign in the column) or negative (minus sign in the column). The biggest negative influence is the time the ladle is empty, and the consumption of electricity and subsequently is the number of vacuuming cycles.

The very negative effect of the empty ladle time can be explained by temperature shocks in the lining during its constant cooling and reheating. These large temperature changes can change the coefficient of thermal expansion in the surface layer with respect to the expansion of the original material, which, even with slight temperature changes, induces shear stress at the interface of the layers sufficient for the formation of cracks. This structural spalling progresses layer by layer and can lead to rapid wear and tear of the lining. The best functioning ladle economy occurs if the lining is kept at high and above all constant temperatures, which requires continuous cycling of the ladle, without the intermediate heating associated with covering the ladle with a lid during transport. Two other factors—the consumption of electrical energy and the number of vacuuming cycles—increase the temperature of the steel in the pan, which accelerates the degradation effects of refractory materials, especially its corrosion. Again, both of these effects could be reduced by cycling the ladle, thereby preserving the overall enthalpy of the lining. The consumption of argon blown into the steel through the bottom plug of the ladle has a positive effect on the life of the lining. We believe this is due to the perfect temperature homogenization of the steel, which causes equalization of temperatures in the lining of the entire ladle. The total amount of blown argon enters analysis as a single constant value for particular heat without differentiation between the blowing phases.

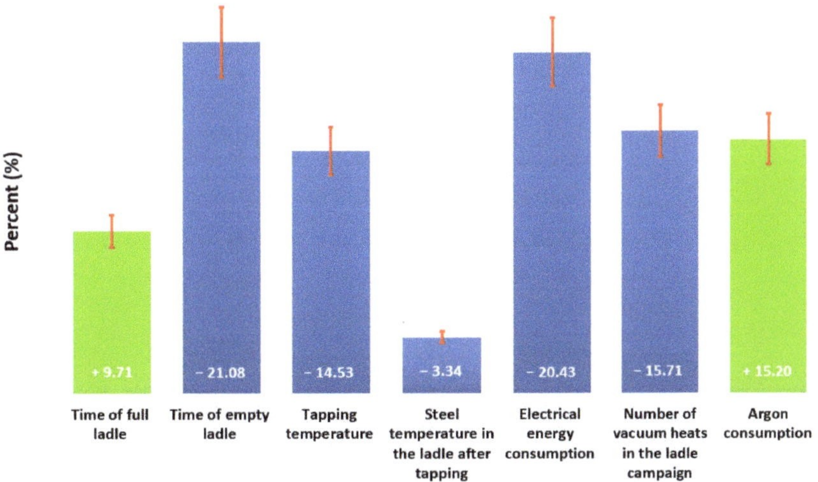

Figure 15. The influence of individual parameters on the life of the lining.

4. Conclusions

As part of the statistical evaluation of the service life of the linings of the casting ladles in the steel plant, application software was created in which data from individual database structures of the control system of the steel plant are processed. The data in this application software is designed to map a ladle campaign from laying to decommissioning. This means that data from individual ladle cycles are always included within the given campaign. These data were subsequently evaluated in neural networks in the Gensym product environment, the so-called NeurOn-Line Studio, a graphical, object-oriented software product for creating neural network applications. For this analysis, a set of input data was created and compared with the output factor, which is the number of castings of the given ladle in the campaign, i.e., the number of cycles of the continuous casting machine, the so-called CCM. The resulting value is the percentage representation of the individual tested parameters on the life of the lining of the casting ladle, which is shown in Figure 15. The most significant adverse effect is the time of the ladle's emptiness, electricity consumption, and, subsequently, the number of vacuuming cycles; moreover, argon blown into the ladle had a very positive effect.

To increase the life of the ladle lining, therefore, it is recommended to cycle it as quickly as possible, if possible, without including high-temperature heating between melts. If the ladle has a sufficiently high enthalpy of the lining before tapping, reduce the time between the end of tapping and the ladle furnace to a minimum, limiting heating by the electrodes on the ladle furnace. The ladle must be covered with a lid at every possible moment to preserve the enthalpy accumulated in the masonry of the lining.

The main innovation of the research lies in the use of neural networks in the field of basin metallurgy. The neural network method has been proven to be a suitable tool for solving the lifetime of casting ladles, especially when compared to classical multiple regression methods. Based on this solution, measures were designed to ensure an increase in the life of the linings of the casting ladles.

Currently, further data collection is taking place, which will expand research in this area. The authors intend to use machine intelligence algorithms to refine and verify the results achieved. The obtained results will be used in the field of predicting the lifetime of the lining of casting ladles in the real operation of a steel plant. The authors also want to use the latest methods of neural networks in other areas of the metallurgical industry.

Author Contributions: Conceptualization, D.J. and P.T.; methodology, D.J., M.M., M.V., P.T., L.K. and J.V.; investigation, M.M., M.V., L.K. and P.T.; writing—original draft preparation, D.J., M.M. and J.V.; writing—review and editing, M.V.; visualization, M.M.; supervision, D.J.; project administration, M.V.; funding acquisition, M.V. All authors have read and agreed to the published version of the manuscript.

Funding: This work is supported by Project No. (FW01010097)—Automated Control Systems in the Field of Ladle Metallurgy, Technology Agency of the Czech Republic (TAČR), No. (SP2022/13)—Low energy processes and materials in industry.

Institutional Review Board Statement: Not applicable.

Informed Consent Statement: Informed consent was obtained from all subjects involved in the study.

Data Availability Statement: The data presented in this study are available on request from the corresponding author.

Conflicts of Interest: The authors declare no conflict of interest.

References

1. Kushnerev, I.V.; Aksel'rod, L.M.; Platonov, A.A. Contemporary Methods for Modeling High-Temperature Systems[1]. *Refract. Ind. Ceram.* **2018**, *59*, 71–79. [CrossRef]
2. Jančíková, Z. Exploitation of Arcificial Intelligence Methods in Material Research. In Proceedings of the Conference Materials, Metallurgy and Interdisciplinary Co—working, Ostrava, Czech Republic, 29–30 September 2008.
3. Branca, T.A.; Fornai, B.; Colla, V.; Murri, M.M.; Streppa, E.; Schröder, A.J. The Challenge of Digitalization in the Steel Sector. *Metals* **2020**, *10*, 288. [CrossRef]
4. Lammer, G.; Lanzenberger, R.; Rom, A.; Hanna, A.; Forrer, M.; Feuerstein, M.; Pernkopf, F.; Mutsam, N. Advanced Data Mining for Process Optimizations and Use of A.I. to Predict Refractory Wear and to Analyse Refractory Behavior. *Iron Steel Technol.* **2017**, *15*, 52–60.
5. Vannucci, M.; Colla, V.; Chini, M.; Gaspardo, D.; Palm, B. Artificial Intelligence Approaches for The Ladle Predictive Maintenance in Electric Steel Plant. *IFAC-PapersOnLine* **2022**, *55*, 331–336. [CrossRef]
6. Li, G.; Jiang, D.; Sun, Y.; Jiang, G.; Tao, B. Arch. Life Prediction Mechanism of Ladle Composite Structure Body Based on Simulation Technology. *Metall. Mater.* **2019**, *64*, 1555–1562. [CrossRef]
7. Mihailov, E.G.; Petkov, V.I.; Doichev, I.V.; Boshnakov, K. Model-Based Approach for Investigation of Ladle Lining Damages. *Int. Rev. Mech. Eng.* **2013**, *7*, 122–130.
8. Petkov, V.; Hadjiski, M.; Boshnakov, K. Diagnosis of Metallurgical Ladle Refractory Lining Based on Non-Stationary On-Line Data Processing. *Cybern. Inf. Technol.* **2013**, *13*, 122–130. [CrossRef]
9. Feuerstein, M. Refractory Wear Modelling using Statistical Methods. Master's Thesis, Graz University of Technology, Graz, Austria, 2016; pp. 1–61.
10. Guo, Y.; Zhang, B.; Yu, S.; Kai, W. Research on an advanced intelligence implementation system for engineering process in industrial field under big data. *Expert Syst. Appl.* **2020**, *161*, 113751. [CrossRef]
11. Boshnakov, K.P.; Petkov, V.I.; Doukovska, L.A.; Vassileva, S.I.; Mihailov, E.G.; Kojnov, S.L. Predictive maintenance model-based approach for objects exposed to extremely high temperatures. *Signal Process. Symp.* **2013**, 1–5. [CrossRef]
12. Chang, W.; Sun, Y.; Li, G.; Jiang, G.; Kong, J.; Jiang, D.; Liu, H. Ladle health monitoring system based on LabVIEW. *Int. J. Comput. Sci. Math.* **2018**, *9*, 566–576. [CrossRef]
13. Tealab, A. Time series forecasting using artificial neural networks methodologies: A systematic review. *Future Comput. Inform. J.* **2018**, *3*, 334–340. [CrossRef]
14. Yemelyanov, V.A.; Yemelyanova, N.Y.; Nedelkin, A.A.; Zarudnaya, M.V. Neural network to diagnose lining condition. In *IOP Conference Series: Materials Science and Engineering*; IOP Publishing: Bristol, UK, 2018. [CrossRef]
15. Yemelyanov, V.; Chernyi, S.; Yemelyanova, N.; Varadarajan, V. Application of neural networks to forecast changes in the technical condition of critical production facilities. *Comput. Electr. Eng.* **2021**, *93*, 107225. [CrossRef]
16. Chernyi, S.; Emelianov, V.; Zinchenko, E.; Zinchenko, A.; Tsvetkova, O.; Mishin, A. Application of Artificial Intelligence Technologies for Diagnostics of Production Structures. *J. Mar. Sci. Eng.* **2022**, *10*, 259. [CrossRef]
17. Zhilenkov, A.; Chernyi, S.; Emelianov, V. Application of Artificial Intelligence Technologies to Assess the Quality of Structures. *Energies* **2021**, *14*, 8040. [CrossRef]

18. Marsland, S. *Machine Learning: An Algorithmic Perspective*, 2nd ed.; Chapman and Hall/CRC: Boca Raton, FL, USA, 2014; pp. 1–458.
19. Joshi, A.V. *Machine Learning and Artificial Intelligence*, 1st ed.; Springer: Cham, Switzerland, 2020; pp. 1–261. [CrossRef]
20. Zou, L.; Zhang, J.; Han, Y.; Zeng, F.; Li, Q.; Liu, Q. Internal Crack Prediction of Continuous Casting Billet Based on Principal Component Analysis and Deep Neural Network. *Metals* **2021**, *11*, 1976. [CrossRef]
21. Panos, L. *Algorithms*; MIT Press: Cambridge, MA, USA, 2020; Volume 99, pp. 1–312. [CrossRef]
22. Aggarwal, C.C. *Neural Networks and Deep Learning*, 1st ed.; Springer: Cham, Switzerland, 2018; pp. 1–497. [CrossRef]
23. Pham, B.T.; Nguyen, M.D.; Bui, K.-T.T.; Prakash, I.; Chapi, K.; Bui, D.T. A novel artificial intelligence approach based on multi-layer perceptron neural network and biogeography-based optimization for predicting coefficient of consolidation of soil. *Catena* **2019**, *173*, 302–311. [CrossRef]
24. Roozbeh, M.; Arashi, M.; Hamzah, N.A. Generalized cross-validation for simultaneous optimization of tuning parameters in ridge regression. *Iran J. Sci. Technol. Trans. Sci.* **2020**, *44*, 473–485. [CrossRef]
25. Gallant, A.R.; White, H. On learning the derivatives of an unknown mapping with multilayer feedforward networks. *Neural Netw.* **1992**, *5*, 129–138. [CrossRef]
26. Roelofs, R.; Shankar, V.; Recht, B.; Fridovich-Keil, S.; Hardt, M.; Miller, J.; Schmidt, L. A meta-analysis of overfitting in machine learning. In Proceedings of the 33rd Conference on Neural Information Processing Systems, Vancouver, BC, Canada, 13–14 December 2019; Curran Associates Inc.: Red Hook, NY, USA, 2019; Volume 32.
27. Song, J.; Li, Y.; Liu, S.; Xiong, Y.; Pang, W.; He, Y.; Mu, Y. Comparison of Machine Learning Algorithms for Sand Production Prediction: An Example for a Gas-Hydrate-Bearing Sand Case. *Energies* **2022**, *15*, 6509. [CrossRef]
28. Zhang, G.; Hu, Y.; Hou, D.; Yang, D.; Zhang, Q.; Hu, Y.; Liu, X. Assessment of Porosity Defects in Ingot Using Machine Learning Methods during Electro Slag Remelting Process. *Metals* **2022**, *12*, 958. [CrossRef]
29. Yan, Z.; Liu, H. SMoCo: A Powerful and Efficient Method Based on Self-Supervised Learning for Fault Diagnosis of Aero-Engine Bearing under Limited Data. *Mathematics* **2022**, *10*, 2796. [CrossRef]
30. Mayet, A.M.; Nurgalieva, K.S.; Al-Qahtani, A.A.; Narozhnyy, I.M.; Alhashim, H.H.; Nazemi, E.; Indrupskiy, I.M. Proposing a High-Precision Petroleum Pipeline Monitoring System for Identifying the Type and Amount of Oil Products Using Extraction of Frequency Characteristics and a MLP Neural Network. *Mathematics* **2022**, *10*, 2916. [CrossRef]
31. Orrù, P.F.; Zoccheddu, A.; Sassu, L.; Mattia, C.; Cozza, R.; Arena, S. Machine Learning Approach Using MLP and SVM Algorithms for the Fault Prediction of a Centrifugal Pump in the Oil and Gas Industry. *Sustainability* **2020**, *12*, 4776. [CrossRef]
32. Zhang, X.; Han, C.; Luo, M.; Zhang, D. Tool Wear Monitoring for Complex Part Milling Based on Deep Learning. *Appl. Sci.* **2020**, *10*, 6916. [CrossRef]
33. Pentoś, K.; Mbah, J.T.; Pieczarka, K.; Niedbała, G.; Wojciechowski, T. Evaluation of Multiple Linear Regression and Machine Learning Approaches to Predict Soil Compaction and Shear Stress Based on Electrical Parameters. *Appl. Sci.* **2022**, *12*, 8791. [CrossRef]
34. Sporka, M. Verification of Effeciency of Ladle Lining Insulation in ArcelorMittal Ostrava a.s. Master's Thesis, VŠB—TU Ostrava, Ostrava, Czech Republic, 2015; pp. 1–65.
35. Tvardek, P. *Využití Expertního Systému G2 pro Operativní Řízení*; Research report; Ispat Nová Huť Ostrava a. s.: Ostrava, Czech Republic, 2003; pp. 1–40.
36. Jančar, D. The Use of Neural Networks for the Analysis of the Lifetime of Casting Ladle Linings. Ph.D. Thesis, VŠB—TU Ostrava, Ostrava, Czech Republic, 2009; pp. 1–150.

Article

Investigation of Molten Metal Infiltration into Micropore Carbon Refractory Materials Using X-ray Computed Tomography

Jakub Stec [1], Jacek Tarasiuk [2], Sebastian Wroński [2], Piotr Kubica [3], Janusz Tomala [3] and Robert Filipek [1,*

[1] Faculty of Materials Science and Ceramics, AGH University of Science and Technology, Al. Mickiewicza 30, 30-059 Kraków, Poland; stec@agh.edu.pl

[2] Faculty of Physics and Applied Computer Science, AGH University of Science and Technology, Al. Mickiewicza 30, 30-059 Kraków, Poland; tarasiuk@agh.edu.pl (J.T.); wronski@fis.agh.edu.pl (S.W.)

[3] Tokai COBEX Polska sp. z o. o., ul. Piastowska 29, 47-400 Racibórz, Poland; piotr.kubica@tokaicobex.com (P.K.); janusz.tomala@tokaicobex.com (J.T.)

* Correspondence: rof@agh.edu.pl

Abstract: The lifetime of a blast furnace (BF), and, consequently, the price of steel, strongly depends on the degradation of micropore carbon refractory materials used as lining materials in the BF hearth. One of the major degradation mechanisms in the BF hearth is related to the infiltration and dissolution of refractory materials in molten metal. To design new and more resilient materials, we need to know more about degradation mechanisms, which can be achieved using laboratory tests. In this work, we present a new investigation method of refractory materials infiltration resistance. The designed method combines a standard degradation test (hot metal penetration test) with X-ray computed tomography (XCT) measurements. Application of XCT measurements before and after molten metal infiltration allows observing changes in the micropore carbon refractory material's microstructure and identifying the elements of the open pore structure that are crucial in molten metal infiltration.

Keywords: blast furnace; carbon refractories; molten metal infiltration; X-ray computed tomography

Citation: Stec, J.; Tarasiuk, J.; Wroński, S.; Kubica, P.; Tomala, J.; Filipek, R. Investigation of Molten Metal Infiltration into Micropore Carbon Refractory Materials Using X-ray Computed Tomography. *Materials* **2021**, *14*, 3148. https://doi.org/10.3390/ma14123148

Academic Editor: Panagiotis (Panos) Tsakiropoulos

Received: 30 April 2021
Accepted: 3 June 2021
Published: 8 June 2021

1. Introduction

While blast furnace (BF) technology remains the main method of obtaining crude iron (known also as pig iron), the cost of steel strongly depends on the lifetime of the BF [1]. Its working time is mainly limited by the degradation of refractory materials, especially when used in the BF hearth area, where liquid metal gathers. It is a consequence of the fact that damage to BF hearth walls cannot be repaired without stopping the work of the whole furnace [2,3]. In that crucial zone, carbon and graphite refractory materials are mainly used, due to their unique properties, such as thermal stability, high thermal conductivity, corrosion resistance, and non-wettability by molten metal and slag [4].

The complexity of the processes occurring inside the BF results in various degradation mechanisms present in each zone of the BF. In general, they can be divided into three groups: mechanical wear (e.g., abrasive wear of descending solid burden, erosion resulting from ascending dust-laden gases and from the molten metal flow), thermal wear (e.g., thermal shocks caused by the tapping cycles and thermomechanical stresses resulting from temperature gradients), and chemical wear, which is a result of the interactions at high temperatures between various compounds present in the BF and lining materials [5]. In the BF hearth, carbon and graphite refractory materials are mainly subjected to chemical wear, which includes various degradation mechanisms, such as alkali [6] and zinc attack [7], carbon monoxide deterioration [8], erosion, dissolution, and infiltration of refractories by molten crude iron [9,10]. All of them do not occur separately as independent processes, but they contribute to the overall degradation of the BF hearth, which changes the temperature distribution and further accelerates the degradation process [11,12].

To design new, more resilient refractory materials, we need to know more about degradation mechanisms. One of the possible sources of such information is postmortem sample analysis. However, this method is limited by sample availability and provides information about the sum of all degradation mechanisms [13]. An alternative approach, which can be used to evaluate new materials and their resistance to various degradation mechanisms are laboratory tests. They are more available than postmortem samples, less expensive, and much quicker, which makes them an essential tool in the process of designing and evaluating new refractory materials. Degradation in contact with molten metal, which is a major degradation mechanism of carbon and graphite materials, can be investigated using the most common tests, e.g., sessile drop, immersion, and crucible, or more complex ones, such as rotatory tests or with induction furnaces [14]. All of these focus on chemical degradation (i.e., dissolution of refractory materials) and/or mechanical erosion, thus on the processes which occur on the surface of refractory materials. However, it is commonly accepted that in the actual BF hearth, molten metal might penetrate refractory bricks up to the 1150 °C isotherm [15]. To investigate the materials resistance to molten metal infiltration, a new test—the hot metal penetration (HMP) test— was designed. It is a modification of the crucible test, in which the carbon-saturated crude iron is forced to infiltrate the tested material by the elevated pressure created inside the crucible by the flow of argon [16]. In the standard HMP procedure, after the infiltration, the crucible is investigated to evaluate if there was a leakage of molten metal (the material can pass or fail the HMP test). To analyze the infiltration zone and metal flow paths, the crucible is cut into smaller pieces and observed via optical or scanning electron microscopy [17]. While this approach provides qualitative information about material infiltration resistance, we are not able to identify the initial elements of material microstructure which are essential in the process of molten metal infiltration.

To better understand the dependence between material microstructure and their degradation resistance, we need to compare the 3D microstructures, before and after the degradation process. This can be achieved by using nondestructive investigation methods, among which X-ray computed tomography (XCT) is a promising candidate. XCT can be used to investigate the microstructure of multiphase materials [18–20] and pore structure [21,22]. Based on the XCT measurements, methods for determining parameters describing pore structure—tortuosity [23] and constrictivity [24]—were developed. Only for carbon and graphite refractory materials, XCT has been used to investigate the pore structure of micropore carbon materials [25], microstructure and tortuosity analysis of carbon anodes for aluminum smelting [26,27] and for porosity investigations of carbon blocks and ramming paste used for silicomanganese production [28].

XCT is a powerful tool, which can be used to investigate the microstructure of composite materials before and after the chemical vapor deposition process [29] or to qualitatively describe the corrosion of steel-fiber reinforced polymer bars [30]. Moreover, XCT can be used in the in situ observation of degradation processes, such as SiC_f/SiC composite fatigue testing [31] or steel bar corrosion in cementitious matrix [32].

In this work, we present a novel investigation method of molten metal infiltration, which combines the hot metal penetration (HMP) test with X-ray computed tomography (XCT) measurements. The new method, the XCT–HMP test, provides a unique opportunity to compare exactly the same volume of refractory material before and after molten metal infiltration. It enables the identification of the preferred infiltration paths, which could be used to correlate the material microstructure with its infiltration and consequently degradation resistance.

2. Materials and Methods

2.1. Materials

The micropore carbon trial material, investigated in our work, was produced on an industrial scale by Tokai COBEX Polska sp. z o.o (Racibórz, Poland). The main raw materials were: artificial amorphous carbon grains containing silicon carbide, synthetic graphite grains, synthetic semigraphite flour, silicon and alumina powder, and coal tar pitch as a binder. The ratio between solid particles and binder was 3.64:1. The prepared raw materials composition was subjected to a two-step mixing process using the Eirich homogenizer [33]. First, the dry compounds were mixed at temperatures below 120 °C for 5 min. After that, the binder was added and the mixing process has been continued for 15 min at 150 °C. The created green paste was further molded into blocks with dimensions of $2500 \times 700 \times 500$ mm^3 using a vibration molding press. Formed blocks were baked in a standard ring furnace under reducing atmosphere. The maximum baking temperature was 1300 °C. From the bottom part of the carbon block, known as the foot, the crucible and sample were cut.

2.2. Hot Metal Penetration Test and Its Modification

2.2.1. Standard HMP Test

The hot metal penetration test was performed using equipment designed and built in collaboration with Tokai COBEX sp. z o.o. (Racibórz, Poland), and Czylok (Jastrzębie-Zdrój, Poland). The scheme of the HMP apparatus is presented in Figure 1a. It consists of a vertical chamber furnace, accessible from the top, lined with alumina and heated by the MoSi$_2$ heating element. The investigated material is machined in the form of a crucible, of which dimensions are presented in Figure 1b. Into the crucible, 150 g of carbon saturated crude iron are placed in the form of granules. The typical composition of crude iron used in the HMP tests is presented in Table 1.

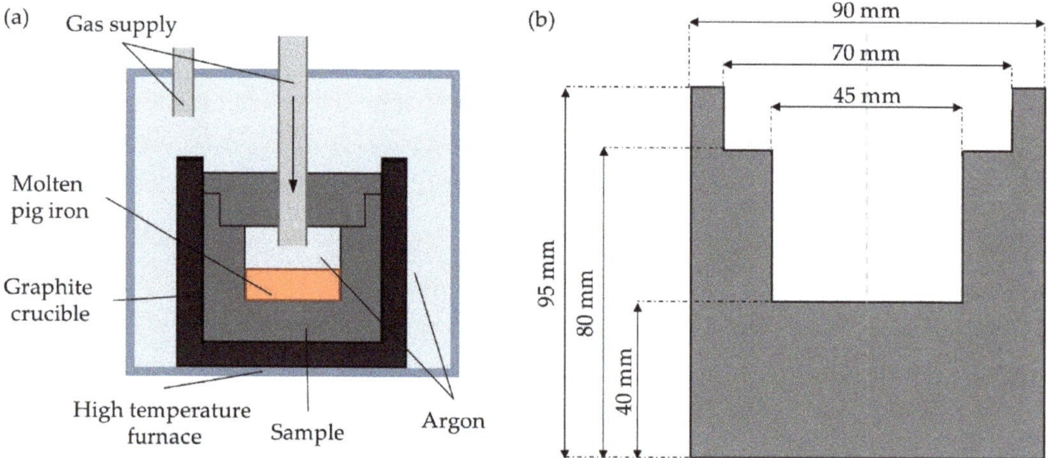

Figure 1. Hot metal penetration test: (**a**) scheme of HMP test and (**b**) dimensions of the HMP crucible.

Table 1. Chemical composition of crude iron used in the HMP test.

Crude Iron Composition [wt.%]							
Fe	C	Si	Mn	P	S	Ti	Zn
94.619	4.780	0.400	0.085	0.073	0.024	0.014	0.005

The tested crucible is tightly closed with a lid made of the same material. The tight connection is ensured by a gasket made of expanded graphite foil. The lid has a hole in the center—a conduit for a pipe that is used to deliver argon inside the crucible. An external graphite crucible is used to protect the furnace heating elements in case of molten metal leakage from the tested crucible. The prepared crucible is placed in the furnace chamber by screwing in a vertical graphite tube, which serves as a gas supply (Figure 2a). The weight of the crucible is continuously monitored, therefore possible metal leaks can be detected during the ongoing process.

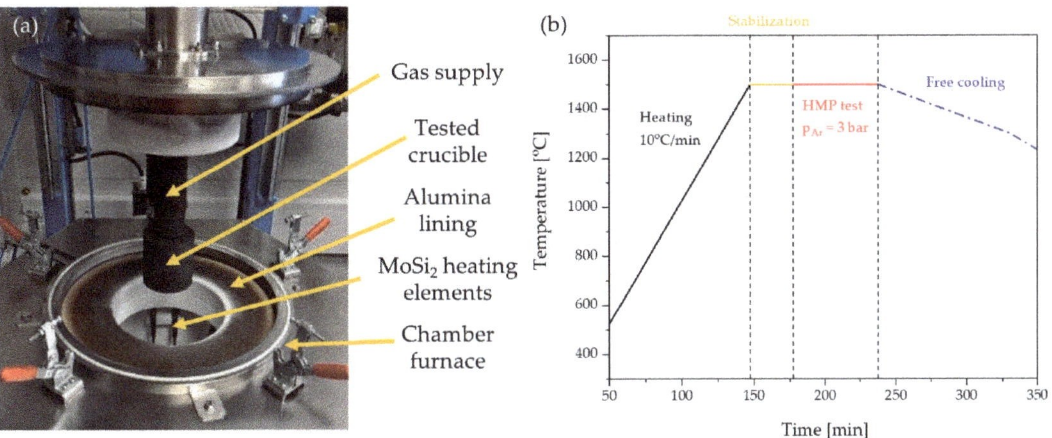

Figure 2. (**a**) Photo of HMP apparatus. (**b**) Heating curve during the HMP test.

The whole system is annealed from room temperature to 1500 °C at a heating rate of 10 °C/min. After reaching the desired temperature, the whole system is maintained for 30 min to stabilize it. Then, an elevated pressure of 3 bars is created inside the crucible using argon flow to force the molten metal infiltration. The crucible was kept at an elevated temperature and pressure for 60 min. After that, the whole system is freely cooled. The heating curve during the standard HMP test is presented in Figure 2b. During all stages of the HMP process, the ambient atmosphere in the furnace is maintained by the flow of argon (10 L/min) to avoid the oxidation of the carbon crucible and gas supply.

2.2.2. Modification—XCT–HMP Test

The modified HMP test (XCT–HMP) was designed based on the idea of observing the 3D microstructure of exactly the same volume before and after molten metal infiltration. One of the major aspects of XCT measurement is its resolution, which strongly depends on the size of the sample. According to Silva et al. [6], molten crude iron can penetrate pores with diameters of up to 5 μm in actual BF hearth. Our previous investigations [25] showed that, for cylindrical samples (Ø = 25 mm, h = 25 mm), we are able to achieve a voxel size of $16 \times 16 \times 16$ μm^3 and observe continuous pore structures through the whole height of the sample. Consequently, the investigation of the whole HMP crucible will result in even lower resolution. To bypass this limitation, we decided to focus on a smaller sample. The scheme of the XCT–HMP test is presented in Figure 3. From the same block of micropore carbon material, a cylindrical sample (Ø = 25 mm, h = 25 mm) and a modified HMP crucible were cut. Compared to the standard HMP crucible, the new one had a hole in the center of its bottom of which the dimensions matched the size of the cylindrical sample. The sample was investigated using XCT, as described in the next section. After the measurement, the sample was inserted into the crucible. The gap between the sample and crucible was filled with sodium silicate sealing paste Firecement HT 1500 °C sealant (SOUDAL, Czosnów, Poland), which ensured a tight connection between them. Then, the modified crucible was used in the HMP procedure described in the previous section. After the test, the cylinder (Ø = 30 mm, h = 40 mm) was cut from the center of the crucible bottom.

This new sample, which included the initial sample and part of the crucible's bottom, was investigated via XCT using the same parameters.

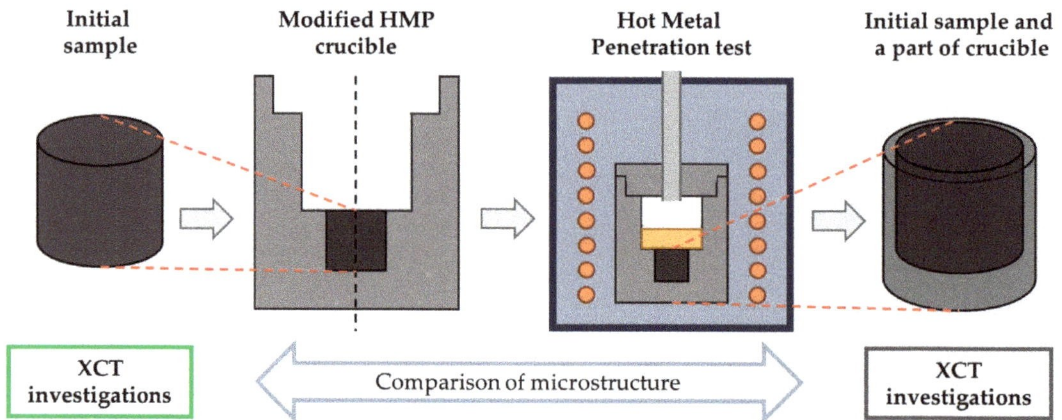

Figure 3. Scheme of the designed new investigation method—the XCT–HMP test.

2.3. X-ray Computed Tomography

2.3.1. X-ray Computed Tomography (XCT) Measurements

The XCT measurements were performed using a Nanotom 180S (GE Sensing & Inspection Technologies GmbH, Wunstorf, Germany). The machine is equipped with a nanofocus X-ray tube with maximum 180 kV voltage. The tomograms were registered on Hamamatsu 2300×2300-pixel detector. During the measurements, a tungsten target was used. The polychromatic beam was filtered using a 0.5-mm copper filter. The working parameters of the X-ray tube were I = 250 µA and V = 70 kV. A total of 1600 projections were taken with 4 integrations for each exposition. The total time of measurement was around 120 min. The reconstructions of the measured objects were done with the aid of the proprietary GE software, datosX ver. 2.1.0, using the Feldkamp algorithm for cone beam X-ray CT [34]. The final resolution of the reconstructed object was 16 µm. The postreconstruction data treatment was performed using VGStudio Max 2.1 software (Volume Graphic GmbH, Heidelberg, Germany) [35].

2.3.2. Processing of XCT Data

The XCT results in the form of 16-bit stacks of 2D cross-section images were imported into open source ImageJ software [36]. In the first step, brightness and contrast were adjusted. Then the porosity of the initial sample and metal inclusions were separated using appropriate thresholds. Next, the continuous pore and metal structures were separated using the Flood Fill 3D tool in ImageJ. Volume fractions of both structures were calculated using Voxel Counter tool for each slice separately and then averaged for the whole volume. Local thicknesses of pore and metal structures were calculated using the thickness option in ImageJ plugin—BoneJ [37]. The 3D visualizations of samples, pore, and metal structures were created using Simpleware ScanIP software (Synopsys, San Jose, CA, USA) [38]. Details of each XCT data processing step were described in [39].

2.4. Microstructure and Chemical Composition Analysis

After HMP test and XCT measurements, the sample (Ø = 30 mm) was cut into 5 mm discs, ground, and polished for microstructure and chemical composition analysis. They were carried out using a scanning electron microscope NOVA NANO SEM 200 (FEI, Acht, The Netherlands) equipped with an energy dispersive spectrometer (EDAX, Tilburg, The

Netherlands). Samples were observed in backscattered electron (BSE) mode. The applied energy of the electron beam was 18 keV.

3. Results

The results of XCT investigations are presented in the form of two types of 2D cross-sections: perpendicular (Figure 4) and parallel (Figure 5) to the molten metal flow direction. The cross-sections of the sample before metal infiltration are presented in the top row, while the cross-sections after the infiltration are presented in the bottom row. In the sample before the HMP test (top row), four phases can be distinguished: pores (black), carbonaceous grains (dark grey), carbonized binder (light grey), and ceramic additives (white). In the sample after the HMP test (bottom row), due to the bigger X-ray beam, the attenuation of heavy solidified metal and consequently its very high brightness [40], the contrast between carbon and ceramics phases in the micropore carbon material is much lower than in the initial sample. As a result, it makes it difficult to distinguish between carbonaceous grains and carbonized binder, as well as between ceramic additives and solidified metal. Four phases can be easily distinguished in the sample after the HMP test: pores (black), carbon phases in the micropore carbon material (dark grey), solidified metal (white), and the sealing paste (light grey) in the gaps between the initial sample and the surrounding crucible. Comparing the sample microstructure before and after HMP test, it can be seen that the major part of the initial porosity was filled with the molten metal. Pores in the lower part of the sample were impregnated with metal to a greater extent than in the upper part of the sample. Portions of molten metal infiltrated below the initial sample in the zone, which presumably was not filled with the sealing paste (Figure 5). The sealing paste was used as a junction material, which provided a tight connection between the crucible and sample in the major part, which prevented the infiltration through the gap between sample and the crucible instead of the sample volume. However, the zones which were not filled with the sealing paste can be observed, especially in the upper parts of the sample (Figure 4b,c)—during the HMP test they were filled with molten metal. Moreover, not only the gap between the sample and crucible was filled with sealing paste, but also a part of the open pore structure in the micropore carbon material (Figure 4c).

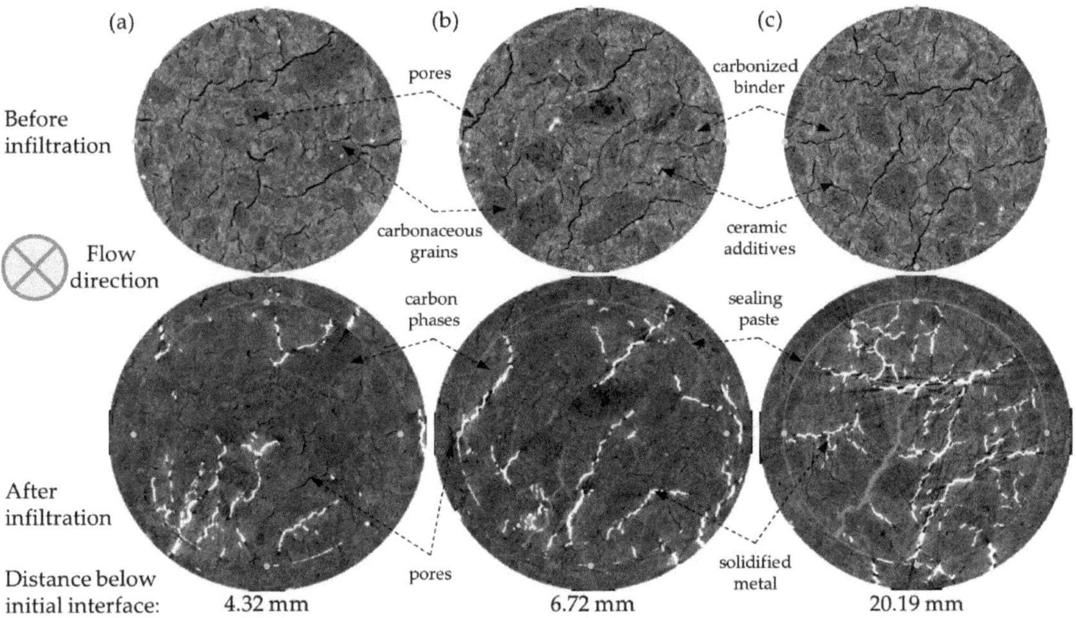

Figure 4. Two-dimensional cross-sections in the plane perpendicular to the molten metal flow of the sample before (top row) and after the HMP test (bottom row) at various distances from the initial molten metal/sample interface: (**a**) 4.32 mm (**b**) 6.72 mm and (**c**) 20.19 mm.

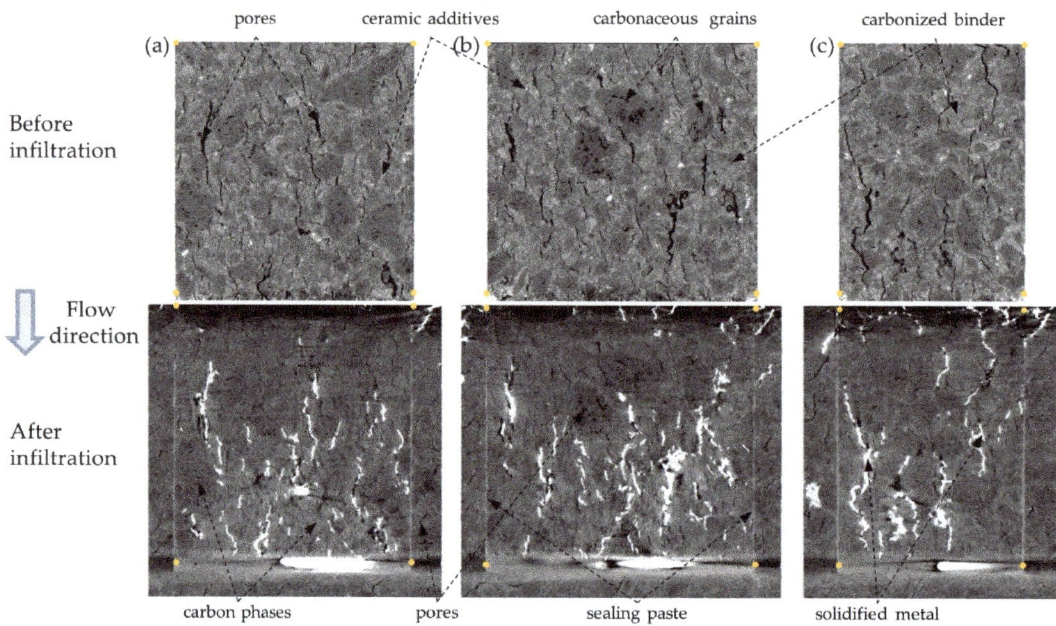

Figure 5. Various 2D cross-sections (a–c) in the plane parallel to the molten metal flow of the sample before (top row) and after the HMP test (bottom row).

Analyzing the 2D cross-sections after the HMP test, the artifacts can be seen both in the upper (Figure 5, dark areas in the top row) and lower part of the sample (Figure 4c), in the zones where there was a higher amount of molten metal. As a result, the phase segmentation of phases other than bright metal in those areas was inaccurate.

To verify if the brighter grey phase in XCT investigation after the HMP test was the actual sodium silicate sealing paste, the sample after metal infiltration was cut into the discs and then further prepared for the scanning electron microscope investigations by grinding and polishing. The microstructure of the crucible sample and the junction between them are presented in Figure 6a. In the observed area, the junction did not fill the whole gap and a part of it was impregnated by the molten metal, which was also visible in the XCT observations (Figure 4, bottom row). The chemical composition analysis of phases in the gap between the sample and crucible were performed in the area marked with an orange rectangle (Figure 6b). Results of the EDS analysis are presented in Figure 6c. In the solidified metal (point 3), we can observe the characteristic peaks for Fe, Si, and C—the three main elements of crude iron used in HMP test (Table 1). The point analysis of the bright grey phase (point 2), shows that it is mainly composed of Si, Al, and O, with small amounts of Na and K, which proves that the bright grey phase observed in the XCT data is the sodium silicate sealing paste.

To evaluate the influence of molten metal on the microstructure of micropore carbon materials, for both XCT measurements, the same volumes, representing the sample before and after molten metal infiltration were cut. The diameters of the analyzed volumes were equal to the diameter of the initial sample (Ø = 25 mm), while their heights were lower, h = 20.8 mm. The chosen samples' height was lower than that of the initial sample (h = 25 mm), due to the fact that part of the crucible bottom was dissolved during the HMP test and the artifacts in the upper part of the sample made the proper phase segmentation of areas close to the initial molten metal/carbon material interface impossible. The 3D visualizations of the chosen volumes for the analysis are presented in Figure 7.

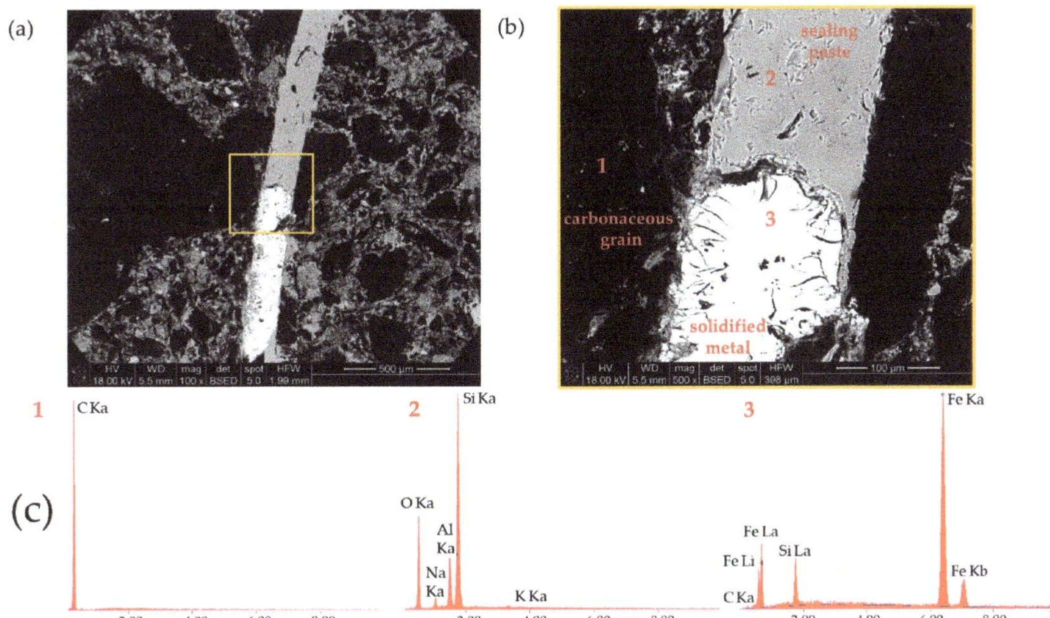

Figure 6. (**a**) SEM image of crucible sample and junction between them after metal infiltration; (**b**) SEM image of junction and solidified metal with marked points for EDS analysis; and (**c**) results of EDS point analysis.

Figure 7. Three-dimensional visualization of samples: (**a**) before and (**b**) after the HMP test.

In the next step, the pores in the initial sample and metal inclusions were separated by applying suitable thresholds to the analyzed images. The separated pores and metal inclusions are presented in the form of 3D visualization in Figure 8a,b. The volume of all pores was 474 mm^3, corresponding to a volume fraction (porosity) of 4.64% (Table 2). The volume of all metal inclusions measured by XCT equaled 366 mm^2, which means that the volume fraction of metal was smaller—3.59% (Table 2). The volume of all metal inclusions was approx. 23% smaller than the volume of all pores. The volumes of pores and metal inclusions were measured for each slice separately, which enabled the presentation of surface fractions as a function of sample height (Figure 9a). The distribution of pores in the initial sample was throughout the whole height of the sample, which resulted in low standard deviation (σ, Table 2). In the case of metal inclusion, the distribution was much more nonuniform. The fraction of metal in the lower part of the sample (between 0 and 10 mm) was more than 4%, while in the upper part was constantly decreasing to 2% at the top sample boundary. This resulted in a much higher standard deviation, approx. 4.5 times higher than for all pores.

Figure 8. Visualization of (**a**) all pores in the initial sample, (**b**) all metallic inclusions in the sample after the test, (**c**) continuous pore structure, and (**d**) continuous metal structure.

Table 2. Volume fraction, standard deviation (σ), and volume (V) of pores and metal inclusions.

All Pores and Metal Inclusions	Volume Fraction [%]	Increase	σ [%]	V [mm³]
Pores	4.64	−23%	0.27	474
Metal	3.59		1.19	366
Continuous Pore and Metal Structures	**Volume Fraction [%]**	**Increase**	**σ [%]**	**V [mm³]**
Pore	2.48	29%	0.42	253
Metal	3.19		1.33	326

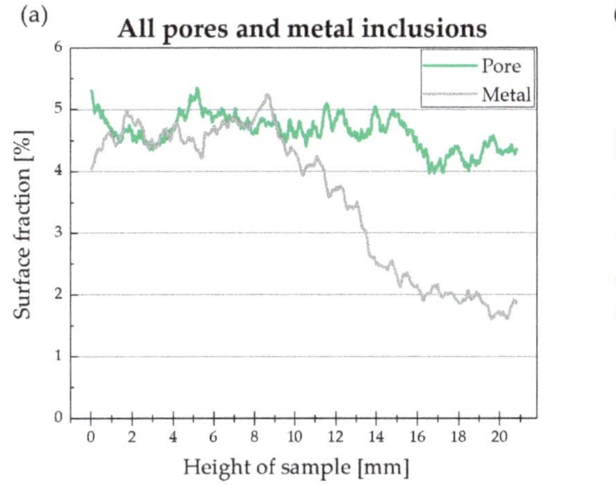

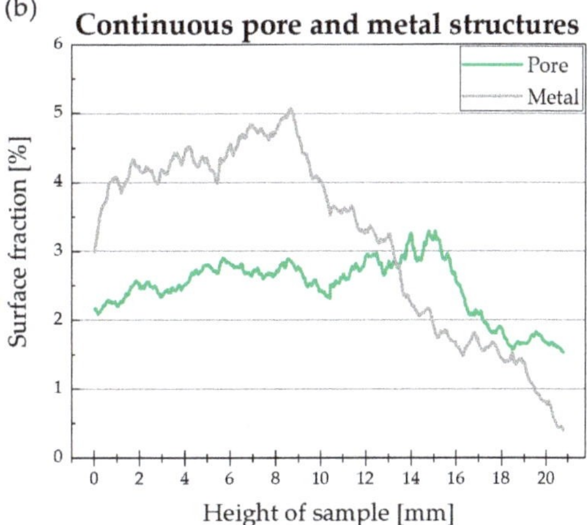

Figure 9. Surface fractions of pores and metal inclusions as a function of the sample height: (**a**) all pores and all metal inclusions and (**b**) continuous pore and metal structure.

Despite the fact that XCT allows measuring the sum of open and closed pores, only open pores contribute to molten metal transport. Thus, in the next stage, the continuous pore and metal structures were separated using the Flood Fill 3D tool. The continuous structures are presented in Figure 8c,d. 3D visualization of continuous pore and metal structures in form of GIF files are presented in Supplementary Materials. The volume of the continuous pore structure was approx. 2 times lower than the volume of all pores, which resulted in a 2.48% volume fraction. Its distribution was less uniform than the distribution of all pores (Figure 9b), which resulted in the higher standard deviation (0.42%). In the case of the continuous metal structure, its volume was also smaller than the volume of all metal inclusions, but the difference was much smaller (approx. 12%). The decrease in metal structure volume was a consequence of the small inclusions close to the top and bottom sample's boundaries which were disconnected from the main metal structure (Figures 8b,d and 9). It resulted in a slightly higher standard deviation (1.33%) than for all metal inclusions (1.19%) The volume fraction of the continuous metal structure was 3.19%, which was 29% larger than the continuous pore structure (Table 2).

The morphology of the pores and metal inclusions was described by the measurements of local thicknesses, presented in Table 3. The average pore thickness of all pores was 115 µm, while for all metal inclusions it was 65% higher (190 µm). The thickness distribution was more uniform for pores (σ = 54 µm) than for the metal inclusion (σ = 84 µm). Comparing the maximum local thicknesses, it can be seen that it was 20% higher for all metal inclusions (843 µm) than for all pores (703 µm). The maximum thicknesses were approx. 6 times higher for pores and 4.4 times higher for metal inclusions than the average thickness.

Table 3. Local thicknesses of the pores and metal inclusions: average local thickness $\bar{d}_{th}$, standard deviation, σ and maximum local thickness d_{th}^{max}.

		Local Thickness [μm]		
		$\bar{d}_{th}$	σ	d_{th}^{max}
All	Pores	115	54	703
	Metal	190	84	843
Increase		65%		20%
Continuous structure	Pore	135	50	410
	Metal	194	85	843
Increase		44%		106%

In the continuous structures, the average thicknesses were higher, especially for the pore structure (135 μm), which resulted in a lower difference between pores and metal—44%. The thickness of the pore structure was slightly more even. The biggest differences were observed in the maximum local thicknesses. While for the continuous metal structure it was the same value like for all metal inclusions, the maximum thickness of the continuous pore structure was much smaller than for all pores (410 μm), and the maximum thickness for continuous metal structure was 2 times higher than for the continuous pore structure.

4. Discussion and Conclusions

Our research showed that XCT can be successfully applied to observe the same volume of refractory material before and after molten metal infiltration. Despite the fact that, according to Silva et al. [6], in the analyzed system (which simulates the conditions inside the actual BF hearth) pores up to 5 μm, which can be infiltrated by molten metal, our investigation with voxel size $16 \times 16 \times 16$ μm^3 allowed us to observe the continuous metal structures. The infiltration paths created during the HMP test were clearly visible in the XCT measurements. The small metal inclusions, disconnected from the main metal structure were observed, which resulted in the difference between the volume fraction of all metal inclusions (3.59%) and the continuous metal structure (3.19%). This difference can be an effect of the applied resolution (larger than the minimum diameter of accessible pores), which as a consequence might result in a lack of ability to observe small connections between the continuous metal structure and separated metal inclusions. However, it is also possible that those crude iron particles were separated from the main metal body. To answer, if they were connected or not, additional XCT measurements with better resolutions are required.

Total and continuous porosities measured by XCT in the initial sample, are much lower than the open porosities measured for this particular micropore carbon material (produced by Tokai COBEX sp. z.o.o., Racibórz, Poland) using other methods, i.e., mercury intrusion porosimetry (16.62 ± 1.09%) or helium porosimetry (18.66 ± 0.70%) [41]. Comparing with the XCT measurements of other carbon refractory materials, the measured total porosity was higher than for the carbon anodes (1.96–3.46%) investigated by Rørvik et al. [27], while the continuous porosity was much lower than for amorphous carbon blocks (6–7%) investigated by the Steenkamp et al. [28]. However, it should be noted that those measurements were performed for XCT investigations with different voxel sizes ($7.3 \times 7.3 \times 7.3$ μm^3 [28] and $27 \times 27 \times 27$ μm^3 [27]) than in this study. Results showed that even using this relatively high resolution ($16 \times 16 \times 16$ μm^3), we are able to observe the elements of continuous pore structure, which serve as a preferred flow path for molten metal infiltration. This observation is in agreement with our previous investigations [25].

The presented modification of hot metal penetration test not only allowed us to observe the penetration of pore structure in micropore carbon materials but also observe the process of carbon material degradation in molten metal. The volume of the continuous metal structure (3.19%) was larger than the volume of the continuous pore structure (2.48%), which indicates that two additional processes occurred during the test: (i) dissolution of carbon material surrounding the continuous pore structure and (ii) creation of a connection between pores separated in the initial sample. Those processes can be observed in local thickness analysis. The dissolution is visible in the increase of the average local thickness from 135 μm (pores) to 194 μm (metal). The creation of the connection is proved by the changes in the maximum local thickness. For solidified metal, the maximum thickness was 843 μm (both for all metal inclusions and continuous metal structure), while for pores there was a significant difference. In total porosity, the maximum thickness was 703 μm and in the continuous pore structure it was 410 μm. Thus, the continuous pore structure was not only filled with molten metal, but it also has become connected with the largest pore present in the investigated sample. The difference in maximum pore (703 μm) and metal (843 μm) thicknesses is a consequence of carbon material dissolution in molten metal.

The application of the sodium silicate sealing paste provided a tight and stable connection between the sample and the crucible. As a result, the major molten metal infiltration occurred through the sample volume, instead of the gap between it and the crucible. However, the entire gap was not filled with sealing paste, and in certain areas the space between the sample and the crucible was filled with molten metal. Simultaneously, part of the sample pore structure was infiltrated by the sealing paste. While it does not have a major influence on the whole HMP test, in the future, the method of applying silicone has to be optimized to ensure an even more tight connection between the sample and the crucible.

In sum, a new method for molten metal infiltration investigations was designed and successfully tested. Thanks to the application of X-ray the computed tomography method enables the investigation of the mechanism of carbon refractory materials degradation by comparison of 3D microstructure before and after infiltration to observe the changes inflicted by the molten metal flow. In the future, we are planning to optimize the infiltration time to observe different infiltration zones. The XCT–HMP test will be used to verify the evolutionary two-phase flow model in 3D geometry coupled with selective dissolution of carbon phases.

Supplementary Materials: The following are available online at https://www.mdpi.com/article/10.3390/ma14123148/s1, Video S1: 3D visualization of continuous pore and metal structures.

Author Contributions: Conceptualization, J.S., J.T. (Janusz Tomala) and R.F.; methodology, J.S., J.T. (Janusz Tomala); software, J.S.; validation, J.S. and R.F.; formal analysis, J.S.; investigation, J.T. (Jacek Tarasiuk), S.W., P.K.; resources, J.T. (Jacek Tarasiuk), J.T. (Janusz Tomala), R.F.; data curation, J.S.; writing—original draft preparation, J.S., P.K.; writing—review and editing, J.S., J.T. (Janusz Tomala), R.F.; visualization, J.S.; supervision, J.T. (Janusz Tomala), R.F.; project administration, J.S.; funding acquisition, J.S., R.F. All authors have read and agreed to the published version of the manuscript.

Funding: This research was funded by the National Science Centre, Poland, grant No. 2019/35/N/ST8/03697.

Institutional Review Board Statement: Not applicable.

Informed Consent Statement: Not applicable.

Data Availability Statement: The data presented in this study are available on request from the corresponding author.

Acknowledgments: Tokai COBEX Polska sp. z o.o. is acknowledged for providing materials and services. Special thanks are addressed to Frank Hiltmann from Tokai COBEX GmbH for fruitful scientific discussions.

Conflicts of Interest: The authors declare no conflict of interest.

References

1. Kawoka, K.; Tsuda, A.; Matsuoka, Y.; Nishioka, K.; Anan, K.; Kakiuchi, K.; Takeshida, H.; Takasaki, H. Latest blast furnace relining technology at Nippon Steel. *Nippon Steel Tech. Rep.* **2006**, *94*, 127–132.
2. Silva, S.N.; Vernilli, F.; Justus, S.M.; Marques, O.R.; Mazine, A.; Baldo, J.B.; Longo, E.; Varela, J.A. Wear mechanism for blast furnace hearth refractory lining. *Ironmak. Steelmak.* **2005**, *32*, 459–467. [CrossRef]
3. Raipala, K. On Hearth Phenomena and Hot Metal Content in a Blast Furnace. Ph.D. Thesis, Helsinki University of Technology, Espoo, Finland, 2003.

4. Joubert, H. Analysis of Blast Furnace Lining/Cooling Systems Using Computational Fluid Dynamics. Master's Thesis, Rand Afrikaans University, Johannesburg, South Africa, 1997.

5. Dzermejko, A.J.; Baret, D.F.; Hubble, D.H. Ironmaking refractory systems. In *Making, Shaping and Treating of Steel–Ironmaking*, 11th ed.; The AISE Steel Foundation: Pittsburgh, PA, USA, 1991; Volume 2, pp. 229–258.

6. Silva, S.N.; Vernilli, F.; Justus, S.M.; Longo, E.; Baldo, J.B.; Varela, J.A.; Lopes, J.M.G. A methodology to investigate the wear of blast furnace hearth carbon refractory lining. *Mater. Corros.* **2012**, *64*, 1032–1038. [CrossRef]

7. Deng, Y.; Lyu, Q.; Zhang, J.; Jiao, K. Erosion of Carbon Brick by Zinc in Hearth of Blast Furnace. *ISIJ Int.* **2020**, *60*, 226–232. [CrossRef]

8. Swarup, D.; Sataravala, R.P. Deterioration of Blast Furnace Linings by the Action of Carbon Monoxide. *Trans. Indian Ceram. Soc.* **1943**, *2*, 15–31. [CrossRef]

9. Deng, Y.; Zhang, J.; Jiao, K. Dissolution mechanism of carbon brick into molten metal. *ISIJ Int.* **2018**, *58*, 815–822. [CrossRef]

10. Xilai, C.; Li, Y.; Li, Y.; Sang, S.; Lei, Z.; Jin, S.; Li, S.; Ge, S. Effect of carbon aggregates on the properties of carbon refractories for a blast furnace. *Metall. Mater. Trans. B* **2013**, *41*, 420–429.

11. Kowalski, W.; Lüngen, H.; Stricker, K. State of the art for prolonging blast furnace campaigns. *Revue Métallurgie* **2000**, *97*, 493–505. [CrossRef]

12. Shinotake, A.; Nakamura, H.; Yadoumaru, N.; Morizane, Y.; Meguro, M. Investigation of Blast-furnace Hearth Sidewall Erosion by Core Sample Analysis and Consideration of Campaign Operation. *ISIJ Int.* **2003**, *43*, 321–330. [CrossRef]

13. De Almeida, B.V.; Neves, E.S.; Silva, S.N.; Junior, F.V. Blast Furnace Hearth Lining: Post Mortem Analysis. *Mater. Res.* **2017**, *20*, 814–818. [CrossRef]

14. Lee, W.E.; Zhang, S. Melt corrosion of oxide and oxide–carbon refractories. *Int. Mater. Rev.* **1999**, *44*, 77–104. [CrossRef]

15. Torrkulla, J.; Saxe, H. Model of the state of the blast furnace hearth. *ISIJ Int.* **2000**, *40*, 438–447. [CrossRef]

16. Internal Tokai COBEX procedures—Hot Metal Penetration test. 2014.

17. Stec, J. Develop/Evaluate a Hot Metal Penetration Test for Furnace Lining Materials. Master's Thesis, AGH University of Science and Technology, Kraków, Poland, 2015.

18. Wejrzanowski, T.; Gluch, J.; Ibrahim, S.H.; Cwieka, K.; Milewski, J.; Zschech, E. Characterization of Spatial Distribution of Electrolyte in Molten Carbonate Fuel Cell Cathodes. *Adv. Eng. Mater.* **2018**, *20*, 1700909. [CrossRef]

19. Chuaypradit, S.; Puncreobutr, C.; Phillion, A.B.; Fife, J.L.; Lee, P.D. Quantifying the Effects of Grain Refiner Addition on the Solidification of Fe-Rich Intermetallics in Al–Si–Cu Alloys Using In Situ Synchrotron X-ray Tomography. In *Proceedings of the International Conference on Martensitic Transformations*; Springer International Publishing: Chicago, IL, USA, 2018; pp. 1067–1073. [CrossRef]

20. Echlin, M.P.; Mottura, A.; Wang, M.; Mignone, P.J.; Riley, D.P.; Franks, G.V.; Pollock, T.M. Three-dimensional characterization of the permeability of W-Cu composites using a new "TriBeam" technique. *Acta Mater.* **2014**, *64*, 307–315. [CrossRef]

21. Ibrahim, S.H.; Skibinski, J.; Oliver, G.; Wejrzanowski, T. Microstructure effect on the permeability of the tape-cast open-porous materials. *Mater. Des.* **2019**, *167*, 107639. [CrossRef]

22. Weber, E.; Fernandez, M.; Wapner, P.; Hoffman, W. Comparison of X-ray micro-tomography measurements of densities and porosity principally to values measured by mercury porosimetry for carbon–carbon composites. *Carbon* **2010**, *48*, 2151–2158. [CrossRef]

23. Shanti, N.O.; Chan, V.W.; Stock, S.R.; De Carlo, F.; Thornton, K.; Faber, K.T. X-ray micro-computed tomography and tortuosity calculations of percolating pore networks. *Acta Mater.* **2014**, *71*, 126–135. [CrossRef]

24. Holzer, L.; Wiedenmann, D.; Münch, B.; Keller, L.; Prestat, M.; Gasser, P.; Robertson, I.; Grobéty, B. The influence of constric-tivity on the effective transport properties of porous layers in electrolysis and fuel cells. *J. Mater. Sci.* **2013**, *48*, 2934–2954. [CrossRef]

25. Stec, J.; Tarasiuk, J.; Nagy, S.; Smulski, R.; Gluch, J.; Filipek, R. Non-destructive investigations of pore morphology of micropore carbon materials. *Ceram. Int.* **2019**, *45*, 3483–3491. [CrossRef]

26. Rørvik, S.; Lossius, L.P. Measurement of tortuosity of anode porosity by 3D micro X-ray computed tomography. *Light Met.* **2017**, 2–5.

27. Rørvik, S.; Lossius, L.P. Characterization of prebaked anodes by micro X-ray computed tomography. *Light Met.* **2017**, 1237–1245.

28. Steenkamp, J.; Joalet, D.; Pistorius, P.C.; Tangstad, M. Wear Mechanisms of Carbon-Based Refractory Materials in Silicomanganese Tap Holes—Part I: Equilibrium Calculations and Slag and Refractory Characterization. *Met. Mater. Trans. A* **2015**, *46*, 653–667. [CrossRef]

29. Baux, A.; Goillot, A.; Jacques, S.; Heisel, C.; Rochais, D.; Charpentier, L.; David, P.; Piquero, T.; Chartier, T.; Chollon, G. Synthesis and properties of macroporous SiC ceramics synthesized by 3D printing and chemical vapor infiltration/deposition. *J. Eur. Ceram. Soc.* **2020**, *40*, 2834–2854. [CrossRef]

30. Zhou, Y.; Zheng, Y.; Pan, J.; Sui, L.; Xing, F.; Sun, H.; Li, P. Experimental investigations on corrosion resistance of innovative steel-FRP composite bars using X-ray microcomputed tomography. *Compos. Part. B Eng.* **2019**, *161*, 272–284. [CrossRef]

31. Quiney, Z.; Weston, E.; Nicholson, P.I.; Pattison, S.; Bache, M.R. Volumetric assessment of fatigue damage in a SiCf/SiC ce-ramic matrix composite via in situ X-ray computed tomography. *J. Eur. Ceram. Soc.* **2020**, *40*, 3788–3794. [CrossRef]

32. Qiu, Q.; Zhu, J.; Dai, J.-G. In-situ X-ray microcomputed tomography monitoring of steel corrosion in engineered cementitious composite (ECC). *Constr. Build. Mater.* **2020**, *262*, 120844. [CrossRef]

33. Hiltmann, F.; Daimer, J.; Hohl, B.; Nowak, R.; Tomala, J. Cathode Quality Improvement by Application of an Intensive Homogenizer for Green Mix Preparation. *Light Met.* **2004**, 593–596.
34. Feldkamp, L.A.; Davis, L.C.; Kress, J.W. Practical one-beam algorithm. *J. Opt. Soc. Am.* **1984**, *1*, 612–619. [CrossRef]
35. Volume Graphic GmbH. Available online: http://www.volumegraphics.com/en/products/vgstudio-max/ (accessed on 16 December 2020).
36. Abràmoff, M.D.; Magalhães, P.J.; Ram, S.J. Image processing with Image. *J. Biophotonics Int.* **2004**, *11*, 36–41.
37. Dougherty, R.; Kunzelmann, K.-H. Computing Local Thickness of 3D Structures with Image. *J. Microsc. Microanal.* **2007**, *13*, 1678–1679. [CrossRef]
38. Simpleware ScanIP. Available online: https://www.synopsys.com/simpleware/software/scanip.html/ (accessed on 21 January 2021).
39. Stec, J. 3D Pore Structure and Infiltration Resistance of Micropore Carbon Materials. Ph.D. Thesis, AGH University of Science and Technology. Wydawnictwo Naukowe AKAPIT, Kraków, Poland, 2020.
40. Stock, S.R. *Microcomputed Tomography. Methodology and Applications*; CRC Press: Boca Raton, FL, USA, 2009.
41. Stec, J.; Smulski, R.; Nagy, S.; Szyszkiewicz-Warzecha, K.; Tomala, J.; Filipek, R. Permeability of carbon refractory materials used in a blast furnace hearth. *Ceram. Int.* **2021**, *47*, 16538–16546. [CrossRef]

Article

Traditional vs. Automated Computer Image Analysis—A Comparative Assessment of Use for Analysis of Digital SEM Images of High-Temperature Ceramic Material

Ilona Jastrzębska [1,*] and Adam Piwowarczyk [2]

[1] Faculty of Materials Science and Ceramics, AGH University of Science and Technology in Cracow, al. A. Mickiewicza 30, 30-059 Cracow, Poland
[2] Faculty of Mechanical Engineering, Cracow University of Technology, al. Jana Pawła II 37, 31-864 Cracow, Poland
* Correspondence: ijastrz@agh.edu.pl

Abstract: Image analysis is a powerful tool that can be applied in scientific research, industry, and everyday life, but still, there is more room to use it in materials science. The interdisciplinary cooperation between materials scientists and computer scientists can unlock the potential of digital image analysis. Traditional image analysis used in materials science, manual or computer-aided, permits for the quantitative assessment of the coexisting components at the cross-sections, based on stereological law. However, currently used cutting-edge tools for computer image analysis can greatly speed up the process of microstructure analysis, e.g., via simultaneous extraction of quantitative data of all phases in an SEM image. The dedicated digital image processing software *Aphelion* was applied to develop an algorithm for the automated image analysis of multi-phase high-temperature ceramic material. The algorithm recognizes each phase and simultaneously calculates its quantity. In this work, we compare the traditional stereology-based methods of image analysis (linear and planimetry) to the automated method using a developed algorithm. The analysis was performed on a digital SEM microstructural image of high-temperature ceramic material from the Cu-Al-Fe-O system, containing four different phase components. The results show the good agreement of data obtained by classical stereology-based methods and the developed automated method. This presents an opportunity for the fast extraction of both qualitative and quantitative from the SEM images.

Keywords: ceramic; spinel; copper; SEM/EDS; digital image; computer analysis; stereology

Citation: Jastrzębska, I.; Piwowarczyk, A. Traditional vs. Automated Computer Image Analysis—A Comparative Assessment of Use for Analysis of Digital SEM Images of High-Temperature Ceramic Material. *Materials* **2023**, *16*, 812. https://doi.org/10.3390/ma16020812

Academic Editor: A. Javier Sanchez-Herencia

Received: 2 November 2022
Revised: 15 December 2022
Accepted: 26 December 2022
Published: 13 January 2023

1. Introduction

Image analysis is becoming an increasingly used tool for various scientific research [1–6], industrial [7,8] and medical [9,10] applications. Often, we are unaware that image analysis accompanies us in our daily lives. Examples of this are traffic analysis, control of vehicle speed, and the detection of license plates [11], or pedestrian detection [12], which all point to smart city development [13]. Other examples of important industrial applications are monitoring the quality of rotors in wind turbines, which allows for early fault detection [7,8], and biometric security [14,15]. The development of image recognition algorithms by automated methods such as in [16] can greatly increase the accuracy of images and obtain plausible results. In particular, the application of image analysis algorithms allows for significantly enhanced diagnostic methods in medicine. An example is the application of a phase retrieval algorithm in an imaging method coupled with computer tomography (CT), which improved the visibility of weakly absorbing objects, thus permitting a lower radiation dose without loss of image quality [10]. Another example is using the 2D phase contrast algorithm, which retrieved better X-ray images with additional information. Moreover, coupling image detection algorithms with methods using high-energy sources, such as synchrotron radiation (e.g., SOLARIS [6]), or facilities with a thermal neutron beam (e.g.,

VIVALDI [17]) have recently allowed profound progress in the diagnostics of tumors in humans [18] and large-scale research facilities have been utilized as neutron research reactor facilities with image-plate detectors for crystallography and biology applications [19,20]. This has an extremely positive impact on both medical and industrial imaging. In this way, image analysis algorithms have been developed to support progress into safer and more sustainable development of nations.

Image analysis is also a powerful tool in materials science. Many instances of its successful use have been reported so far. Digital image processing in joining technology permitted detection of the border between the melting pool of liquid metal and adjacent non-molten base metal despite their temperature being the same, as they are at the phase transformation stage [1]. This research afforded the opportunity to develop a smart welding filter equipped with augmented reality, which displays additional information to the operator, making the joining process more precise, operator-friendly, and effective.

In materials engineering, microstructure mostly determines materials' properties. Modifications on the microstructural scale enable tailoring of the desired properties of a material. Scanning electron microscopy equipped with energy (SEM/EDS) or electron microprobe analysis with weave dispersive spectroscopy (EPMA/WDS) are common and effective tools that permit the extraction of spectroscopic qualitative information [21]. The obtained microstructural images can be subjected to image analysis to retrieve more data on materials, such as the quantity of the individual phase components. Traditional methods based on stereology rules can be applied, but they are time-consuming and less accurate due to the occurrence of systematic errors. The application of automated, computer-based methods has been recently developed for different kinds of materials. Binarization was used to detect the pores in cement pastes with different W/C ratios and their further geometrical characterization (diameter, roundness) to investigate the formation of capillary pores [4]. Image recognition and analysis were conducted on Al_2O_3/WS_2 coatings on Al alloys to determine the volume share of areas of distinguishing filamentous morphology [5]. Kubinova et al. [22] compared several stereological and digital methods for estimating the surface area and volume of cells on confocal microscopy images and discussed their pros and cons in the context of the isotropy of the studied material.

Image analysis can be performed on different types of photographs. Different attributes of every individual phase on an image can be used for the development of a robust algorithm for color, shape, or area. The image analysis covers object detection (qualitative information) as well as the measurement of the amounts of specific objects (quantitative information). The human is the "link" in the analysis process and has control over the program action, so also impacts the accuracy of the results. If input data (an image or a series of images) are prepared inadequately in the initial stage, the output data will be erroneous, inaccurate, and overestimated or underestimated.

Comparison of traditional (manual/computer-aided) to automated computer measurements pros and cons can be performed. A computer will not count anything on its own, while a human must control a machine. Based on knowledge and experience, a human must precisely determine what shall be counted and assess whether the result is correct. As measurements are performed after binarization of the original image (Figure 1), preceded by various operations, a representative and good-quality image is of crucial significance.

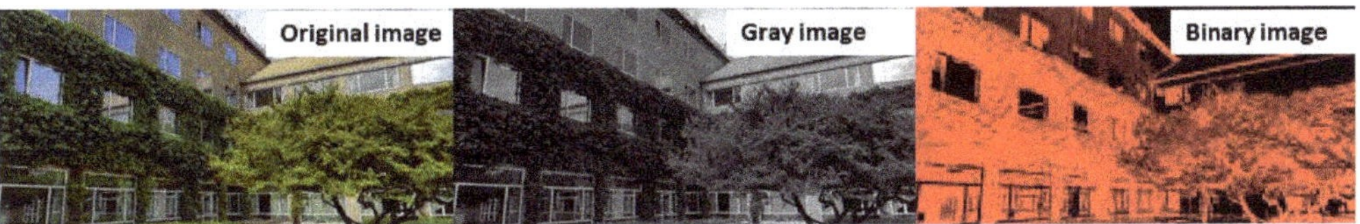

Figure 1. Comparison of an original color image with grey and binary images. Aarhus University campus, during 15th International Congress for Stereology and Image Analysis, Denmark 2019.

One of the most common problems in image analysis is uneven background, which derives from the unequal lightening of the objects. This occurs especially in optical microscopy and may be a serious problem in appropriate object recognition, especially for low-contrast images. Often, it is not visible to the human eye, which leads to unreliable quantification and misinterpretation. A new method for shade correction in optical microscopy images, developed by Gądek et al. [23], was based on the simulation of the image background in which pixel values represent smooth grey-level changes. In contrast, Biżantowicz [24] developed a focus stacking algorithm for SEM images, which eliminates the adverse features of SEM photos, such as drifts or changes in geometry, and allowed for an increase in the depth of field (DoF) via digital image correction combined with assembling a series of photographs into one image.

Currently, in the characterization of ceramic materials, the most common technique of microstructural analysis is the SEM/EDS technique, which is widely available both at universities and in numerous R&D departments of industrial companies. The method requires a small sample of about 1 cm^2 area and provides a wide range of enlargements. However, mostly qualitative information is retrieved from these images, while quantitative data are determined from manual or computer-aided calculations based on stereology laws, or, using complementary methods, e.g., refinement of X-ray patterns; these are both, in fact, laborious and time-consuming. In this work, we compare classical stereology-based vs. automated computer analysis of grey SEM microstructural images of high-temperature ceramic material from the Cu-Al-Fe-O system. Starting from showing versatile transformations, which enable increased image quality, we finally present the algorithm permitting for retrieving reliable quantitative data on the amounts of the microstructural phase objects present in the analyzed material. The application of this kind of algorithm can greatly enhance the extraction of additional data from a single SEM image; thus, it can make R&D activities more effective and sustainable.

2. Methods and Materials

The material taken for the image analysis was ceramic oxide material, produced via the arc melting technique [25] that was previously used successfully for the synthesis of numerous high-temperature spinel compounds [26–28] and high-temperature materials [29,30]. The starting materials were analytical grade powders of Fe_2O_3, CuO, and Al_2O_3 (Sigma Aldrich) mixed in the molar proportions 0.25:0.5:1 and homogenized in a ball mill. The sample for arc-melting was prepared in the form of a cylindrical-shaped disc of 20 mm diameter and 10 mm height. The arc-melted material was observed using a scanning electron microscope (Nova NanoSEM200 (FEI)) equipped with an energy-dispersive spectrometer (EDX). The imaging was performed in back-scattered electron (BSE) mode, using an accelerating voltage of 18 kV and 2000× magnification.

Most SEM microscopes detect two types of electrons, namely back-scattered electrons (BSE) and secondary electrons (SE). The former are produced due to elastic scattering of beam electrons from the atomic nucleus of the sample and are derived from the deeper parts of materials (100–1000 Å); the latter come from the sample and are produced as a result of bombarding of the surface atoms (10–300 Å) by beam electrons [21,31]. The detectors used for BE (passive detectors, scintillation detectors, semiconductor BSE detector) and SE (Everhart–Thornley detector, through-the-kens electron detectors) were described in [21]. Detection of the BSE signal produces images, in which contrast depends on relative differences in atomic numbers, in contrast to secondary electrons detection mode, which provides information only about the topography of the specimen. In general, the brighter the microarea, the higher its atomic mass. In contrast, the darker microareas indicate lower atomic mass, or resin-filled porosity. Thanks to these signals, the determination of phase composition can be postulated.

The cross-section sample for SEM observation was prepared by a ceramographic technique. First, the sample was embedded in an ambient cure two-component epoxy resin. Then, the resin-embedded cross-sectioned sample proceeded with rough grounding and fine polishing. The impregnated sample was coated with carbon in a vacuum to enable the charge transfer of SEM beam electrons at the ceramic, electrically isolated specimen. Finally, the sample was mounted on the SEM microscope table using carbon tape.

For the quantitative analysis of the image, proper preparation of the sample is crucial [32]. Additionally, the number of images captured should be enough to retrieve the representative information, and will be greater for higher-magnification photographs. Based on experience, the approximate number of images that should be taken to obtain reliable information on a sample is 20–40, which, specifically, depends on the analyzed sample (larger for anisotropic materials). The average time of such analysis, together with EDS measurements, is 1–2 h.

Figure 2a presents a greyscale SEM microstructure image of the studied, melted ceramic material, which is composed of four distinct phases, determined based on the chemical EDS analysis. The darkest phase, which created a specific pattern, is Al_2O_3 with a slight amount of Fe (0.9, point 1). Between alumina grains, three phases of different intensities of grey are observed. The lightest-color phase is copper oxide, CuO_x (point 2). The rest phases, lighter-grey (point 3) and darker-grey (point 4), are spinel compounds of different stoichiometry, namely, Cu-rich and Fe-rich spinels, respectively. The EDS spectra for the discussed points are presented in Figure 2b, and the results are included in Table 1.

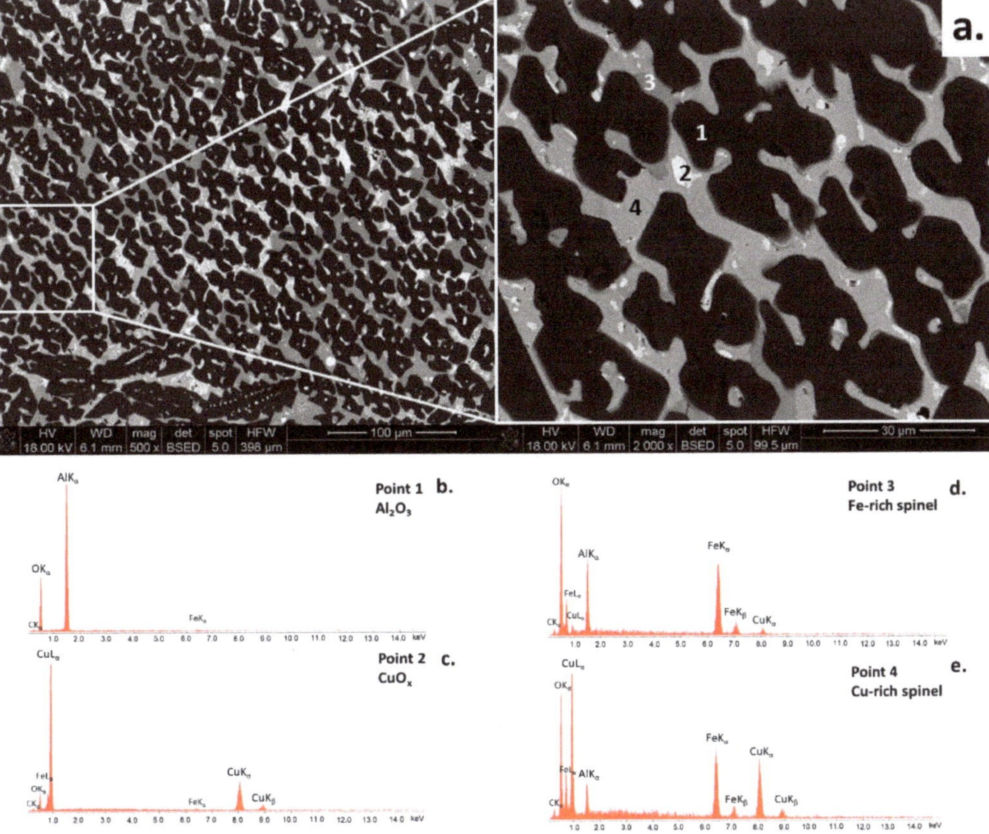

Figure 2. (**a**). SEM microstructure image of ceramic high-temperature material together with EDS spectra in points 1–4 (**b–e**).

Table 1. EDS analysis of microareas of high-temperature ceramic material corresponding to Figure 2.

Point	Phase	Chemical Composition, mol. % *		
		Cu	Fe	Al
1	Alumina Al_2O_3	-	0.9	47.0
2	Copper oxide CuO_x	67.3	1.8	0.6
3	Fe-rich spinel $(Fe,Cu)(Fe,Al,Cu)_2O_4$	2.6	31.9	13.0
4	Cu-rich spinel $(Cu,Fe)(Cu,Fe,Al)_2O_4$	30.7	19.4	7.7

* The rest (to 100%) is oxygen.

The SEM microphotograph, saved in lossless *tiff* format, was subjected to image analysis by using traditional methods based on stereology as well as an automated method based on the developed algorithm. For the traditional methods of linear and planimetry, the free and open-code software *ImageJ* was used [9]. The automatic algorithm for the simultaneous recognition and measurements of phases present in the studied sample was developed using *Aphelion* software ver. 4.4.0.

3. Traditional Methods of Image Analysis

Stereology covers the number of methods developed for the description of 3D objects based on 2D images [33]. The selected methods, especially planimetry owing to its usefulness, were included in standards, e.g., American Society for Testing and Materials [34] for the estimation of average grain size in all single-phase materials, or International Organization for Standardization [35] for the estimation of grain size in Cu and Cu alloys, and other applications [36]. Using stereological-based methods, the volumetric proportions between coexisting phases can be determined using only fragmental information based on a flat cross-sectioned sample. A reliable analysis requires a random area, using material that is located uniformly within the volume of the material and without the privileged direction of orientation [37].

Traditional stereological methods, including planimetry, linear analysis, or point counting, are used to obtain quantitative information about the objects distributed in the material. Specifically, the global parameters are determined. They describe relations between selected features and the entire analyzed space, including the volume share of the selected element in the material (V_V), the surface area share of cross-sections (A_A), and linear its share (L_L). According to the Cavalieri–Hecquert principle, the global parameters are Equation (1) [33,38].

$$V_V = A_A = L_L \tag{1}$$

3.1. Linear Analysis

The linear analysis is based on the secant of known length on the analyzed microstructure image followed by the determination of the sum of the chords belonging to this secant, covering the interest phase. The linear share is the ratio of the sum of chords to the length of the secant. A specific number of secants is applied for the analyzed image in order to reduce the uncertainty. The quantity information is obtained by dividing the sum of the chords cutting out the interest phase by the total length of the secant (2).

$$L_L(P) = \frac{\sum_{i=1}^{n} \sum_{j=1}^{m} c_{ij}}{n \cdot l} \tag{2}$$

where:

c—the length of the chord,

n—the number of secants,

m—the number of the cut phase in the following *i*-measurement,

l—the length of the secant.

Based on the Cavalieri–Hecquert principle (1) and the calculated surface area of the interest phase (P1, P2, P3. or P4), the volume share of the specific phase in the material is equal to its linear share at the cross-section, namely $V_V = L_L$ [33,38].

3.2. Planimetry

Planimetry is based on the measurement of the surface area of the interest phase, via summing the pixels corresponding to this phase, extracted out of the rest phases by binarization. The quantity information is obtained by dividing the surface of the phase by the total surface of the image (3).

$$A_A(P) = \frac{\sum_{i=1}^{n} A(P)}{A(A)} \tag{3}$$

where:

A(P)—the total surface of the phase at the image,

A(A)—the total surface of the analyzed image.

Based on the Cavalieri–Hecquert principle (1) and the calculated surface area of the interest phase (P1, P2, P3, or P4), the volume share of the specific phase in the material is equal to its surface share at the cross-section, namely, $V_V = A_A$ [33,38].

4. Operations in Computer Image Analysis

Different color models, also called color spaces, are commonly used in image analysis. Using a color space, it is possible to define a specific combination of color models and their representation functions. The identification of a color space permits the automatic identification of the associated color models. During image processing, one of the important steps is the selection of the color model. Several color models were developed, as presented in Table 2. They can be used independently of the desired results, as shown by Figure 3 (RGB) and Figure 4 (HSI).

Table 2. Selected color models and their designations.

Name	Color Space Definitions
RGB	Red, Green, and Blue
HSI	Hue, Saturation, and Intensity
HSV	Hue, Saturation, and Value
YUV	Luminance and Chrominance

One method of image processing is converting an original color image into a grey one (0–255 grayscale [39]; Figure 1, middle), followed by analysis. This simple and quick transformation can extract information that can be used in the subsequent steps of the algorithm. However, it simultaneously causes a loss of information in the individual components. Alternatively, binarization of the color image can be performed for one individual component of the image, which permits obtaining the binary image directly from the color image (Figure 1, right).

Figure 3. Comparison of (**a**) the color image in the RGB model, and images formed by decomposing the image into (**b**) R, (**c**) G, (**d**) B components.

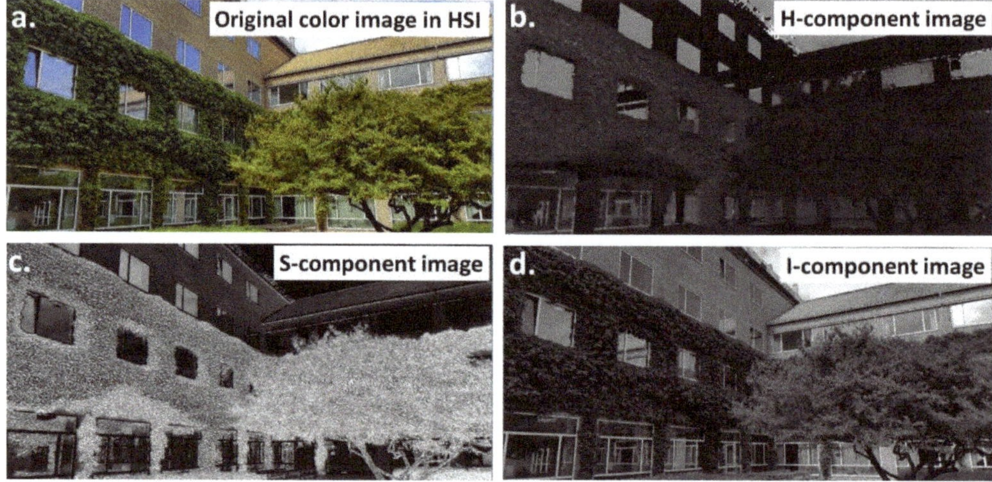

Figure 4. Comparison of the (**a**) color image in the HSI model and images formed by decomposing the image into (**b**) hue (H), (**c**) saturation (S), and (**d**) intensity (I) components.

If object detection on a grey image is not recommended or if a determination of the binarization thresholds for a color image is problematic, the RGB image can be split into components. This transformation produces three multi-shade images, representing the red (R), green (G), and blue (B) components, respectively (Figure 3a–d). Then, an individual image can be subjected to further transformations, such as object detection.

In some cases, the RGB model does not permit the detection of objects effectively. This results from insufficient color differences between the objects and the background, in terms of the values of the individual color components. Therefore, it may be useful to perform a color transformation from the RGB model to the HSI model. The result of this transformation is an image stored in the HSI model, as shown in Figure 4a–d. In this method, the color image is split into its components and the differences between the images are analyzed. Even if there is no significant difference in the share of the individual RGB components, the differences will be visible in the share of the components of saturation, intensity, and brightness of colors. The most suitable component can be selected for further transformation and detection.

Noise is commonly encountered during image acquisition. This can make analysis difficult or, in some cases, impossible. For this purpose, several filter tools have been developed to reduce the presence of noise. However, special caution should be taken during using a filter for noise reduction to avoid the introduction of undesired changes to the sourced image. Otherwise, this can cause irreversible losses in the information contained in the image [39]. Filters are mainly used for sharpening, blurring, or edge detection [40], as presented in Figure 5a–d.

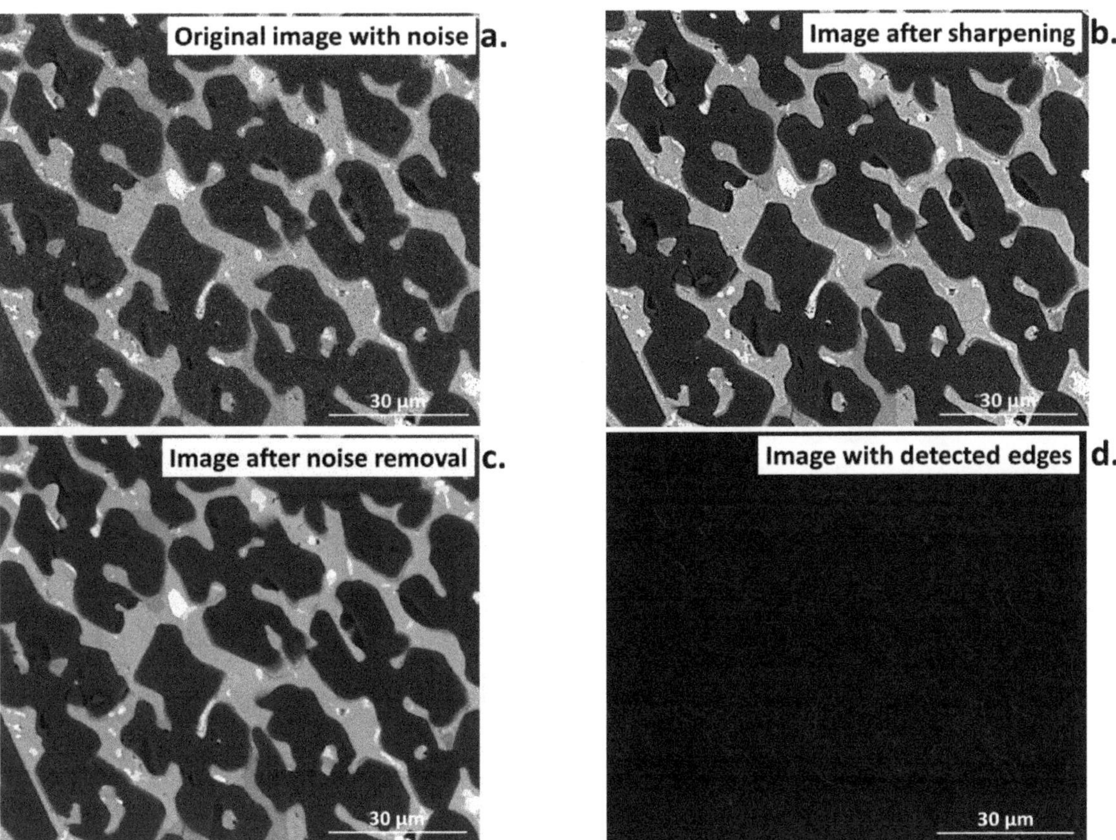

Figure 5. Comparison of the (**a**) input image with added noise, and images (**b**) after sharpening, (**c**) after noise removal, (**d**) with detected edges in the image without noise.

Filters can help significantly in image preparation, e.g., if the image was registered with a signal. This can be evidenced by comparing the profile of the image with noise (Figure 6a,b) to the image after noise reduction (Figure 6c,d) using the median filter [39]. The profile after filtering is significantly smoothed compared to the image with noise. This is an ideal operation for inhomogeneous noise with varying degrees of grey. Sharpening filters should be avoided when performing quantitative image analysis as they can introduce additional noise, despite the improvement in the quality of the image.

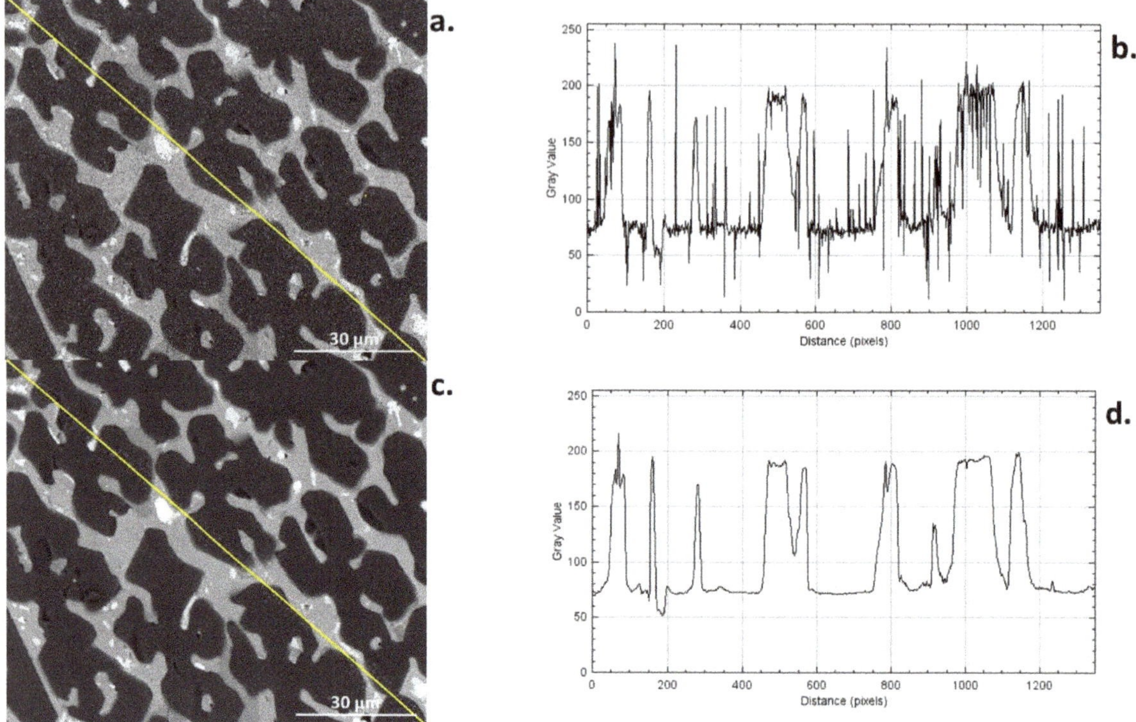

Figure 6. Comparison of the image and noise with its profile (**a**,**b**) vs. the image after noise reduction and its profile (**c**,**d**); the yellow line shows the test site for creating the profile.

Morphological operations belong to the most important operations in the analysis of the digital image, as, combined with other operations, they allow for complex image transformations to extract desired information. Morphological operations can be described as filters that are distinguished from classical filters by their selectivity [3,33,39]. In this case, the specific points are subjected to transformation. A structural element, called a pattern or a template, plays a role in every morphological operation. The general scheme of the algorithm for morphological operations can be divided into three stages. First, a center point to each point in the image is applied. Then, the configuration of points is checked to confirm that it is identical to one in the pattern. Finally, the operation according to the given transformation is performed [3]. Four morphological operations are mostly used, namely erosion, dilation, opening, and closing (Figure 7a–d).

Erosion permits the removal of isolated points, small particles, and narrow spears from an image, as shown in Figure 7b. In addition, it smooths the edges of objects in an image and reduces their surface edges. Erosion can also lead to the division of the objects into several smaller ones, so it can be used to divide connected particles.

Dilation is the inverse transformation to erosion, also called a maximal filter. The characteristic feature of dilation is that it closes small holes (fills the gaps in objects) [3,33]. As result, the area of each object in the transformed image is greater than in the input image, as shown in Figure 7c.

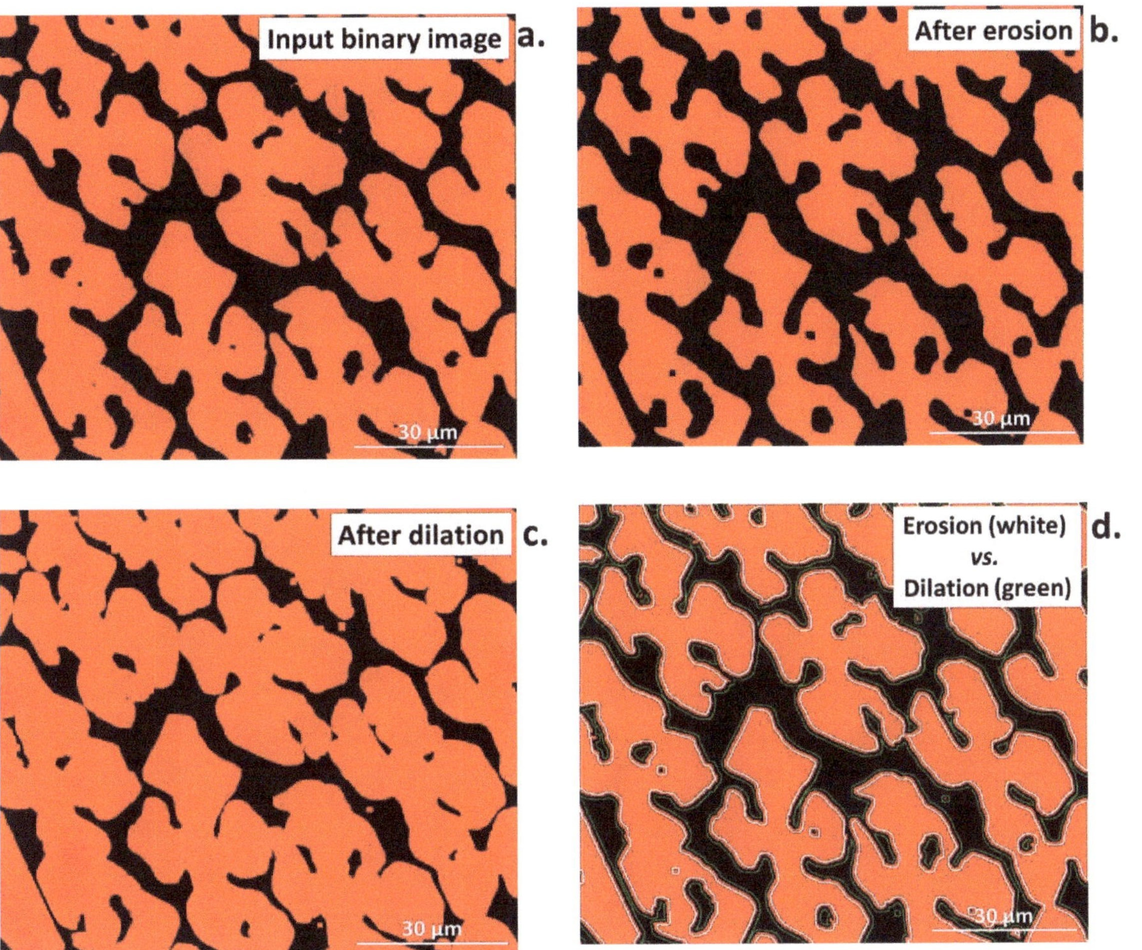

Figure 7. Comparison of the (**a**) input image after Otsu binarization to images after (**b**) erosion, (**c**) dilation, and (**d**) erosion (white lines) and dilation (green lines) superimposed on the input binary image.

Opening and closing are more complex operations, combining erosion with dilation (Figure 8a–d). In the opening operation, the erosion is followed by dilation. However, it uses the same pattern (structural element) and maintains the same operation size in both steps. First, erosion breaks up the thin lines in the input image that connect objects. Then, dilation permits the approximate recreation of these connections [3,33]. As a result, the separation of elements will be obtained in the output image, as shown in Figure 8b. The characteristic feature of the opening operation is that it removes small elements and details without changing the size of the main part of the image.

The closing operation is the reverse of the opening one. It includes dilation followed by erosion. The main role of closing is to fill in the narrow elements in the image, such as indentations or small details. Similar to opening, closing does not change the size of the image [33]. The results of the closing operation can be seen in Figure 8c and the comparison between opening and closing is presented in Figure 8d.

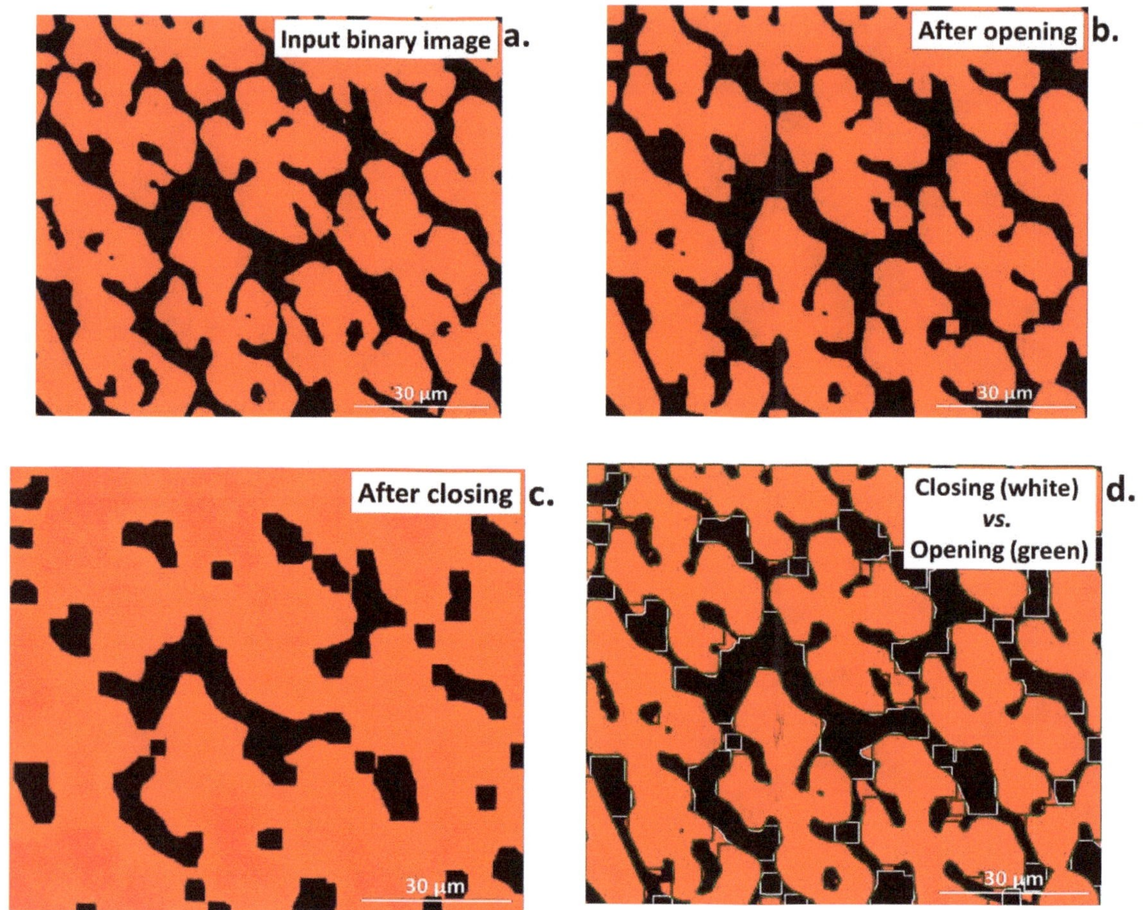

Figure 8. Comparison of the (**a**) input image after Otsu binarization to images after (**b**) opening, (**c**) closing, and (**d**) closing (while lines) and opening (green lines) superimposed on the input binary image.

Binarization is one of the most significant operations in image analysis [3,39,41]. It facilitates the analysis and conducting of measurements at the images, especially in terms of percentage quantification of the objects relative to the rest of the image, determination of the object parameters (areas, perimeters, etc.), and displaying them on screen. Numerous binarization methods have been developed so far, e.g., applying an upper threshold, a lower threshold, and two thresholds. Moreover, various automatic binarization methods are used that determine threshold values based on the analysis of selected image features. The Otsu or entropy thresholding method is an example of the automatic threshold algorithm, based on the histogram of grey-level distribution.

As can be seen from Figure 9a–c, every individual method of binarization produces different results. Otsu binarization (Figure 9b), which is a histogram-based method, transforms a grayscale input image into a binary image by minimizing the weighted sums of the variances of two classes, namely foreground objects and background [41]. This method belongs to simplest and most successful; however, it may lose some important details. The entropy threshold method (Figure 9c) provides a set of regions from the input image, by applying automatic thresholding based on the entropy computation of the histogram. This method, due to fixed threshold values, has problems with images of unimodal histograms and may lose information for small, low-contrast target objects [41]. In all cases, to acquire information about the reliability of the selected binarization method as well as to avoid errors, the object of interest in the binary image (Figure 10b,c) should be compared to the original, before binarization (Figure 10a).

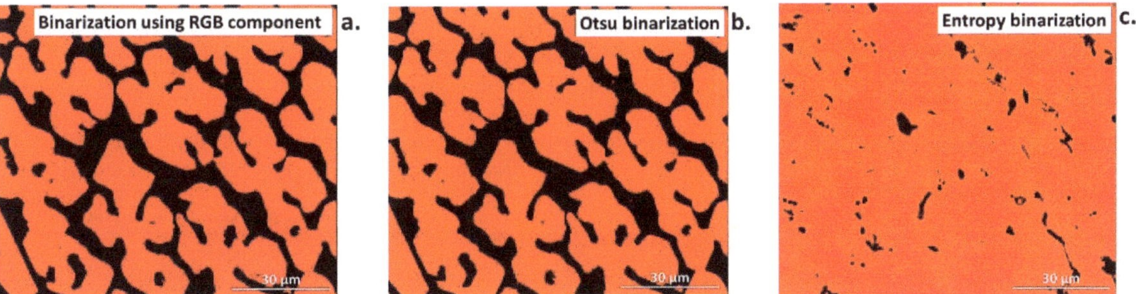

Figure 9. Comparison of various binarization methods, (**a**) using the blue component for RGB, and using automatic threshold methods (**b**) Otsu, (**c**) entropy.

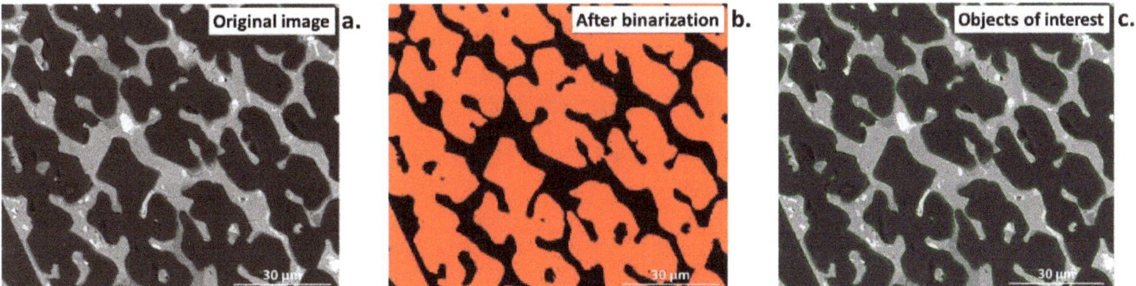

Figure 10. Comparison of (**a**) the original microstructural image of ceramic material with the image (**b**) after binarization, (**c**) comparison of images with marked objects of interest (green) permits for the visual assessment of the quality of the binarization.

Binarization by automatic thresholding has both pros and cons compared with manual thresholding. The advantage is that automatic binarization does not require the selection of the range of threshold for a particular object, as it occurs automatically, releasing object and background distribution. In the case of entropy automatic binarization, for each threshold, the t-value (between image-min and image-max) and two probability distributions (object and background distribution) are derived from the original grey-level distribution image [41]. As result, two entropy values are calculated. For a small number of studied objects and a large series of photographs, the use of the automated method is an appropriate solution if the same image quality is provided. However, when the number of interest phases increases or there are slight differences in the quality of the images, it may result in different detection of objects, e.g., for sample 1 the range needed for detection of a studied phase is 0–100, while for sample 2 it is 0–120 for the same object. Overall, the choice of binarization method shall be adapted to the studied image and object of interest; however, in the former case, the manual (interactive) method will be more appropriate as it permits precise adjusting of the range of threshold [39]. Automatic binarization works best for images with that which significantly differ in contrast, such as bright objects on dark backgrounds or the opposite [33]. Nevertheless, binarization alone is not a sufficient operation before image analysis, and in most cases requires additional operations, such as morphological transformations.

5. Results of the Quantification Analysis of the SEM Image

5.1. Traditional Stereology-Based Methods

In this work we used linear and planimetry methods. The measurements were performed on a digital SEM image. The digital image was analyzed using *ImageJ* software. Twenty-five lines of the same length were applied in the linear analysis. For planimetry analyses, before performing the binarization, the image was first filtered using a median filter, then transformed to an 8-bit image (256 of grey), normalized, and finally transformed to a 1-bit image using interactive thresholding.

The results of linear analysis on a grey SEM image of the studied material are presented in Figure 11a,b, while the result of the planimetry is presented in Figure 12a–d. The quantitative information extracted from both analyses are included in Table 3.

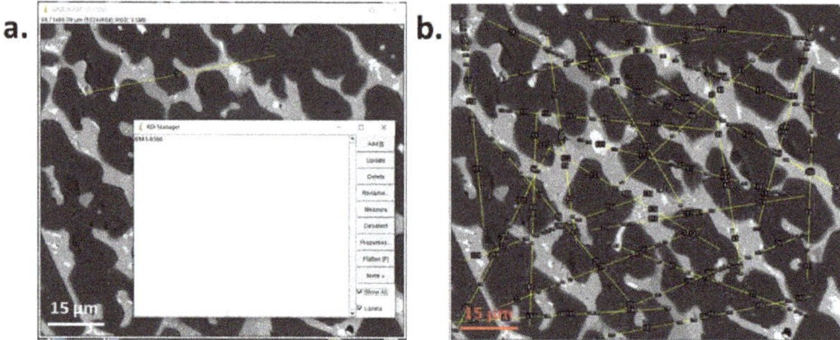

Figure 11. Windows of *ImageJ* during the linear analysis of the image, (**a**) using ROI Manager for adding the secants and chords, (**b**) with all secants and chords.

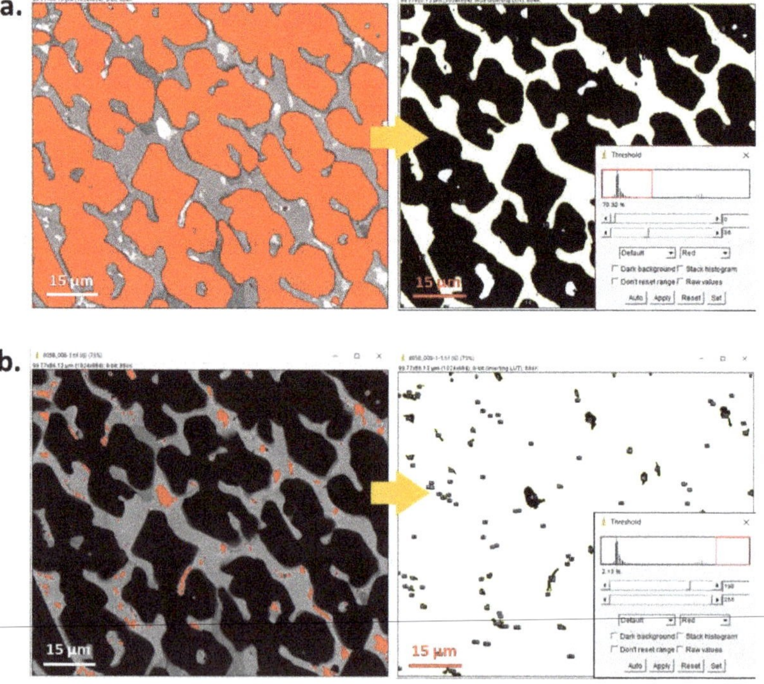

Figure 12. *Cont.*

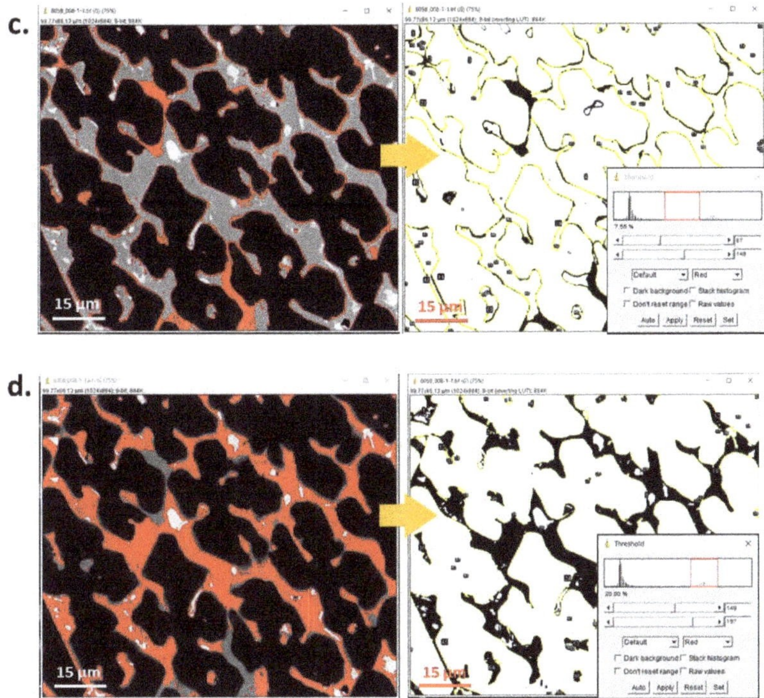

Figure 12. Window of *ImageJ* during binarization (by manual thresholding) and planimetry analysis of phases (**a**) P1, (**b**) P2, (**c**) P3, (**d**) P4.

Table 3. Results of traditional image analysis corresponding to Figure 2 (linear) and Figure 3 (planimetry).

Phase Name	Phase Amount, %	
	Linear	Planimetry
P1 (darkest)	70.6	71.4
P2 (lightest)	2.7	2.1
P3 (dark grey)	1.9	8.6
P4 (light grey)	24.6	21.6

Estimated error: linear ± 2.0, planimetry ± 1.0.

As can be seen from Table 3, both methods produced similar results. The greatest agreement was obtained for the P2 phase, which was CuO_x, randomly distributed in the matrix, at the level of 2.7% and 2.1% for the linear and planimetry methods, respectively. Additionally, a similarly small difference between results was determined for the P1 phase, Al_2O_3, which constituted 70.6% and 71.4% for the linear and planimetry methods, respectively. The light-grey Cu-rich spinel phase P4, which occurred as a continuous phase, showed a higher discrepancy in both methods, of 24.6% (linear) and 21.6% (planimetry). The greatest difference was determined for the dark-grey Fe-rich spinel phase P3 of 1.9% (linear) and 8.6% (planimetry). The difference results from detecting the edges of the Al_2O_3 grains as the P3 phase, due to the same pixel values.

5.2. Automated Method

Aphelion software was used for the development of the algorithm for automated image analysis. The user interface of *Aphelion* during analysis is presented in Figure 13. The software is equipped with the necessary modules to perform individual operations and permits the simultaneous detection and percentage amount measurements of the objects present in the SEM image. In addition, all the data produced during analysis can be saved as a macro command in the Visual Basic language.

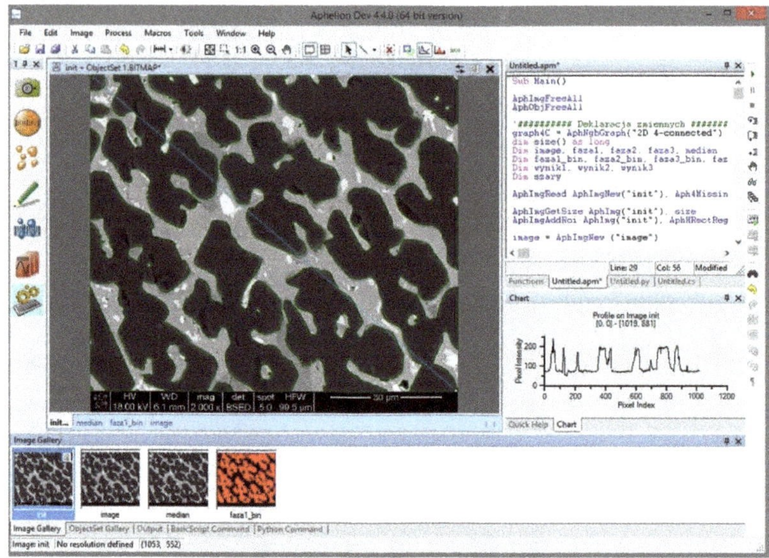

Figure 13. User interface of *Aphelion* during the automated analysis of an SEM image of high-temperature ceramic material.

The goal of the developed algorithm was to automatically detect the four phases present in the multiphase material, namely P1, P2, P3, and P4 (previously determined using SEM/EDS on cross-sections), color every individual phase as separate, and finally, measure its percentage 2D surface area relative to the total image area. A schematic presentation of the developed algorithm is depicted in Figure 14.

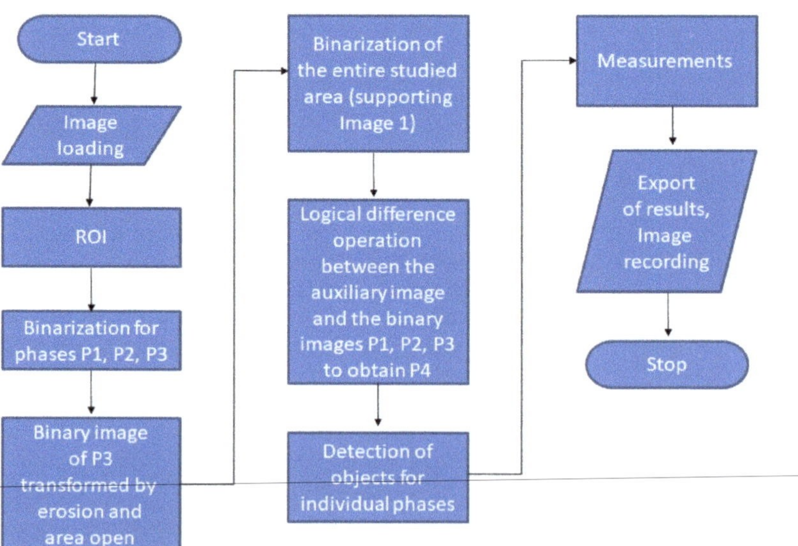

Figure 14. Flowchart summarizing the developed algorithm for image analysis.

Figure 15a–j show the output images from the following steps of the image analysis. In the first stage, the input image (Figure 15a) is marked with the region of interest (ROI. Figure 15b) to remove the black bar along the bottom that contains metadata of the SEM analysis. Only the ROI is taken for the image analysis, which is then subjected to median filtering to remove minimal noise (Figure 15c). Subsequently, the binary images obtained by thresholding are generated for every individual object of interest, namely phases marked as P1, P2, and P3 (threshold selected using a profile; Figure 15d–f). The fourth object—P4—is determined by first detecting the ROI analyzed area and subsequently applying a log difference between ROI and the previously detected objects P1, P2, and P3 (Figure 15g). Additionally, the alternating operations of erosion, areaopen (a built-in function in *Aphelion*), erosion, and areaopen were applied, which allowed the removal of the lines of P1 phase grain boundaries (treated the same as P3) and final detection of the P3 phase. The next step was the presentation of every detected phase in a separate image (Figure 15h–k). Finally, all the detected phases were superimposed on the same image, marked in different colors for better and complete visualization of the results (Figure 15l).

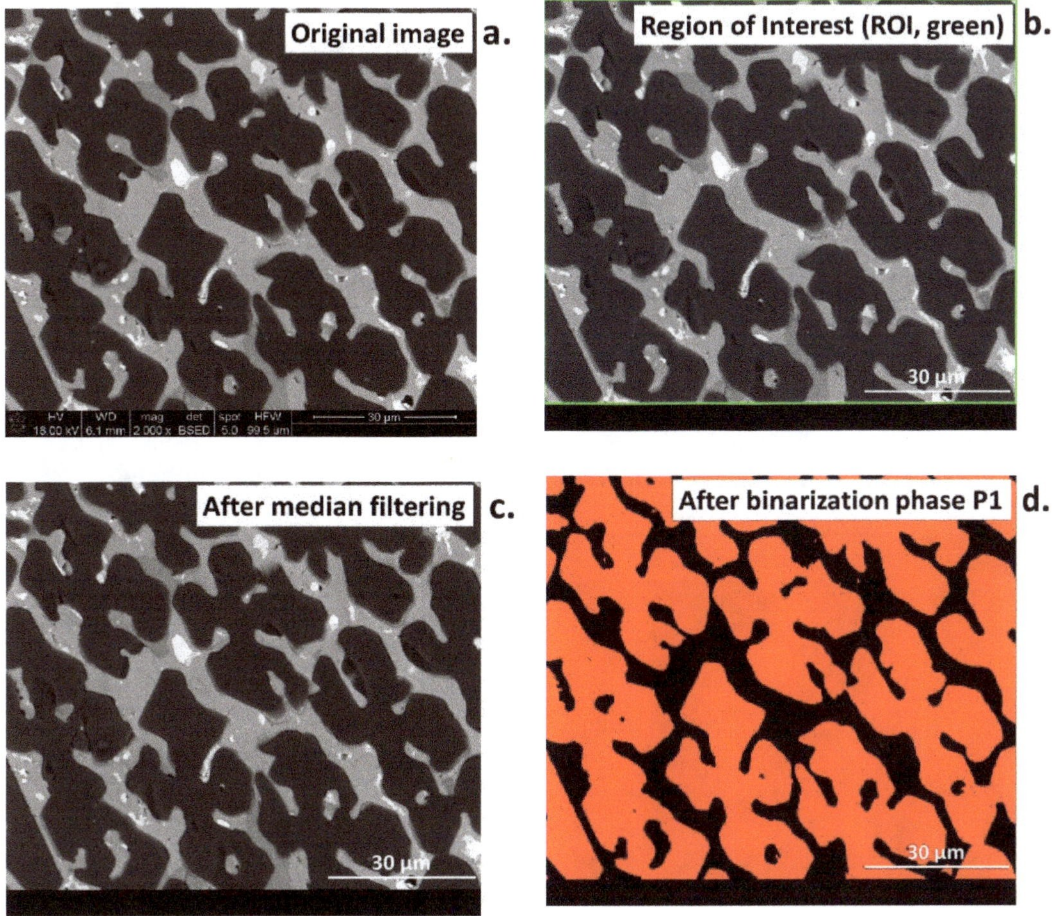

Figure 15. *Cont.*

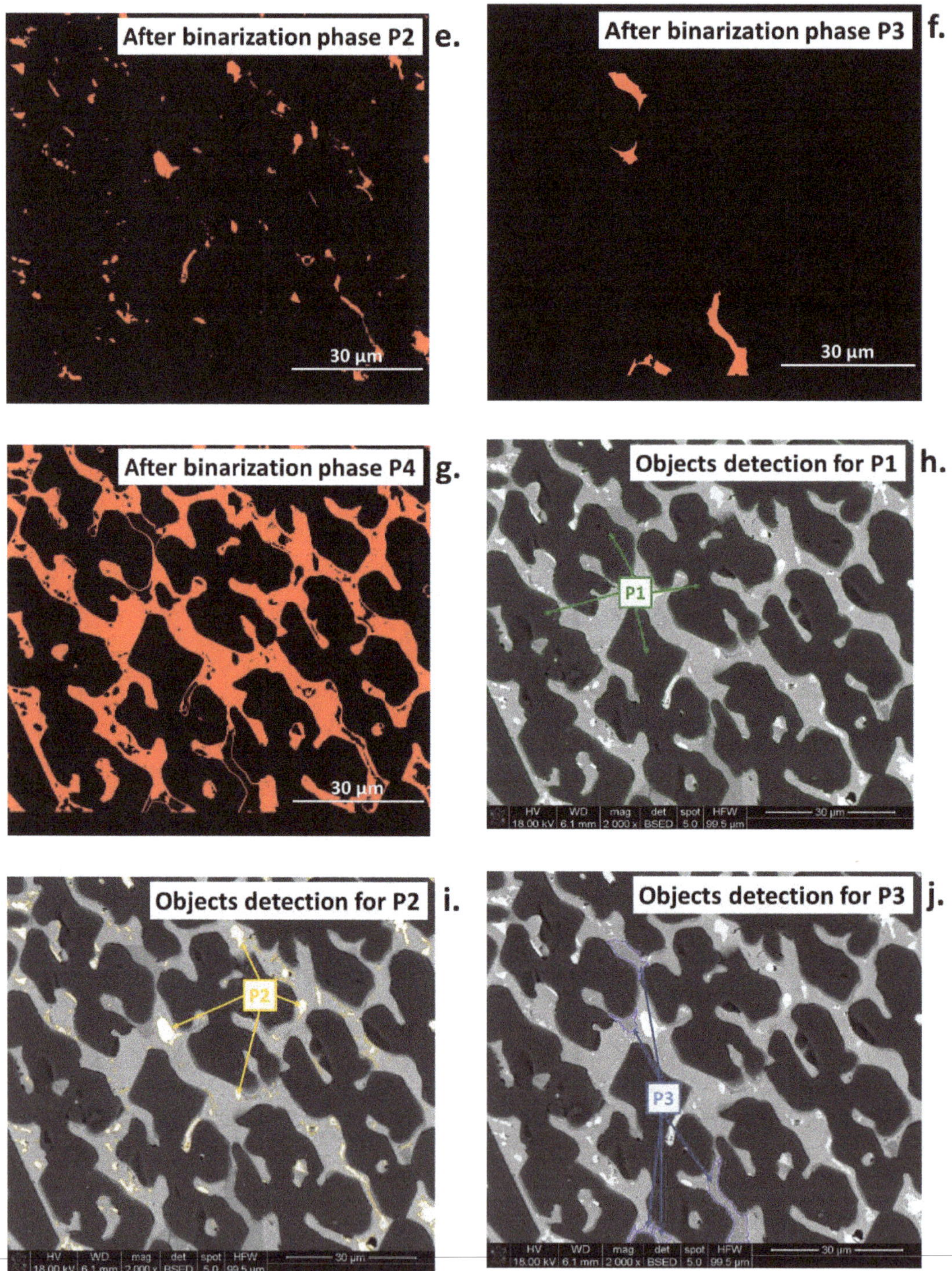

Figure 15. *Cont.*

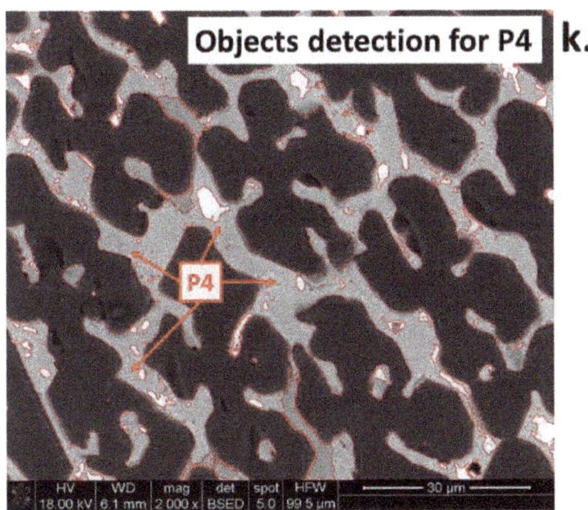

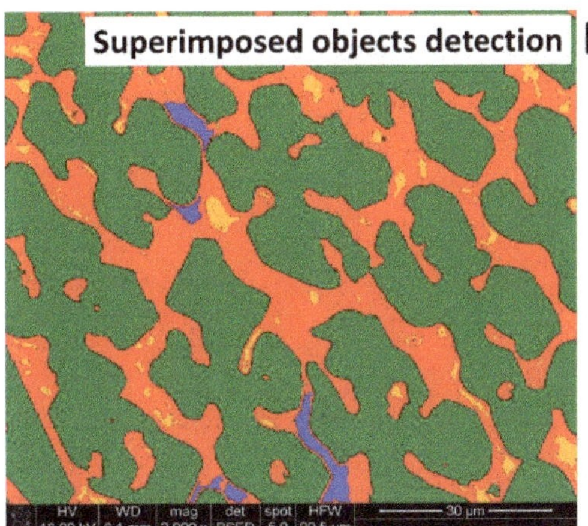

Figure 15. (**a–l**) Images presenting the steps of the algorithm for image analysis of high-temperature ceramic material.

The percentage content of each phase was calculated based on the corresponding surface area with reference to the total surface area of the analyzed image, according to Equation (4), and was presented in Table 4.

$$Vv(P) = A(P)/A(A) \tag{4}$$

where:

A(P)—the surface area of the analyzed object P1, P2, P3, or P4;

A(A)—the surface area of the total sample.

The window of the software during the image analysis was showed in Figure 16. The superimposed image with all the detected phases, in magnification 500× was demonstated in Figure 17. The results of image analysis can be exported to CSV and Excel formats. The parameters given by the software involve area, volume fraction, Crofton perimeter, etc., are attached in Table 5, and can be further used for statistical purposes. The Crofton perimeter is an estimate of object perimeter (more complex than the four-point connections-based neighbor-analysis *<Perimeter>* and results in high accuracy. Additionally, it provides a more accurate estimate of the Euclidean object perimeter and is less sensitive to object orientation compared to *Perimeter* [39].

Table 4. Percentage distribution of analyzed phases P1, P2, P3, and P4 in the image, corresponding to Figure 15l (2000×) and Figure 17 (500×).

Phase Name *	Amount of Phase, %	
Magnification	2000×	500×
P1 (green)	66.3 ± 0.1 *	69.3 ± 0.1
P2 (orange)	10.7 ± 0.4	2.8 ± 1.3
P3 (blue)	9.9 ± 0.2	1.2 ± 2.1
P4 (red)	13.1 ± 0.2	26.7 ± 0.1

* relative error.

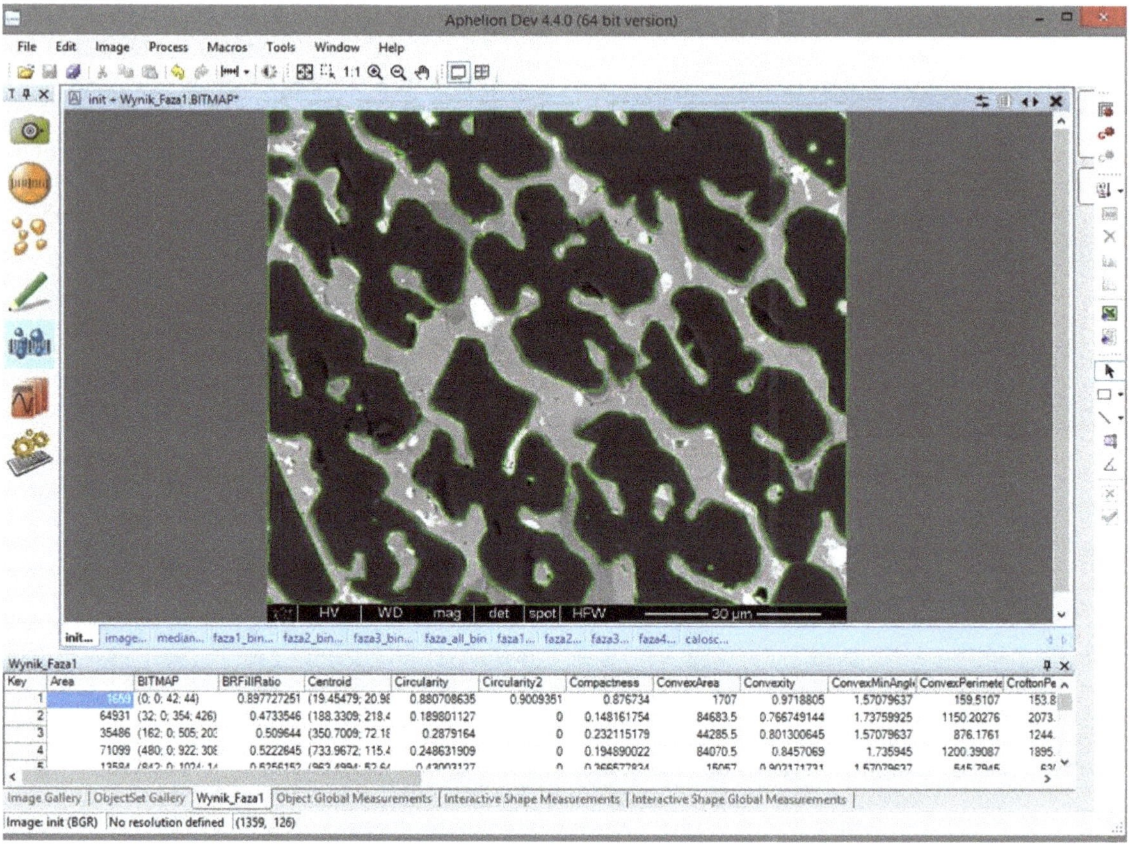

Figure 16. User interface of *Aphelion* during measurements of the P1 phase.

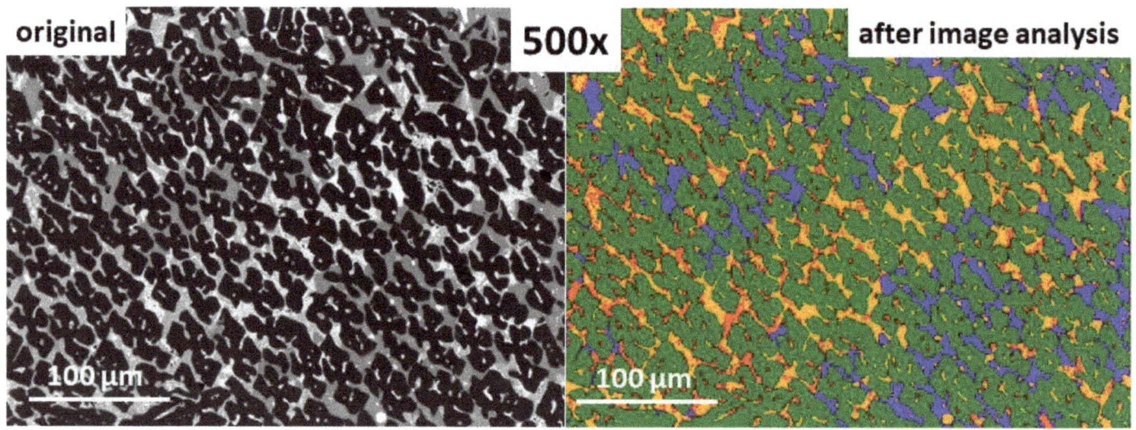

Figure 17. 500× magnified original SEM image (**left**), and image the after analysis using the developed algorithm (**right**).

Table 5. Two-dimensional parameters obtained as a result of the automated image analysis using *Aphelion* software.

Parameter	Description
Pixel Count	Number of pixels making up the object
Height	The difference between an object's highest Y coordinate and its lowest Y coordinate
Width	The difference between an object's right X coordinate and its left X coordinate
Centroid	The average position of all pixels in an object expressed as a pair of x, y coordinates (i.e., the center of mass of the object)
Major Axis	Angle in radians from the X-axis of the principal axis of inertia. This object attribute gives the main orientation of the object to the X-axis.
BR Fill Ratio	The ratio between the area of an object and the area of its bounding rectangle. The bounding rectangle has the same orientation as the X.Y coordinate system of the image.
Perimeter	An estimate of the object perimeter based on the number of 4-connected neighboring pixels along the object boundary
Crofton Perimeter	Facility circuit estimate based on a more complex analysis than 4-connectivity
Compactness	An object attribute that is equal to $16.\mathrm{Area}/\mathrm{Perimeter}^2$
Bounding Rectangle To Perimeter	The ratio between the perimeter of an object and the perimeter of its bounding rectangle, where the latter is oriented along the X, and Y axis. The perimeter measure used for this ratio is Perimeter, as described above.
Number of Holes	The number of holes in an object. A hole is one or more connected background pixels completely contained within an object.
Area	Facility area
Elongation	The absolute value of the difference between the inertia of the major and minor axes is divided by the sum of these inertias. The minor axis is defined as the axis perpendicular to the major axis.
Circularity	For a given object this attribute is equal to: $\frac{4\pi \times Area}{Crofton\ Perimeter^2}$
Intercepts	Several transitions from background to object in $0°, 45°, 90°$, and $135°$ directions
Equivalent Diameter	Specifies the diameter of a circle whose area is equal to the area of the object
Convexity	This attribute is equal to the area of the object divided by the area of its convex hull

Table 5. *Cont.*

Parameter	Description
Perimeter Variation	The sum of the changes in direction between the boundary pixels where a change of 45 degrees equals 1, a change of 90 degrees equals 2, and a change of 135 degrees equals 3
Convex Min Angle	The minimum of the angles formed by adjacent pairs of line segments comprising a polygonal object boundary is given in radians
Symmetry Mean Difference	The average of the absolute values of the difference in length between the centroid and the two opposite boundary points of the object
Convex Area	The convex hull area of the object
Convex Perimeter	Circumference of the object's convex hull using the Perimeter measure
Holes Area	A vector containing the surface area of the holes in the object
Holes Total Area	The total area of facility openings

6. Discussion

In this work, we showed the results of image analysis conducted using two approaches: traditional stereology-based and automated techniques. Linear analysis and planimetry were applied as a traditional route, while an algorithm was developed to automatically process and analyze the SEM image of ceramic multiphase material.

Image analysis is most reliable for the non-porous, homogeneous material. However, common real materials are inhomogeneous and contain defects such as intergranular or inside-grain porosity. An example is slag samples from non-ferrous metallurgy processes [42]. The traditional stereology-based image analysis for such materials would need analysis of multiple images and would be extremely time-consuming. Thus, for this kind of material computer, image analysis utilizing algorithms for the automated detection and measurements of the material components is most promising and needed.

The results of the SEM image analysis conducted in this work using traditional and automated methods were found to be in good agreement, as seen in Figure 18. The dispersion between phase amounts using traditional (linear) and automated methods was below 1.5% for P1–P3 and 2% for P4 (Table 3 vs. Table 4). The dispersion of results for phase P3 (dark-grey phase in the matrix) between linear and planimetry methods results from counting the pixels of P1 phase at grain borders, which were equal as it comes to their value to pixels for the phase P3 (Figure 19). The analyzed image contained phases of different mechanical properties, with the most abundant high-hardness Al_2O_3 phase (9 on the Mohs scale) distributed among softer spinel phases (8 on the Mohs scale) [43] and tenorite CuO (3.5 on the Mohs scale). This produced blurred grain boundaries of the P1 phase, which appeared brighter and had pixel values the same as the P3 phase, causing the P1 boundaries to be treated by the algorithm as P3.

The challenge of the algorithm in distinguishing the binary image between the P3 phase and the grain boundaries of the P1 phase (Figure 20a–f) was resolved using the combined action of morphological operations. The following sequence of morphological operations was applied to solve the issue, namely erosion, area open, erosion and, again, area open. This resulted in the correct P3 phase detection. For the detection of only the Flines representing the P1 grain boundaries, the algorithm used the logical difference between the original binary image (Figure 20a) with the previously detected P3 phase (Figure 20e).

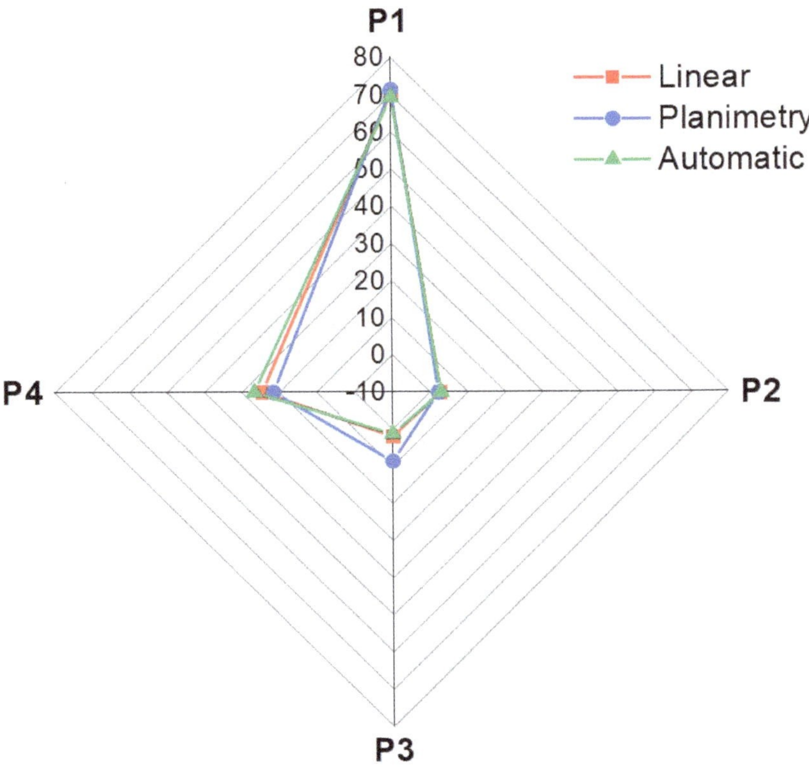

Figure 18. Spider chart presenting the results of image analysis obtained using traditional (linear and planimetry) and automated methods.

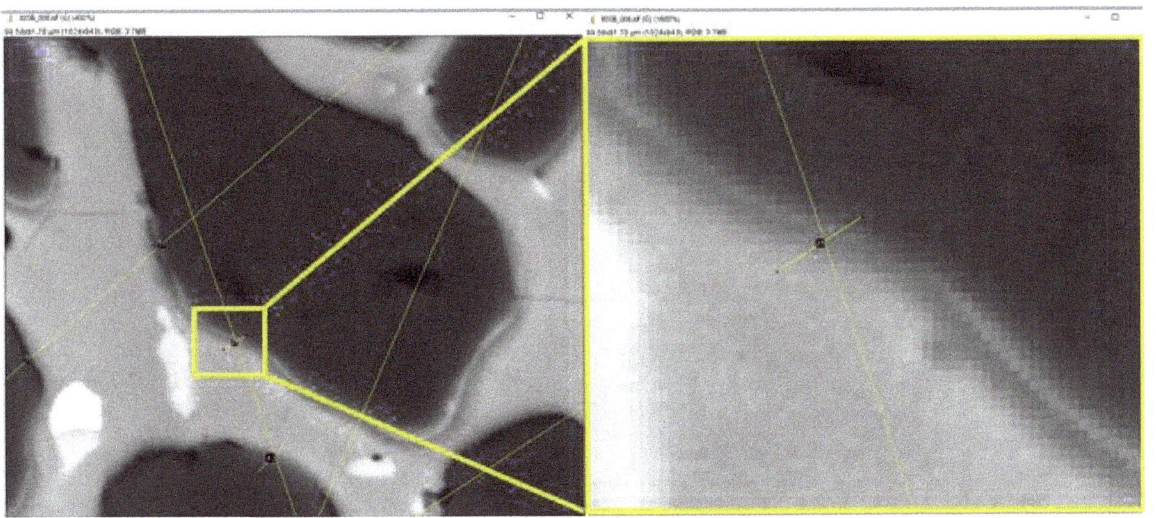

Figure 19. Magnified SEM image during linear analysis showing the source of the error resulted from blurred grain boundary.

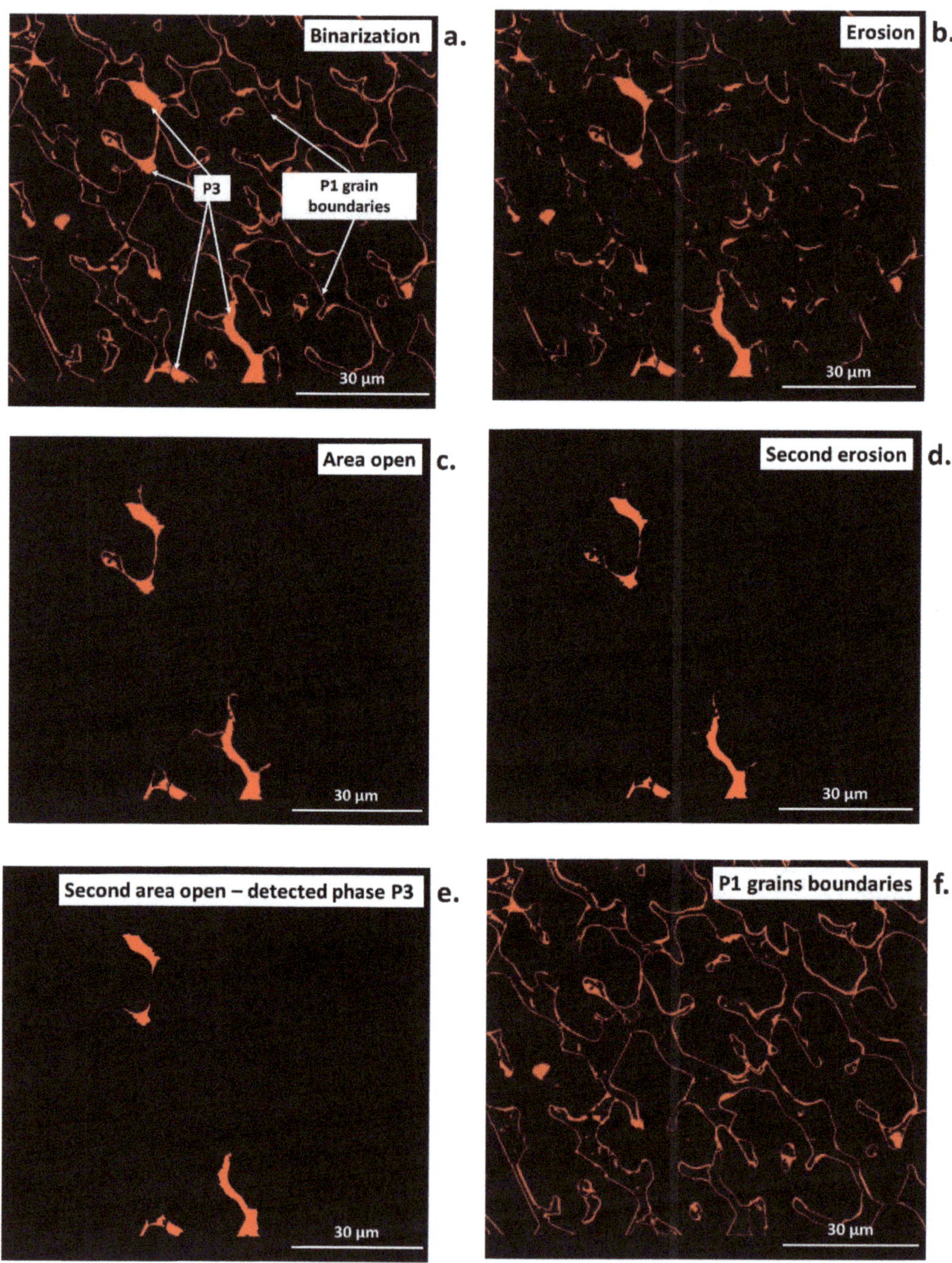

Figure 20. SEM image during automated analysis for P3 phase: (**a**) after thresholding, (**b**) after thresholding followed by morphological transformations (**c**–**f**).

The automated computer image analysis enables the simultaneous phase detection of multiple-phase material on the analyzed digital SEM image, and allows for measurements of their amounts. The quantity measurements are conducted on the binary images, so images are reduced to 1-bit color depth via thresholding. Thus, binarization plays a key role in image analysis and may be the source of large errors. The results obtained in this work by the automated method confirm those obtained by linear and planimetry belonging to the traditional methods. Often, converting an initial color image in the RGB model into another one, such as HSI, helps to make the algorithm simpler and the results more satisfactory.

Numerous obstacles may arise during development of the algorithm for automated image analysis. Frequently, noise causes difficulties in proper object detection since small elements are treated as searched objects (Figure 4a). Another obstacle is a shadow, defined as uneven background illumination. A shadow can cause uneven illumination of objects and is produced during image acquisition, especially in the case of optical microscopy. As a result, the image is brighter in the middle and darker at the border of the field of view. This is a common problem in microscopic images and leads to incorrect qualitative descriptions of the objects being analyzed at a given time. The human eye does not notice subtle and minimal differences in background brightness. Approximately 40–50 shades of grey are recognized by the human eye [39,44]. The shading effect leads to difficulties in detecting objects correctly, especially using algorithms with automatic thresholding [23,45].

Another aspect that can affect automated object detection is the choice of image analysis method, i.e., algorithms based on human knowledge and experience or based on machine learning or neural networks. Automated detection is faster and more accurate, while human-made ones, based on experience in creating image analysis solutions, are simultaneously more able to detect errors and provide better control over the final result [46,47]. Nevertheless, the code should be transparent and as simple as possible, built by applying the best initial knowledge and experience of a human.

Performing reliable image analysis strongly depends on the sample preparation procedure as well as the conditions of SEM analysis, which both should be repeatable to obtain sharp boundaries between microstructural objects. Numerous factors impact the quality of SEM images and the conditions of their repeatability. The depth of field of a scanning electron microscope is dependent on many factors of both physical and microscope construction nature, including the diffraction of electrons, the diameter of the aperture, magnification, and the working distance (which impacts spot size and aperture angle). These wider aspects can be developed in future works.

One of the most important aspects, in terms of the industrial application of automated image analysis based on dedicated algorithms, is the time required for image analysis. Traditional techniques for the quantitative determination based on stereological principles (using parameters V_V, A_A, L_L) are universal and efficient but, simultaneously, time-consuming. In this work, the time to perform the computer-aided linear analysis together with the analysis of the results was 170 min, but 60 min for the planimetry (the classical manual method would take at least 50% longer). In contrast, the total time for the analysis after the development of the algorithm was less than 5 s. Thus, the application of modern software for image analysis, such as *Aphelion*, facilitates the development of dedicated algorithms for precise, more user-friendly, and faster image analysis.

7. Conclusions

In this work, a digital SEM image of four-phase high-temperature ceramic material from the Cu-Al-Fe-O system was subjected to image analysis to obtain quantitative data on the coexisting phases using traditional (linear, planimetry) and automated method. The results of stereology-based methods and automated methods utilizing the developed algorithm of image analysis produced comparable results with a maximum 2% dispersion, requiring 5 s for the automated method compared to 60 min for classical planimetry. This shows that automated methods of image analysis can be successfully used for precise, effective, and fast SEM image analysis, and can be widely applied for the characterization of microstructure and tailoring of the properties of ceramic high-temperature materials and other similar materials.

Author Contributions: Conceptualization: I.J., Writing—original draft preparation: I.J., A.P., Design of the work and providing the material for analysis: I.J., Methodology, I.J., A.P., Software: I.J., A.P., Validation: I.J., Formal analysis: I.J., A.P., Investigation: I.J., Data curation: I.J., Data visualization: I.J., A.P., Project administration: I.J., Funding acquisition: I.J. All authors have read and agreed to the published version of the manuscript.

Funding: This research was supported through a grant from The National Centre for Research and Development, LIDERXII, Grant no. LIDER/14/0086/L-12/20/NCBR/2021, 2022-2025 (Principal Investigator: I. Jastrzębska).

Data Availability Statement: The data presented in this study are available on request from the corresponding author.

Acknowledgments: The authors wish to thank Aleksandra Kalęba for technical support in image analysis using traditional methods.

Conflicts of Interest: The authors declare no conflict of interest.

References

1. Prusak, Z.; Tadeusiewicz, R.; Jastrzębski, R.; Jastrzębska, I. The Advances and Perspectives in Using Medical Informatics for Steering Surgical Robots in Welding and Training of Welders Applying Long-Distance Communication Links. *Weld. Technol. Rev.* **2020**, *92*, 15–26.
2. Konovalenko, I.; Maruschak, P.; Prentkovskis, O. Automated method for fractographic analysis of shape and size of dimples on fracture surface of high-strength titanium alloys. *Metals* **2018**, *8*, 161. [CrossRef]
3. Tadeusiewicz, R.; Korohoda, P. *Computer Analysis and Image Processing*; (In Polish: Komputerowa Analiza i Prztwarzanie Obrazów); Publisher of the Foundation for the Advancement of Telecommunications (In Polish: Wydawnictwo Fundacji Postępu Telekomunikacji): Kraków, Poland, 1997.
4. Lyu, K.; She, W.; Miao, C.; Chang, H.; Gu, Y. Quantitative Characterization of Pore Morphology in Hardened Cement Paste via SEM-BSE Image Analysis. *Constr. Build. Mater.* **2019**, *202*, 589–602. [CrossRef]
5. Korzekwa, J.; Gądek-Moszczak, A. Application of the Image Analysis Methods for the Study of Al_2O_3 Surafce Coatings. *Qual. Prod. Improv.* **2019**, *1*, 406–411. [CrossRef]
6. Zając, M.; Giela, T.; Freindl, K.; Kollbek, K.; Korecki, J.; Madej, E.; Pitala, K.; Kozioł-Rachwał, A.; Sikora, M.; Spiridis, N.; et al. The first experimental results from the 04BM (PEEM/XAS) beamline at Solaris. *Nucl. Instrum. Methods Phys. Res. Sect. B Beam Interact. Mater. At.* **2021**, *492*, 43–48. [CrossRef]
7. Berg, J.C.; Barone, M.F.; Yoder, N. *SMART Wind Turbine Rotor: Data Analysis and Conclusions*; SAND2014-0712; Sandia National Laboratories: Albuquerque, CA, USA, 2014.
8. Winstroth, J.; Schoen, L.; Ernst, B.; Seume, J.R. Wind Turbine Rotor Blade Monitoring Using Digital Image Correlation: A Comparison to Aeroelastic Simulations of a Multi-Megawatt Wind Turbine. *J. Phys. Conf. Ser.* **2014**, *524*, 012064. [CrossRef]
9. Schneider, C.A.; Rasband, W.S.; Eliceiri, K.W. NIH Image to ImageJ: 25 Years of Image Analysis. *Nat. Methods* **2012**, *9*, 671–675. [CrossRef]
10. Kitchen, M.J.; Buckley, G.A.; Gureyev, T.E.; Wallace, M.J.; Andres-Thio, N.; Uesugi, K.; Yagi, N.; Hooper, S.B. CT Dose Reduction Factors in the Thousands Using X-Ray Phase Contrast. *Sci. Rep.* **2017**, *7*, 15953. [CrossRef]
11. Mangala, A.G.; Balasubramani, R.A. Review On Vehicle Speed Detection Using Image Processing. *Int. J. Curr. Eng. Sci. Res.* **2017**, *4*, 23–28.
12. Liu, B.; Su, S.; Wei, J. The Effect of Data Augmentation Methods on Pedestrian Object Detection. *Electronics* **2022**, *11*, 3185. [CrossRef]
13. Tundys, B.; Bachanek, K.; Puzio, E. *Smart City. Models, Generations, Measuremet and Directions of Development*; (In Polish: Modele, generacje, pomiar i kierunki rozwoju); Edu-Libri: Kraków, Poland, 2022.
14. Soldek, J.; Shmerko, V.; Phillips, P.; Kukharev, G.; Rogers, W.; Yanushkevich, S. Image Analysis and Pattern Recognition in Biometric Technologies 1997. In Proceedings of the International Conference on the Biometrics: Fraud Prevention, Enchanced Service, Las Vegas, NV, USA, 1 January 1997; pp. 270–286.
15. Shao, Z.; Zhou, Y.; Zhu, H.; Du, W.L.; Yao, R.; Chen, H. Facial Action Unit Recognition by Prior and Adaptive Attention. *Electronics* **2022**, *11*, 3047. [CrossRef]
16. Wang, Y.; Yu, M.; Jiang, G.; Pan, Z.; Lin, J. Image registration algorithm based on convolutional neural network and local homography transformation. *Appl. Sci.* **2020**, *10*, 732. [CrossRef]
17. Wilkinson, C.; Cowan, J.A.; Myles, D.A.A.; Cipriani, F.; McIntyre, G.J. VIVALDI—A thermal-neutron laue diffractometer for physics, chemistry and materials science. *Neutron News* **2002**, *13*, 37–41. [CrossRef]
18. Pacilè, S.; Baran, P.; Dullin, C.; Dimmock, M.; Lockie, D.; Missbach-Guntner, J.; Quiney, H.; McCormack, M.; Mayo, S.; Thompson, D.; et al. Advantages of breast cancer visualization and characterization using synchrotron radiation phase-contrast tomography. *J. Synchrotron Radiat.* **2018**, *25*, 1460–1466. [CrossRef]

19. Popov, A.I.; Zimmermann, J.; McIntyre, G.J.; Wilkinson, C. Photostimulated luminescence properties of neutron image plates. *Opt. Mater.* **2016**, *59*, 83–86. [CrossRef]

20. McIntyre, G.J.; Mélési, L.; Guthrie, M.; Tulk, C.A.; Xu, J.; Parise, J.B. One picture says it all—High-pressure cells for neutron Laue diffraction on VIVALDI. *J. Phys. Condens. Matter* **2005**, *17*, S3017–S3024. [CrossRef]

21. Goldstein, J.I.; Newbury, D.E.; Michael, J.R.; Ritchie, N.W.M.; Scott, J.H.J.; Joy, D.C. *Scanning Electron Microscopy and X-Ray Microanalysis*, 4th ed.; Springer: Berlin/Heidelberg, Germany, 2018; EBook.

22. Wolszczak, P.; Kubínová, L.; Janáček, J.; Guilak, F.; Opatrný, Z. Comparison of Several Digital and Stereological Methods for Estimating Surface Area and Volume of Cells Studied by Confocal Microscopy. *Cytometry* **1999**, *36*, 85–95.

23. Gądek-Moszczak, A.; Wojnar, L.; Piwowarczyk, A. Comparison of Selected Shading Correction Methods. *System Safety: Hum. Tech. Facil. Environ.* **2019**, *1*, 819–826.

24. Profile, S.E.E. Focus Stacking Algorithm for Scanning Electron Microscopy (In Polish: Algorytm cyfrowej korekcji głębi ostrości w elektronowej mikroskopii skaningowej). *Mechanik* **2016**, *87*, 59–68.

25. Jastrzębska, I.; Jastrzębski, A. Smart Equipment for Preparation of Ceramic Materials by Arc Plasma Synthesis (APS). In Proceedings of the Electronic Materials and Applications EMA2022, Orlando, FL, USA, 18–20 January 2017; Available online: https://ceramics.org/meetings-events/acers-meeting-archives/electronic-materials-and-applications-ema-2017-archive (accessed on 1 December 2022).

26. Jastrzębska, I.; Szczerba, J.; Stoch, P.; Błachowski, A.; Ruebenbauer, K.; Prorok, R.; Śniezek, E. Crystal Structure and Mössbauer Study of FeAl2O4. *Nukleonika* **2015**, *60*, 45–47. [CrossRef]

27. Jastrzebska, I.; Bodnar, W.; Witte, K.; Burkel, E.; Stoch, P.; Szczerba, J. Structural Properties of Mn-Substituted Hercynite. *Nukleonika* **2017**, *62*, 95–100. [CrossRef]

28. Jastrzębska, I.; Szczerba, J.; Stoch, P. Structural and microstructural study on the arc-plasma synthesized (APS) $FeAl_2O_4$-$MgAl_2O_4$ transitional refractory compound. *High-Temp. Mater. Process.* **2017**, *36*, 299–303. [CrossRef]

29. Śnieżek, E.; Szczerba, J.; Stoch, P.; Prorok, R.; Jastrzębska, I.; Bodnar, W.; Burkel, E. Structural properties of MgO-ZrO2 ceramics obtained by conventional sintering, arc melting and field assisted sintering technique. *Mater. Des.* **2016**, *99*, 412–420. [CrossRef]

30. Szczerba, J.; Śniezek, E.; Stoch, P.; Prorok, R.; Jastrzebska, I. The role and position of iron in 0.8CaZrO3−0.2CaFe2O4. *Nukleonika* **2015**, *60*, 147–150. [CrossRef]

31. Słowik, G. Fundamentals of Electron Microscopy and Its Selected Applications in Characterization of Carrier Catalysts. In *Adsorbents and Catalysts. Selected Technologies and the Environment*; (In Polish: Podstawy mikroskopii elektronowej i jej wybrane zastosowania w charakterystyce katalizatorów nośnikowych. Rozdział 12 Adsorbenty i katalizatory. Wybrane technologie a środowisko); Ryczkowski, J., Ed.; University of Rzeszów: Rzeszów, Poland, 2012; Chapter 12; pp. 219–243.

32. Trzciński, J. Combined SEM and Computerized Image Analysis of Clay Soils Microstructure: Technique & Application. Advances in Geotechnical Engineering: The Skempton Conference. In Proceedings of the Advances in Geotechnical Engineering, London, UK, 29–31 March 2004; pp. 654–666.

33. Wojnar, L.; Kurzydłowski, K.J.; Szala, J. *Practice of Image Analysis*; (In Polish: Praktyka Analizy Obrazu); Polish Steorological Sciety: Kraków, Poland, 2002; EBook.

34. *ASTM E112-10*; Standard Test Methods for Determining Average Grain Size, American Society for Testing and Materials. Endolab: Riedering, Germany.

35. *ISO 2624*; Copper and Copper Alloys-Estimation of Average Grain Size. ISO: Geneva, Switzerland.

36. Makowski, K. New Requirements for Ergonomic Parameters of Respiratory Protection Equipment. (In Polish: Nowe Wymagania w Zakresie Parametrów Ergonomicznych Sprzętu Ochrony Układu Oddechowego). *Occup. Saf.* **2020**, *10*, 22–25.

37. Russ, J.; Neal, F. Segmentation and Thresholding. In *The Image Processing Handbook*, 7th ed.; CRC Press: Boca Raton, FL, USA, 2016; pp. 381–437.

38. Grabowski, G. *Quantitative Microstructure Analysis of Ceramic Materials*; (In Polish: Ilościowa Analiza Mikrostruktury Materiałów Ceramicznych); Akapit: Kraków, Poland, 2022.

39. Wojnar, L. *Image Analysis How Does It Work ?* (In Polish: Analiza Obrazu Jak to działa?); Cracow University of Technology Kraków: Kraków, Poland, 2020.

40. Liu, Z.; Hong, H.; Gan, Z.; Wang, J.; Chen, Y. An Improved Method for Evaluating Image Sharpness Based on Edge Information. *Appl. Sci.* **2022**, *12*, 6712. [CrossRef]

41. Malik, M.; Spurek, P.; Tabor, J. Cross-Entropy Based Image Thresholding. *Schedae Inform.* **2015**, *24*, 21–29.

42. Ludwig, M.; Śnieżek, E.; Jastrzębska, I.; Piwowarczyk, A.; Wojteczko, A.; Li, Y.; Szczerba, J. Corrosion of Magnesia-Chromite Refractory by PbO-Rich Copper Slags. *Corros. Sci.* **2022**, *109949*, 1–22. [CrossRef]

43. The Mohs Hardness Scale and Chart For Select Gems. Available online: https://www.gemsociety.org/article/select-gems-ordered-mohs-hardness/ (accessed on 18 October 2022).

44. Tadeusiewicz, R.; Jastrzębska, I.; Jastrzębski, R. The Possibility of Creating a Welding Mask with Computer Processing of Spatial Image Instead of Welding Filters. *Weld. Technol. Rev.* **2010**, 2–15.

45. Gądek-Moszak, A. Problem of the Objects Detection on Low Quality 3D Image—Example Solution. *Qual. Prod. Improv.* **2017**, *6*, 131–141. [CrossRef]

46. Piwowarczyk, A.; Wojnar, L. Machine Learning Versus Human-Developed Algorithms in Image Analysis of Microstructures. *Qual. Prod. Improv.* **2019**, *1*, 412–416. [CrossRef]
47. Chan, H.; Cherukara, M.; Loeffler, T.D.; Narayanan, B.; Sankaranarayanan, S.K.R.S. Machine Learning Enabled Autonomous Microstructural Characterization in 3D Samples. *npj Comput. Mater.* **2020**, *6*, 1–9. [CrossRef]

*materials*

MDPI

Article

Analysis of the Quasi-Static and Dynamic Fracture of the Silica Refractory Using the Mesoscale Discrete Element Modelling

Aleksandr S. Grigoriev [1,*], Andrey V. Zabolotskiy [2], Evgeny V. Shilko [1,*], Andrey I. Dmitriev [1] and Kirill Andreev [3,4]

[1] Institute of Strength Physics and Materials Science of Siberian Branch of the Russian Academy of Sciences (ISPMS SB RAS), 2/4, pr. Akademicheskii, 634055 Tomsk, Russia; dmitr@ispms.ru
[2] Magnezit Group, 34, St. Solnechnaya, 456910 Satka, Russia; azabolotskiy@magnezit.com
[3] The State Key Laboratory of Refractories and Metallurgy, Wuhan University of Science and Technology (WUST), 947 Heping Ave., Wuhan 430081, China; kpandreev@gmail.com
[4] Ceramic Research Centre, Tata Steel, 211 Rooswijkweg, 1951 MD Velsen Noord, The Netherlands
[*] Correspondence: grigoriev@ispms.ru (A.S.G.); shilko@ispms.ru (E.V.S.)

Abstract: Computer modelling is a key tool in the optimisation and development of ceramic refractories utilised as insulation in high-temperature industrial furnaces and reactors. The paper is devoted to the mesoscale computer modelling of silica refractories using the method of homogeneously deformable discrete elements. Approaches to determine the local mechanical properties of the constituents from the global experimental failure parameters and respective crack trajectories are considered. Simulations of the uniaxial compressive and tensile failure in a wide range of quasi-static and dynamic loading rates (10^2 s^{-1}) are performed. The upper limit of the dynamic loading rates corresponds to the most severe loading rates during the scrap loading on the refractory lining. The dependence of the strength, fracture energy, and brittleness at failure on the loading rate is analysed. The model illustrates that an increase in the loading rate is accompanied by a significant change in the mechanical response of the refractory, including a decrease in the brittleness at failure, a more dispersed failure process, and a higher fraction of the large grain failure. The variation of the grain–matrix interface's strength has a higher impact on the static compressive than on the static tensile properties of the material, while the material's dynamic tensile properties are more sensitive to the interface strength than the dynamic compressive properties.

Keywords: refractories; dynamic loading; fracture; mesoscale computer simulation; discrete element method (DEM)

Citation: Grigoriev, A.S.; Zabolotskiy, A.V.; Shilko, E.V.; Dmitriev, A.I.; Andreev, K. Analysis of the Quasi-Static and Dynamic Fracture of the Silica Refractory Using the Mesoscale Discrete Element Modelling. *Materials* **2021**, *14*, 7376. https://doi.org/10.3390/ma14237376

Academic Editors: Jacek Szczerba and Ilona Jastrzębska

Received: 27 September 2021
Accepted: 27 November 2021
Published: 1 December 2021

Publisher's Note: MDPI stays neutral with regard to jurisdictional claims in published maps and institutional affiliations.

1. Introduction

Refractories are ceramic materials utilised to construct industrial furnaces and high-temperature reactors [1–8]. They have a composite microstructure, which typically features large filler grains, a matrix of smaller grains, pores, and micro-cracks. There exist refractories of different chemical compositions and morphology of microstructure to match different service conditions. Their mechanical behaviour is characterised by the quasi-brittle failure and sensitivity of strength to hydrostatic stress [2,3]. Due to similar microstructural set-up and mechanical behaviour, they are related to civil engineering materials, rocks, and soils. In service, refractories are exposed to mechanical loads of diverse types and intensity [2,3], featuring compressive and tensile stresses resulting from the interactions between different parts of the refractory structure and under the influence of the process conditions. The majority of loads are characterised by quasi-static rates. In specific instances, e.g., under the impact of the hydraulic hammer repair works or during the loading of scrap [1,9], dynamic loading occurs. The refractories' ability to sustain different loads is critical for the reliability and productivity of the high-temperature industrial equipment.

For refractories, microstructural aspects of crack formation and propagation have been attracting major research attention [3,10–13]. One of the key factors influencing fracture behaviour is the interface between microstructural constituents. The results obtained for several types of refractories depict that the brittleness at failure correlates with an increase of the relative crack length along the grain/matrix interface [14]. In model civil engineering concrete, the samples with weaker interfaces had lower bending strength [15] and brittleness at failure [16]. For composites in general, the strength of the interface between the matrix and reinforcing constituents (large grains, fibres) positively correlates with the material's strength [17–19]. For fracture energy and the brittleness at failure, this relationship has a maximum [17] extremely strong interface that leads to brittle failure. Such observations are made for quasi-static failure. A few existing publications on the dynamic failure of refractories only address the macrostructural aspects of failure [9,20]; thus, little is known about its meso-structural mechanisms.

Apart from classical materials' science analysis, computer modelling is an efficient tool for the mesoscale analysis of fracture [10,11,21–26]. The advantage of numerical computer analysis is the ability to freely vary the properties of different constituents and correlate those with macroscopic mechanical properties. The challenge of mesoscale modelling is the accurate representation of the properties of the constituents forming the material [27,28]. Most studies on refractories are based on generic representation of the microstructure [10,21,22,25,26]. The computer modelling of refractories' mechanical behaviour is realised using two alternative approaches. The first one is a continuum-based approach, such as finite element method (FEM) [10,21,22,29,30], which allows the representation of complex shapes of constituents and their non-linear behaviour. However, this method has a limitation in the modelling of crack growth [27,31,32]. The widely used algorithm of the "smearing" of real discontinuity (crack), including the phase-field models and smeared crack band models, does not always permit the appropriate simulation of crack localisation, especially under the conditions of unstable crack growth. The advanced XFEM implementations [33,34] require rather higher computational costs and still experience difficulties with crack weaving and branching. Contrary to the continuum-based methods, the discrete approach is perfectly suited to model formation and growth of micro- and macro-cracks [27,35–38]. Here, the material is represented by the ensemble of bonded finite-sized particles. The formation of a discontinuity is modelled as an extremely localised process by means of separating the surfaces of adjacent particles. For refractories, the discrete element method (DEM) has been applied in a limited number of studies [11,25], in which its classical implementation, treating the particles as undeformable (rigid) volumes, has been used. To the authors' best knowledge, the computer modelling of the failure of refractories under dynamic loads has not yet been performed.

In the current paper, the fracture of silica refractories was analysed using the mesoscale DEM model. The advanced method of homogeneously deformable DEM was used [39–41], which assents to the modelling of crack formation and growth both under quasi-static and dynamic loading. The macroscopic parameters of failure and mesoscopic processes of crack formation under uniaxial compression and tension were examined within the broad interval of strain rates from quasi-static values up to the dynamic rates of 10^2 s^{-1}, specific focus points being the accurate representation of the properties of constituents and analysis of the grain–matrix interface's influence on the failure of the material.

2. Materials and Methods

2.1. Material and Laboratory Tests

Computer modelling was used to simulate the commercially available silica brick (Figure 1). Such bricks are used to construct coke ovens, heat exchangers of blast furnaces, and glass-making furnaces. Properties of the studied material were broadly discussed in a previous publication [42]. SiO_2 comprises more than 96 wt % of the material. The main mineralogical phases are tridymite and cristobalite. Tridymite forms the grains of the matrix and the rims of larger grains [12,42]. Cristobalite is predominantly found in large grains. The grain size distribution, as indicated by the supplier, is as follows: 1–2 mm grains are 2 wt %, 0.5–1 mm grains make up 13 wt %, and 0.1–0.5 mm grains comprise 29 wt %; the rest of the grains are finer than 0.088 mm. The bulk and true densities of the brick are 1.84 g/cm^3 and 2.35 g/cm^3, respectively. The total porosity is 22%.

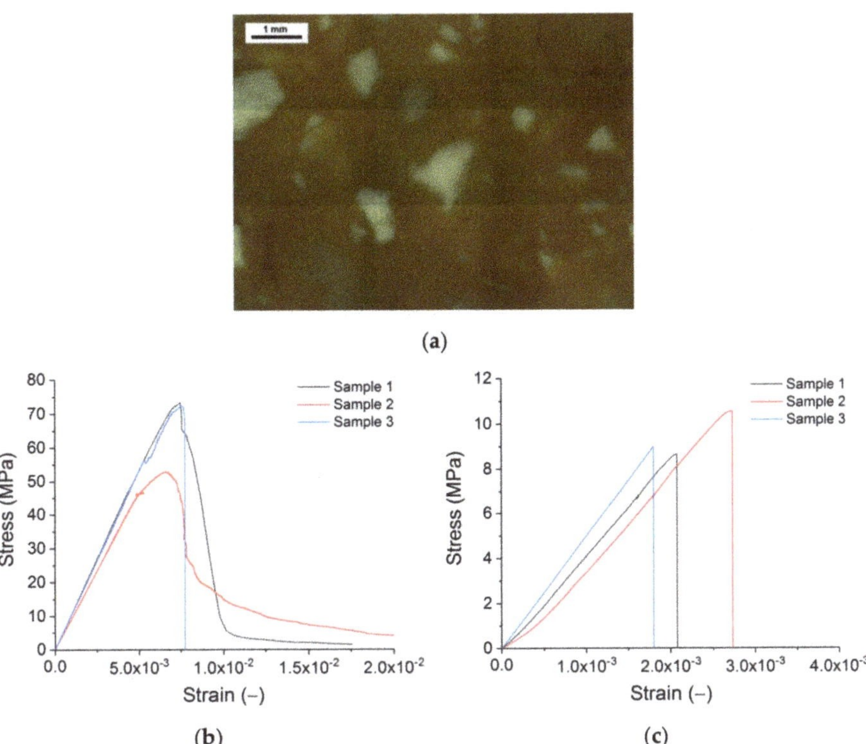

Figure 1. A typical microstructure of silica refractories as seen in light optical microscopy conducted in cross-parallel light (**a**), and experimentally obtained diagrams of a uniaxial compression test (**b**) and a three-point bending test (**c**).

Room temperature mechanical properties are described in the following sentences. Dynamic Young´s modulus is 12 GPa [42]. Static Young´s modulus depends on the test set-up. Typically, it can be 2–3 times lower than the dynamic value (Figure 1b,c). The compressive strength is 40–50 MPa (Figure 1b). The typical value of three-point bending strength is 8–10 MPa (Figure 1c). Specific fracture energy under tension is 60 N/m [42]. The number of laboratory samples used to obtain different mechanical properties varied between 3 and 10. For mechanical properties, the typical statistical spread is 10–20% of the average value.

In the paper, images of cracks obtained by us in different laboratory experiments are used to validate the DEM results. In the compressive test, the prismatic samples had a height of 50 mm and a cross-section of 30 × 30 mm^2. In the three-point bending tests, the geometry was 40 × 40 × 120 mm^3 with a span of 100 mm. The loading rate was 0.1 mm/min and 0.3 mm/min for the compressive and bending tests, respectively. For the wedge splitting test, the typical geometry [12] and the loading rate of 0.5 mm/min were used. The width, depth, and height of the samples were 100 mm, 65 mm, and 100 mm, respectively. Due to notches, the failed cross-section had the width and height of 55 mm and 66 mm, respectively. The vertical notch was 12 mm. The wedge angle was 8°. In the analysis of the crack propagation through different constituents of the microstructure,

6 crack trajectories of three wedge splitting test samples were analysed. Per sample, two polished sections exposing crack at different depths of the sample were used.

2.2. DEM Models

Two-dimensional (2D) mesoscopic DEM samples of a representative material volume were constructed (Figure 2), and refractory was considered a particle-reinforced composite. The microstructure is modelled by the discrete element's ensembles with appropriate mechanical characteristics. The discrete element's diameter (35 µm) was chosen to ensure that the minimum size of grains exceeded the element diameter by at least three times. A special study for the convergence of the results presented that a decrease in the element diameter does not lead to a significant change in the fracture pattern and integral characteristics of the mechanical response of the samples. The plane stress approximation was used.

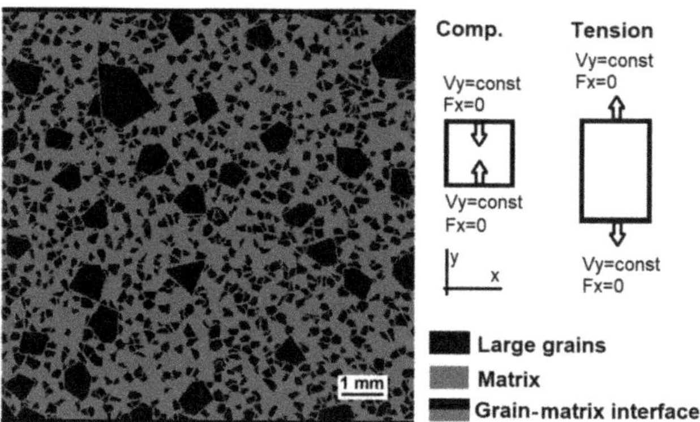

Figure 2. Schematic of DEM models of uniaxial compression and tension. The picture shows the model sample 1×1 cm^2.

The model explicitly takes into consideration the particles of three grain fractions larger than 0.1 mm (1–2 mm, 0.5–1 mm, and 0.1–0.5 mm grains), which were considered non-porous inclusions; their shape qualitatively corresponds to that of the corresponding fractions of a real sample (Figure 1a). The volumetric content of fractions conforms to the volumetric content of similar fractions in a real refractory. The sample regions formed by finer fractions and microscale pores were modelled as homogeneous continual, forming the matrix of the modelled material. We assumed that all the porosity of the refractory is concentrated in the matrix. The spatial placement and determination of the large grains' geometry in the model were executed out using the Voronoi tessellation algorithm [43]. Representing by the model of the grain size distribution as during the pressing of the bricks ignores the changes of the grain size and shape to possibly occur during the sintering of the bricks.

Uniaxial tension and compression of the samples were modelled. For the compressive and tensile tests, the geometry of the model was 1×1 cm^2 and 1×1.5 cm^2, respectively (Figure 2). For every loading mode, we considered three "statistically equivalent" samples with different spatial distributions of inclusions but the same percentage of different fractions. The failure at room temperature was modelled.

The specimens were loaded by setting the constant velocity V_y to the elements of the corresponding end surfaces along the Y-axis. Zero external forces were applied to these end surfaces in the horizontal direction to fulfil the boundary condition $\sigma_{xx} = \sigma_{xy} = 0$. For quasi-static loading, $V_y = 2$ mm/s was used, corresponding to the strain rates of 0.4 s^{-1} and 0.25 s^{-1} for compression and tension, respectively. Dynamic loading was performed with the strain rates in the range 5 s^{-1} up to 100 s^{-1}. The upper value corresponds to the characteristic strain rate under the dynamic loading of lining by falling scrap.

2.3. Homogeneously Deformable Discrete Elements

Computer modelling was carried out using the in-house software MCA 2D Load Test (version 3.0.6.5, ISPMS SB RAS, Tomsk, Russia), which implements the method of homogeneously (simply) deformable discrete elements. In this section, the keystones of this method and the mechanical model of quasi-brittle refractory materials are described.

In comparison with classical DEM, the method of homogeneously deformable discrete elements allows accounting for the stress–strain state of the elements, including their volume strains [39,40,44,45]. It also allows for more accurate implementation of mechanical strength criteria, including the Drucker–Prager criterion. This implementation of DEM is based on the following approximations.

First, the evolution of the ensemble of elements is determined by solving the Newton–Euler system of equations of motion for equivalent balls with central and tangential interaction [37,46].

The second approximation is the interaction of discrete elements through flat faces. For the interacting pair of unstrained elements, the initial value of the area of such a face equals the area of the common face of corresponding polyhedrons, which account for both for the equivalent balls and the "as if filled" voids between them [40]. The introduction of the contact area allows operating with central (normal) and tangential (shear) stresses.

The third approximation is a uniform distribution of strains and stresses in the volume of the element. The averaged values of the stress tensor components in the volume of the element $\overline{\sigma}_{\alpha\beta}$ are determined according to [37,39]:

$$
\overline{\sigma}^{\,i}_{\alpha\beta} = \frac{R_i}{\Omega^0_i} \sum_{k=1}^{N_i} \left(\vec{n}_{ik}\right)_\alpha \left(\vec{F}_{ik}\right)_\beta = \frac{R_i}{\Omega^0_i} \sum_{k=1}^{N_i} \left(\vec{n}_{ik}\right)_\alpha \left(\vec{F}^{centr}_{ik} + \vec{F}^{tang}_{ik}\right)_\beta =
$$
$$
= \frac{R_i}{\Omega^0_i} \sum_{k=1}^{N_i} S^0_{ik}\left(\vec{n}_{ik}\right)_\alpha \left(\vec{T}_{ik}\right)_\beta = \frac{R_i}{\Omega^0_i} \sum_{k=1}^{N_i} S^0_{ik}\left(\vec{n}_{ik}\right)_\alpha \left(\sigma_{ik}\left(\vec{n}_{ik}\right)_\beta + \tau_{ik}\left(\vec{t}_{ik}\right)_\beta\right)'
$$

$$(1)$$

where R_i is the radius of the equivalent sphere approximating the element i, Ω^0_i is the volume of unstrained element i, S^0_{ik} is the contact area in unstrained pair i–k, $\vec{F}_{ik}$ is the vector of interaction force, $\vec{F}^{centr}_{ik}$ and $\vec{F}^{tang}_{ik}$ are central and tangential contributions to the total force ($\vec{F}_{ik} = \vec{F}^{centr}_{ik} + \vec{F}^{tang}_{ik}$), $\vec{T}_{ik}$ is the traction vector at the area of contact of elements i and k, σ_{ik} and τ_{ik} are central and tangential components of T_{ik} in the pair i–k, $\vec{n}_{ik}$ is the unit normal vector directed along the line connecting the mass centres of the elements, $\vec{t}_{ik}$ is the unit tangent vector directed in the tangential plane, and $\left(\vec{W}\right)_\beta$ is the projection of some vector $\vec{W}$ onto the β-axis of lab coordinates. The components of the average strain tensor $\overline{\varepsilon}^{\,i}_{\alpha\beta}$ are calculated incrementally with the use of the material's constitutive law and calculated average stresses [40].

The fourth approximation is that spatial parameters of element–element interaction are divided into the contributions of both elements:

$$
\begin{cases} \Delta h_{ik} = \Delta q_{ik} + \Delta q_{ki} = \Delta\varepsilon_{ik}R_i + \Delta\varepsilon_{ki}R_k \\ \Delta l_{ik} = \Delta\gamma_{ik}R_i + \Delta\gamma_{ki}q_{ki} \end{cases}.
$$

$$(2)$$

Hereinafter, the symbol Δ denotes an increment of the corresponding parameter per time step Δt, h_{ij} is element–element overlap, q_{ik} and q_{ki} are the distances from mass centres of elements i and k to the centre of interaction area ($q_{ik} + q_{ki} = r_{ik}$; r_{ik} is the distance between mass centres of the elements; $q_{ik} = R_i$ and $q_{ki} = R_k$ for the unstrained elements), l_{ij} is the relative shear displacement of elements in the pair i–k [37], ε_{ik} and ε_{ki} are the central strains of the elements i and k in the pair i–k, γ_{ik} and γ_{ki} are the corresponding shear angles of the elements in the pair.

The fifth approximation is that the relation for the central force of element–element interaction takes into account the volumetric part of internal stresses in the interacting elements (mean stresses):

$$
\begin{cases}
\sigma_{ik} = \sigma_{ik}^{pair}(\varepsilon_{ik}) + A_i\overline{\sigma}_{mean}^{i} = \sigma_{ki} = \sigma_{ki}^{pair}(\varepsilon_{ki}) + A_k\overline{\sigma}_{mean}^{k} \\
\tau_{ik} = \tau_{ik}^{pair}(\gamma_{ik}) = \tau_{ki} = \tau_{ki}^{pair}(\gamma_{ki})
\end{cases}
, \tag{3}
$$

where the index *"pair"* denotes pair-wise function, $\overline{\sigma}_{mean}^{i} = \left(\overline{\sigma}_{xx}^{i} + \overline{\sigma}_{yy}^{i} + \overline{\sigma}_{zz}^{i}\right)/3$ is the mean stress in the element i, and A_i is the material parameter. The tangential interaction is formulated in pair-wise form as in classical DEM formalism. Note that Equation (3) takes into account the necessity to satisfy Newton's third law ($\sigma_{ik} = \sigma_{ki}$ and $\tau_{ik} = \tau_{ki}$).

The last two approximations (Equations (2) and (3)) are the consequences of the deformability of the discrete element and the uniform (homogeneous) distribution of stresses and strains in the volume of the element. Another consequence of the above approximations is that the particular expressions for central and tangential element–element interaction forces (Equation (3)) replicate the expressions for the constitutive relations of the modelled material.

2.3.1. The Linear–Elastic Behaviour of Discrete Elements

The mechanical response of all components of the model is assumed to be isotropic and linear–elastic prior to fracture. The corresponding constitutive equation is Hooke's law:

$$
\begin{cases}
\Delta\sigma_{\alpha\alpha} = 2G\Delta\varepsilon_{\alpha\alpha} + (1 - \tfrac{2G}{K})\Delta\sigma_{mean} \\
\Delta\tau_{\alpha\beta} = 2G\Delta\varepsilon_{\alpha\beta}
\end{cases}
, \tag{4}
$$

where α, $\beta = x,y,z$; $\sigma_{\alpha\alpha}$ and $\varepsilon_{\alpha\alpha}$ are the diagonal components of stress and strain tensors; $\tau_{\alpha\beta}$ and $\varepsilon_{\alpha\beta}$ are the off-diagonal components; $\sigma_{mean} = \left(\sigma_{xx} + \sigma_{yy} + \sigma_{zz}\right)/3$ is the mean stress; K is the bulk modulus; and G is the shear modulus. The corresponding formulation of the relationships for specific forces σ_{ik} and τ_{ik} (response of element i to the action of the neighbour k) is similar [39,40,44]:

$$
\begin{cases}
\Delta\sigma_{ik} = 2G_i\Delta\varepsilon_{ik} + \left(1 - \tfrac{2G_i}{K_i}\right)\Delta\overline{\sigma}_{mean}^{i} \\
\Delta\tau_{ik} = 2G_i\Delta\gamma_{ik}
\end{cases}
, \tag{5}
$$

where G_i and K_i are corresponding elastic moduli for the material of the element i. Equations (4) and (5) are written in the hypo-elastic form. Then, Equations (2) and (3) take the following form:

$$
\begin{cases}
\sigma_{ik}^{cur} = \sigma_{ik}^{pre} + \Delta\sigma_{ik} = \sigma_{ik}^{pre} + 2G_i\Delta\varepsilon_{ik} + \left(1 - \tfrac{2G_i}{K_i}\right)\Delta\overline{\sigma}_{mean}^{i} = \\
= \sigma_{ki}^{cur} = \sigma_{ki}^{pre} + \Delta\sigma_{ki} = \sigma_{ki}^{pre} + 2G_k\Delta\varepsilon_{ki} + \left(1 - \tfrac{2G_k}{K_k}\right)\Delta\overline{\sigma}_{mean}^{k} \\
\Delta h_{ik} = \Delta\varepsilon_{ik}R_i + \Delta\varepsilon_{ki}R_k
\end{cases}
, \tag{6}
$$

$$
\begin{cases}
\tau_{ik}^{cur} = \tau_{ik}^{pre} + \Delta\tau_{ik} = \tau_{ik}^{pre} + 2G_i\Delta\gamma_{ik} = \tau_{ki}^{cur} = \tau_{ki}^{pre} + \Delta\tau_{ki} = \tau_{ki}^{pre} + 2G_k\Delta\gamma_{ki} \\
\Delta l_{ik} = \Delta\gamma_{ik}R_i + \Delta\gamma_{ki}R_k
\end{cases}
. \tag{7}
$$

Here, the upper indexes *"cur"* and *"pre"* denote values of specific forces at the current and previous time steps of integrating the motion equations.

A pair of discrete elements can be in a bound or unbound state. The connected pair of elements resists tension and compression as well as shear. An unbound pair resists only compression, and the value of shear resistance is limited from above by the value of the dry friction force: $|\tau_{ik}| \leq \mu|\sigma_{ik}|$, where μ is the local value of the friction coefficient [37].

2.3.2. Quasi-Static Model of Local Fracture

Local failure is modelled as a bonded-to-unbonded transition. We considered two alternative failure criteria, namely the Drucker–Prager criterion and the combined Drucker–Prager–Rankine criterion.

The Drucker–Prager criterion takes into account the dependence of shear strength on local pressure, which is essential for all brittle and quasi-brittle materials [47,48]:

$$\sigma_{fract} = 1.5(\beta - 1)\sigma_{mean} + 0.5(\beta + 1)\sigma_{eq} = \sigma_c \tag{8a}$$

where $\beta = \sigma_c/\sigma_t$, σ_c, and σ_t are the values of the uniaxial tensile strength (UTS) and uniaxial compressive strength (UCS) for the component or the interface.

It is known that the Drucker–Prager failure criterion may overestimate the failure stress in the parametric area of tensile volumetric stresses. Thus, this criterion is often supplemented by the Rankine condition:

$$\sigma_1 = \sigma_t, \tag{8b}$$

where σ_1 is maximum principal stress. Rankine criterion limits the main tensile stress to the tensile cut-off.

The model of local fracture includes not only the condition for bonded-to-unbonded transition (fracture criterion) but also the separation of the surfaces. In this way, when the failure criterion is met, the cohesion/adhesion of the elements is gradually lost. Loss of cohesion/adhesion is regarded as a decrease in the proportion of the bonded part of the contact surface over time. It is controlled by the dimensionless parameter k_{bond}, which is the ratio of the bonded part of the contact surface to the total value of the contact surface ($0 \leq k_{bond} \leq 1$). Two limiting values are $k_{bond} = 0$ (totally unbonded pair) and $k_{bond} = 1$ (totally bonded pair). The bonded part of the contact area in the i–k pair is $S^{ik}_{bond} = S_{ik}k^{ik}_{bond}$.

For the elastic–brittle large grains, the model of the dynamic separation of the surfaces at a constant rate was applied:

$$\frac{dk_{bond}}{dt} = -\upsilon_s < 0. \tag{9}$$

Here, t is time, and υ_s is the rate of bond breaking. This model assumes that not only the beginning but also the continuation of the separation of surfaces at time t takes place when the fracture criterion is met at a given time. The dynamic surface separation model is based on describing a brittle bond break as the unstable crack propagation along the contact surface. For convenience, we use the velocity of propagated crack $V_s = \upsilon_s R$. Here, R is the element radius. The parameter V_s is the input parameter of the model. The value of this parameter is physically limited by Rayleigh wave velocity V_R.

For the porous matrix, a surface separation model is based on the approximation of stable crack growth. The surface separation rate is variable, depending on the dynamics of deformation of the pair. The current rate of separation of surfaces is proportional to the sum of the current values of the rate of expansion and the rate of relative shear of the elements of the pair under consideration:

$$\frac{dk_{bond}}{dt} = -\left(\frac{1}{\varepsilon_{max}} \frac{d\varepsilon_{pair}}{dt} + \frac{1}{\gamma_{max}} \frac{d\gamma_{pair}}{dt} \right) < 0. \tag{10}$$

Here, ε_{max} and γ_{max} are dimensionless input parameters of the model; $d\varepsilon_{pair}$ and $d\gamma_{pair}$ are changes in the central strain and the shear strain in the couple of elements during the time dt. For the pair of elements i–k, these changes are defined as follows:

$$
\begin{aligned}
d\varepsilon^{ik}_{pair} &= \begin{cases} dh_{ik}/(R_i + R_k), & if\ dh_{ik} > 0 \\ 0, & if\ dh_{ik} \leq 0 \end{cases} \\
d\gamma^{ik}_{pair} &= \begin{cases} d|l_{ik}|/(R_i + R_k), & if\ d|l_{ik}| > 0 \\ 0, & if\ d|l_{ik}| \leq 0 \end{cases}
\end{aligned}
\tag{11}
$$

The physical meaning of ε_{max} is the normalized change of distance between the elements when under the condition of uniaxial tension, the cohesion is fully lost. The physical meaning of γ_{max} is analogous and relates to the condition of pure shear.

The continuation of the separation of surfaces at time t is applied when the fracture criterion is met at a given time. The described "deformation" model is in many respects similar to well-known non-potential cohesive zone models, which are used in finite element modelling. The validity of using such a model for a porous matrix is justified by the fact that highly porous materials often exhibit a rough crack trajectory.

2.3.3. Modelling of the Grain–Matrix Interfaces

The interface zones connecting the matrix and inclusions were modelled in the approximation of an infinitely small interface thickness. The approximation is valid because the matrix and the rims of the grains have similar chemical and mineralogical compositions. Zones of apparent increase of porosity at the interface are not properly quantified. Thus, the only characteristics of the interfaces are the parameters of the Drucker–Prager strength criterion and the model of adhesion loss. The model of stable crack growth (Equations (10) and (11)) was used.

2.3.4. Dynamic Formulation of Fracture Criterion

The classical fracture criteria, including Drucker–Prager failure criterion (Equation (8a)), are quasi-static, since they do not take into account the finite time of damage and crack formation. This can lead to a significant underestimation of the dynamic strength, which is seen under the loading strains of $10\ \mathrm{s}^{-1}$ and higher [49]. The general dynamic formulation of the Drucker–Prager failure criterion was used:

$$\sigma_{fract}(t) = 1.5\left(\beta_{dyn} - 1\right)\sigma_{mean}(t) + 0.5\left(\beta_{dyn} + 1\right)\sigma_{eq}(t) = \sigma_c^{dyn}(t - t_0). \tag{12a}$$

Here, $\beta_{dyn} = \sigma_c^{dyn}(t - t_0)/\sigma_t^{dyn}(t - t_0)$; $\sigma_c^{dyn}(t - t_0)$ and $\sigma_t^{dyn}(t - t_0)$ are dynamic values of compressive and tensile strength, respectively, t_0 is the starting time of incubation of fracture when the parameter σ_{fract} reaches the static strength $\sigma_c = \sigma_c^{st}$; $T_{fract} = t - t_0$ is the time parameter of local failure. The dynamic Rankine condition is defined as follows:

$$\sigma_1(t) = \sigma_t^{dyn}(t - t_0). \tag{12b}$$

The input functions $\sigma_c^{dyn}\left(T_{fract}\right)$ and $\sigma_t^{dyn}\left(T_{fract}\right)$ can be determined from standard laboratory dynamic tests [49]. Those functions are material properties.

It is known that despite significant differences in the absolute values of the dynamic strength of various brittle materials, the dependence of their normalized strength on the strain rate can be approximated with acceptable accuracy by a single curve [50–53]. There is the following set of unifying dependences [49]:

$$\frac{\sigma_c^{dyn}(T_{fract})}{\sigma_c^{st}} = \frac{1}{f_{scale}}\begin{cases} 1 - 0.37\ln\left(\frac{T_{fract}}{F_c T_1} - 0.01\right) & \text{at } T_{fract} \le F_c \cdot 3.762 \cdot 10^{-4} s \\ 1 - 0.83\ln\left(\frac{T_{fract}}{F_c T_2} - 0.0129\right) & \text{at } T_{fract} > F_c \cdot 3.762 \cdot 10^{-4} s \end{cases}, \tag{13}$$

$$\frac{\sigma_t^{dyn}(T_{fract})}{\sigma_t^{st}} = \frac{1}{f_{scale}}\begin{cases} 1 - 1.66\ln\left(\frac{T_{fract}}{F_t T_3} - 0.238\right) & \text{at } T_{fract} \le F_t \cdot 4.623 \cdot 10^{-5} s \\ 1 - 0.19\ln\left(\frac{T_{fract}}{F_t T_4} - 0.164\right) & \text{at } T_{fract} > F_t \cdot 4.623 \cdot 10^{-5} s \end{cases}. \tag{14}$$

Here, $T_1 = 10^{-3}$ s, $T_2 = 1.55 \times 10^{-2}$ s, $T_3 = 5.25 \times 10^{-5}$ s, and $T_4 = 2.5 \times 10^{-4}$ s. Nondimensional parameters $F_c = \frac{\sigma_c^{st}}{E} / \frac{(\sigma_c^{st})^{ref}}{E^{ref}}$ and $F_t = \frac{\sigma_t^{st}}{E} / \frac{(\sigma_t^{st})^{ref}}{E^{ref}}$ characterize the ratios between Young's modulus (E), the static strengths (σ_c^{st} and σ_t^{st}) of the studied material, and the properties of the known reference material. Here, the expressions were derived using the properties of consolidated sandstone ($E^{ref} = 16$ GPa, $(\sigma_c^{st})^{ref} = 70$ MPa, $(\sigma_t^{st})^{ref} = 10$ MPa).

The parameter f_{scale} is a dimensionless scale factor that characterizes the ratio of the scale of macroscopic samples ($H_{macro} \approx 10^{-2}$ m) to the considered scale of fracture. For particle-reinforced composite, as the modelled refractory material, such parameter is the characteristic distance between the centres of the modelled large grains $H_{heter} = H_{meso}/\sqrt{N_{inc}}$, where H_{meso} is the sample size, and N_{inc} is the number of inclusions in the model composite sample. For the modelled refractory, f_{scale} is estimated as the ratio $f_{scale} = H_{macro}/H_{heter} \approx 25$. For grain–matrix interfaces, the Young's modulus used in Equations (13) and (14) was set as equal to that of the matrix, as it is the softest constituent.

2.4. Model Parameters

The properties of the microstructure's constituents were estimated from the experimental data (Table 1). The grains' density was made equal to the true density of the material, while the density of the matrix was calculated from the total porosity, apparent, and true density—see Section 2.1. In the calculation, it was assumed that all the porosity of the material is localised in the matrix. The Young's modulus of the grains was set equal to the dynamic modulus of cristobalite [54]. The Young's modulus of the matrix was calculated using Counto's equation for the bi-phasal materials. The reverse analysis was performed for the known parameter of the grains, aiming for the total Young's modulus of 6 GPa. The latter value was chosen regarding the fact that the Young's modulus found in bending and compressive tests was 4–6 GPa and 6–7 GPa, respectively. As the first approximation, the strength of the grains was estimated by extrapolating the correlations of the sample strength and portion of the grain through the grain fracture trajectory [12]. The tensile strengths of the matrix and interface between the large grain and matrix were obtained by scaling those to the strength of the large grains. For that, Knoop hardness tests (Shimadzu HMV-2, Kyoto, Japan) performed on different constituents of the microstructure of the modelled material were used [12]. The ratio of the strength between the different constituents was equal to the differences between average Knoop hardness values for the respective constituents.

Table 1. Initial values of mechanical characteristics of mesoscale components.

Properties	Grains	Matrix	Interfaces
Density (kg/m^3)	2350	1670	Not applicable
Young's modulus (GPa)	65	2.75	Not applicable
Poisson's ratio (-)	0.2	0.2	Not applicable
Tensile strength σ_t^{st} (MPa)	8.5	5.4	3.6
Compressive strength σ_c^{st} (MPa)	34	21.6	14.4

The ratio between the compressive and tensile strengths of the constituents was calculated from the relationship of $\sigma_c^{st}/\sigma_t^{st} = \left(\sqrt{3}+\tan\varphi\right)/\left(\sqrt{3}-\tan\varphi\right)$ [55]. The experimental data on refractories [56,57] reveal that the angle of friction φ can vary in the wide range of, at least, between 35–45° and 60–70°. For the given range, the ratio can vary between 2 and 9. Regarding these data, the initial assumption for the grains, matrix, and interfaces was $\sigma_c^{st}/\sigma_t^{st} = 4$ (Table 1).

The following parameter was used to model the separation of the surfaces upon failure (Table 2). For the grains, $V_s = 0.5V_R$, which is representative of the dense, brittle materials. For the matrix and interfaces, the initial values of ε_{max} and γ_{max} are the average values of the post-peak strain seen in compression and bending tests (Figure 1b,c). In general, those parameters are 10% of the ratio σ_t^{st}/E.

The value of the local coefficient of friction μ was set equal to 0.1. It was seen that varying the value of μ in the range from 0 to 0.3 does not lead to a significant change in the fracture pattern and mechanical characteristics of the samples under unconfined uniaxial loading.

Table 2. Defined parameters controlling the separation of elements.

Parameter	Grains	Matrix	Interfaces
V_s (m/s)	1700	Not applicable	Not applicable
ε_{max}	Not applicable	0.00025	0.00025
γ_{max}	Not applicable	0.00025	0.00025

3. Results and Discussion

3.1. Validation of the Model and the Adjustment of the Local Properties

The reference characteristics that were utilised in assessing the accuracy of the mechanical model were the quasi-static values of the ultimate tensile (UTS) and compressive (UCS) strengths, post-peak behaviour, and parameters of the crack trajectory. The procedure for evaluating and refining model parameters described below is a sequence of logically connected stages.

3.1.1. Quasi-Static Fracture Criterion

First, an examination of the alternative failure criteria and values of their parameters (Table 1) was carried out. Figure 3 illustrates a typical example of a fracture pattern under uniaxial tension for the criteria (8a) and (8b). The fracture occurs by the formation of a transverse main crack in the middle of the specimen. For the model with the Drucker–Prager criterion, the main crack trajectory has many branches and small internal cracks on both sides of the main crack's line (Figure 3a). This type of fracture under tensile conditions is more likely to occur in loosely bound materials [58] but is not typical for silica refractory. Experimental data [12] indicate that tensile cracks in silica refractory are localised. The combined Drucker–Prager–Rankine criterion provides the required localised fracture pattern (Figure 3b), which is qualitatively appropriate with the experimental data (Figure 3c). The Rankine criterion had little effect on the fracture pattern and strength in the compressive models. The results displayed in the following sections are obtained using the combined Drucker–Prager–Rankine criterion.

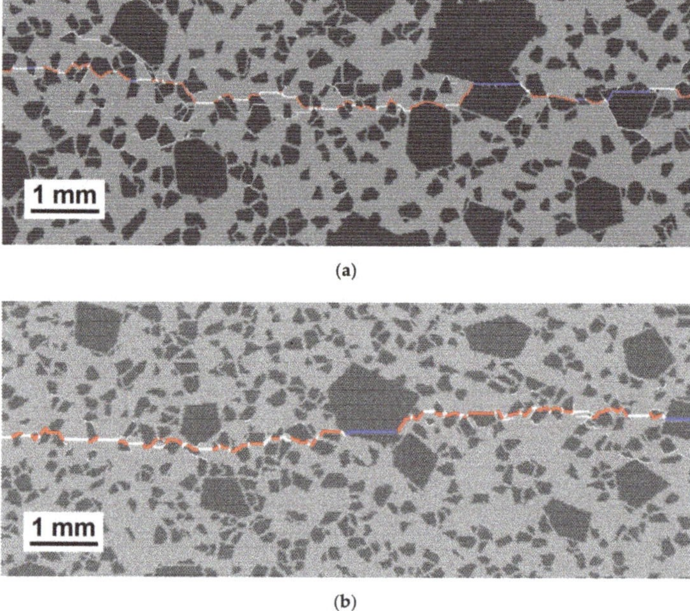

(a)

(b)

Figure 3. *Cont.*

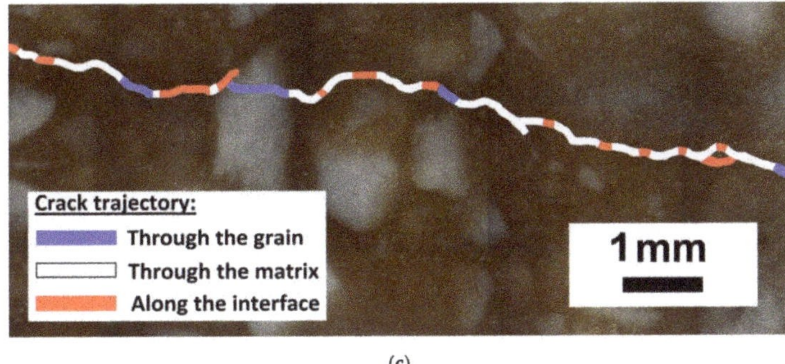

(c)

Figure 3. Examples of the fracture pattern obtained with Drucker–Prager criterion (**a**), combined Drucker–Prager–Rankine criterion (**b**), and a tensile crack formed in a sample of silica refractory under monotonic wedge splitting loading (**c**). The thick curved line in (**c**) shows the crack trajectory. White, blue, and red portions of the crack lines indicate crack propagation through matrices, through grains, and along interfaces, respectively. The width of each image represents 1 cm of the sample.

3.1.2. Strength of Constituents

Under tension and compression, the modelled material had a Young's modulus of 5.5 GPa (Figure 4). Thus, the selected elastic characteristics of the constituents (Table 1) produce an acceptable representation of Young's modulus, as seen in the lab tests. However, the obtained strength was an order of magnitude lower than the experimental values. Furthermore, the numerically obtained strength ratio UCS/UTS $\approx$ 3.8 is lower than the experimental value of approximately 5 (Figure 1).

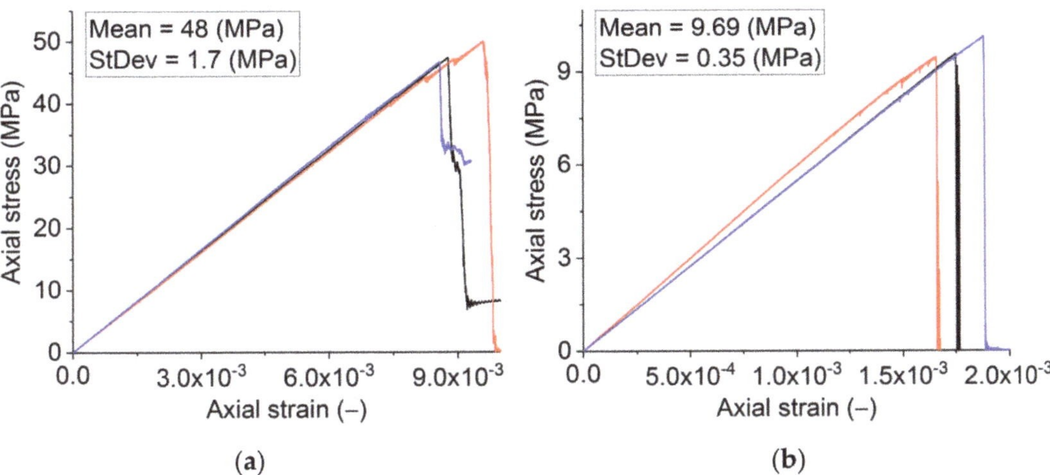

(a)   (b)

Figure 4. Numerically obtained uniaxial compression (**a**) and tension (**b**) diagrams with the use of the final strength parameters for mesoscopic structural elements. Diagrams are shown for the three statistically equivalent samples.

To match the laboratory data of macroscopic strength, the strength of large grains, matrix, and interfaces was increased. It was assumed that although the strengths of the constituents were underestimated (Table 1), their relative values are correct—this assumption is summarised in Table 3. Note that this implicitly assumes the same ratio σ_c / σ_t for all structural elements.

Table 3. Strength ratios.

Strength/Strength	Grains/Matrix	Grains/Interfaces	Matrix/Interfaces
σ_t / σ_t	1.574	2.361	1.5
σ_c / σ_c	1.574	2.361	1.5

The determination of the values of the local strength of structural elements was executed in two stages. In the first stage, we varied the σ_c / σ_t ratio for the structural elements to obtain the required UCS/UTS. We increased the values of σ_c of the structural elements while maintaining constant values of σ_t. The required ratio of macroscopic strengths UCS/UTS = 5 is achieved at $\sigma_c / \sigma_t = 10$. In the second stage, we jointly increased σ_c and σ_t for structural elements while keeping their σ_c / σ_t ratio unchanged, thus rendering it possible to achieve the required values of uniaxial compressive and tensile strength of the samples (UCS $\approx$ 48 MPa, UTS $\approx$ 9.7 MPa, UCS/UTS = 5). The stress–strain curves after the final tuning are presented in Figure 4a,b. Corresponding values of local strength are shown in Table 4. One should note that the values of σ_c and σ_t (UCS and UTS) obtained for the grains are similar to those of dense silica [59]. However, unlike in the model, the real stress–strain curves exhibit some non-linearity at the peak. This effect will be addressed further in Section 3.1.4.

Table 4. Final strength values.

Strength	Grains	Matrix	Interfaces
σ_c^{st} (MPa)	680	432	288
σ_t^{st} (MPa)	68	43.2	28.8

The analysis of the tensile stress–strain curves indicated the fracture energy of $\approx$90 N/m. The stress–strain curve demonstrates the brittle failure. For quasi-brittle materials, this was observed when the elastic energy released upon the crack formation exceeds the energy to be consumed by the forming crack. This balance is influenced not only by the material properties but also by the shape of the loaded body [60]. For the same reason, brittle failure is observed on the laboratory three-point bending curves (Figure 1c). The energy from the stress–strain curves showing brittle failure indicates the elastic energy released, which exceeds the fracture energy. Therefore, it is not surprising that the fracture energy obtained by us exceeds the fracture energy of 60 N/m obtained for this material in the stable regime of crack propagation enabled by the wedge-splitting test [42].

Here, it should be noted that the compared virtual and real samples have different shapes and boundary conditions. For instance, due to a higher height to width ratio, the real compressive sample is expected to have somewhat lower strength and more brittle failure. However, the friction between the loading plates and sample, which is absent in the model, can reduce these effects. The lab data of purely uniaxial tensile and compressive tests could be used to further improve the model.

The fracture patterns for the "statistically equivalent" models with final values of local strength parameters (Table 4) are compared with those registered in the refractory samples. For uniaxial compression, one can see that all DEM samples display similar fracture patterns, namely the vertical main cracks split the sample into several large parts (Figure 5a–c). The trajectories of tensile cracks have a localised nature (Figure 6a–c). The modelled failure is rather similar to that in real samples (Figures 5d and 6d–f).

3.1.3. Properties of the Interface

To validate the initial estimate of the interface strength (Table 4), the analysis of the crack trajectory was performed (Figure 7). Calculations for the interface's strength between 65% and 100% of the strength of the matrix were conducted. The initially set ratio of the matrix strength to interface strength of 1.5 corresponds to the value of 66.7%. In all cases, σ_c and σ_t for interfaces were varied synchronously, keeping the ratio $\sigma_c / \sigma_t = 10$ as constant.

In modelled crack trajectories, the portions of the crack propagating through the interface, large grains, and matrix were measured. Additionally, the ratio of the crack length (L2) to its projection on the plane parallel to the direction of the crack propagation (L1) was calculated. The latter ratio quantified the crack's non-linearity. Averaged results for the three model samples were weighed against the experimental crack propagation analysis for the same grade of silica refractories, for which the experimental cracks were obtained in the wedge-splitting test [12]. In the fractographic analysis, six crack trajectories of three wedge-splitting tests were analysed. Per sample, two polished sections exposing cracks at different depths of the sample were used.

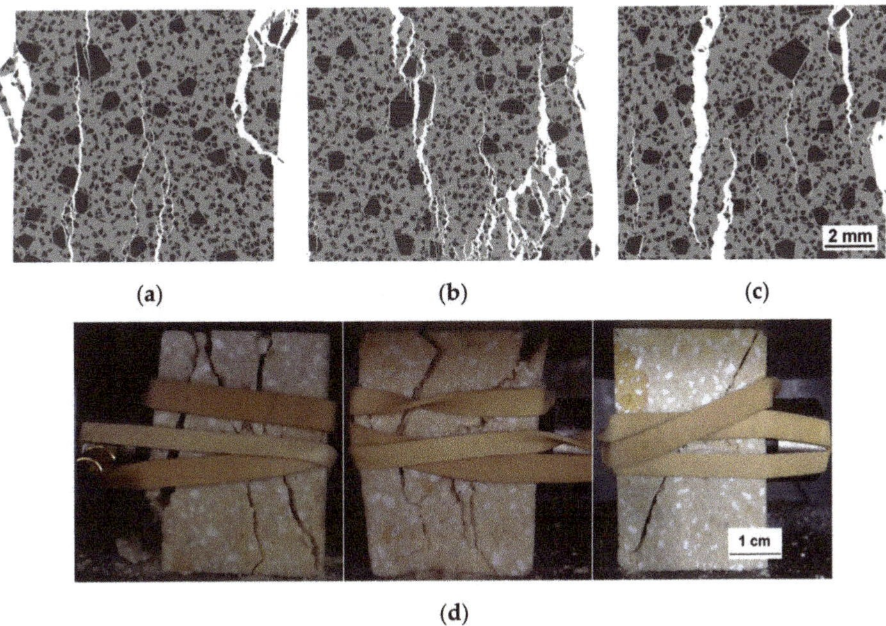

Figure 5. Crack patterns for uniaxial compression tests (**a–c**) models with final values of local strength parameters (1×1 cm^2) (**a–c**), typical crack patterns in the real samples (**d**).

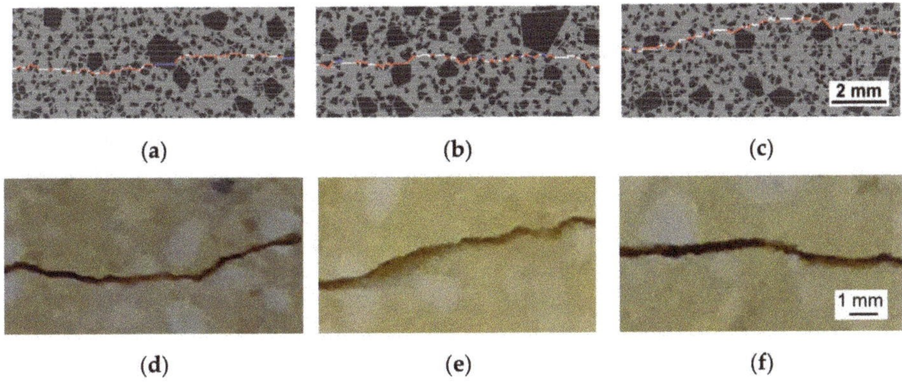

Figure 6. Crack patterns of the model samples under uni-axial tension (**a–c**) and typical crack patterns in the real samples under three-point bending (**d–f**). White, blue, and red portions of the crack lines in (**a–c**) indicate crack propagation through matrices, through grains, and along interfaces, respectively. The width of each image represents 1 cm of the sample. See also Supplementary Materials with full-size (**a–c**).

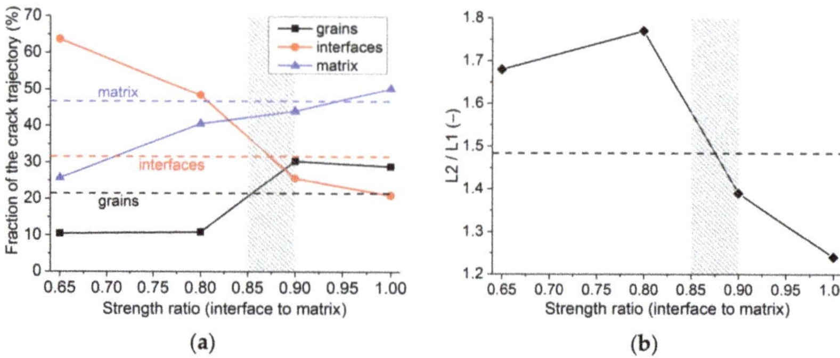

Figure 7. Comparison of the virtual and real crack trajectories for crack portions through different constituents of the model samples (**a**) and the crack non-linearity parameter L2/L1 (**b**). Horizontal dashed lines denote experimental values for silica refractory [12]. The rectangular shaded area marks the interval where the model is considered to agree with the experiment.

For the models with low interface strength, the portion of the crack trajectory through the interface approaches 0.7 (Figure 7a). The crack jumps from the interface of one large grain to another, producing a trajectory of high non-linearity (L2/L1). For the models where the interface's strength approaches that of the matrix, the preferred crack propagation route is through the matrix. When a crack meets a grain, it splits it, causing the non-linearity of the crack to become lower (Figure 7b). The best agreement between the parameters of the modelled and experimental crack trajectory is achieved for the interface strength which is 85–90% of the matrix strength.

An increase of the interface strength from 65% to 100% produces the average increase of UTS of $\approx$10% (from 9.7 to 10.7 MPa) and UCS of $\approx$25% (from 48 to 59.7 MPa). The strength ratio increases approximately by 10% (UCS/UTS $\approx$ 5.5). In any case, the strength values and their ratio are in the characteristic range these parameters have in the real silica refractory (Figure 1b,c).

The observations of different crack trajectories and almost identical strength values, which is observed here for samples with different interface strengths, is in agreement with lab test results [12]. There, different participation of the interface failure was detected for two groups of samples tested under monotonic and cyclic wedge splitting. However, the average strength values for both groups were similar. Apart from the global strength, the interface strength is to influence the fracture energy. The correlation of the fracture energy and the interface strength is to have the maximum. DEM modelling may be used to find the optimum enabling best combination of the global strength and the fracture energy.

3.1.4. Influence of Finite Time of Fracture Incubation

Obtained curves (Figure 4), especially for compression, exhibit unrealistically brittle failure. To explore the possibility of controlling the length of the sample softening section of the stress–strain curve under compression, the parameters ε_{max} and γ_{max} were varied. The values of these parameters decreased and increased within five times relative to the "basic" values (Table 2). As above, we assumed $\varepsilon_{max} = \gamma_{max}$. In all cases, the values were the same for the matrix and interfaces. The variation of the parameters had a limited effect on the softening. Particularly, the parameter ε_{max} does influence the length of the softening slope. However, this effect is limited to crack formation. The failure remains brittle.

Further adaptation of the model was to account for the finite time of incubation of damage, i.e., to use the dynamic model with combined criteria (12)–(14). The observed brittle compressive failure of the model samples is believed to be a consequence of the strain rate ($\approx$10^{-1} s^{-1}) far beyond the typical quasi-static range (10^{-3}–10^{-5} s^{-1}) in experimental studies of refractory samples. The relatively large strain rate was used in the simulation to reduce the analysis time. At such characteristic values of loading rates ($\geq$10^{-1} s^{-1}), elastic waves from emerging damages do not have time to completely dissipate before new damages appear and thereby affect the conditions of new damage formation.

To accurately simulate the fracture of the sample at strain rates $\approx 10^{-1}$ s^{-1}, the calculations were implemented using the dynamic model of fracture. The final quasi-static mechanical characteristics of the constituents (Table 4) were used in the dynamic model. The static strength of the interface was set as 90% of the strength of the matrix ($\sigma_c = 390$ MPa, $\sigma_t = 39$ MPa).

The dynamic model captures the dynamic hardening of materials even at the considered relatively low strain rates ($\approx 10^{-1}$ s^{-1}). Consequently, as a result, the strength of the sample is higher than in the model with static formulation (Figures 8 and 9). The difference is $\approx 7\%$ and 5–6% for compressive and tensile strengths, respectively. The fracture of the model becomes less brittle. The quasi-brittle nature of the fracture seen under compression (Figure 8) is due to the gradual formation of mesoscopic internal cracks and their coalescence into macrocracks. The length of the softening section for compression and tension is $\approx 1.5 \times 10^{-3}$ and $\approx 1.2 \times 10^{-4}$, respectively. The value for compression quantitatively agrees with the data for samples of the real material. One can note that the ratio of the lengths of the softening sections for tension and compression corresponds to that of macroscopic strengths with a proportionality coefficient of ≈ 2.

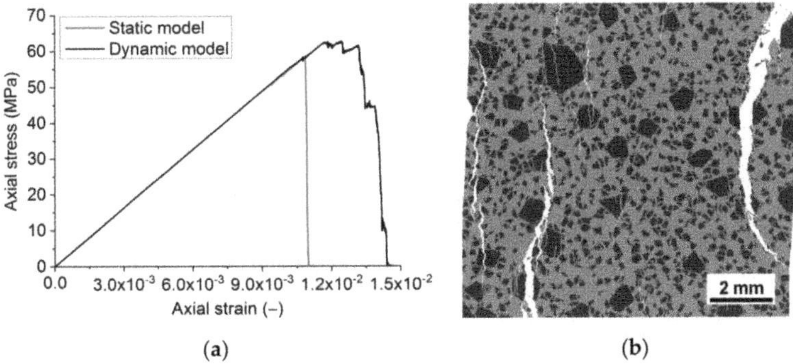

Figure 8. Compression diagrams obtained using the static and dynamic fracture models (**a**) and the broken sample in the dynamic fracture model (**b**). Strain rate 4×10^{-1} s^{-1}.

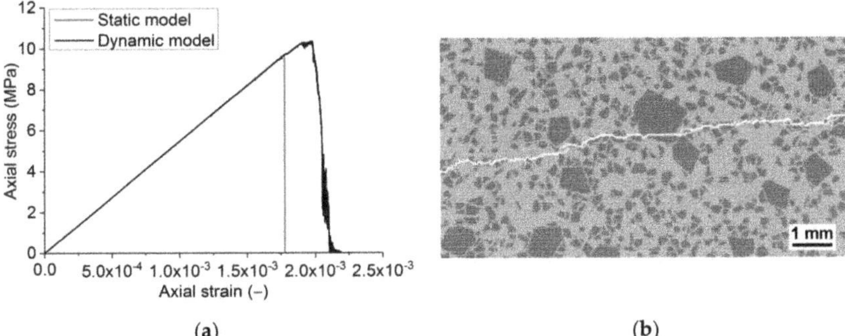

Figure 9. Uniaxial tension diagrams obtained using the static and dynamic fracture models (**a**) and the broken sample in the dynamic fracture model (**b**). Strain rate 2.5×10^{-1} s^{-1}.

3.2. Dynamic Mechanical Behaviour

Traditionally, two characteristic strain rate intervals are distinguished: "low" ($\dot{\varepsilon} < 10^{-3}$ s^{-1}) and "large" strain rates ($\dot{\varepsilon} > 10^{0}$ s^{-1}). In the first interval, the majority of brittle solids can be roughly assumed to be strain rate insensitive [61–63]. In the second, the increase of the strain rate causes the non-linear increase of strength. For concrete, for example, the dynamic increase factor (DIF) approaches 2 at $\dot{\varepsilon} \approx 10^2$ s^{-1}. DIF is the normalised increase of strength due to the increase of the strain rate from quasi-static to a given dynamic value. From the three-point bending and wedge splitting tests reported for this material [12], strain rates typical for mechanical tests on refractories can be estimated as 3–7 $\times$ 10^{-5} s^{-1}. Regarding the fact that the height from which the scrap and other solid particles are loaded may be as high as 5–7 m, the velocity at the impact on the face of the refractory lining will exceed the loading velocity in the lab tests by many orders of magnitude. By scaling the velocities of the falling scrap and loading piston of the lab experiments [12], one can estimate the maximal strain rates experienced by refractories as 50–150 s^{-1}. Similar magnitudes of strain rates are expected from the impacts of the hydraulic hammer during the repair of the lining.

Dynamic modelling was performed with the material properties, as presented in Table 4, and the interface strength of 90% of the matrix strength. From the analysis, one can observe a non-linear increase of the strength with the increasing rate of loading (Figure 10), which is typical for the vast majority of materials [50–53]. It should also be noted that the slope of the softening section becomes flatter at high strain rates. This is due to the fact that at high strain rates, the fracture time of the sample becomes comparable to the loading time, as a result of which there is a constant "additional load" of the breaking sample even after the loss of stability.

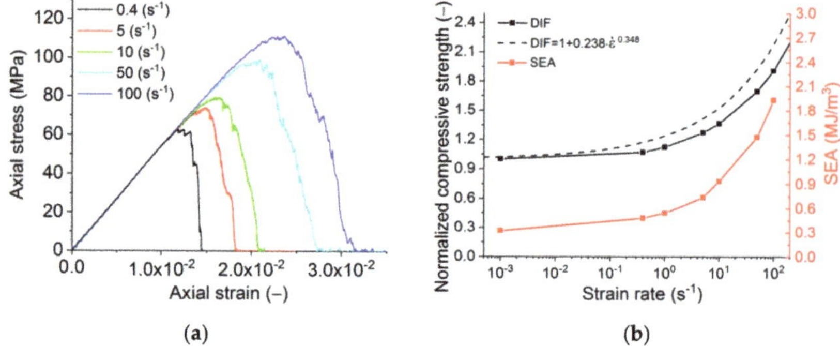

Figure 10. Results for dynamic compressive loading: (**a**) stress–strain diagrams, (**b**) effect of the strain rate on normalized strength and specific energy absorption (SEA). The dashed line is the averaging analytical approximation of experimental data for various brittle and quasi-brittle materials [49].

The energy absorption (or fracture energy) is also an important characteristic of the material. In the simplest case, specific energy absorption (SEA) can be estimated as the area under the loading diagram (σ–ε). In this paper, SEA is determined by the area under the loading curve. The red line in Figure 10b illustrates the change in SEA with an increasing loading rate. Its strain-rate dependence is a non-linear (power-law) function, and SEA's absolute value in the considered range of strain rates increases by an order of magnitude owing to distributed cracking. To translate it to more conventional means for refractories units of N/m (specific fracture energy), the reported values of SEA should be multiplied by the height of the sample. One should note that this approach neglects the non-linearity of the crack (and multiple cracking under dynamic loading). However, the fracture energy measured for refractories in lab tests also neglects the crack waviness.

For reference, one can consider the SEA of civil engineering concrete measured at 100 s^{-1} in a split Hopkinson pressure bar set-up. SEA was found to be about 1 MJ/m^3 [64]. The material has a quasi-static compressive strength of 50 MPa. The SEA's order of magnitude agrees with our modelling results.

Multiple increases in SEA and fracture energy at high strain rates are reflected in the fracture pattern (Figure 11). Increased fragmentation of the sample is seen for the increasing rate of loading. This is a typical regularity of the change in fracture pattern for the majority of brittle and quasi-brittle materials under dynamic impact [65].

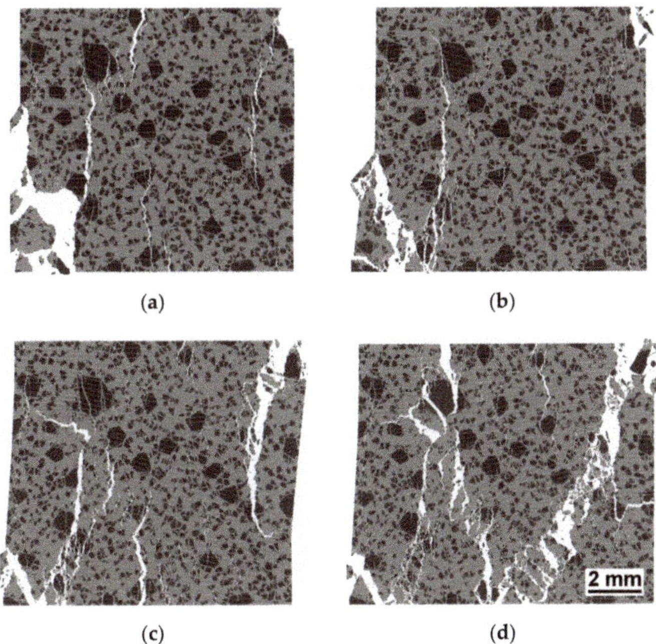

Figure 11. Fracture patterns under dynamic compression for the strain rates of 5 s^{-1} (**a**), 10 s^{-1} (**b**), 50 s^{-1} (**c**), and 100 s^{-1} (**d**).

The predicted compression and tension strength values broadly agree with the averaged experimental data for brittle materials [50–53] (Figures 10b and 12b). In the figures, the latter is shown as the dashed line. The deviations lie within the range of the strength scatter for various brittle and quasi-brittle materials. For refractories, the typical values for the properties scatter are 10–20% of the average value. The deviation is higher for the tensile failure than for compressive failure, especially if one considers the change of slope in the experimental and numerical curves. This feature of the tensile behaviour may indicate that with an increase in the strain rate, the governing mechanism of fracture changes from separation to a mixed mechanism including shear.

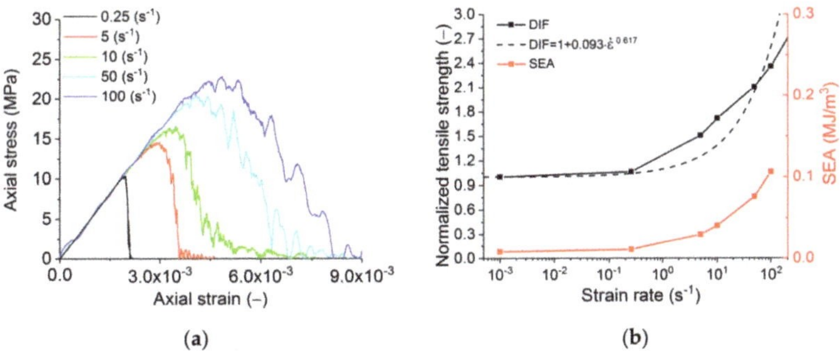

Figure 12. Results for dynamic tensile loading: (**a**) stress–strain diagrams, (**b**) effect of the strain rate on normalised strength and specific energy absorption (SEA). The dashed line is the analytical approximation reported elsewhere [49].

Note that the presented averaged dependence was employed to determine the fracture time of the components and parameters of relationships (13) and (14). The differences between the dynamic strength curve of the refractory under consideration and averaged curve are obviously associated with the influence of the internal structure's features (size and shape of grains, the characteristic distance between them, interfaces, etc.).

For dynamic tensile tests, one can perceive that the increasing loading rate produces more developed strain softening (Figure 12a) on account of the transition from the extremely localised fracture at low strain rates to the formation of a system of mesoscopic cracks (Figure 13). The fracture under dynamic loading is characterised by the presence of several centres of growth of transverse mesoscopic cracks (the initial stage of fracture), following which these initial cracks are combined into one main fracture. Due to several fracture sites, the main crack under dynamic loading is much more tortuous than under quasi-static loading.

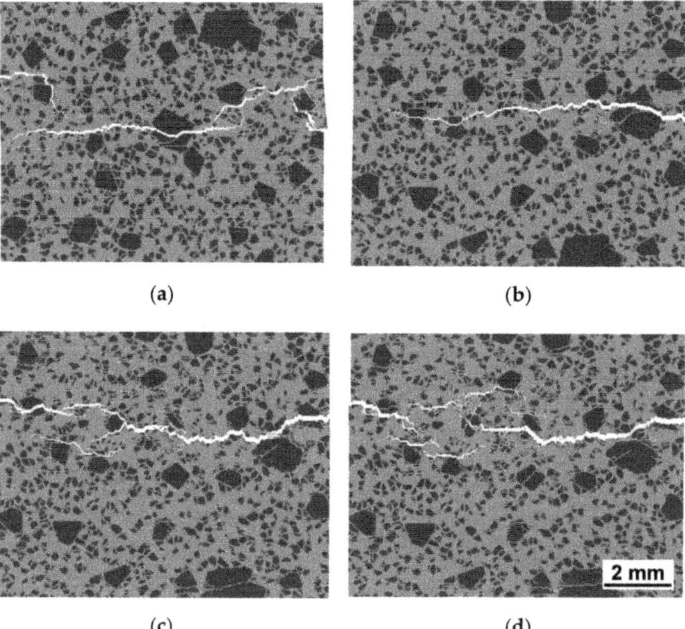

(a) (b) (c) (d)

Figure 13. Fracture patterns under dynamic tension for the strain rates of 5 s^{-1} (**a**), 10 s^{-1} (**b**), 50 s^{-1} (**c**), and 100 s^{-1} (**d**).

For refractories, the critical material property is the brittleness at failure. One of the indexes of this property is the ratio of fracture energy to strength: the lower the ratio, the higher the brittleness. With increasing strain rate, the refractory's strength is identified to vary by approximately two times. For the same interval of strain rates, the fracture energy is identified to vary by ten times. Therefore, increasing of the loading rate causes the decrease of brittleness (Figure 14). As the fracture energy increases, the decreased brittleness should result from a more distributed cracking.

Regarding the complex nature of the dynamic failure, the damage development in the process of loading has been analysed. The sample damage was characterised by the number of broken bonds in pairs of discrete elements normalised to the initial number of bonds in the sample. The contribution of each constituent to the sample damage (fraction of damage) was characterised by the ratio of the number of broken bonds in this constituent to the total number of broken bonds in the sample. Figure 15 illustrates how the constituents (grains, matrix, and interfaces) contribute to the damage accumulation in the sample under dynamic loading. The damage starts in the interfaces and grains, after which the contribution of the matrix to fracture gradually increases. This is associated with the coalescence of cracks in neighbouring grains. At the end of loading (sample failure), the contribution of the grains, matrix, and the interface is ≈50%, ≈25%, and ≈25%, respectively. Hence, the contribution of grains to the total fracture of samples under dynamic loading is much greater than the contributions of the matrix and interfaces.

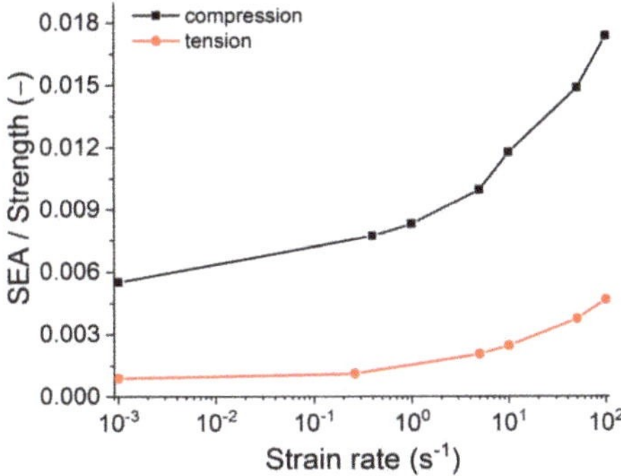

Figure 14. The dependence of the ratio of specific energy absorption to strength on the strain rate.

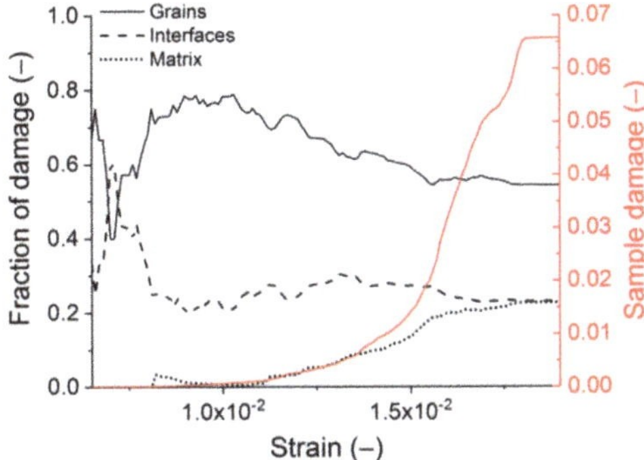

Figure 15. Contributions of constituents to the sample damage accumulation during dynamic compressive loading. The strain rate is $5\ \mathrm{s}^{-1}$. The red curve is the absolute sample damage.

Figure 16 reveals how the contributions of constituents to the fracture of refractory samples alter with a strain rate increase. Here are the "final" values of fractions corresponding to fractured samples.

From experimental studies [12], the values of the quasi-static tensile crack path fractions per grains, matrix, and interfaces are notable for the considered refractory material. When comparing these data with the results of numerical simulation of dynamic tension (Figure 16a), one can witness significant changes in the fracture mechanisms during the transition from a quasi-static loading mode to a dynamic one. In the range of quasi-static strain rates ($<10^{-2}\ \mathrm{s}^{-1}$), the dominant contribution to fracture is made by damage in the matrix. Under dynamic loading, according to the results of numerical modelling, most of the damage is concentrated in large grains. The increasing contribution of the large grain damage with increasing strain rate is systematically seen for tension (Figure 16a) and compression (Figure 16b). Grain failure is observed throughout the sample. The effect is much more pronounced under tension. This intragranular damage is expected to be caused by secondary elastic waves from the forming main cracks.

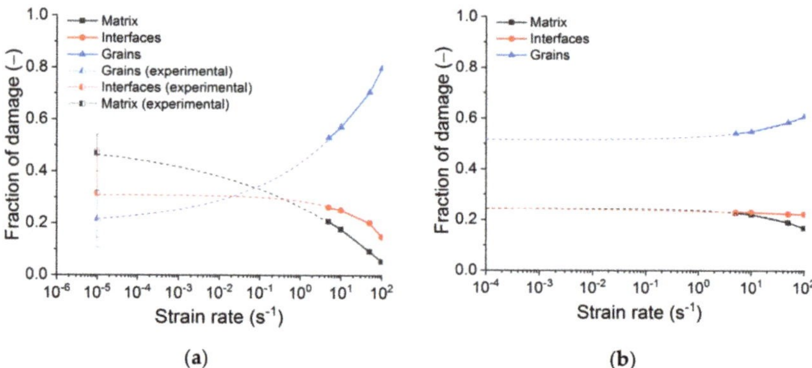

Figure 16. The dependences of the contributions of constituents to the fracture of refractory samples on the strain rate: tension (supplemented by experimental data) (**a**) and compression (**b**).

To analyse the effect of the interface strength on the macroscopic dynamic strength of silica refractory, we carried out an additional numerical study with an interface strength that was 65% of the matrix strength (Table 4). The results confirmed that the difference between DIF for the models with the interface strength of 65% and 90% is less than 10% for compression and within 20% for tension—it is nearly constant at $\dot{\varepsilon} > 1$ s^{-1}. Thus, the obtained estimates of compressive DIF for refractory with interface strength 90% of the matrix strength (Figures 10b and 12b) can be used as the "master curves". Indeed, as the strength of the refractory interfaces changes, it is sufficient to measure the new UCS and UTS values of the samples under quasi-static loading. The dependences of the dynamic strength of the new material on the strain rate can be estimated with acceptable accuracy based on the use of the "master curves".

4. Conclusions

The method of deformable discrete elements has been applied to model the meso-structural aspects of the failure of silica refractories. It has been discovered that the dynamic form of the combined Drucker–Prager–Rankine criterion is suitable to model the failure of quasi-brittle refractories under dynamic and quasi-static loading. This is explained by the importance of dynamic strengthening as well as for rather low rates of loading. A promising algorithm for determining the properties of microstructural constituents has been proposed. The algorithm is based on the scaling from the constituent of the best-known properties. The scaling factors are based on the Knoop hardness of the grains, matrix, and their interface. The validation and fine tuning of the properties were executed using the information on the fractions of the constituents in the real crack trajectories.

Compressive and tensile numerical tests were performed in the range of loading rates spanning the values typical for quasi-static laboratory tests, respective quasi-static in-service loads, and maximal dynamic in-service loads. From the numerical analysis, the following was concluded:

- An acceptable agreement was seen between the loading rate dependencies predicted by the model and average experimental trend typical for brittle and quasi-brittle materials. Particularly, the simulation results showed a non-linear increase of compressive and tensile strength of refractory with increasing strain rate. In both cases, the dynamic strength curves can be approximated by power functions with exponents significantly less than 1. The exponent for dynamic tensile strength is 1.4 times greater than the exponent for dynamic compressive strength. These results correlate well with experimental data on diverse heterogeneous ceramic-based materials, including concretes and rocks. Experimental data show that, first, the dependences DIF ($\dot{\varepsilon}$) are power functions with exponents usually around 0.5, and, second, the exponent for tensile DIF is up to two times greater than for compressive DIF [50–53]. The simulation results also indicate a transition from localised fracture (single macrocracks) to multiple fractures (branching network of mesocracks) and a progressive increase in the degree of sample fragmentation with increasing strain rate. The effect is especially pronounced at $\dot{\varepsilon} > 10^1$ s^{-1}. This is also in good agreement with the dynamic test results of diverse quasi-brittle porous materials. Finally, it is pertinent to mention the increase in the number of broken mesoscopic hard inclusions at high strain rates observed in experiment and simulation (the trend from crack propagation around the inclusions to cutting them).

- The disproportional (power law) increase of the strength and fracture energy with the loading rates signals a reduced brittleness for the refractories under higher dynamic loading rates. On the microstructural level, this is seen by a more dispersed failure process and a higher fraction of large grain failure typically observed for higher rates of loading. A detailed experimental and computational study of this effect is the subject of further research.
- The compressive properties are more sensitive to the interface strength than the tensile properties. Under tension, a significant variation of the crack trajectory is observed despite the limited sensitivity of the global strength on the strength of the interface.
- The variation of the interface strength has a quantitatively similar effect on the quasi-static and dynamic compressive strength of the material.
- For tensile loading, the dependencies of strength on the loading rate for models with different interface strengths are less similar than those for compressive loading. This can indicate that the change in the contributions of different constituents to the sample damage with an increase in the tensile strain rate substantially depends on the static strength of the interfaces. A detailed analysis of this effect can be the subject of further research.

Refractory materials are designed to operate at high temperatures and temperature gradients. Therefore, thermal shock resistance is among the key service characteristics of refractories. The advantages of the method of uniformly deformable discrete elements make it possible to implement not only mechanical but coupled thermomechanical (for example, thermoelastic) models of quasi-brittle materials and take into account the thermal expansion of components and temperature dependence of component's (local) strength. The development of such a model for silica refractory material and its application to study micro-scale mechanisms of formation of thermally induced cracks during thermal shocks is a priority topic for further research. The far-reaching goal of such research is the computer design of the multiscale internal structure of the refractory material to achieve the specified mechanical and thermophysical properties at high operating temperatures. The presented methodology of the development of a mesoscale mechanical model is a necessary initial stage of that research.

Supplementary Materials: The following are available online at https://www.mdpi.com/article/10.3390/ma14237376/s1, Figure S1: The full-size images of Figure 6a–c.

Author Contributions: Conceptualization, K.A. and A.V.Z.; software, E.V.S. and A.I.D.; simulation, A.S.G. and E.V.S.; writing—original draft preparation, A.S.G., E.V.S., A.I.D. and K.A.; visualization, A.S.G. and A.V.Z.; supervision, K.A.; project administration, E.V.S. and A.I.D.; funding acquisition, K.A. All authors have read and agreed to the published version of the manuscript.

Funding: This research was funded by the Government research assignment for ISPMS SB RAS, project FWRW-2021-0002.

Conflicts of Interest: The authors declare no conflict of interest.

References

1. Shinohara, Y. *Refractories Handbook*; Japanese Association of Refractories: Tokyo, Japan, 1998.
2. Schacht, C. *Refractory Linings: Thermomechanical Design and Applications*; CRC Press: Boca Raton, FL, USA, 2019.
3. Harmuth, H.; Bradt, R.C. Investigation of refractory brittleness by fracture mechanical and fractographic methods. *Interceram Int. Ceram. Rev.* **2010**, *1*, 6–10.

4. Ludwig, M.; Śnieżek, E.; Jastrzębska, I.; Prorok, R.; Sułkowski, M.; Goławski, C.; Fischer, C.; Wojteczko, K.; Szczerba, J. Recycled magnesia-carbon aggregate as the component of new type of MgO-C refractories. *Constr. Build. Mater.* **2021**, *272*, 121912. [CrossRef]

5. Madej, D.; Szczerba, J. Detailed studies on microstructural evolution during the high temperature corrosion of SiC-containing andalusite refractories in the cement kiln preheater. *Ceram. Int.* **2017**, *43*, 988–1996. [CrossRef]

6. Yao, H.; Chen, H.; Ge, Y.; Wei, H.; Li, Y.; Saxén, H.; Wang, X.; Yu, Y. Numerical Analysis on Erosion and Optimization of a Blast Furnace Main Trough. *Materials* **2021**, *14*, 4851. [CrossRef]

7. Gómez-Rodríguez, C.; Antonio-Zárate, Y.; Revuelta-Acosta, J.; Verdeja, L.F.; Fernández-González, D.; López-Perales, J.F.; Rodríguez-Castellanos, E.A.; García-Quiñonez, L.V.; Castillo-Rodríguez, G.A. Research and Development of Novel Refractory of MgO Doped with ZrO_2 Nanoparticles for Copper Slag Resistance. *Materials* **2021**, *14*, 2277. [CrossRef]

8. Tang, H.; Li, C.; Gao, J.; Touzo, B.; Liu, C.; Yuan, W. Optimization of Properties for Alumina-Spinel Refractory Castables by CMA (CaO-MgO-Al_2O_3) Aggregates. *Materials* **2021**, *14*, 3050. [CrossRef] [PubMed]

9. Ratle, A.; Allaire, C. A new method for impact testing of refractories. *J. Can. Ceram. Soc.* **1996**, *65*, 263–269.

10. Gruber, D.; Sistaninia, M.; Fasching, C.; Kolednik, O. Thermal shock resistance of magnesia spinel refractories—Investigation with the concept of configurational forces. *J. Eur. Cer. Soc.* **2016**, *36*, 4301–4308. [CrossRef]

11. Asadi, F.; André, D.; Emam, S.; Doumalin, P.; Huger, M. Numerical modelling of the quasi-brittle behaviour of refractory ceramics by considering microcracks effect. *J. Eur. Ceram. Soc.* **2021**, in press. [CrossRef]

12. Andreev, K.; Yin, Y.; Luchini, B.; Sabirov, I. Failure of refractory masonry material under monotonic and cyclic loading—Crack propagation analysis. *J. Con. Build. Mat.* **2021**, *299*, 124203. [CrossRef]

13. Zabolotskiy, A.V.; Turchin, M.Y.; Khadyev, V.T.; Migashkin, A.O. Numerical investigation of refractory stress-strain condition under transient thermal load. *AIP Conf. Proc.* **2020**, *2310*, 020355. [CrossRef]

14. Harmuth, H. Characterisation of the Fracture Path in 'Flexible' Refractories. *Adv. Sci. Technol.* **2010**, *70*, 30–36. [CrossRef]

15. Rossello, C.; Elices, M. Fracture of model concrete: 1. Types of fracture and crack path. *Cem. Concr. Res.* **2004**, *34*, 1441–1450. [CrossRef]

16. Rossello, C.; Elices, M.; Guinea, G.V. Fracture of model concrete: 2. Fracture energy and characteristic length. *Cem. Concr. Res.* **2006**, *36*, 1345–1353. [CrossRef]

17. Brandt, A.J. *Cement-Based Composites*; CRC Press: London, UK, 2014. [CrossRef]

18. Mao, Y.; Coenen, J.W.; Riesch, J.; Sistla, S.; Almanstötter, J.; Jasper, B.; Terra, A.; Höschen, T.; Gietl, H.; Linsmeier, C.; et al. Influence of the interface strength on the mechanical properties of discontinuous tungsten fiber-reinforced tungsten composites produced by field assisted sintering technology. *Compos. A Appl. Sci. Manuf.* **2018**, *107*, 342–353. [CrossRef]

19. Sabirov, I.; Kolednik, O. Local and global measures of the fracture toughness of metal matrix composites. *Mater. Sci. Eng. A* **2010**, *527*, 3100–3110. [CrossRef]

20. Smirnov, I.; Konstantinov, A. Strain Rate Dependencies and Competitive Effects of Dynamic Strength of Some Engineering Materials. *Appl. Sci.* **2020**, *10*, 3293. [CrossRef]

21. Henneberg, D.; Ricoeur, A.; Judt, P. Multiscale modeling for the simulation of damage processes at refractory materials under thermal shock. *Comput. Mater. Sci.* **2013**, *70*, 187–195. [CrossRef]

22. Makarian, K.; Santhanam, S. Micromechanical modeling of thermo-mechanical properties of high volume fraction particle-reinforced refractory composites using 3D Finite Element analysis. *Ceram. Int.* **2020**, *46*, 4381–4393. [CrossRef]

23. Šavija, B.; Smith, G.E.; Liu, D.; Schlangen, E.; Flewitt, P.E.J. Modelling of deformation and fracture for a model quasi-brittle material with controlled porosity: Synthetic versus real microstructure. *Eng. Fract. Mech.* **2019**, *205*, 399–417. [CrossRef]

24. Šavija, B.; Zhang, H.; Schlangen, E. Assessing hydrated cement paste properties using experimentally informed discrete models. *J. Mater. Civ. Eng.* **2019**, *31*, 04019169. [CrossRef]

25. Andreev, K.; Sinnema, S.; Stel, J.; Allaoui, S.; Blond, E.; Gasser, A. Effect of binding system on the compressive behaviour of refractory mortars. *J. Eur. Ceram. Soc.* **2014**, *34*, 3217–3227. [CrossRef]

26. Luchini, B.; Sciuti, V.F.; Angélico, R.A.; Canto, R.B.; Pandolfelli, V.C. Critical inclusion size prediction in refractory ceramics via finite element simulations. *J. Eur. Ceram. Soc.* **2017**, *37*, 315–321. [CrossRef]

27. Moreira, M.H.; Cunha, T.M.; Campos, M.G.G.; Santos, M.F.; Santos, T., Jr.; Andrfe, D.; Pandolfelli, V.C. Discrete element modelling—A promising method for refractory microstructure design. *Am. Ceram. Soc. Bull.* **2020**, *99*, 22–28.

28. Nguyen, T.T.; André, D.; Huger, M. Analytical laws for direct calibration of discrete element modelling of brittle elastic media using cohesive beam model. *Comput. Part. Mech.* **2019**, *6*, 393–409. [CrossRef]

29. Samadi, S.; Jin, S.; Harmuth, H. Combined damaged elasticity and creep modeling of ceramics with wedge splitting tests. *Ceram. Int.* **2021**, *47*, 25846–25853. [CrossRef]

30. Moreira, M.H.; Ausas, R.F.; Dal Pont, S.; Pelissari, P.I.; Luz, A.P.; Pandolfelli, V.C. Towards a single-phase mixed formulation of refractory castables and structural concrete at high temperatures. *Int. J. Heat Mass Transf.* **2021**, *171*, 121064. [CrossRef]

31. Li, W.; Zhou, X.; Carey, J.W.; Frash, L.P.; Cusatis, G. Multiphysics lattice discrete particle modeling (M-LDPM) for the simulation of shale fracture permeability. *Rock Mech. Rock Eng.* **2018**, *51*, 3963–3981. [CrossRef]

32. Sadowski, T.; Pankowski, B. Numerical modelling of two-phase ceramic composite response under uniaxial loading. *Compos. Struct.* **2016**, *143*, 388–394. [CrossRef]

33. Moës, N.; Dolbow, J.; Belytschko, T. A finite element method for crack growth without remeshing. *Int. J. Numer. Methods Eng.* **1999**, *46*, 131–150. [CrossRef]
34. Mohammadnejad, T.; Khoei, A. An extended finite element method for hydraulic fracture propagation in deformable porous media with the cohesive crack model. *Finite Elem. Anal. Des.* **2013**, *73*, 77–95. [CrossRef]
35. Cusatis, G.; Pelessone, D.; Mencarelli, A. Lattice discrete particle model (LDPM) for failure behavior of concrete. I: Theory. *Cem. Concr. Compos.* **2011**, *33*, 881–890. [CrossRef]
36. Rezakhani, R.; Zhou, X.; Cusatis, G. Adaptive multiscale homogenization of the lattice discrete particle model for the analysis of damage and fracture in concrete. *Int. J. Solids Struct.* **2017**, *125*, 50–67. [CrossRef]
37. Potyondy, D.O.; Cundall, P.A. A bonded-particle model for rock. *Int. J. Rock Mech. Min. Sci.* **2004**, *41*, 1329–1364. [CrossRef]
38. Jebahi, M.; Andre, D.; Terreros, I.; Iordanoff, I. *Discrete Element Method to Model 3D Continuous Materials*; Wiley: London, UK, 2015.
39. Psakhie, S.G.; Shilko, E.V.; Grigoriev, A.S.; Astafurov, S.V.; Dimaki, A.V.; Smolin, A.Y. A mathematical model of particle–particle interaction for discrete element based modeling of deformation and fracture of heterogeneous elastic–plastic materials. *Eng. Fract. Mech.* **2014**, *130*, 96–115. [CrossRef]
40. Psakhie, S.G.; Shilko, E.V.; Schmauder, S.; Popov, V.L.; Astafurov, S.V.; Smolin, A.Y. Overcoming the limitations of distinct element method for multiscale modeling of materials with multimodal internal structure. *Comput. Mater. Sci.* **2015**, *102*, 267–285. [CrossRef]
41. Dmitriev, A.I.; Österle, W.; Wetzel, B.; Zhang, G. Mesoscale modeling of the mechanical and tribological behavior of a polymer matrix composite based on epoxy and 6 vol.% silica nanoparticles. *Comput. Mater. Sci.* **2015**, *110*, 204–214. [CrossRef]
42. Andreev, K.; Tadaion, V.; Zhu, Q.; Wang, W.; Yin, Y.; Tonnesen, T. Thermal and mechanical cyclic tests and fracture mechanics parameters as indicators of thermal shock resistance—Case study on silica refractories. *J. Eur. Cer. Soc.* **2019**, *39*, 1650–1659. [CrossRef]
43. Starinshak, D.P.; Owen, J.M.; Johnson, J.N. A new parallel algorithm for constructing Voronoi tessellations from distributed input data. *Comput. Phys. Commun.* **2014**, *185*, 3204–3214. [CrossRef]
44. Psakhie, S.; Shilko, E.; Smolin, A.; Astafurov, S.; Ovcharenko, V. Development of a formalism of movable cellular automaton method for numerical modeling of fracture of heterogeneous elastic-plastic materials. *Frattura Integr. Strutt.* **2013**, *24*, 26–59. [CrossRef]
45. Shilko, E.V.; Konovalenko, I.S.; Konovalenko, I.S. Nonlinear mechanical effect of free water on the dynamic compressive strength and fracture of high-strength concrete. *Materials* **2021**, *14*, 4011. [CrossRef]
46. Jing, L.; Stephansson, O. *Fundamentals of Discrete Element Method for Rock Engineering: Theory and Applications*; Elsevier: Amsterdam, The Netherlands, 2007.
47. Drucker, D.C.; Prager, W. Soil mechanics and plastic analysis for limit design. *Q. Appl. Math.* **1952**, *10*, 157–165. [CrossRef]
48. Öztekin, E.; Pul, S.; Hüsem, M. Experimental determination of Drucker-Prager yield criterion parameters for normal and high strength concretes under triaxial compression. *Constr. Build. Mater.* **2016**, *112*, 725–732. [CrossRef]
49. Grigoriev, A.S.; Shilko, E.V.; Skripnyak, V.A.; Psakhie, S.G. Kinetic approach to the development of computational dynamic models for brittle solids. *Int. J. Impact Eng.* **2019**, *123*, 14–25. [CrossRef]
50. Xu, H.; Wen, H.M. Semi-empirical equations for the dynamic strength enhancement of concrete-like material. *Int. J. Impact Eng.* **2013**, *60*, 76–81. [CrossRef]
51. Bischoff, P.H.; Perry, S.H. Compressive behavior of concrete at high strain rates. *Mater. Struct.* **1991**, *24*, 425–450. [CrossRef]
52. Ramesh, K.T.; Hogan, J.D.; Kimberley, J.; Stickle, A. A review of mechanisms and models for dynamic failure, strength, and fragmentation. *Planet. Space Sci.* **2015**, *107*, 10–23. [CrossRef]
53. Kimberley, J.; Ramesh, K.T.; Daphalapurkar, N.P. A scaling law for the dynamic strength of brittle solids. *Acta Mater.* **2013**, *61*, 3509–3521. [CrossRef]
54. Gregorova, E.; Pabst, W. Elastic properties of silica polimorphs—A review. *Ceramics-Silikáty* **2013**, *57*, 167–184.
55. Kovrizhnykh, A.M. Plane stress equations for the von Mises–Schleicher yield criterion. *J. Appl. Mech. Tech. Phys.* **2004**, *45*, 894–901. [CrossRef]
56. Andreev, K.; Harmuth, H.; Gruber, D.; Presslinger, H. Thermo-mechanical behaviour of the refractory lining of a BOF converter—A numerical study. *Taikibutsu* **2004**, *56*, 478–482.
57. Jin, S.; Harmuth, H.; Gruber, D.; Sidi Mammar, A. Advanced measures to characterize mechanical properties of refractories. *China's Refract.* **2018**, *27*, 8–13. [CrossRef]
58. André, D.; Levraut, B.; Tessier-Doyen, N.; Huger, M. A discrete element thermo-mechanical modelling of diffuse damage induced by thermal expansion mismatch of two-phase materials. *Comput. Methods Appl. Mech. Eng.* **2017**, *318*, 898–916. [CrossRef]
59. AZO MATERIALS. Available online: https://www.azom.com/properties.aspx?ArticleID=1114 (accessed on 16 September 2021).
60. Nakayama, J.; Abe, H.; Bradt, R.C. Crack Stability in the Work-of-Fracture Test: Refractory Applications. *J. Am. Ceram. Soc.* **1981**, *64*, 671–675. [CrossRef]
61. Zhang, Q.B.; Zhao, J. A review of dynamic experimental technique and mechanical behavior of rock materials. *Rock Mech. Rock Eng.* **2014**, *47*, 1411–1478. [CrossRef]
62. Petrov, Y.V.; Smirnov, I.V.; Volkov, G.A.; Abramian, A.K.; Bragov, A.M.; Verichev, S.N. Dynamic failure of dry and fully saturated limestone samples based on incubation time concept. *J. Rock Mech. Geotech. Eng.* **2017**, *9*, 125–134. [CrossRef]

63. Cai, X.; Zhou, Z.; Zang, H.; Song, Z. Water saturation effects on dynamic behavior and microstructure damage of sandstone: Phenomenon and mechanisms. *Eng. Geol.* **2020**, *276*, 105760. [CrossRef]
64. Zhao, X.; Li, H.; Wang, C. Quasi-static and dynamic compressive behavior of ultra-high toughness cementitious composites in dry and wet conditions. *Constr. Build. Mater.* **2019**, *227*, 117008. [CrossRef]
65. Radchenko, P.A.; Batuev, S.P.; Radchenko, A.V. Numerical analysis of concrete fracture under shock wave loading. *Phys. Mesomech.* **2021**, *24*, 40–45. [CrossRef]

materials

Article

Synthesis of Niobium-Alumina Composite Aggregates and Their Application in Coarse-Grained Refractory Ceramic-Metal Castables

Tilo Zienert [1,*], Dirk Endler [1], Jana Hubálková [1], Gökhan Günay [2], Anja Weidner [2], Horst Biermann [2], Bastian Kraft [3], Susanne Wagner [3] and Christos Georgios Aneziris [1]

[1] TU Bergakademie Freiberg, Institute of Ceramics, Refractories and Composite Materials, Agricolastr. 17, 09599 Freiberg, Germany; dirk.endler@ikfvw.tu-freiberg.de (D.E.); Jana.Hubalkova@ikfvw.tu-freiberg.de (J.H.); aneziris@ikfvw.tu-freiberg.de (C.G.A.)

[2] TU Bergakademie Freiberg, Institute of Materials Engineering, Gustav-Zeuner-Str. 5, 09599 Freiberg, Germany; goekhan.guenay@iwt.tu-freiberg.de (G.G.); weidner@ww.tu-freiberg.de (A.W.); biermann@ww.tu-freiberg.de (H.B.)

[3] Karlsruhe Institute of Technology, Institute for Applied Materials-Ceramic Materials and Technology, Haid-und-Neu Straße 7, 76131 Karlsruhe, Germany; bastian.kraft@kit.edu (B.K.); susanne.wagner@kit.edu (S.W.)

* Correspondence: tilo.zienert@licoo.de or tilo.zienert@ikfvw.tu-freiberg.de; Tel.: +49-3731-39-1546

Citation: Zienert, T.; Endler, D.; Hubálková, J.; Günay, G.; Weidner, A.; Biermann, H.; Kraft, B.; Wagner, S.; Aneziris, C.G. Synthesis of Niobium-Alumina Composite Aggregates and Their Application in Coarse-Grained Refractory Ceramic-Metal Castables. *Materials* **2021**, *14*, 6453. https://doi.org/10.3390/ma14216453

Academic Editors: Jacek Szczerba and Ilona Jastrzebska

Received: 1 September 2021
Accepted: 13 October 2021
Published: 27 October 2021

Publisher's Note: MDPI stays neutral with regard to jurisdictional claims in published maps and institutional affiliations.

Abstract: Niobium-alumina aggregate fractions with particle sizes up to 3150 µm were produced by crushing pre-synthesised fine-grained composites. Phase separation with niobium enrichment in the aggregate class 45–500 µm was revealed by XRD/Rietveld analysis. To reduce the amount of carbon-based impurities, no organic additives were used for the castable mixtures, which resulted in water demands of approximately 27 vol.% for the fine- and coarse-grained castables. As a consequence, open porosities of 18% and 30% were determined for the fine- and coarse-grained composites, respectively. Due to increased porosity, the modulus of rupture at room temperature decreased from 52 MPa for the fine-grained composite to 11 MPa for the coarse-grained one. However, even the compressive yield strength decreased from 49 MPa to 18 MPa at 1300 °C for the fine-grained to the coarse-grained composite, the latter showed still plasticity with a strain up to 5%. The electrical conductivity of fine-grained composite samples was in the range between 40 and 60 S/cm, which is fifteen magnitudes above the values of pure corundum.

Keywords: refractory composite; aggregate synthesis; castable

1. Introduction

Refractory ceramics are typically used in metallurgy applications. They are characterised by good mechanical properties, good thermal shock behaviour as well as producibility of large, coarse-grained components by water-based casting processes, such as pressure slip casting [1] or castable technology [2].

All physical and mechanical properties of refractory castables are depending on the firing temperature. The higher the firing temperature of the castables the more pronounced is the ceramic bonding of the fine matrix. Hence, the high degree of sintering results in lowering porosity and enhancing both modulus of rupture as well as Young´s modulus [3]. The open porosity of conventional alumina castables bonded with high alumina cement fired at 1000 °C varies from 20% to 30% depending on the type of aggregates and cement used [4]

After firing at 1600 °C, the low cement castables (LCC) and cement-free castables (NCC) based on alumina have an open porosity in the range from 15 to 20% [5]. The Young's moduli of alumina at ambient temperature of such castables range between 60 GPa and 100 GPa depending on the chemical and granulometric composition [6]. For the industrial application of refractory castables, a detailed evaluation of predominant stress conditions is essential. If a high erosion resistance is required, the microstructure of the castables should be highly sintered with a low degree of porosity. Nevertheless, the high degree of sintering involves a pure brittle behaviour accompanied by a high susceptibility to cracking and subsequent spalling. Otherwise, if high thermal shock resistance is of prime importance, a low Young´s modulus and relative high porosity are beneficial, provided that the mechanical stability, i.e., sufficient modulus of rupture, is ensured. The shrinkage of coarse-grained castable during the sintering should not exceed a value of 2% to prevent crack formation.

Ceramic-matrix and metal-matrix composites have been investigated since several decades [7]. In particular, the combination of ceramics with refractory metals is of interest for high-temperature and wear applications, e.g., corundum-molybdenum [8–11]; mullite-molybdenum [12–14]; zirconia-(tantalum, niobium) [15,16]. In that works, fine-grained powders were used to produce dense sintered composites focusing on improving the mechanical properties of the ceramic part by the combination of the ceramic's high strength and the metal's toughness and grain fining ability [17]. High shrinkage on sintering, limited thermal shock ability and/or the sintering technology as hot isostatic pressing are restricting the maximum size of fine-grained components in fabrication.

The concept of coarse-grained refractory composites based on tantalum-alumina and niobium-alumina was recently introduced [18] combining the fields of powder metallurgy and castable technology. Shrinkage values of $\approx$1% and good mechanical properties at ambient temperature as well as in the temperature range of 1300–1500 °C [19] were obtained. Low shrinkage on sintering causes also low values of residual stresses in the refractory ceramic, which enables one to fabricate large refractory components. Refractory composite castables based on coarse-grained tabular alumina and fine-grained metal of 11 vol.% and 21 vol.% substituing the alumina matrix showed the best results regarding physical and mechanical properties. It was suggested to produce pre-synthesised composite aggregates for further composite castables.

The approach of refractory ceramic-metal castables aims to leverage the synergy effects of both refractory ceramics and refractory metals and hence to develop smart key components for metallurgical industry with targeted functionalisation, such as integrated electrodes for non-wetting behaviour [20], ladle sliding gate plates with enhanced thermal shock properties, special casting nozzles with anti-clogging properties or integrated heating elements.

The wetting behaviour of a liquid metal/ceramic system depends on the bonding characteristics of both system components as well as on the magnitude of interactive forces at the interface. It is characterised by the contact angle of the liquid drop formed on the solid substrate indicating the chemical affinity between the two materials and by the work of adhesion as a measure of interface adhesion strength [21]. Speaking about industrial applications of refractory ceramics, it is necessary to know the interfacial properties (surface/interfacial tension and surface segregation) of melts and characteristics of solids (surface roughness, impurities, porosity) including the operating conditions such as temperature, time, gas composition and pressure [22,23].

The aim of the present work is to investigate the feasibility to produce refractory niobium-alumina composite material in a novel two-step castable procedure. The first step comprises the synthesis of targeted niobium-alumina composite aggregrates while the second step involves the production of coarse-grained refractory castables utilising the pre-synthesised composite aggregates. Such a fractal design is needed in order to ensure a coherent composition at different aggregate size scales and hence to achieve an electrical conductivity and proper mechanical properties (thermal shock resistance, sufficient high-temperature strength). For this purpose, the metallic phase should constitute 65 vol.% of the composite material.

2. Methods

2.1. Sample Preparation

Initially, the preliminary fine-grained material had to be synthesised and subsequently crushed and sieved in fractions to obtain refractory metal-ceramic aggregates of different sizes. The composite niobium-alumina aggregates were then used to produce coarse-grained refractory composite castables.

The base material for the production of aggregates with an intended niobium amount of 65 vol.% was synthesised from raw materials (particle sizes and chemical information are listed in Table 1) as follows. Powders of niobium (EWG Wagner, Weissach, Germany), alumina (Martoxid, Martinswerke, Germany), a hydratable alumina binder (Alphabond 300, Almatis, Ludwigshafen, Germany) and dispersing alumina (ADS-1, Almatis, Ludwigshafen, Germany) were first dry mixed for one minute using an Eirich mixer (Gustav-Eirich Maschinenfabrik, Hardheim, Germany) followed by wet mixing for four minutes with step-wise water addition until almost self-flowability was achieved. The composition of the fine-grained composite material is given in Table 2. This mixture was then filled in steel molds with dimensions of 25 mm × 25 mm × 150 mm by vibrational assisted casting. After setting at room temperature for 48 h, the samples were demolded and dried for 24 h at 130 °C in air atmosphere. Afterwards, the prisms were sintered at a temperature of 1600 °C for 4 h in a corundum tube furnace under flowing argon atmosphere, which was purified by a titanium getter. This material is hereinafter referred to as *fine-grained composite*.

Table 1. Determined particle sizes of the used raw materials and their chemistry.

Material	Particle Size in µm				Purity	Main Impurities
	d_{10}	d_{50}	d_{90}	d_{max}		
niobium				75	99.95 wt.% Nb	O, C, Ta
Martoxid	0.13	0.63	3.57		99.8 wt.% Al_2O_3	Na, Si, Ca, Fe
CT9FG	1.96	5.53	20.63		99.5 wt.% Al_2O_3	Na, Fe, Si
CL370	0.17	0.54	5.77		99.7 wt.% Al_2O_3	Na, Si, Fe, Ca
Alphabond300		4-8		30	88 wt.% Al_2O_3 (min)	Ca, Na, Si

Table 2. Castable recipe for the fine-grained composite.

Material	mass in g	Amount	
		in mass%	in vol.%
niobium	1644.9	79.95	64.31
Martoxid (alumina)	395.1	19.21	33.60
Alphabond300 (binder)	12.2	0.59	1.47
ADS-1 (dispersant)	5.2	0.25	0.62
water	112.9	5.20	27.0

In a further step, the synthesised fine-grained prisms were broken and crushed using a jaw crusher (BB50, Retsch, Haan, Germany) with jaws made from hard metal (92% WC–8% Co). The crushed material was sieved into the four aggregate classes 0–45 µm, 45–500 µm, 500–1000 µm and 1000–3150 µm.

The particle size distribution (Bettersizer S3 plus, 3P Instruments GmbH & Co. KG, Odelzhausen, Germany), density and phase assemblage using X-ray powder diffraction (XRD) (Empyrean, Malvern Panalytical GmbH, Kassel, Germany) in combination with Rietveld refinement using HighScore Plus 4.8 [24] of the synthesised composite aggregates were determined. XRD measurements were done using Cu-K_{α_1} radiation between 15–140° 2Θ with 0.0143° step size and an exposion time of 160 s/step.

Densities were measured using mercury intrusion porosimetry (AutoPore V 9600, Micromeritics Instrument Corp., Norcross, GA, USA) on particles according to DIN ISO

15901-1:2019-03 and gas pycnometry with helium (AccuPyc 1340TEC, Micromeritics Instrument Corp., Norcross, GA, USA) on fine-powdered material according to DIN 66137-2:2019-03 of each aggregate class. In addition, true densities of each aggregate class were calculated based on the results of the XRD/Rietveld analysis. Particle morphology of the synthesised aggregates was investigated based on laser-scanning microscopy (VK-X1000, Keyence Deutschland GmbH, Neu-Isenburg, Germany).

2.2. Design of Coarse-Grained Castables

Particle size distributions of castables can be mathematically described using particle packing models expressing the cumulative sum curve or *cumulative percent finer than particle size d* CPFT(d). A commonly used model is the Dinger–Funk model [25], which allows to take the minimum and maximum particle sizes (d_{min} and d_{max}) into account. Recently, this model was modified by Fruhstorfer [26] to

$$CPFT_{\text{mod-DF}}(d) = 100\% \cdot \frac{d^{n(d)} - d_{min}^{n(d)}}{d_{max}^{n(d)} - d_{min}^{n(d)}} \tag{1}$$

using the particle-size dependent distribution modulus

$$n(d) = n_{min} + d \cdot \frac{n_{max} - n_{min}}{d_{max}}, \tag{2}$$

where n_{min} and n_{max} are the minimum and the maximum distribution modulus, respectively. The flowability of a castable can be related to its particle distribution and hence, n_{min} and n_{max} can be used as parameters to describe a chosen material mixture. In case of tabular alumina with a maximum particle size of 3150 µm, best properties in terms of flowability, density and pore sizes of the castable were found for $n_{min} = 0.28$ and $n_{max} = 0.8$ [27].

The synthesised niobium-alumina aggregates were wet mixed with the hydratable alumina binder Alphabond 300 and the reactive alumina CL370 (Almatis, Ludwigshafen, Germany) according to the recipe as discussed in Section 3.3 using a concrete laboratory mixer (ToniMAX, Toni Baustoffprüfsysteme GmbH, Berlin, Germany). The coarse-grained mixture was then casted into the prismatic molds, set, dried and sintered following the same procedure as the preliminary, fine-grained material. The sintered castable samples will be referred to as *coarse-grained composite* hereinafter.

2.3. General Sample Characterisation

The fine-grained as well as the coarse-grained composites (four prisms each) were characterised in terms of shrinkage, envelope density and open porosity according to DIN EN 993-1:2019-03; elastic constants according to DIN EN 843-2:2006 by the ultrasonic procedure (UKS-D device, GEOTRON-ELEKTRONIK, Pirna, Germany).

The cold modulus of rupture (MOR) of the fine-grained composite (two prisms) and the coarse-grained composites (three prisms) was determined using an universal testing machine TIRAtest 28100 (TIRA GmbH, Schalkau, Germany). In order to achieve a uniform distribution of bending moment, four point bending setup with a support span of 125 mm, a load span of 62.5 mm, a pre-load of 20 N and a loading-rate of 150 N/s was applied.

On cut slices, the interior microstructure and phase assemblage of the prisms were studied by scanning electron microscopy (SEM) using back-scattered electron (BSE) and secondary electron (SE) contrast together with energy dispersive X-ray spectroscopy (EDS) (Philips XL30 ESEM FEG, Amsterdam, the Netherlands). Some cut slices were also crushed and powdered using an agate mortar and analysed by XRD/Rietveld refinement.

2.3.1. CT Measurements

The macrostructure of the composite material was analysed using a microfocus X-ray tomograph CT-ALPHA (ProCon X-ray GmbH, Sarstedt, Germany) equipped with a transmission X-ray tube FXE-160.20/25 (Feinfocus, Garbsen, Germany) and a flat panel X-ray detector Dexela 1512 (Perkin Elmer, Rodgau, Germany).

The µ-CT was operated at 150 kV and 120 µA using a 0.6 mm copper filter. The exposure time was set to 2 s. The volume data were reconstructed by means of the software Volex 6.0 (Fraunhofer EZRT, Fürth, Germany) with a voxel size of 9.8 µm. The reconstructed volume data were visualised using the software VG Studio MAX 2.2 (Volume Graphics, Heidelberg, Germany) and quantified with the software MAVI 1.5.3 (Fraunhofer ITWM, Kaiserslautern, Germany). The image processing comprised a cropping step in order to cut out a defined volume of the reconstructed data ($350 \times 350 \times 1050$ voxels) followed by a segmentation step in order to transform the grey scale into binary image. The binarisation was performed using Otsu´s method [28] being based on a global thresholding strategy. Subsequently, the field features (volume density and total porosity) were determined.

2.3.2. High-Temperature Compressive Strength Measurements

Quasi-static compression tests were performed on both, fine-grained and coarse-grained composites (one specimen each). The cylindrical specimens with 12 mm in diameter and 20 mm in height were drilled from sintered prisms. The compression tests were conducted at an initial strain rate of 7.5×10^{-4} 1/s at 1300 °C up to a possible maximum strain of 30% using an electro-mechanical, high-temperature testing machine (Z020, Zwick Roell, Haan, Germany) with a protective argon gas chamber (Maytec, Olching, Germany) integrated into the testing machine. To prevent the oxidation of the specimens, the test chamber was evacuated to 0.8 mbar vacuum and then filled with argon twice. Thus, the tests were performed under argon atmosphere and ambient pressure. The presence of oxygen was controlled by an oxygen sensor (Stange Elektronik, Gummersbach, Germany).

Before the tests, the lower piston was moved upwards with a speed of 0.1 mm/min to apply a pre-load of 5 N to the specimens. The heating of the specimens was carried out inductively via a medium-frequency induction generator HF 5010 (TRUMPF Hüttinger, Ditzingen, Germany) with a heating rate of 30 K/s and a water-cooled copper coil. The temperature was detected by a pyrometer Metis MS09 (Sensortherm GmbH, Steinbach/Ts, Germany) with a wavelength of 0.9 µm and an emission coefficient of 0.93. In case of the coarse-grained composite, a metallic susceptor made of Mo-alloy TZM was used, since the material could not be heated by induction. The specimens were kept 20 min under the pre-load after reaching the test temperature to achieve a homogeneous temperature field over the entire height of the specimens. More details of the test procedure were given elsewhere [19].

2.3.3. Electrical Conductivity Measurements

Two samples for electrical conductivity measurements were prepared as follows. First a powder mixture of 65 vol.% niobium and 35 vol.% alumina (calcinated alumina CT9FG, Almatis, Ludwigshafen, Germany) was dry mixed and homogenised, pressed to cylinders and finally sintered under argon atmosphere at a temperature of 1600 °C for 4 h.

To remove the formed sinter skins and to create a plane surface, the circular faces of the samples were grinded using SiC paper (P400 and P600). Then the surfaces were sputter-coated with gold, using a Quorum Q150T ES (Quantum Design GmbH, Darmstadt, Germany). Due to surface roughness, the circular faces were coated with a layer of silver paste to ensure that the whole surface area of the sample was connected by one continuous electrode.

The investigations were performed using a four-point-measurement setup, which is schematically shown in Figure 1. During the tests, currents of 1 mA, 10 mA and 100 mA were applied to the samples working with a Keithley 220 Programmable Current Source (Keithley Instruments, Cleveland, OH, USA) and the resulting voltages were measured using a Keithley 2000 Multimeter (Keithley Instruments, Cleveland, OH, USA). The resistance R of the samples was calculated with

$$R = \frac{U}{I},$$ (3)

where I is the applied current and U is the resulting voltage [29]. The specific resistance ρ was calculated according to

$$\rho = \frac{R \cdot A}{l}$$ (4)

where A is the surface area of the prepared electrode and l the distance between the prepared sample surfaces.

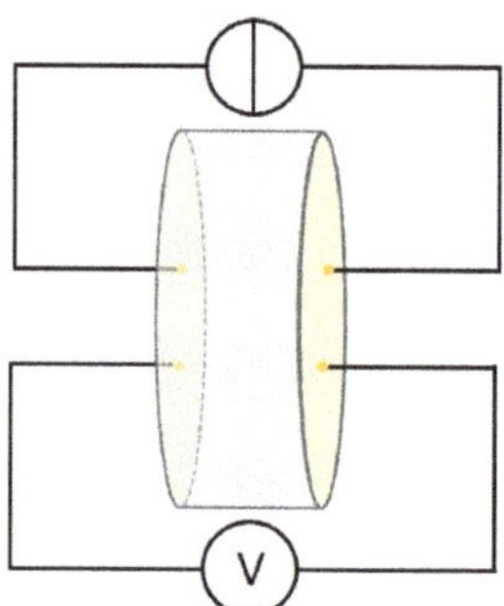

Figure 1. Schematic illustration of electrical conductivity measurement setup.

3. Results and Discussion

3.1. Fine-Grained Composite

SEM micrographs of the sintered fine-grained composite can be seen in Figure 2. The niobium particles are homogeneously distributed within a fine-grained matrix of alumina (Figure 2a). Pores with a diameter up to $\approx 100\,\mu m$ were only randomly distributed within the sample volume and, therefore, less visible in the presented micrographs. The impurity phase NbO identified by EDS measurements was frequently present within niobium particles as it is shown in Figure 2b. Ternary oxide formation, e.g., of $AlNbO_4$ as reported in [19], at the interface alumina/niobium oxide was not detected by SEM/EDS. Furthermore, the niobium material showed only a less pronounced tendency for surface diffusion along the alumina grains in comparison to the previous case of a mixture with a tabular alumina material [18].

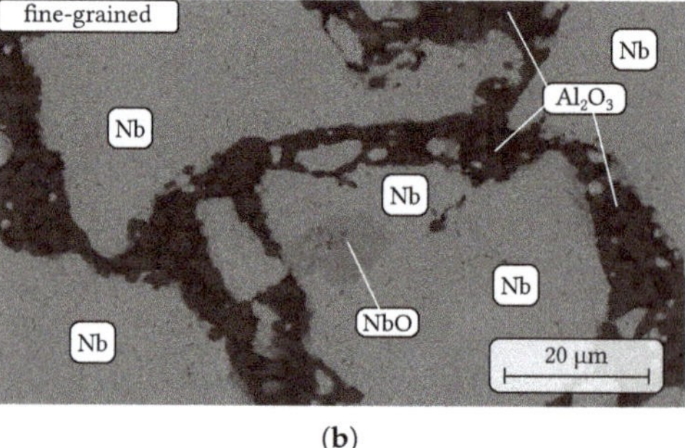

Figure 2. SEM micrographs of the fine-grained composite (**a**) SE contrast, overview and in detail (**b**) BSE contrast, NbO impurity within Nb grains surrounded by a finer grained alumina matrix.

3.2. Synthesised Composite Aggregates

3.2.1. Morphology and Particle Size Distributions

Photographs of the aggregate fractions 500–1000 µm and 1000–3150 µm are presented in Figure 3. The crushing process produced mainly plate-like shaped particles. The particle height is much smaller than the length and width. Quantitatively spoken, the ratio of height to the maximal length is approximately 0.5–0.7. Within this overall morphology, two types can be distinguished. First, roughly two third of the particles approximate an imaginary plate where a relatively flat plateau in height-direction was formed. For the other one third of particles, a more or less pronounced maximum peak can be observed. These two shape categories are exemplary illustrated in Figure 4 determined on particles from the 1000–3150 µm aggregate fraction.

The characteristic particle sizes d_{10}, d_{50} and d_{90} of the synthesised composite aggregates measureable with laser granulometry are listed in Table 3. A continuous uniform particle size distribution was assumed for the aggregate class 1000–3150 µm.

Figure 3. Photographs of the synthesised aggregate fractions (**a**) 500–1000 µm and (**b**) 1000–3150 µm.

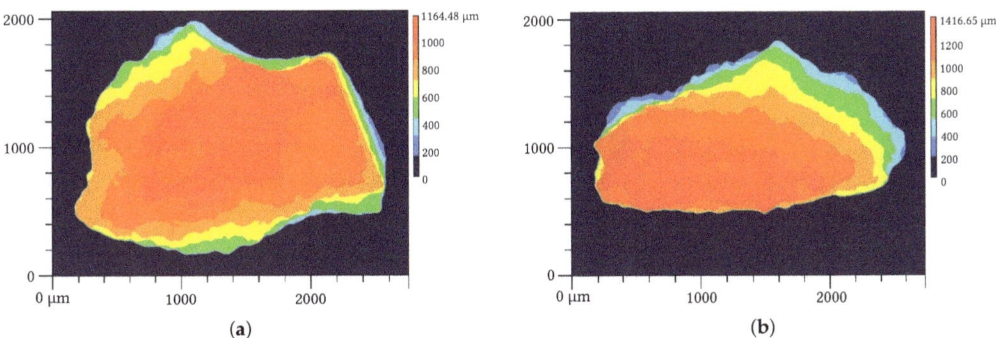

Figure 4. Height maps of particles from the aggregate fraction 1000–3150 µm determined with laser-scanning microscopy showing two types of plate-like particle morphologies: (**a**) relatively flat surface plane and (**b**) a more or less pronounced maximum in the vertical direction.

Table 3. Characteristic particle sizes of three synthesised composite aggregate classes measureable by laser granulometry.

Aggregate	Particle Size in μm		
Class	d_{10}	d_{50}	d_{90}
0–45 μm	0.2	11.7	50.4
45–500 μm	1.5	83.6	324.6
500–1000 μm	199.7	461.2	701.4

3.2.2. Density and Porosity

The envelope density of the fine-grained composite was determined with the Archimedes principle to 5.66 g/cm³ including 18% open porosity. A skeleton density of 6.73 g/cm³ was determined with mercury intrusion porosimetry (MIP) resulting in an envelope (bulk) density of 5.78 g/cm³ and hence an (open) porosity of 16.5% as presented in Table 4. The total porosity of the fine-grained composite was estimated based on the CT image as shown in Figure 5 to 32.60%, which gives an estimated closed porosity in the range of 10–15%. However, the estimated porosity must be interpreted with caution. There are several constraints affecting the quantitative CT analysis. Due to the limited resolution, e.g., reconstructed voxel size, pores smaller than approximately 20 μm were unevitably neglected. Furthermore, uneven thickness of the scanned sample resulted in an irregular grey value distribution along the width of the sample, as it can be seen in Figure 5. Therefore, the image processing, particulary the binarisation step, is a subject of errors.

Table 4. Determined values of envelope density ρ_b and open porosity π_a according to DIN EN 993-1:2019-03, the skeleton density ρ_S and porosity ε according to DIN ISO 15901-1:2019-03, and the true density ρ_F according to DIN 66137-2:2019-03 of the synthesised aggregates as well as the sintered fine-grained and coarse-grained composites. If applicable, values are given as mean value $\pm$ standard deviation (number of samples $n = 4$).

	Archimedes		MIP			He	XRD
	ρ_b in g/cm³	π_a in %	ρ_S in g/cm³	ε in %	ρ_b in g/cm³	ρ_F in g/cm³	ρ in g/cm³
fine-grained composite	5.66 ± 0.05	18.0 ± 0.7	6.73	16.48	5.78		
aggregate fraction							
0–45 μm							5.899
45–500 μm						7.068	6.886
500–1000 μm						6.917	6.641
1000–3150 μm			6.88	15.83	5.79	6.918	6.661
coarse-grained composite	4.57 ± 0.04	30.4 ± 0.3	6.49	25.55	5.17		

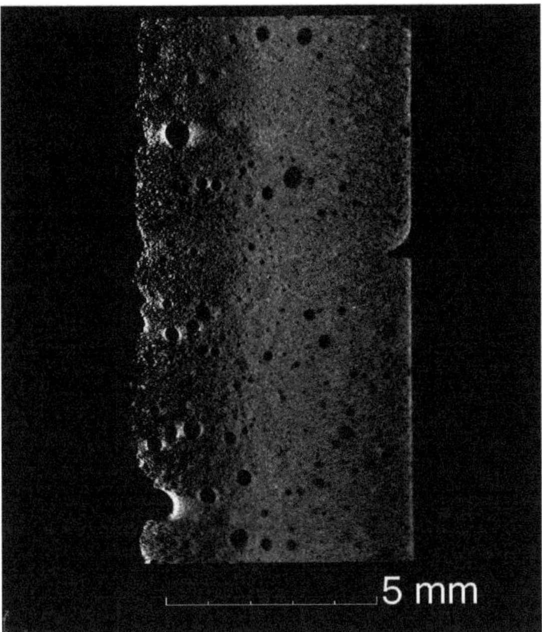

Figure 5. CT image of the fine-grained composite showing inhomogeneously distributed pores with a diameter larger than 100 μm.

3.2.3. XRD Analysis

Detailed results of the Rietveld refinement of the aggregate fractions are listed in Table 5. The crushing procedure of the sintered fine-grained composite induced phase separation resulting in alumina enrichment in the finest aggregate fraction 0–45 μm and niobium enrichment in the aggregate fraction of 45–500 μm. For the aggregate fractions larger than 500 μm, the volume ratio of niobium to corundum is around 55:40, whereas for the aggregate fractions 0–45 μm and 45–500 μm this ratio is 40:57 and 60:35, respectively. Phase separation due to crushing is also confirmed by the results of the density measurements using helium pycnometry as given in Table 4. The largest density was determined for the 45–500 μm aggregate fraction, which is in good agreement with the calculated true densities based on the XRD results. However, the true densities determined by helium pycnometry are 2.5–4% larger than the calculated ones, which reflects the error of the quantitative phase analysis.

In all fractions, the impurity phases β-Nb$_2$C [30] (adopted from #ICSD 33575) and NbO [31] (#ICSD 27574) were detected each with an amount between $\approx$1–2 vol.%. It was found that the volume content V_{impurity} of NbO and β-Nb$_2$C in each aggregate fraction is linearly depending on the corresponding niobium content V_{Nb}, which can be described by

$$V_{\text{impurity}} = f_{\text{impurity}} \cdot V_{\text{Nb}} \, . \tag{5}$$

For NbO and β-Nb$_2$C, the factors f_{NbO} and $f_{\beta\text{-Nb}_2\text{C}}$ were obtained by least-square fitting to 0.0368 and 0.0369, respectively, resulting in $f_{\text{NbO}+\beta\text{-Nb}_2\text{C}}$ of 0.0737. Such linear dependency means that the grains of NbO and β-Nb$_2$C are always connected to the niobium crystals. Crushing did not result in a separation of the formed Nb-based impurities from the niobium raw material.

As an example, the result of the XRD measurement and the Rietveld refinement of the aggregate fraction 1000–3150 μm can be seen in Figure 6a. The niobium-related reflections were described using two niobium phases each with a different UVW profile and lattice parameter. For the 1000–3150 μm aggregate fraction, lattice parameter of $a = 3.31377$ Å and $a = 3.30858$ Å were obtained for the phases marked as 'Niobium-1' and 'Niobium-2' in Figure 6, respectively. Using two niobium phases during Rietveld refinement led to a large drop in the resulting weighted profile R-values R_{wp} in comparison to using only one phase. For a detailed view, the refinement result of the niobium (132) reflection is shown in Figure 6b. The calculated profiles of the two niobium phases are different, which can be related to a difference in crystallite size. Our material has at least two different morphologies of metal grains. One are the particles of the raw material, which are connected to NbO particles as shown in Figure 2b. The second type of metal grains are the ones that originate from diffusion along the surface of the corundum particles (compare also with the reported microstructures in [18]). The difference in the lattice parameter of the two niobium phases can be related to a different chemistry. It can be assumed that the niobium diffused along the corundum grains has a larger amount of dissolved impurity elements (such as O, K, Na) in comparison to the relatively large particles of the niobium raw material. However, it should be mentioned that no differences in chemistry were observed by SEM/EDS measurements. Most likely such differences are below the detection limit of the EDS method. The spread of lattice parameters could be also explained by the presence of residual stress in niobium caused by the crushing process.

Figure 7 shows the determined lattice parameters of the phases Niobium-1 and Niobium-2 for all synthesised aggregate fractions. The dependence of the lattice parameter of Niobium-1 on the particle size shows a pronounced maximum for the aggregate fraction 45–500 µm. Contrarily, the lattice parameter of Niobium-2 reveal an opposite tendency with a minimum for the aggregate fraction 45–500 µm. It can be assumed that the chemically inhomogeneous niobium entails local differences in hardness and toughness leading to the observed phase separation during crushing.

Table 5. Results of the Rietveld refinement of the four aggregate fractions 0–45 µm, 45–500 µm, 500–1000 µm and 1000–3150 µm. The overall fit quality of the Rietveld refinement can be judged with the presented weighted profile R-values R_{wp}.

Aggregate Fraction	Phase (Amount in mass/vol.%) Lattice Parameter in Å			Remarks
0–45 µm	**Nb-1 (28.8/20.0)** $a = 3.31178$	**α-Al$_2$O$_3$ (38.8/57.3)** $a = 4.75892$ $c = 12.99214$	**β-Nb$_2$C (2.1/1.5)** $a = 5.36096$ $c = 4.96013$	$R_{wp} = 7.22$ 39.9 vol.% Nb
	Nb-2 (28.8/19.9) $a = 3.31045$		**NbO (1.6/1.3)** $a = 4.21061$	
45–500 µm	**Nb-1 (33.1/26.9)** $a = 3.31575$	**α-Al$_2$O$_3$ (20.3/35.0)** $a = 4.75862$ $c = 12.9912$	**β-Nb$_2$C (2.6/2.2)** $a = 5.36064$ $c = 4.9614$	$R_{wp} = 7.84$ 60.3 vol.% Nb
	Nb-2 (41.4/33.4) $a = 3.30833$		**NbO (2.6/2.4)** $a = 4.2103$	
500–1000 µm	**Nb-1 (28.3/22.1)** $a = 3.31557$	**α-Al$_2$O$_3$ (24.7/40.9)** $a = 4.75885$ $c = 12.99186$	**β-Nb$_2$C (2.5/2.1)** $a = 5.36271$ $c = 4.96004$	$R_{wp} = 7.00$ 55.0 vol.% Nb
	Nb-2 (42.3/32.9) $a = 3.30913$		**NbO (2.2/2.0)** $a = 4.21104$	
1000–3150 µm	**Nb-1 (37.6/29.5)** $a = 3.31377$	**α-Al$_2$O$_3$ (24.1/40.2)** $a = 4.75866$ $c = 12.99121$	**β-Nb$_2$C (2.4/2.0)** $a = 5.36314$ $c = 4.95887$	$R_{wp} = 7.49$ 55.8 vol.% Nb
	Nb-2 (33.7/26.3) $a = 3.30858$		**NbO (2.2/2.0)** $a = 4.21097$	

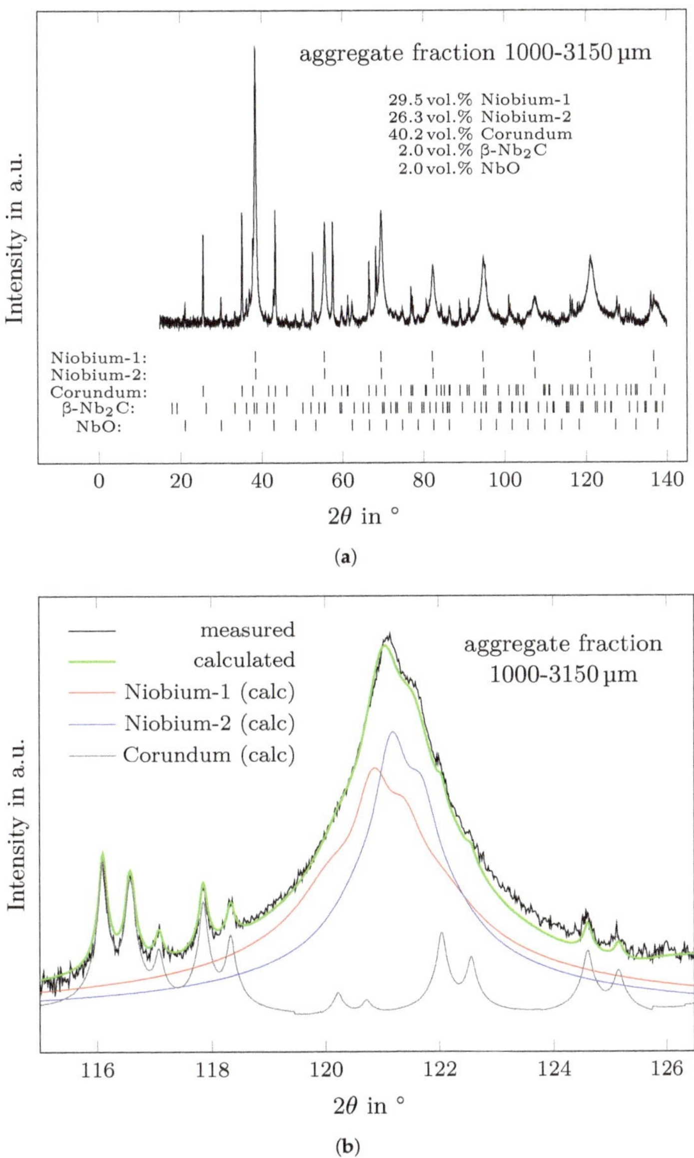

Figure 6. Results of XRD measurement and Rietveld refinement of the 1000–3150 µm aggregate fraction. (**a**) Determined diffraction pattern and calculated positions of each phase reflections between 15 and 140° 2θ and (**b**) enlarged view showing the calculated intensities of the niobium (132) reflection.

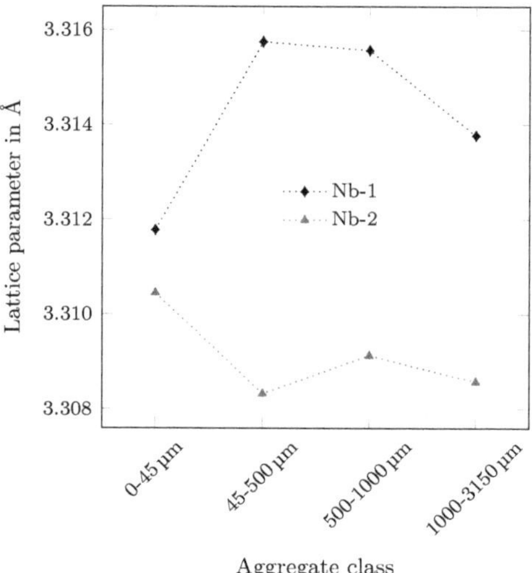

Figure 7. Lattice parameters of the two niobium phases used for Rietveld refinement obtained for all synthesised aggregate fractions.

3.3. Coarse-Grained Castable Recipe

Using $n_{min} = 0.28$ and $n_{max} = 0.8$ in Equation (1) as mentioned in Section 2.2, a volume ratio of the aggregate classes 0–45 µm:45–500 µm:500–1000 µm:1000–3150 µm of 27.6:23.2:9.05:40.1, which is 0.69:0.58:0.23:1 normalised to the largest aggregate class, should be utilised in the castable mixture.

The amount of raw material was a strongly limiting factor for obtaining the intended volume ratios of the synthesised aggregate fractions. Nevertheless, the actual yields of the single aggregate fractions during the crushing procedure did not satisfy the requirements on the adequate volume amounts. Therefore, it was necessary to adjust the volume ratio of the aggregate classes. Finally, a mass ratio of 5.7:33.6:28.2:32.5 of the aggregate classes 0–45 µm:45–500 µm:500–1000 µm:1000–3150 µm was approved for producing the coarse-grained castable. Using the above discussed densities of the synthesised aggregate fractions, a normalised volume ratio of 0.18:1.03:0.87:1 of the aggregate fractions was realised.

For the castable mixture, approximately 15 vol.% of alumina-based fines and binder were additionally added to the synthesised aggregates. The castable component's particle size distributions are shown in Figure 8 according to the derived recipe (see Table 6). Finally, the normalised volume ratio of the castable mixture was 0.95:1.23:0.43:1, resulting in an increased amount of particles of the aggregate classes 45–500 µm and 500–1000 µm in comparison to the optimal values shown above. The particle distribution of the coarse-grained composite is characterised by Equation (1) using $d_{min} = 0.01$ µm, $d_{max} = 3150$ µm with $n_{min} = 0.2716(30)$ and $n_{max} = 0.295(24)$.

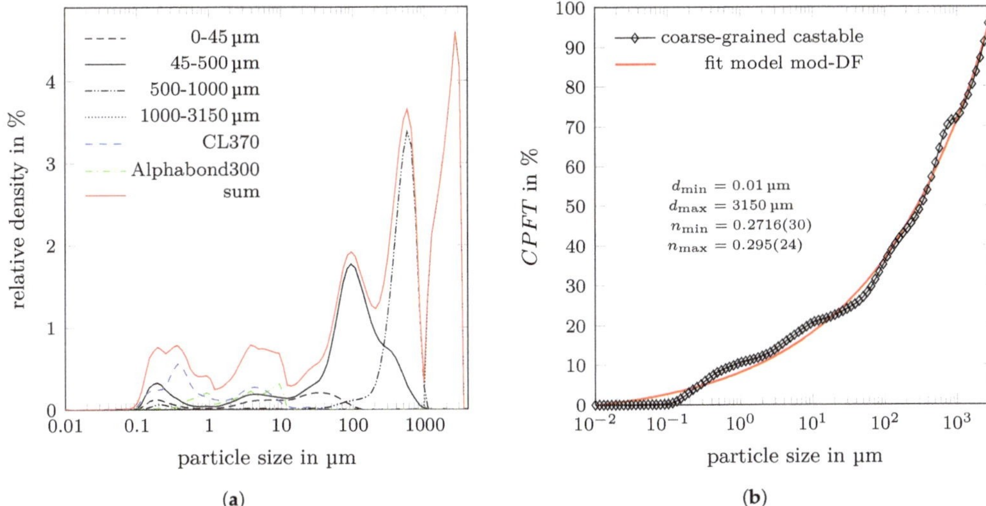

Figure 8. (a) Determined particle size distributions of the used castable components as they contribute to the castable's particle size distribution (calculated sum distribution); (b) CPFT of the particles of the coarse-grained castable together with the results of the fit using the modified Dinger-Funk distribution model (see Equation (1)).

Table 6. Coarse-grained castable recipe.

Material	Density	mass	Amount	
	in g/cm³	in g	in mass-%	in vol.%
0–45 µm	5.79	89.98	5.18	4.92
45–500 µm	5.79	533.02	30.69	29.12
500–1000 µm	5.79	447.5	25.77	24.45
1000–3150 µm	5.79	514.97	29.65	28.14
CL370 (alumina)	4.013	110.0	6.33	8.67
Alphabond300 (binder)	2.781	41.38	2.38	4.71
water	1	120	6.46	27.52

3.4. Coarse-Grained Composite

SEM micrographs of the sintered coarse-grained composite are shown in Figures 9 and 10. In comparison to the sintered fine-grained composites (Figure 2), the niobium particles are inhomogenuously distributed as the niobium-rich pre-sintered coarse-grained particles are now embedded within an alumina-rich matrix based on CL370 and the de-hydrated Alphabond300 binder as it can be seen in Figure 9. In addition, large pores with a few hundred micrometers in diameter were observed regularly within the microstructure (Figure 9a). In Figure 9b, the indicated particle boundaries of the pre-synthesised aggregates are drawn for more clarity. However, the phase assemblage of the niobium particles did not change during the second sintering process in comparison to the sintered fine-grained composite. Still, niobium oxide is present within some of the niobium particles without any SEM/EDS-detectable chemical reaction with alumina (see Figure 10).

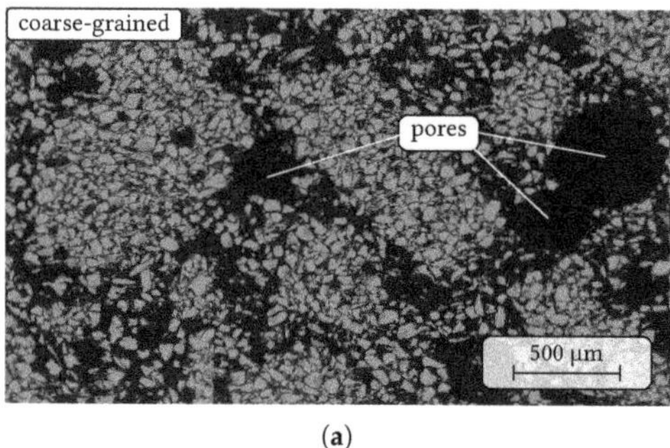

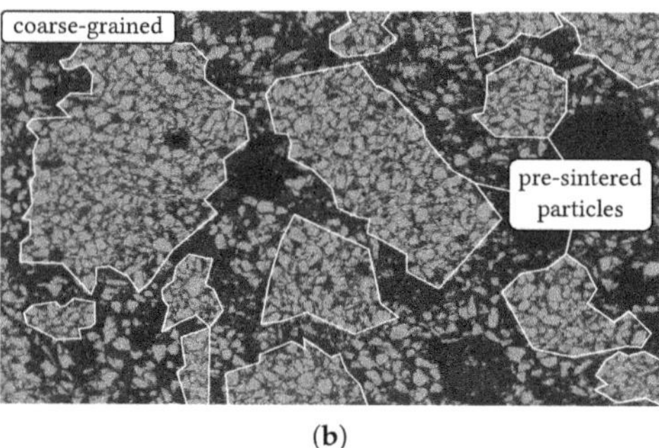

(**a**)   (**b**)

Figure 9. SEM micrograph (BSE contrast) of the coarse-grained composite with niobium (light grey) and alumina (dark grey). (**a**) overview with pores up to a few hundreds micrometers size, (**b**) the same micrograph with indicated particle boundaries of the pre-sintered aggregates.

Figure 10. SEM micrograph in BSE contrast showing similar phase assemblage as the fine-grained composite with niobium particles containing niobium oxide impurity within a fine-grained alumina matrix.

The envelope density of the coarse-grained composites was determined to $4.57\,g/cm^3$ with 30.4% open porosity, whereas MIP measurements on fragments of the coarse-grained prisms give an envelope (bulk) density of $5.17\,g/cm^3$ including 25.6% (open) porosity. These differences can indicate surface effects. However, both measurements point up that the coarse-grained composite exhibit a lower density than the fine-grained one (see Table 4).

3.5. Comparison of Mechanical Properties and Elastic Constants of the Fine- and Coarse-Grained Composites

The determined shrinkage of the sintered fine-grained composite is $4.27 \pm 0.42\%$ whereas the coarse-grained composite evinces a significant reduced shrinkage values of $1.61 \pm 0.13\%$.

The determined values of Young's modulus E, shear modulus G and Poisson's ratio ν of the fine-grained and the coarse-grained composites are listed in Table 7. The Young's modulus of the coarse-grained composite is only $\approx 50\%$ of the fine-grained one, whereas the shear modulus and Poisson's ratio dropped by roughly one third. It can be assumed that the relatively low elastic constants originate from the high porosity of the coarse-grained composite. As it is shown in Table 4, the determined open porosity increased from 18 to 20% in case of the fine-grained composite to 25–30% for the coarse-grained samples.

The four-point bending strength also decreased from 52 MPa to 11 MPa comparing the fine- and coarse-grained composites (see Table 7). The fine-grained composite has a similar strength as a formerly produced fine-grained niobium-alumina composite containing 11 vol.% metal [18].

Table 7. Determined values of the elastic constants Young's modulus E, shear modulus G and Poisson's ratio ν and the four-point bending strength σ of the sintered fine-grained and coarse-grained composites given as mean value $\pm$ standard deviation.

	Elastic Constants			MOR
	E **in GPa**	G **in GPa**	ν	σ **in MPa**
fine-grained	97.8 ± 5.0	38.0 ± 2.1	0.287 ± 0.014	52.0 ± 7.0
coarse-grained	46.0 ± 1.8	24.5 ± 2.2	0.185 ± 0.021	11.3 ± 0.4

3.6. High-Temperature Compressive Strength

Figure 11a shows the high-temperature compressive behaviour of fine-grained and coarse-grained composite materials at 1300 °C. The fine-grained composite has a significantly higher strength compared to coarse-grained composite with a compressive yield strength of 49 MPa, while the coarse-grained specimen achieved 18 MPa only. In addition, pronounced plastic deformation occurred in the fine-grained composite, as it is visible from Figure 11b. After reaching 10% of strain, the stress on the fine-grained composite slightly increased due to strain hardening within the material. Furthermore, the effective cross-section of the sample was increased. The specimen was then compressed well up to 30% of strain. On the other hand, the coarse-grained composite showed more brittle behaviour with limited plasticity and failed after approximately 4–5% of strain.

Figure 11b compares the appearence of the specimens before and after the tests. As it can be seen, some cracks occurred on the fine-grained specimen during the test, however, in spite of these cracks, the specimen showed good plastic behaviour and did not fail. The coarse-grained composite showed fracture to larger fragments as it is clearly visible in Figure 11b. Thus, the specimen showed limited plasticity and more brittle behaviour and the specimen failed. The presence of higher porosity is considered to be the fundamental reason for the brittle behaviour.

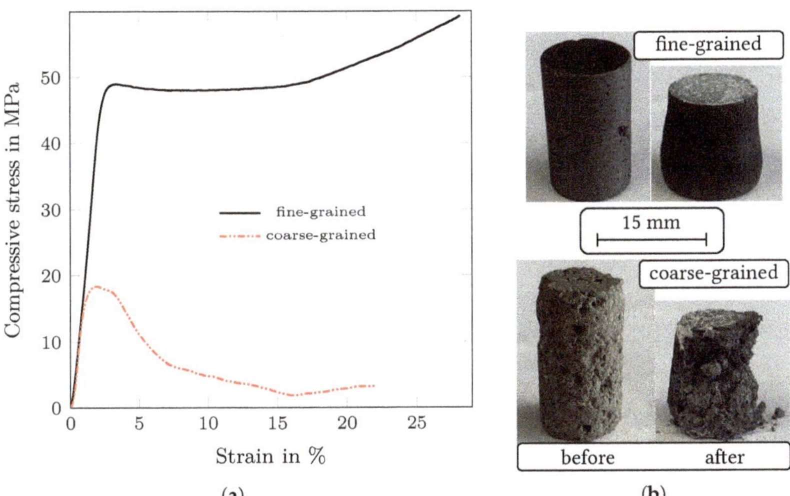

Figure 11. (**a**) Results of compression tests at a temperature of 1300 °C performed on the fine-grained and coarse-grained composites and (**b**) photographs of the corresponding specimens before and after the tests.

3.7. Electrical Conductivity

Two samples were produced with a density of 5.3 g/cm³ and 5.2 g/cm³, and a determined electrical conductivity of 40.58 S/cm and 59.14 S/cm, respectively. The high electrical conductivity is caused by the formation of a conducting network of the metallic component through the composite material. According to [32], the specific electrical resistance of pure niobium is in the range of $1.46 \cdot 10^{-5} - 1.70 \cdot 10^{-5}\,\Omega \cdot$ cm, resulting in an electrical conductivity in the range of 68.49–58.82 kS/cm. A comparison of these values with the results of the investigated niobium-alumina composites shows a decrease of the electrical conductivity in the order of three magnitudes.

The lower electrical conductivity is caused by the composition of the material, made from 65 vol.% electrically conductive niobium and 35 vol.% isolating alumina. Furthermore, the sample density has to be considered, since the samples exhibited a geometric density of about 75 % theoretical density. The influence of porosity on electrical conductivity has been studied elsewhere and is reported to decrease with increasing porosity [33–35].

4. Conclusions

Niobium-alumina aggregates were successfully synthesised and their application in coarse-grained refractory castables was demonstrated. It was found that the crushing process of the fine-grained 65/35 vol.% niobium-alumina composite induced phase separation resulting in niobium contents between 40 and 60 vol.%, depending on the aggregate class. The porosity of the fine-grained composite might play a role here as well as the amount of diffusion of the niobium along the grain boundaries of the alumina material. Such interactions must be understood in more detail for a better designing of refractory composite aggregates and the related coarse-grained castables.

Obviously, the physical properties and composition of the aggregates are depending on the aggregate fractions, i.e., on the aggregate size. In addition, the particle morphology and the porosity of the aggregates play an important role in packing density, and water demand for self-flowability of the castable mixture. It was shown that the particle morphology variety can not be ignored. Therefore, a systematic analysis of particle morphologies should be part of further works.

A relatively high water content of almost 30 vol.% was necessary for achieving a vibration-assisted flowability of the fine- and coarse-grained castables. Therefore, the porosity of the coarse-grained composite increased due to aggregate properties (roughness, relatively high open porosity) in comparison to the fine-grained composite, which is also reflected by lower values of elastic constants and mechanical strength. However, a strain value of almost 5% was achieved for the coarse-grained composite in compression at a temperature of 1300 °C demonstrating good thermal shock properties in refractory ceramic applications.

For the coarse-grained composite, shrinkage of 1.6% was determined after sintering. In combination with the high metal content, the material can be used to produce large components without the risk of shrinkage cracks on sintering.

The electrical conductivity of the refractory composite strongly depends on chemistry and porosity. The investigated pressed and sintered fine-grained composites had a similar porosity as the casted ones. Thus, the values of electrical conductivity should be in the same order of magnitude for the two materials. With 40–60 S/cm, the electrical conductivity of our fine-grained composite is only one magnitude below the one of silicon carbide [36] but fifteen magnitudes above the electrical conductivity of pure alumina [37]. Thus, the (coarse-grained) material is a candidate for high-temperature heating elements or for refractory linings with voltage-controlled metal melt wetting. For such a purpose, the composite must be used under protective atmosphere or the surface must be coated, e.g., with alumina, for oxidation resistance. In further works, the electrical conductivity of the refractory composite should be investigated on densely sintered samples to study the influence of chemistry as well as on samples containing different levels of closed and open porosity.

Mechanical and functional properties of coarse-grained refractory composite are mainly depending on the properties of the used aggregates. Despite the presented splintery-shaped particles produced by crushing, circular-shaped, porous aggregates based on alginate gelation [38] or almost dense particles based on pelletising can improve thermomechanical properties of coarse-grained refractory composites. It is a challenging task to find the best-fitting material solution for high temperature application simultaneously offering a good thermal-shock behaviour and sufficient mechanical properties [39]. By assembling of pre-synthesised dense and porous aggregates in different size-dependent volume ratios in a castable offset could be a promising way to achieve this objective, particularly for large refractory components.

Author Contributions: Conceptualization, T.Z. and C.G.A.; data curation, T.Z.; formal analysis, J.H.; methodology, T.Z., D.E., J.H. and C.G.A.; validation, T.Z., A.W., H.B., S.W. and C.G.A.; investigation, T.Z., D.E., J.H., G.G., B.K.; resources, S.W., H.B. and C.G.A.; writing—original draft preparation, T.Z., G.G. and B.K.; writing—review and editing, T.Z., J.H., A.W., H.B., S.W. and C.G.A.; visualization, T.Z.; project administration, J.H., A.W., H.B., S.W. and C.G.A.; funding acquisition, J.H., A.W., H.B., S.W. and C.G.A. All authors have read and agreed to the published version of the manuscript.

Funding: This research was funded by the German Research Foundation (DFG) within the Research Unit FOR 3010 (Project number: 416817512). The X-ray diffractometer was acquired through the "Major Research Instrumentation" funding programme of the German Research Foundation (DFG), reference number: INST 267/157-1 FUGG (Project number: 395259190).

Institutional Review Board Statement: Not applicable.

Informed Consent Statement: Not applicable.

Data Availability Statement: The data presented in this study are available on request from the corresponding author.

Acknowledgments: We are especially grateful for the suggestions of reviewer 2 for improving the manuscript.

Conflicts of Interest: The authors declare no conflict of interest. The funders had no role in the design of the study; in the collection, analyses, or interpretation of data; in the writing of the manuscript, or in the decision to publish the results.

References

1. Moritz, K.; Gerlach, N.; Hubálková, J.; Aneziris, C.G. Pressure slip casting of coarse-grained alumina-carbon materials. *Int. J. Appl. Ceram. Technol.* **2019**, *16*, 14–22. [CrossRef]
2. Luz, A.P.; Braulio, M.A.; Pandolfelli, V.C. *Refractory Castable Engineering*; F.I.R.E. Compendium Series; Göller Verlag: Baden-Baden, Germany, 2015; Volume 1.
3. Long, B.; Xu, G.; Buhr, A.; Jin, S.; Harmuth, H. Fracture behaviour and microstructure of refractory materials for steel ladle purging plugs in the system Al_2O_3-MgO-CaO. *Ceram. Int.* **2017**, *43*, 9679–9685. [CrossRef]
4. Lee, W.; Vieira, W.; Zhang, S.; Ghanbari Ahari, K.; Sarpoolaky, H.; Parr, C. Castable refractory concretes. *Int. Mater. Rev.* **2001**, *46*, 145–167. [CrossRef]
5. Routschka, G.; Wuthnow, H. *Handbook of Refractory Materials*, 4th ed.; Vulkan-Verlag GmbH: Essen, Germany, 2012.
6. Ghassemi Kakroudi, M.; Yeugo-Fogaing, E.; Huger, M.; Gault, C.; Chotard, T. Influence of the thermal history on the mechanical properties of two alumina based castables. *J. Eur. Ceram. Soc.* **2009**, *29*, 3197–3204. [CrossRef]
7. Rodriguez-Suarez, T.; Bartolomé, J.; Moya, J. Mechanical and tribological properties of ceramic/metal composites: A review of phenomena spanning from the nanometer to the micrometer length scale. *J. Eur. Ceram. Soc.* **2012**, *32*, 3887–3898. [CrossRef]
8. McHugh, C.; Whalen, T.; Humenik, M., Jr. Dispersion-Strengthened Aluminum Oxide. *J. Am. Ceram. Soc.* **1966**, *49*, 486–491. [CrossRef]
9. Nawa, M.; Sekino, T.; Niihara, K. Fabrication and mechanical behaviour of Al_2O_3/Mo nanocomposites. *J. Mater. Sci.* **1994**, *29*, 3185–3192. [CrossRef]
10. Wei, W.C.J.; Wang, S.C.; Cheng, F.H. Characterization of Al_2O_3 composites with fine Mo particulates, I. Microstructural development. *Nanostruct. Mater.* **1998**, *10*, 965–981. [CrossRef]
11. Zhou, Y.; Gao, Y.; Wei, S.; Pan, K.; Hu, Y. Preparation and characterization of Mo/Al_2O_3 composites. *Int. J. Refract. Met. Hard Mater.* **2016**, *54*, 186–195. [CrossRef]
12. Moya, J.; Bartolome, J.; Diaz, M.; Requena, J. Mullite/molybdenum composites. *Compos. Interfaces* **1998**, *6*, 325–342. [CrossRef]
13. Diaz, M.; Bartolomé, J.; Requena, J.; Moya, J. Wet processing of mullite/molybdenum composites. *J. Eur. Ceram. Soc.* **2000**, *20*, 1907–1914. [CrossRef]
14. He, Q.; Wang, Y.; Fu, Z.; Wang, R.; Wang, H.; Wang, W.; Zhang, J.; Zhang, F. Low-temperature sintering and microstructure control of mullite-Mo composites. *Ceram. Int.* **2016**, *42*, 5339–5344. [CrossRef]

15. Smirnov, A.; Bartolomé, J. Microstructure and mechanical properties of ZrO_2 ceramics toughened by 5–20 vol% Ta metallic particles fabricated by pressureless sintering. *Ceram. Int.* **2014**, *40*, 1829–1834. [CrossRef]
16. Smirnov, A.; Beltrán, J.; Rodriguez-Suarez, T.; Pecharromán, C.; Munoz, M.; Moya, J.; Bartolomé, J. Unprecedented simultaneous enhancement in damage tolerance and fatigue resistance of zirconia/Ta composites. *Sci. Rep.* **2017**, *7*, 44922. [CrossRef]
17. Yeomans, J. Ductile particle ceramic matrix composites-Scientific curiosities or engineering materials? *J. Eur. Ceram. Soc.* **2008**, *28*, 1543–1550. [CrossRef]
18. Zienert, T.; Farhani, M.; Dudczig, S.; Aneziris, C.G. Coarse-grained refractory composites based on Nb-Al_2O_3 and Ta-Al_2O_3 castables. *Ceram. Int.* **2018**, *44*, 16809–16818. [CrossRef]
19. Weidner, A.; Ranglack-Klemm, Y.; Zienert, T.; Aneziris, C.G.; Biermann, H. Mechanical High-Temperature Properties and Damage Behavior of Coarse-Grained Alumina Refractory Metal Composites. *Materials* **2019**, *12*, 3927. [CrossRef]
20. Aneziris, C.G.; Hampel, M. Microstructured and Electro-Assisted High-Temperature Wettability of MgO in Contact with a Silicate Slag-Based on Fayalite. *Int. J. Appl. Ceram. Technol.* **2008**, *5*, 469–479. [CrossRef]
21. Novakovic, R.; Ricci, E.; Muolo, M.; Giuranno, D.; Passerone, A. On the application of modelling to study the surface and interfacial phenomena in liquid alloy-ceramic substrate systems. *Intermetallics* **2003**, *11*, 1301–1311. [CrossRef]
22. Eustathopoulos, N.; Nicholas, M.; Drevet, B., Eds. *Wettability at High Temperatures*, 1st ed.; Pergamon Materials Series; Pergamon Press: Oxford, UK, 1999; Volume 3.
23. Giuranno, D.; Bruzda, G.; Polkowska, A.; Nowak, R.; Polkowski, W.; Kudyba, A.; Sobczak, N.; Mocellin, F.; Novakovic, R. Design of refractory SiC/$ZrSi_2$ composites: Wettability and spreading behavior of liquid Si-10Zr alloy in contact with SiC at high temperatures. *J. Eur. Ceram. Soc.* **2020**, *40*, 953–960. [CrossRef]
24. Degen, T.; Sadki, M.; Bron, E.; König, U.; Nénert, G. The HighScore suite. *Powder Diffr.* **2014**, *29*, S13–S18. [CrossRef]
25. Dinger, D.; Funk, J. Particle Packing II - Review of Packing of Polydisperse Particle Systems. *Interceram* **1992**, *41*, 176–179.
26. Fruhstorfer, J. Continuous gap-graded particle packing designs. *Mater. Today Commun.* **2019**, *20*, 100550. [CrossRef]
27. Fruhstorfer, J.; Aneziris, C. The Influence of the Coarse Fraction on the Porosity of Refractory Castables. *J. Ceram. Sci. Technol.* **2014**, *5*, 155–166.
28. Otsu, N. A Threshold Selection Method from Gray-Level Histograms. *IEEE Trans. Syst. Man, Cybern.* **1979**, *9*, 62–66. [CrossRef]
29. Orazem, M.E.; Tribollet, B. *Electrochemical Impedance Spectroscopy*, 2nd ed.; John Wiley & Sons, Inc.: Hoboken, NJ, USA, 2017.
30. Terao, N. Structure des Carbures de Niobium. *Jpn. J. Appl. Phys.* **1964**, *3*, 104–111. [CrossRef]
31. Brauer, G. Die Oxyde des Niobs. *Z. Für Anorg. Und Allg. Chem.* **1941**, *248*, 1–31. [CrossRef]
32. Shabalin, I.L. *Ultra-High Temperature Materials I*; Springer Science+Business Media: Dodrecht, The Netherlands, 2014.
33. Probst, N.; Grivei, E. Structure and electrical properties of carbon black. *Carbon* **2002**, *40*, 201–205. [CrossRef]
34. Kennedy, L.J.; Vijaya, J.J.; Sekaran, G. Electrical conductivity study of porous carbon composite derived from rice husk. *Mater. Chem. Phys.* **2005**, *91*, 471–476. [CrossRef]
35. El Khal, H.; Cordier, A.; Batis, N.; Siebert, E.; Georges, S.; Steil, M. Effect of porosity on the electrical conductivity of LAMOX materials. *Solid State Ionics* **2017**, *304*, 75–84. [CrossRef]
36. Pai, C.h.; Koumoto, K.; Yanagida, H. Effects of Sintering Additives on the Thermoelectric Properties of SiC Ceramics. *J. Ceram. Soc. Jpn.* **1989**, *97*, 1170–1175. [CrossRef]
37. Salmang, H.; Scholze, H. *Keramik*, 7 ed.; Springer: Berlin/Heidelberg, Germany, 2007.
38. Storti, E.; Neumann, M.; Zienert, T.; Hubálková, J.; Aneziris, C.G. Metal-Ceramic Beads Based on Niobium and Alumina Produced by Alginate Gelation. *Materials* **2021**, *14*, 5483. [CrossRef]
39. Cristante, Å.; Nascimento, L.A.; Neves, E.S.; Vernilli, F. Study of the castable selection for blast furnace blowpipe. *Ceram. Int.* **2021**, *47*, 19443–19454. [CrossRef]

Article

Metal-Ceramic Beads Based on Niobium and Alumina Produced by Alginate Gelation

Enrico Storti *, Marc Neumann, Tilo Zienert, Jana Hubálková and Christos Georgios Aneziris

Institute of Ceramics, Refractories and Composite Materials, TU Bergakademie Freiberg, Agricolastraße 17, 09599 Freiberg, Germany; Marc.Neumann@ikfvw.tu-freiberg.de (M.N.); Tilo.Zienert@ikfvw.tu-freiberg.de (T.Z.); Jana.Hubalkova@ikfvw.tu-freiberg.de (J.H.); aneziris@ikfvw.tu-freiberg.de (C.G.A.)
* Correspondence: enrico.storti@ikfvw.tu-freiberg.de; Tel.: +49-3731-39-2176

Abstract: Full metal-ceramic composite beads containing different amounts of niobium and alumina, particularly 100 vol% alumina, 100 vol% niobium, and 95/5 vol% niobium/alumina, were produced by the alginate gelation process. The suspension for bead fabrication contained sodium alginate as gelling agent and was added dropwise into a calcium chloride solution to trigger the consolidation process. After debinding in air, sintering of the composite beads was performed under inert atmosphere. Samples in green and sintered state were analyzed by digital light microscopy and scanning electron microscopy equipped with energy dispersive X-ray spectroscopy. Investigations by mercury intrusion porosimetry revealed that pure alumina beads featured smaller pores compared to composite beads, although the open porosities were comparable. The fracture strength was evaluated on single beads. Contrary to the pure alumina, the composite beads showed a clear plastic deformation. Pure niobium beads showed a ductile behavior with very large deformations. XRD analyses revealed the presence of calcium hexaluminate and beta-alumina as minor phases in the alumina beads, while the composite ones contained about 25 wt% of impurities. The impurities comprised NbO arising from the oxidation, and β-Nb$_2$C, from the reaction with the residual sodium alginate.

Keywords: metal-ceramic composites; alginate gelation; refractory metals; computed tomography; niobium

Citation: Storti, E.; Neumann, M.; Zienert, T.; Hubálková, J.; Aneziris, C.G. Metal-Ceramic Beads Based on Niobium and Alumina Produced by Alginate Gelation. *Materials* **2021**, *14*, 5483. https://doi.org/10.3390/ma14195483

Academic Editors: Jacek Szczerba and Ilona Jastrzębska

Received: 16 August 2021
Accepted: 17 September 2021
Published: 22 September 2021

Publisher's Note: MDPI stays neutral with regard to jurisdictional claims in published maps and institutional affiliations.

1. Introduction

Ceramic-metal composites benefit from the combination of high melting point, hardness and chemical stability of ceramics and high toughness and ductility of metals. Usually, the application temperature of such composites is limited by the relatively low melting point of the used metals or by the reaction between the metal and ceramic phase and/or environment. Regarding the melting point, so-called refractory metals, such as Zr, Mo, Nb, W, and Ta (to name the most abundant), can help extending the application to higher temperatures. In particular, niobium and tantalum exhibit a similar thermal expansion coefficient as alumina in a wide temperature range, which allows to produce composites with a high thermal shock resistance [1–4]. However, most studies on Nb-alumina composites are using very fine powders.

In a recent work, Zienert et al. reported for the first time the production of coarse-grained refractory composites consisting of refractory metals and refractory ceramic materials [5]. Refractory metals with a melting point above 2000 °C, namely Nb and Ta, were chosen and combined with alumina. Composites showed plastic deformation behavior between 1300 °C and 1500 °C [6]. The shrinkage of the coarse-grained refractory composites was clearly reduced in comparison to fine-grained materials, with the additional benefit of low residual stresses.

Especially in the case of ceramic materials, the ability to produce near-net shaped parts is very beneficial since it reduces the need for final machining, which is usually limited for sintered bodies (due to their brittleness) or often results in rejected green pieces (due to their weakness). Among these shaping methods, aqueous gel casting is one of the most common technique. In general, a ceramic suspension with high solid contents is mixed with an additive allowing for a direct sol-gel transition. Originally, toxic chemicals, such as acryl amide, were utilized as gelling agents [7–10]. Due to the toxicity of acryl amide, natural compounds are preferred today [11]. They are effective at low concentration, inexpensive, and environmentally friendly. Santacruz et al. obtained green densities higher than 60% of theoretical density and high green strength by gel casting of alumina suspensions, using concentrated agarose solutions [12]. It was demonstrated that gel casting with agar is also a good forming technique to produce net-shaped bodies in a very short time (consolidation takes <1 min) and with similar density values to those obtained by slip casting of nanosuspensions [13].

A particular class of natural compounds commonly used in gel casting is that of alginates, which are derived from brown seaweeds. They are widely applied as thickeners, stabilizers and gelling agents, especially in the food and pharmaceutical industry [14]. Alginates are water-dispersible salts of polysaccharides consisting of two essential monomers, 1,4-β-D-mannuronate (M) and 1,4-α-L-guluronate (G). Due to their biaxially linked structure, the combination of neighboring G monomers (so-called "G-blocks") is of special interest for the gel-casting process. In particular, G-blocks enable the transition from an alginate sol to a stiff gel structure based on their buckled chain structure [15]. The process is based on the substitution of monovalent cations by divalent alkaline earth ions, such as Ca^{2+} or Ba^{2+}, which results in the bonding between two G-blocks of different polysaccharide chain segments and leads to a 3D crosslinking and, thus, in the gelation. This phenomenon occurs stepwise and depends on the ratio between divalent ion (M^{2+}) and G monomers [14]. The added alginates also provide a stabilizing effect. On the one hand, the viscosity increase and thickening effect result in a kinetic stabilization effect, hindering the sedimentation of dispersed particles in the suspension. On the other hand, alginates are natural polyelectrolytes that arrange around the solid particles leading to a steric stabilization effect. When free alkaline earth ions are supplied, the gelation takes place rather instantly, which can be exploited to manufacture granules or components with high specific surface area [16]. The shape and cross section of the manufactured product is determined by the nozzle through which the suspension is released and the gelation progress depends on the diffusion of alkaline earth ions into the structure core, limiting the achievable thickness. The suspension is usually pumped, poured, or dropped into a solution of alkaline earth salts, such as calcium chloride. This approach is especially useful in the food or pharmaceutical industry, for example, to encapsulate materials in alginate beads. However, technical applications, such as the gel casting of "spaghetti filters" for steel melt filtration, are possible, as well [17,18]. Alternatively, it can be beneficial to delay the gelation process by adding a chelating agent, which forms a complex with the calcium ions. By using an acid, calcium ions are then released from the complex and react with the alginate, finally forming a three-dimensional network. Silicon carbide bodies with high bending strength, good surface quality, and homogeneous microstructure were obtained with this approach [19].

Recently, hollow and full beads based on zirconia and steel were fabricated using the alginate gelation technique [20]. In particular, the authors combined austenitic stainless steel with transformation induced plasticity (TRIP) and/or twinning induced plasticity (TWIP) and partially stabilized zirconia reinforcing particles. The resulting samples exhibited high yield strength and high energy absorption capability, especially under compression, making them excellent candidates as crash absorber material. In addition, beads were spray coated in order to further improve the energy absorption in comparison to pure steel and uncoated MMC (metal-matrix composite) beads [21]. In the present work, a similar approach was used to produce full metal-ceramic beads based on alumina and niobium fine powders as starting materials. Such beads should serve as basis material for the manufacturing of coarse-grained refractory components with special functional properties, such as high electrical conductivity and improved thermal shock resistance, in comparison to traditional refractories.

2. Materials and Methods

2.1. Raw Materials

The main raw material for the beads production was niobium powder (Nb 99.95%, EWG Sondermetalle GmbH, Weissach-Flacht, Germany) with measured d_{10}, d_{50}, and d_{90} of 9.2 , 32.1, and 67.1 µm, respectively. Furthermore, fine calcined alumina (CT9FG, Almatis GmbH, Ludwigshafen, Germany) with measured d_{10}, d_{50} and d_{90} of 2.0, 5.5 and 20.6 µm, respectively, was also used. Sodium alginate powder (Carl Roth GmbH, Karlsruhe, Germany) worked as binder and enabled the gelation of the slurry. Liquid additives KM 1001 and KM 2000 (Zschimmer & Schwarz GmbH, Lahnstein, Germany) were required to reduce the water content of the slurry and provide sufficient stability. Finally, calcium chloride dihydrate $CaCl_2 \cdot 2H_2O$ ($\geq$99%, Carl Roth GmbH, Germany) was used as solidifying agent.

2.2. Bead Production

The fabrication process was based on the gelation of alginate in contact with bivalent ions in an aqueous solution as solidifying agent. The raw powders (niobium powder, alumina and sodium alginate) were dry mixed for 5 min using a homogenizer DIAX 600 (Heidolph Instruments, Schwabach, Germany). Similarly, the liquid additives were dissolved in deionized water. Next, the two mixtures were filled into a polypropylene barrel together with alumina milling balls (3 cm in diameter) and mixed for 4 h. The formulations of the different slurries are given in Table 1. In the Nb_95 slurry, 5 vol% fine alumina was used in order to promote sintering of the samples, since the applied sintering temperature was relatively low compared to the melting point of Nb.

Table 1. Composition of the slurries for beads production (values in mass%).

	$Al_2O_3\_100$	Nb_95	Nb_100
Materials			
CT9FG alumina	62.4	1.65	-
Nb powder	-	67.75	69.4
Sodium alginate	0.6	0.6	0.6
Deionized water	37	30	30
Additives (relative to solids)			
KM 1001	0.6	0.6	0.6
KM 2000	1	1	1

After mixing, the suspension was loaded into a 50 mL syringe and poured dropwise into a water solution (1 mass%) of calcium chloride from a height of approximately 3 cm. The bead formation occurred instantly. The process was automated by means of an infusion pump Perfusor Secura FT (Braun Melsungen AG, Melsungen, Germany), consisting of a computer-controlled motor turning a screw that pushes the plunger on the syringe. Different nozzle gauges allowed to control the size of the produced beads, as shown in Table 2. For simplicity, the three different sizes were labeled "L" (large), "M" (medium) and "S" (small), respectively. In general, thinner nozzles required a lower infusion rate in order not to exceed the maximum pressure applicable by the pump. It was not possible to use nozzles smaller than 21G (0.8 mm in diameter) without modifying the slurries, i.e., reducing the solids content.

Table 2. Properties of the beads before and after thermal treatment. "L" = Large; "M" = Medium; "S" = Small.

Property	Unit	Batch						
		$Al_2O_3\_100$-L	$Al_2O_3\_100$-M	$Al_2O_3\_100$-S	Nb_95-L	Nb_95-M	Nb_95-S	Nb_100-L
Nozzle diameter	mm	no nozzle	1.2	0.8	no nozzle	1.2	0.8	no nozzle
Bead diameter (dried)	mm	3.01 ± 0.10	2.59 ± 0.11	2.42 ± 0.09	2.71 ± 0.15	2.12 ± 0.05	1.79 ± 0.06	2.69 ± 0.16
Bead diameter (sintered)	mm	2.91 ± 0.09	2.42 ± 0.09	2.15 ± 0.06	2.47 ± 0.11	1.97 ± 0.10	1.75 ± 0.06	2.60 ± 0.13
Shrinkage	%	3.3	6.7	11.4	8.6	6.9	2.1	3.5
Open porosity	%	47.1	46.0	45.5	43.1	42.2	42.5	42.5
Apparent density	g/cm^3	3.98	3.95	3.93	6.58	6.85	6.54	7.24

After gelation, the wet green beads were removed from the calcium chloride solution with the aid of a sieve, washed with deioinzed water to remove any solution residues, and, finally, dried at 50 °C for 24 h. It was reported that the decomposition of sodium alginate should be completed at approximately 680 °C, with three main exothermic peaks [20]. However, in the case of metal-ceramic composite beads, it was not possible to perform the debinding step up to this temperature, due to the high affinity of Nb to oxygen [5]. Instead, the following debinding program was chosen: 0.5 K/min heating rate up to 220 °C, 30 min dwell time at 220 °C, then 0.1 K/min up to 275 °C and 0.5 K/min to 300 °C. The furnace was then left to cool freely. With this schedule, only the first exothermic peak of sodium alginate was met. However, a preliminary debinding test showed some oxidation of Nb already at 300 °C. After debinding, the composite beads were sintered under inert atmosphere (Ar 5.0) inside an XGraphit furnace (XERION ADVANCED HEATING Ofentechnik GmbH, Berlin, Germany) with a heating rate of 5 K/min up to 1600 °C and a dwell time of 120 min at this temperature. Regarding the pure ceramic beads, a combined debinding and sintering program inside the same furnace was applied. In this case, three intermediate dwell steps were used, along with a final dwell time of 120 min at 1600 °C (as for the composite samples). The full schedule is shown in Figure 1.

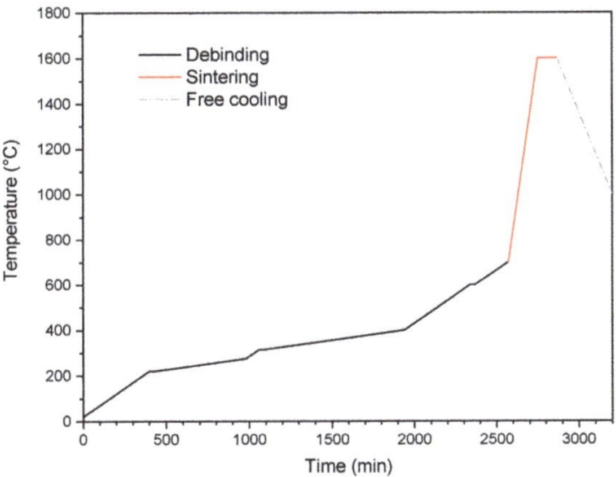

Figure 1. Combined debinding and sintering program for the pure alumina beads.

2.3. Characterization

The rheological behavior of the slurries was evaluated by means of a rheometer HAAKE MARS 60 (Thermo Scientific, Waltham, MA, USA), using a coaxial cylinder measuring system with rotor CC38/Ti/SE and cup Z40. The measurements were performed under controlled-rate/steady-state conditions at a temperature of 20 °C. The shear rate was increased from 0 to 500 s^{-1}, and it was then kept constant for 90 s and, finally, decreased again.

Mercury intrusion porosimetry was used to evaluate the open porosity and pore size distribution (PSD) of the sintered beads according to the International standard ISO/DIS 15901-1. The morphology and fracture surfaces of the samples were analyzed with aid of a digital light microscope (VHX-200 D, Keyence, Neu-Isenburg, Germany). The average diameter of each sample batch in dried, as well as in sintered, condition and the resulting shrinkage were estimated through image analysis on the micrographs. In addition, scanning electron microscopy coupled with EDS (ESEM XL30, FEI/Philips, Mainz, Germany) was used.

The fracture strength of the beads in sintered state was estimated with the aid of an universal testing machine TT2420 (TIRA GmbH, Germany) and a measuring test device for single beads equipped with a load cell of 1 kN. Thirty beads per batch were loaded to fracture in a diametral compression mode between two parallel steel plates at a displacement rate of 0.05 mm/s. The fracture stress was estimated from the force-displacement data according to the method reported by Kschinka et al., who investigated the mechanical behavior of glass spheres [22]:

$$\sigma_f = \frac{2.8P}{\pi d_f{}^2},$$

(1)

where P is the load at fracture, and d_f is the distance between the loading points at the instant of failure. d_f was calculated for each single bead from the initial diameter, the crosshead speed, and the time to fracture. It should be pointed out here that the calculated strength is not equivalent to the stress obtained by a standard compression test on cylindrical samples. It was demonstrated that the tensile strength of brittle samples can be indeed determined by the compression test of irregular specimens [23]. After calculating the fracture strength, the failure probability was estimated for each sample, and the Weibull analysis was performed according to the European Standard DIN EN 843-5. Finally, the results were plotted in the double logarithmic coordinate system.

The 3D-macrostructure of selected large (L) beads was analyzed before and after mechanical loading using microfocus X-ray computed tomography (µ-CT). The analyses were performed with a CT-ALPHA (ProCon X-Ray GmbH, Sarstedt, Germany) equipped with a transmission X-ray tube FXE-160.20/25 (Feinfocus, Garbsen, Germany) and a flat panel X-ray detector Dexela 1512 (Perkin Elmer, Solingen, Germany). The µ-CT was operated with at 150 kV and 80 µA using a 0.3 mm copper filter in order to reduce beam hardening artifacts. The exposure time was set to 1 s for alumina samples and to 2 s for niobium containing samples. The volume data were reconstructed using the software Volex 6.0 (Fraunhofer EZRT, Fürth, Germany). The resulting voxel size was 5.5 µm for all investigated samples. The reconstructed volume data were visualized using the software VG Studio MAX 2.2 (Volume Graphics, Heidelberg, Germany).

Beads in sintered state were manually ground into fine powder and analyzed by means of XRD. The XRD diffractometer Empyrean (Malvern Panalytical GmbH, Kassel, Germany) was operated in Bragg-Brentano geometry, with standard Cu Kα radiation (λ = 1.54 Å) and a 2D detector. The measurements took place in the a 2θ-range from 10 to 140° with a scan step size of 0.0143° and a holding time of 160 s per step. The X-ray source was operated at 40 kV and 40 mA. The ICSD database was used to determine the phase composition of the samples in the HighScore Plus software version 4.8 (Malvern Panalytical GmbH, Germany). Rietveld refinement was applied for the quantitative analysis.

3. Results

The results of the rheological investigation are plotted in Figure 2. As expected, the slurries showed a clear shear thinning behavior up to 500 s^{-1}, together with a slight thixotropy. The shear thinning behavior is a fundamental feature for slurries which need to be pumped or sprayed through a nozzle. Despite the lower solids content, the Al$_2$O$_3$_100 slurry showed an overall higher dynamic viscosity compared to the others, hence requiring a higher shear stress to be stirred. This behavior was likely due to the particle size and morphology of the raw materials used in the preparation of the slurries: the tabular alumina had a d_{90} of 20 µm, against the d_{90} of 67 µm for the niobium powder. Due to the very similar composition, the Nb_95 and Nb_100 slurries showed almost identical rheological behavior.

Figure 3 presents the large (L) beads in green state, after drying. It can be observed that samples from all batches had very similar size and shape. The use of nozzles allowed to decrease the average diameter and obtain almost perfect spheres. In addition, the metal-ceramic beads were comparatively smaller than the pure alumina ones, regardless of the nozzle used, as shown in Table 2. Since the syringes and nozzle-to-solution distance were not changed during the experiments, the size difference can be entirely attributed to the different rheological properties of the slurries. All beads showed a regular surface without defects, regardless of the composition. Large beads in sintered conditions are presented in Figure 4. Compared to the green state, a limited shrinkage was detected. Defects, such as pores or cracks, were not observed, indicating that the sintering and, especially, debinding schedules were effective. During sintering at 1600 °C, some niobium diffused from the samples into the alumina crucibles, which turned from white to gray. This is confirmed by the lighter gray shade of the composite beads after the thermal treatment. A strong diffusion of niobium along alumina grains was recently reported by Zienert et al. [5].

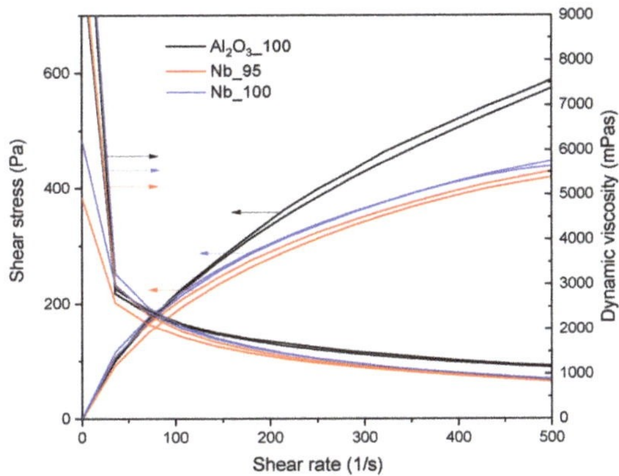

Figure 2. Rheological curves for all slurries.

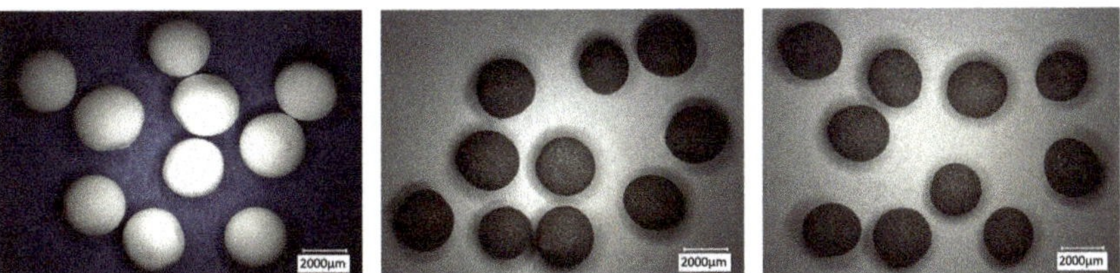

Figure 3. Beads in green state, after drying. Al₂O₃_100-L (**left**), Nb_95-L (**center**), and Nb_100-L (**right**).

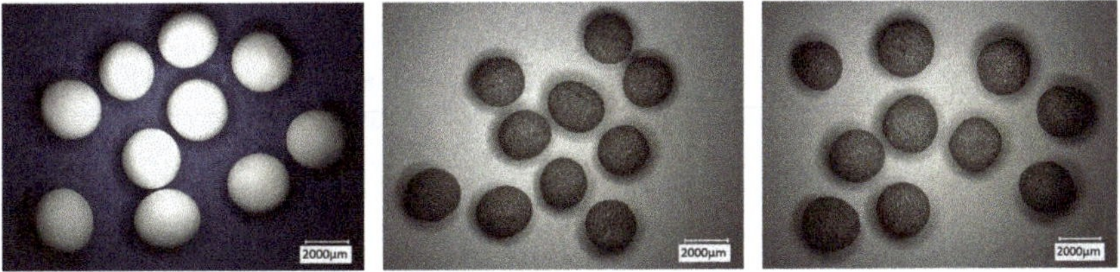

Figure 4. Beads in sintered state. Al₂O₃_100-L (**left**), Nb_95-L (**center**), and Nb_100-L (**right**).

The main results from mercury intrusion porosimetry are reported in Table 2. All batches showed an open porosity above 40% after sintering, which is quite high in comparison to common refractory ceramics. In addition, the larger beads generally had even higher porosity, as high as 47.1% for the Al₂O₃_100-L batch. Larger beads have a smaller surface-to-volume ratio, which might result in less effective debinding and sintering steps. The apparent density obtained through mercury intrusion was similar for batches from the same slurry and with different sizes. As expected, the Nb_100-L batch showed a higher value of apparent density compared to the Nb_95-L one, due to the lack of alumina in the corresponding slurry. The mercury intrusion curves for all batches are presented in

Figure 5. From these, several phenomena can be observed. First, alumina beads clearly featured smaller pores than Nb-based samples in general. The average pore size was about 1.8 μm for alumina batches and one order of magnitude higher (≈13 μm) for the Nb-containing ones. Since a diffusion rate close to zero can be assumed for niobium even at 1600 °C (which is only 0.68·T_m of Nb), it is obvious that sintering was much more effective for pure alumina beads. Second, the bead size had practically no impact on the pore size distribution of any batch. In particular, the smaller beads only showed slightly larger pores on average. Finally, the 5 vol% Al_2O_3 addition entailed no remarkable effect in the pore size distribution of large composite beads.

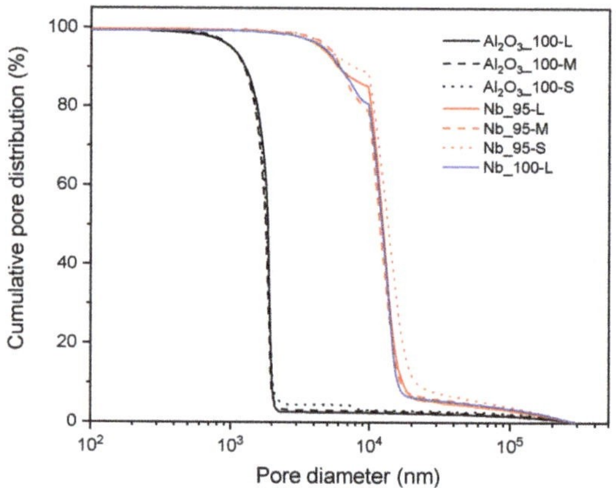

Figure 5. Normalized mercury intrusion curves for all batches.

Some selected load-displacement curves for the different batches are presented in Figure 6. The fracture behavior of alumina beads was purely linear elastic, i.e., the curves were almost straight lines up to the point of fracture, at which the specimen broke into two hemispheres. In contrast, the composite beads did not separate at the point of fracture due to their high metallic content (see CT analysis below)). For comparison, only the linear elastic part of the force-displacement curves was taken into account for these batches. The confidence interval (not shown) was much narrower for the smaller alumina beads compared to the larger ones. In case of the composite beads, the confidence interval was quite narrow for any size. Observing Figure 6a,b, it is evident that the alumina beads broke at higher loads than the composite ones. On the other hand, the Nb-based samples reached much higher displacements, indicating a higher elastic compliance. Taking the size difference into account, all beads had similar strength levels (see Table 3). Finally, Figure 6c presents a selected load-displacement curve from the Nb_100-L batch. In this case, the samples showed a plastic behavior, with a clear plateau following the linear elastic region. Since it was not possible to identify a fracture point, no Weibull analysis was performed for the Nb_100-L batch.

The Weibull diagrams for all batches are presented in Figure 7. It was observed that the strength values were quite similar throughout all samples, regardless of the composition or size. As shown in Table 2, the bead diameters were all within the same order of magnitude; hence, no remarkable size effect was expected. In general, the Weibull parameter m was slightly higher for the alumina samples, as presented also in Table 3. The calculated values for m were within the expected range of partially sintered alumina [24]. The lower modulus m of the composite beads can be attributed to the interaction of two separate phases within the samples, which possess significantly different mechanical properties. However, the main factor was related to the average particle size and particle size distribution of the used raw materials. As mentioned in Section 2.1 and shown in the SEM images below, the alumina was much finer than niobium powder. In addition, the particle size distribution was not optimized to obtain the best possible packing density in the composite samples. As a result, the pure alumina beads showed better mechanical properties simply due to the finer particle size and better sintering activity of the material at 1600 °C.

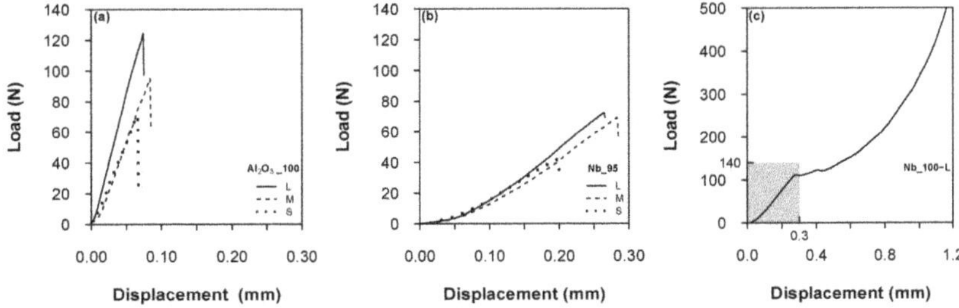

Figure 6. Selected load-displacement curves (compressive loading): (**a**) Al$_2$O$_3$_100 batches; (**b**) Nb_95 batches; (**c**) Nb_100-L batch. The gray box represents the scale regions of (**a**), (**b**).

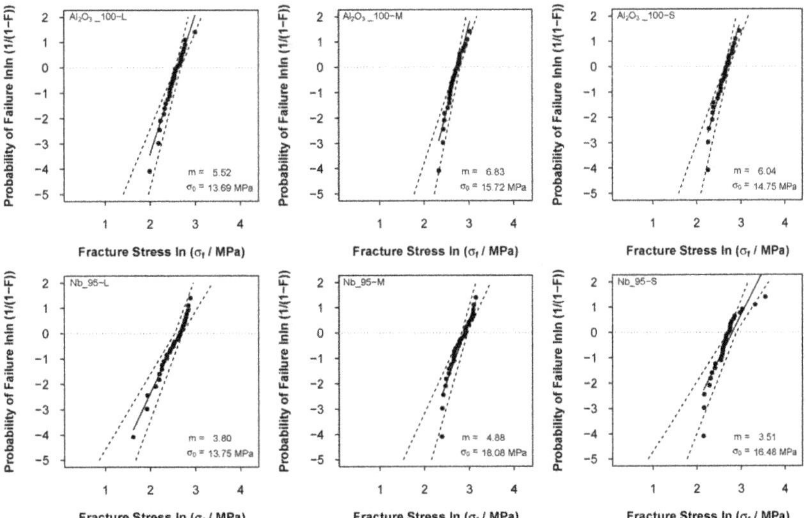

Figure 7. Weibull diagrams from the strength data (30 beads per batch). The dashed lines identify the 95% confidence interval.

The parameters of the fracture strengths obtained from the Weibull analysis are given in Table 3, according to the the European Standard DIN EN 843-5. In particular, C_l and C_u are the lower and upper limit of the 95% confidence interval for σ_0 (characteristic Weibull strength of the samples), respectively. Similarly, D_l and D_u are defined as the lower and upper limit of the 95% confidence interval for the Weibull modulus m.

Table 3. Characteristics of the Weibull distribution, as defined in the European Standard DIN EN 843-5.

	m	D_u	D_l	σ_0 (MPa)	C_u (MPa)	C_l (MPa)
Al$_2$O$_3$_100-L	5.52	7.05	4.33	13.69	14.52	12.92
Al$_2$O$_3$_100-M	6.83	8.73	5.36	15.72	16.48	15.00
Al$_2$O$_3$_100-S	6.04	7.72	4.74	14.75	15.56	13.99
Nb_95-L	3.80	4.86	2.98	13.75	14.97	12.64
Nb_95-M	4.88	6.24	3.83	18.08	19.31	16.93
Nb_95-S	3.51	4.48	2.75	16.48	18.07	15.05

Reconstructed volume images from the CT analyses on the sintered beads are presented in Figure 8. The shape was quite close to that of a sphere, regardless of the composition. Pores can be observed on the Nb_95-L sample. On the other hand, in case of the pure alumina beads, the voxel size was larger than the average pore size (see Figure 5); hence, the samples appeared completely dense in this analysis. Three-dimensionally-rendered images of beads after the compression test are shown in Figure 9. As mentioned above, the alumina beads showed a clear linear elastic behavior: the Al$_2$O$_3$_100-L was clearly split into two hemispheres, typical for a brittle sample. From the analysis of multiple samples, it was confirmed that the beads were actually full. Hollow beads are also interesting and will be investigated in a future work, for instance, to decrease the total mass of the final components. The beads containing Nb behaved in a different way under load, and it was observed that they did not break catastrophically in two or more parts. Instead, after fracture, they kept deforming under compression until the test was manually terminated. The Nb_95-L bead showed a large deformation and vertical cracks, but its parts were still connected together. In case of the Nb_100-L sample, the deformation exceeded 50% without fragmentation. The use of such composite beads for the production of refractory components would likely result in a highly ductile character, as opposed to traditional ceramic parts.

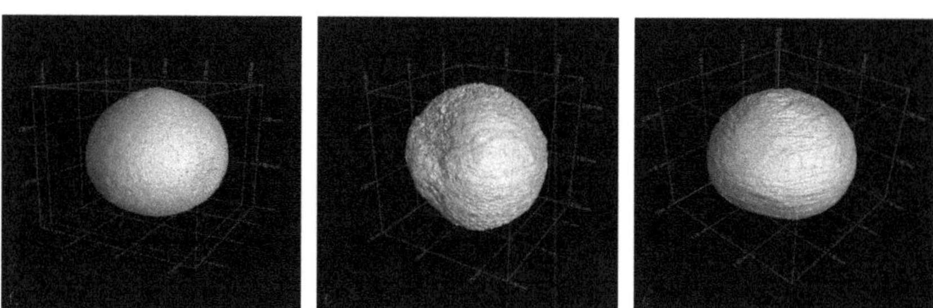

Figure 8. CT-rendered images of beads before compression test. Al$_2$O$_3$_100-L (**left**), Nb_95-L (**center**), and Nb_100-L (**right**).

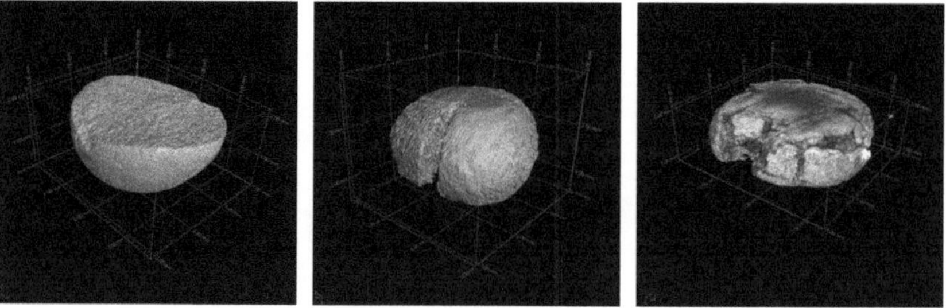

Figure 9. CT-rendered images of beads after compression test. Al$_2$O$_3$_100-L (**left**), Nb_95-L (**center**), and Nb_100-L (**right**).

SEM micrographs of the surface of pure alumina beads are presented in Figure 10. Samples of different sizes had very comparable microstructure; hence, only a $Al_2O_3\_100$-S bead is shown here. A lot of pores were detected, as expected from the mercury intrusion porosimetry results. Overall, a regular microstructure with fine particles was observed. For the composite beads, a Nb_95-S bead is presented in Figure 11. The used magnifications are the same as in Figure 10, so the particle size difference is evident. Here, the use of a BSE detector allowed to clearly distinguish Nb and Nb-containing phases (such as carbides and oxides) from alumina, due to the remarkable difference in atomic number. In the micrograph, bright particles consisted mainly of Nb, while dark gray ones were identified as Al_2O_3 or similar. Overall, the alumina phase was well distributed within the Nb matrix as intended. In addition, it was observed that the Nb particles were only poorly sintered together. This was expected since the sintering temperature for the beads was only $0.68 \cdot T_m$ of Nb. The microstructure of Nb_100-L beads was similar, without the contrast from alumina grains. For this reason, micrographs of such samples are omitted here. EDS analysis on all composite beads revealed the presence of Nb, Al, O, and Ca. Calcium does not originate from the raw materials, but it was incorporated from the calcium chloride solution into the beads during the gelation phase.

Figure 10. SEM (BSE mode) micrographs of $Al_2O_3\_100$-S beads after sintering: overview (**left**) and detail (**right**).

Figure 11. SEM (BSE mode) micrographs of Nb_95-S beads after sintering: overview (**left**) and detail (**right**).

XRD analysis of the corundum powder revealed 98.9 mass% α-Al_2O_3 (ICSD 73725) with lattice parameters $a = 4.7594$ Å and $c = 12.9921$ Å, and 1.1 mass% of $Na_2O \cdot 11Al_2O_3$ (depicted as β-alumina) [25] (ICSD 60635) with lattice parameters $a = 5.6026$ Å and $c = 22.5166$ Å. XRD results of $Al_2O_3\_100$-M are shown in Figure 12. Only traces of β-alumina were found (≈ 0.2 mass%) in the $Al_2O_3\_100$-M sample with lattice parameters $a = 5.5729$ Å and $c = 22.4955$ Å, whereas no β-alumina was observed for the $Al_2O_3\_100$-L and -S beads. However, in all three samples, ≈ 3 mass% $CaO \cdot 6Al_2O_3$ [26] (ICSD 202616, depicted as CA6) was formed with lattice parameters $a = 5.5587$ Å and $c = 21.90$ Å. Based on the smaller lattice parameters of β-alumina of the $Al_2O_3\_100$-M sample in comparison to the corundum raw material, it can be assumed that this β-alumina underwent a chemical change during sintering and was incorporated into the lattice of CA6, which has a similar crystal structure.

Besides niobium [27] (ICSD 76416) with lattice parameter $a = 3.3058$ Å, ≈ 1.1 mass% β-Nb$_2$C [28,29] (adopted from ICSD 33575) with lattice parameters $a = 5.38493$ Å, and $c = 4.96897$ Å, 2.3 mass% NbO [30] (ICSD 40318) with lattice parameter $a = 4.21246$ Å and traces of at least one other phase were detected in the niobium raw material. The XRD pattern of the samples Nb_95-M and Nb_100-L are shown in Figure 13. After sintering the Nb_95 and Nb_100 samples, the lattice parameter of niobium slightly increased to $a = 3.3069$ Å, which could be related to oxygen incorporation. In addition, the amount of β-Nb$_2$C and NbO increased strongly to values between 13–21 mass% and 5–10 mass%, respectively. For the formation of niobium carbide, carbon was available from the alginate residues still present after the debinding step and also from the graphite crucibles in which the samples were enclosed during sintering. Lower amounts of secondary phases were found for the Nb_100-L sample. In all niobium-containing samples, the lattice parameters of β-Nb$_2$C were lowered to $a \approx 5.375$ Å and $c \approx 4.958$ Å. The chemical composition of NbO was only slightly changed during annealing as the lattice parameter was with $a = 4.2104$ Å always close to the original value. Only weak reflexes corresponding to 0.5–2 mass% α-Al$_2$O$_3$ and 0.5–4 mass% CaO·6Al$_2$O$_3$ were observed in the XRD patterns of the Nb_95 samples. Here, the lattice parameter a of CA6 remained almost the same as in the corundum raw material, whereas c increased to ≈ 21.92 Å.

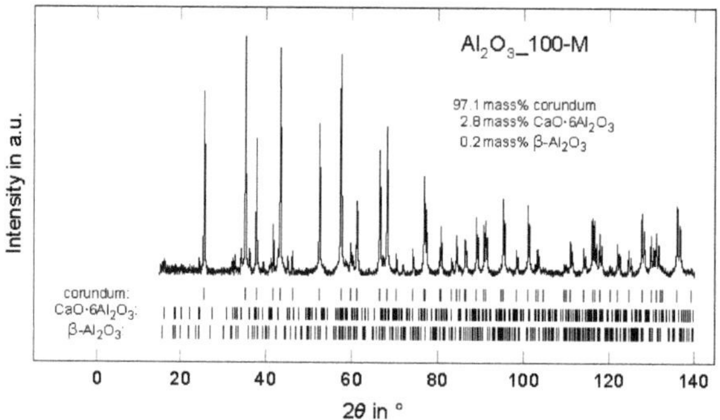

Figure 12. XRD pattern of Al$_2$O$_3$_100-M.

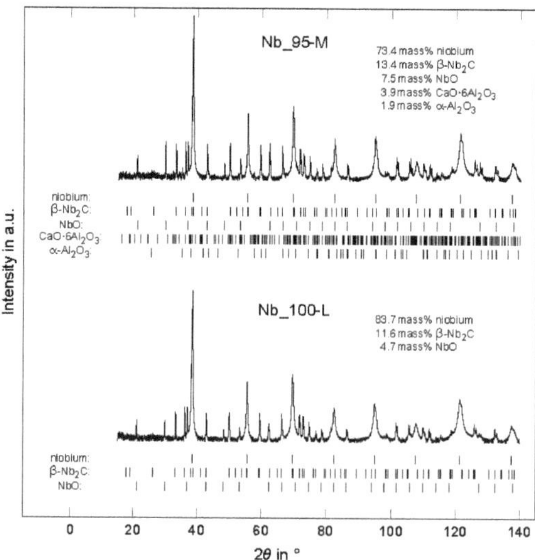

Figure 13. XRD patterns of Nb_95-M and Nb_100-L.

4. Conclusions

In this study, metal-ceramic composite beads containing niobium and alumina, particularly 100 vol% alumina, 100 vol% niobium, and 95/5 vol% niobium/alumina, were produced by means of gel casting with sodium alginate as binder. The slurries were poured dropwise through nozzles with different diameters into a calcium chloride solution in order to obtain beads with distinct sizes. The samples containing Nb were carefully debound in air and subsequently sintered at 1600 °C under inert atmosphere to prevent the metal from oxidating.

Full beads with good spherical shape and average final diameters in the range 1.75–2.9 mm were obtained. The linear shrinkage after debinding and sintering was quite high, reaching more than 10% for the $Al_2O_3\_100$-S batch. The open porosity obtained by mercury intrusion on the sintered beads amounted to more than 40% for all batches, with slightly higher values for the pure alumina samples. In addition, the intrusion curves revealed that pure alumina beads had a much smaller average pore size compared to the Nb-containing beads. This was likely related to the smaller grain size and, thus, higher sintering activity of alumina in contrast to the poor diffusivity of Nb at 1600 °C, since this sintering temperature was only equal to $0.68 \cdot T_m$. SEM investigations confirmed a more pronounced sintering of the pure alumina beads, with a more regular surface and finer pores. Under compression, the composite beads clearly showed a plastic behavior and achieved very high deformations without splitting, while the pure alumina ones failed catastrophically as expected. The compression tests allowed to estimate the fracture stress of the samples and to produce a Weibull plot. This was not possible in case of the pure Nb beads, due to the lack of a clear peak in the curves. The XRD analysis revealed the presence of calcium hexaluminate and of β-alumina in all batches, while NbO and β-Nb_2C were found in the composite beads due to the reaction with the residual carbon and oxygen from the alginate binder.

This study demonstrated that the gel casting process allows to reliably produce spherical grains with defined properties, which can be used as aggregate fraction in new coarse-grained refractory castables, among others. Such products are being currently investigated and will be the object of an upcoming study. In addition, the particle size distribution of the slurries will be optimized to obtain the best possible particle packing, in order to improve the material densification during thermal treatment.

Author Contributions: Conceptualization, E.S., J.H., and C.G.A.; methodology, E.S.; software, M.N.; validation, E.S., M.N., T.Z., and J.H.; formal analysis, M.N. and T.Z.; investigation, E.S., M.N., T.Z., and J.H.; resources, C.G.A.; data curation, M.N.; writing—original draft preparation, E.S.; writing—review and editing, M.N., T.Z., and J.H.; visualization, E.S., M.N., T.Z., and J.H.; project administration, C.G.A.; funding acquisition, J.H. and C.G.A. All authors have read and agreed to the published version of the manuscript.

Funding: This research was funded by the German Research Foundation (DFG) within the Research Unit FOR 3010 (Project number: 416817512).The X-ray diffractometer was acquired through the "Major Research Instrumentation" funding program of the German Research Foundation (DFG), reference number: INST 267/157-1 FUGG (Project number: 395259190).

Institutional Review Board Statement: Not applicable.

Informed Consent Statement: Not applicable.

Data Availability Statement: Not applicable.

Acknowledgments: The authors would like to thank C. Ludewig for the sample preparation, G. Schmidt for the SEM investigations, and M. Müller for the mercury intrusion porosimetry measurements.

References

1. Beals, J.T.; Nardone, V.C. Tensile behaviour of a niobium/alumina composite laminate. *J. Mater. Sci.* **1994**, *29*, 2526–2530. [CrossRef]
2. Shaw, L.; Miracle, D.; Abbaschian, R. Microstructure and mechanical properties of metal/oxide and metal/silicide interfaces. *Acta Metall. Mater.* **1995**, *43*, 4267–4279. [CrossRef]
3. García, D.E.; Schicker, S.; Janssen, R.; Claussen, N. Nb- and Cr-Al2O3 composites with interpenetrating networks. *J. Eur. Ceram. Soc.* **1998**, *18*, 601–605. [CrossRef]
4. Thomson, K.E.; Jiang, D.; Yao, W.; Ritchie, R.O.; Mukherjee, A.K. Characterization and mechanical testing of alumina-based nanocomposites reinforced with niobium and/or carbon nanotubes fabricated by spark plasma sintering. *Acta Mater.* **2012**, *60*, 622–632. [CrossRef]
5. Zienert, T.; Farhani, M.; Dudczig, S.; Aneziris, C.G. Coarse-grained refractory composites based on Nb-Al2O3 and Ta-Al2O3 castables. *Ceram. Int.* **2018**, *44*, 16809–16818. [CrossRef]
6. Weidner, A.; Ranglack-Klemm, Y.; Zienert, T.; Aneziris, C.G.; Biermann, H. Mechanical High-Temperature Properties and Damage Behavior of Coarse-Grained Alumina Refractory Metal Composites. *Materials* **2019**, *12*, 3927. [CrossRef]
7. Young, A.C.; Omatete, O.O.; Janney, M.A.; Menchhofer, P.A. Gelcasting of Alumina. *J. Am. Ceram. Soc.* **1991**, *74*, 612–618. [CrossRef]
8. Janney, M.A.; Omatete, O.O.; Walls, C.A.; Nunn, S.D.; Ogle, R.J.; Westmoreland, G. Development of low-toxicity gelcasting systems. *J. Am. Ceram. Soc.* **1998**, *81*, 581–591. [CrossRef]
9. Ortega, F.S.; Sepulveda, P.; Pandolfelli, V.C. Monomer systems for the gelcasting of foams. *J. Eur. Ceram. Soc.* **2002**, *22*, 1395–1401. [CrossRef]
10. Zhou, L.; Huang, Y.; Xie, Z. Gelcasting of concentrated aqueous silicon carbide suspension. *J. Eur. Ceram. Soc.* **2000**, *20*, 85–90. [CrossRef]
11. Ortega, F.S.; Valenzuela, F.A.; Pandolfelli, V.C. Gelcasting ceramic foams with alternative gelling agents. *Mater. Sci. Forum* **2003**, *416-418*, 512–518. [CrossRef]
12. Santacruz, I.; Baudín, C.; Moreno, R.; Nieto, M.I. Improved green strength of ceramics through aqueous gelcasting. *Adv. Eng. Mater.* **2004**, *6*, 672–676. [CrossRef]
13. Santacruz, I.; Nieto, M.I.; Binner, J.; Moreno, R. Gel casting of aqueous suspensions of BaTiO3 nanopowders. *Ceram. Int.* **2009**, *35*, 321–326. [CrossRef]
14. Funami, T.; Fang, Y.; Noda, S.; Ishihara, S.; Nakauma, M.; Draget, K.I.; Nishinari, K.; Phillips, G.O. Rheological properties of sodium alginate in an aqueous system during gelation in relation to supermolecular structures and Ca2+ binding. *Food Hydrocoll.* **2009**, *23*, 1746–1755. [CrossRef]
15. Draget, K.I.; Skjåk Bræk, G.; Smidsrød, O. Alginic acid gels: the effect of alginate chemical composition and molecular weight. *Carbohydr. Polym.* **1994**, *25*, 31–38. [CrossRef]
16. Oppelt, M.; Wenzel, C.; Aneziris, C.G.; Berek, H. Processing and Characterization of MMC Beads Based on Zirconia and TRIP Steel. *Metall. Mater. Trans. B Process. Metall. Mater. Process. Sci.* **2014**, *45*, 2000–2008. [CrossRef]
17. Wetzig, T.; Schmidt, A.; Dudczig, S.; Schmidt, G.; Brachhold, N.; Aneziris, C.G. Carbon-Bonded Alumina Spaghetti Filters by Alginate-Based Robo Gel Casting. *Adv. Eng. Mater.* **2020**, *22*, 1900657. [CrossRef]
18. Wu, X.; Wetzig, T.; Aneziris, C.G.; Weidner, A.; Biermann, H. Compression Behavior of Carbon-Bonded Alumina Spaghetti Filters at Room and High Temperatures. *Adv. Eng. Mater.* **2021**, 2100613. [CrossRef]
19. Wang, X.; Xie, Z.P.; Huang, Y.; Cheng, Y.B. Gelcasting of silicon carbide based on gelation of sodium alginate. *Ceram. Int.* **2002**, *28*, 865–871. [CrossRef]
20. Oppelt, M.; Aneziris, C.G. Analysis and evaluation of different influencing factors in processing of hollow and full beads based on TRIP steel. *J. Alloy. Compd.* **2015**, *634*, 43–49. [CrossRef]
21. Oppelt, M.; Leißner, T.; Berek, H.; Baumgart, C.; Krüger, L.; Peuker, U.; Aneziris, C.G. Processing and Characterization of Beads with Graded Layer Compositions Based on Zirconia and TRIP-Steel. *Adv. Eng. Mater.* **2019**, *21*, 1–9. [CrossRef]
22. Kschinka, B.A.; Perrella, S.; Nguyen, H.; Bradt, R.C. Strengths of Glass Spheres in Compression. *J. Am. Ceram. Soc.* **1986**, *69*, 467–472. [CrossRef]
23. Hiramatsu, Y.; Oka, Y. Determination of the tensile strength of rock by a compression test of an irregular test piece. *Int. J. Rock Mech. Min. Sci. Geomech. Abstr.* **1966**, *3*, 89–90. [CrossRef]
24. Nanjangud, S.C.; Brezny, R.; Green, D.J. Strength and Young's Modulus Behavior of a Partially Sintered Porous Alumina. *J. Am. Ceram. Soc.* **1995**, *78*, 266–268. [CrossRef]
25. Yamaguchi, G.; Suzuki, K. On the Structures of Alkali Polyaluminates. *Bull. Chem. Soc. Jpn.* **1968**, *41*, 93–99. [CrossRef]
26. Utsunomiya, A.; Tanaka, K.; Morikawa, H.; Marumo, F.; Kojima, H. Structure refinement of CaO·6Al2O3. *J. Solid State Chem.* **1988**, *75*, 197–200. [CrossRef]
27. Edwards, J.W.; Speiser, R.; Johnston, H.L. High Temperature Structure and Thermal Expansion of Some Metals as Determined by X-Ray Diffraction Data. I. Platinum, Tantalum, Niobium, and Molybdenum. *J. Appl. Phys.* **1951**, *22*, 424–428. [CrossRef]
28. Terao, N. Structure des Carbures de Niobium. *Jpn. J. Appl. Phys.* **1964**, *3*, 104–111. [CrossRef]

29.   Smith, J.; Carlson, O.; De Avillez, R. The niobium-carbon system. *J. Nucl. Mater.* **1987**, *148*, 1–16.
30.   Brauer, G. Die Oxyde des Niobs. *Z. Anorg. Allg. Chem.* **1941**, *248*, 1–31. [CrossRef]

materials

MDPI

Article

Effect of Hollow Corundum Microspheres Additive on Physical and Mechanical Properties and Thermal Shock Resistance Behavior of Bauxite Based Refractory Castable

Rimvydas Stonys, Jurgita Malaiškienė *, Jelena Škamat and Valentin Antonovič

Institute of Building Materials, Vilnius Gediminas Technical University, 08217 Vilnius, Lithuania;
rimvydas.stonys@vilniustech.lt (R.S.); jelena.skamat@vgtu.lt (J.Š.); valentin.antonovic@vgtu.lt (V.A.)
* Correspondence: jurgita.malaiskiene@vilniustech.lt; Tel.: +370-8-5-2512329

Abstract: This paper analyses the effect of hollow corundum microspheres (HCM) on both physical-mechanical properties (density, ultrasonic pulse velocity, modulus of elasticity, and compressive strength) and thermal shock resistance behavior of refractory medium cement castable with bauxite aggregate. Moreover, the scanning electron microscopy (SEM) results of HCM and refractory castable samples are presented in the paper. It was found that the replacement of bauxite of 0–0.1 mm fraction by HCM (2.5%, 5%, and 10% by weight of dry mix) had no significant effect on the density and compressive strength of castable, while the modulus of elasticity decreased by 15%. Ultrasonic pulse velocity (*Vup*) values and the visual analysis of the samples after thermal cycling showed that a small amount of HCM in composition of refractory castable could reduce the formation and propagation of cracks and thus increase its thermal shock resistance.

Keywords: refractory castable; hollow corundum microspheres; bauxite aggregate; thermal shock resistance

Citation: Stonys, R.; Malaiškienė, J.; Škamat, J.; Antonovič, V. Effect of Hollow Corundum Microspheres Additive on Physical and Mechanical Properties and Thermal Shock Resistance Behavior of Bauxite Based Refractory Castable. *Materials* **2021**, *14*, 4736. https://doi.org/10.3390/ma14164736

Academic Editor: Christos G. Aneziris

Received: 12 July 2021
Accepted: 20 August 2021
Published: 22 August 2021

Publisher's Note: MDPI stays neutral with regard to jurisdictional claims in published maps and institutional affiliations.

1. Introduction

Most high temperature processes in industrial furnaces are combined with cycles of heating and cooling. In these applications, refractory materials are exposed to temperature gradient during operation, which causes thermal stresses and damage to the material. Therefore, the durability of the lining significantly depends on the thermal shock resistance of refractory materials used in thermal units. The thermal shock resistance of castable depends on many factors: chemical composition, microstructure, phase transformation at high temperature during firing process, etc. [1,2].

The authors [3] point out that about one-third of refractories fail due to insufficient thermal shock resistance. The destruction of refractory material is regarded as a two-stage process that includes crack formation, its further growth and propagation. Based on the thermoelastic theory of crack nucleation, the stress-to-elastic modulus ratio (σ/E) is used to assess the ability of a material to resist crack nucleation [4,5]. This means that a stronger material with a lower modulus of elasticity will have a higher thermal resistance. At the same time, the characteristics of the material's microstructure make it possible to control the propagation of cracks [6–8]. Thus, the formation of a heterogeneous structure by combining components with rather different coefficients of thermal expansion makes it possible to obtain a system of microstructurally small cracks that reduce the modulus of elasticity and effectively inhibit the propagation of larger cracks [5,9]. It has been noted that thermal shock resistance of a material can be improved by creating a fragmentary structure [10] and by reinforcing it with fibers of different nature: metallic, ceramic, and carbon [11,12]. A structure of controlled fragmentation is more mobile and the grains and crystals in such a structure can expand freely without causing additional stresses. As reinforcing inclusions often have a higher strength than the refractory matrix, they not only inhibit crack propagation in the material, but also inhibit the composite's failure.

The presence of pores not only improve the thermal resistance of the material but also reduce stress levels and slow down the propagation of micro-cracks [13,14]. This positive role of pores is achieved if they are round and of quite a big size. Pores of a regular spherical shape can significantly reduce the level of stress concentration, as the propagation of the crack that enters the pore is stopped or suspended.

In terms of increasing the thermal shock resistance of refractories by inhibiting cracks, a certain positive effect can be achieved by using additives with particle of a spherical shape, such as hollow corundum microspheres (HCM), which, also, have a low thermal conductivity and low sintering activity [15,16]. Authors of [17] found that HCM improve mullite castables' thermal shock resistance and explain this effect by the strong bonding at the microsphere and matrix interface, when the crack propagation is stopped at the corundum microsphere. It is also found that the addition of HCM has a positive effect on the strength of various types of materials. Authors of [18] found that the addition of microporous corundum contributes to increase the strength of refractory castable with spinel. The study of corundum modified refractories [19] revealed that the smaller the size of corundum spheres (spheres ranging in size from 100 to 2000 microns were studied) the stronger their effect on increasing the strength of refractory.

Thermal shock resistance of refractories is determined using tests in which the material is heated and cooled, and the number of cycles that a material can withstand prior to failure and without spalling, is taken as its resistance to thermal shock. However, experimental method of thermal shock resistance evaluation is both expensive and time-consuming [20]. The level of degradation of the samples can also be monitored before and during testing by nondestructive test, such as ultrasonic pulse velocity technique [21]. The formation of the cracks has a significant impact on the ultrasonic velocity and the Young's modulus of the refractories. The revealed correlation between thermomechanical properties, Vup, microstructure, and the nature of crack propagation makes it possible to predict the thermal shock resistance of the material without performing a large number of experimental cycles.

In the experiments described in this paper, the addition of corundum microspheres was used to improve both the physical and mechanical properties as well as the thermal resistance of medium cement castable with bauxite aggregate.

2. Materials and Methods

The following materials were used to make the samples.

High alumina cement Gorkal-70 (G70) (chemical composition, mass %: Al_2O_3—71.0; CaO—28.0; SiO_2—0.5, and Fe_2O_3—0.4. Blaine surface area 450 m^2/kg, bulk density 1100 kg/m^3) manufactured by Górka Cement Sp. zo.o. (Trzebinia, Poland).

Microsilica (MC) RW-Fuller (chemical composition, mass %: SiO_2—96.1, Al_2O_3—0.2, Fe_2O_3—0.1, C—0.6, CaO—0.3, MgO—0.4, K_2O—1.2, Na_2O—0.1, and SO_3—0.3) manufactured by RW Silicium GmbH (Pocking, Germany).

Reactive alumina (RA) CTC 20 (chemical composition, mass %: Al_2O_3—99.7; Na_2O—0.1; Fe_2O_3—0.03; SiO_2—0.03; and CaO—0.02. Blaine surface area 2100 m^2/kg) manufactured by Almatis (Ludwigshafen, Germany).

Calcined alumina (CA) CT 19 (chemical composition, mass %: Al_2O_3—99.8; Na_2O—0.1; Fe_2O_3—0.02; SiO_2—0.01; and CaO—0.03. Blaine surface area 400 m^2/kg) manufactured by Almatis (Ludwigshafen, Germany).

Bauxite (chemical composition, mass %: Al_2O_3—81.7; SiO_2—10.0; TiO_2—4.4; Fe_2O_3—2.4; CaO—0.5; P_2O_5—0.3; K_2O—0.3; MgO—0.2; ZrO_2—0.2; Na_2O—0.04; and SO_3—0.02) of different fractions was used as a filler by Stanchem (Niemce, Poland).

The mixtures were prepared using the following chemical additives: polycarboxylate ether (PCE) Castament FS 20 manufactured by BASF Construction Solutions GmbH (Trotsberg, Germany) and sodium tripolyphosphate (TP) $Na_5P_3O_{10}$. Bauxite of different fractions was used as a filler. In experimental compositions 25%, 50%, and 100% of bauxite of 0–0.1 mm fraction (bulk density 1770 kg/m^3) were replaced with hollow corundum microspheres (Kit-Stroi SPb, Saint Petersburg, Russia) with a particle size ranging from 5 to

100 μm (Figure 1a), which corresponds to 2.5%, 5%, and 10% of the dry mixture mass. The bulk density of the HCM is 1750 kg/m^3.

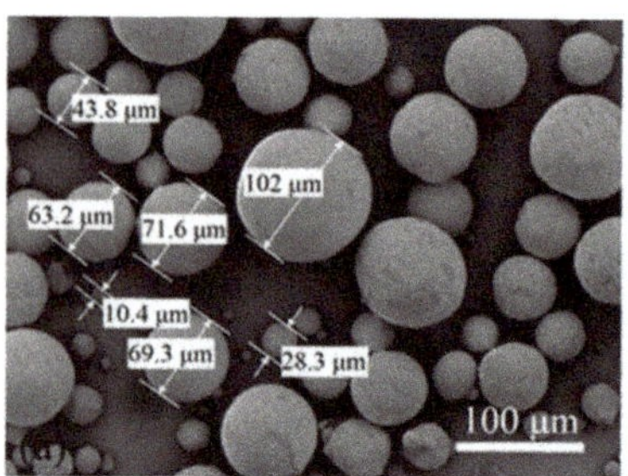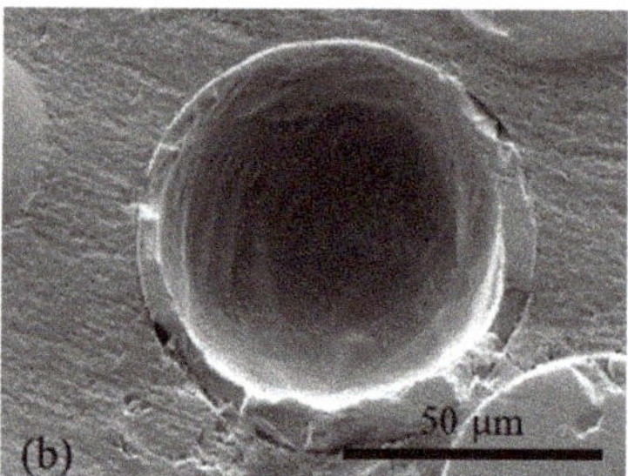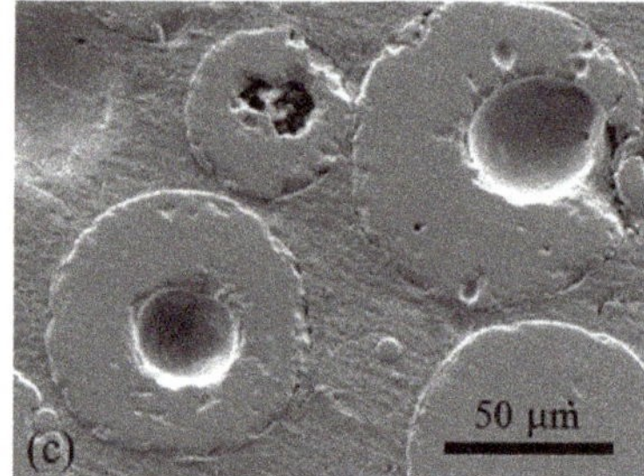

Figure 1. External (**a**) and internal (**b,c**) morphology of HCM (SEM images).

SEM studies revealed that some of the HCM used in this work were solid. The hollow spheres had the wall thickness varying over a wide range (Figure 1b,c).

According to the results of chemical analysis (by X-ray fluorescence spectroscopy (XRF)) HCM consists of 99.0 % of Al_2O_3; the rest are SiO_2, Na_2O, and ZrO_2 (~0.65% in total) and such compounds as MgO, P_2O_3, SO_3, K_2O, CaO, Fe_2O_3, ZnO, Ga_2O_3, and Y_2O_3 (~0.35% in total). X-ray diffraction analysis (XRD) showed that the main phase of HCM is α-Al_2O_3 of trigonal crystal structure, which is a stable form of aluminum oxide. The diffraction curve also shows reflections that can be attributed to cubic (γ-Al_2O_3) and monocline (θ-Al_2O_3) modifications of aluminum oxide and reflections typical of SiO_2 (Figure 2).

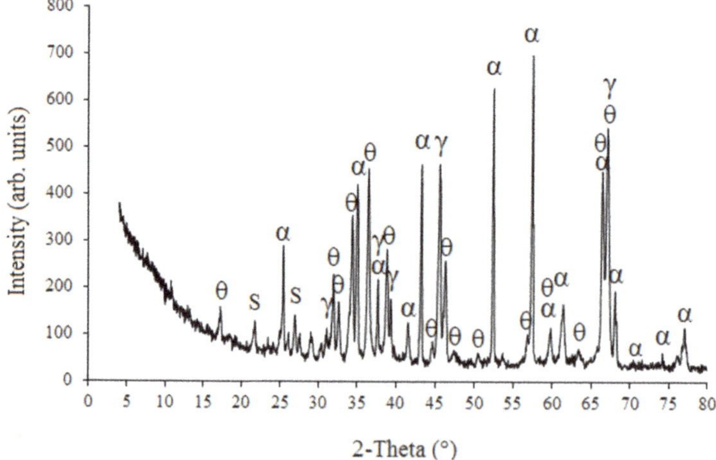

Figure 2. XRD pattern of HCM diffraction curve: α—α-Al_2O_3; γ—γ-Al_2O_3; θ—θ-Al_2O_3; S-SiO_2.

The castable mixes were prepared using potable water. Compositions of castables are given in Table 1.

Table 1. Composition of castables.

Grade	G70, g	MC, g	RA, g	CA, g	Bauxite 0–0.1 mm, g	Bauxite 1–3 mm, g	Bauxite 0–1 mm, g	PCE, g *	TP, g *	HCM, g *	Water, g *
K0	1440	360	600	600	1200	5400	2400	12	12	0	840
K2.5	1440	360	600	600	900	5400	2400	12	12	300	840
K5	1440	360	600	600	600	5400	2400	12	12	600	840
K10	1440	360	600	600	0	5400	2400	12	12	1200	840

*—above 100%, calculated according mass of dry materials.

Castable dry materials were mixed for 5 min, then water was added and the mixing continued for another 5 min. Then, 70 mm × 70 mm × 70 mm samples were formed from the prepared mixture and kept in the mold for 24 h. Afterwards the samples were taken out, they were conditioned for 2 days at 20 ± 1 °C and dried at 110 ± 5 °C.

Samples (3 of each composition) for the tests of physical and mechanical properties according to LST EN ISO 1927-6:2013 were fired at 1100 ± 5 °C and 1300 ± 5 °C. The compressive strength was determined using the press ALPHA3-3000S (Riedlingen, Germany).

The samples for thermal shock resistance tests were fired at 950 °C. Thermal shock resistance of castable was established for 3 samples of each composition in accordance with DIN 51068, according to which the samples were kept for 15 min at 950 °C and then cooled for 3 min in water (cycle 1).

The destruction of the material was estimated by means of ultrasonic tests with the samples subjected to thermal cycling (3 of each composition). The ultrasonic pulse velocity (Vup, m/s) was calculated according to literature [22].

The compressive strength degradation was calculated according to Equation (1):

$$\sigma = \sigma_0 \cdot \left(\frac{\mathrm{Vup}_0}{\mathrm{Vup}_{30}} \right)^n \tag{1}$$

where: σ_0 is the compressive strength of the sample before exposure of the material to the thermal shock testing, MPa; Vup_0 is the longitudinal velocity before testing, m/s; Vup_{30} is the longitudinal velocity after testing (30 cycles), m/s; n is the material constant (0.488) [23,24].

The modulus of elasticity was calculated from the equation [25] for 3 samples of each composition:

$$E = \mathrm{Vup}^2 \cdot \rho \frac{(1 + \mu)(1 - 2\mu)}{1 - \mu} \tag{2}$$

where: Vup is ultrasonic pulse velocity, m/s; ρ is density, kg/m^3; and μ is Poisson's ratio of 0.17 for all castables.

Also, for the evaluation of material thermal shock resistance, the σ/E ratio was calculated.

Microstructure analysis of the materials was done with a scanning electron microscope SEM JEOL JSM-7600F (Tokyo, Japan) on splitting surfaces pre-coated with a conductive gold layer (QUORUM Q150R ES vacuum sputtering coating machine, Quorum technologies, Laughton, UK).

3. Results and Discussion

The replacement of the bauxite part by having the same bulk density HCM was not found to have a significant effect on the density of castables after hardening at 20 °C (~2730 kg/m^3), drying at 110 °C (~2650 kg/m^3), and after firing at different temperatures (at 1100 °C—~2590 kg/m^3 and at 1300 °C—~2620 kg/m^3). The difference in the average density values did not exceed 1%.

The mechanical testing of castable specimens showed (Figure 3) that up to 5% addition of HCM had no significant effect on the strength of the castable. A slight increase in compressive strength (CS) at 5% HCM after drying and firing at 1100 °C can be noted, but the difference with the control sample does not exceed 5%. In specimens containing up to

10% of HCM, a more expressed tendency of castable strength reduction was observed both after hardening and drying, as well as after firing at 1100 and 1300 °C. This reduction can be related to a clustering of HCM particles and formation of more voids in the sample matrix.

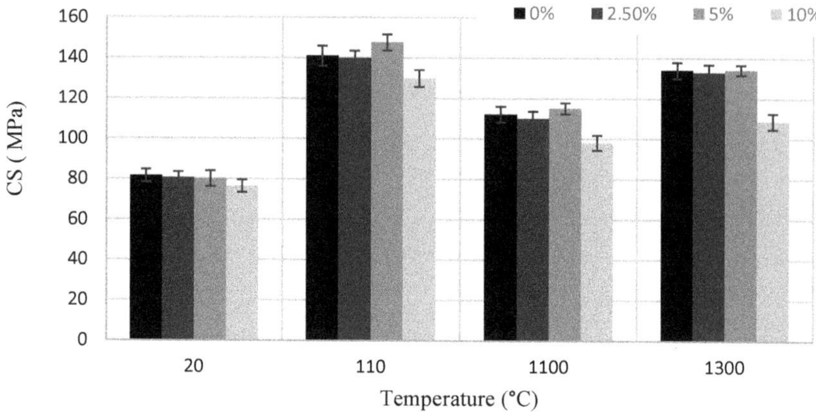

Figure 3. Dependence of compressive strength (CS) of the castable on the amount of HCM and temperature.

The castable compositions investigated in this work withstood 30 cycles without significant signs of fracture. However, the formation of micro and macro cracks on the surface of the specimens was observed during thermal cycling and the length and width of the cracks increased with the number of thermal shocks. The difference in the damage of the different samples is well noticeable after 15 cycles and after 30 cycles the pattern of crack formation did not change much. The largest number of macro-cracks (Figure 4) was detected on the surfaces of control specimens (K0), as well as on specimens K10 containing the largest amount of HCM (10 wt. % of dry mix). The specimens with a lower HCM content of 2.5% and 5% (K2.5 and K5) were damaged the least.

Both the material structural continuity and density determine ultrasonic wave velocity in materials: the structural damages disturb propagation of ultrasonic wave and decrease its velocity. The ultrasonic wave in material with defects reflect, refract, and diffract, which will lead to changes (decreases) in the propagation velocity of the ultrasonic waves [26,27].

The drop in Vup after firing at high temperatures is typical for refractory concretes: the discontinuity of microstructure increases due to destroying hydraulic bonds and evaporation of chemically combined hydraulic phases. The curves of ultrasonic pulse velocity variation DV (%) in the tested castable samples during thermal cycling are shown in Figure 5. The bigger decrease in ultrasonic pulse velocity corresponds to a bigger amount of microcracks and other discontinuities developed in material during thermal cycling and, respectively, lower material capability to withstand thermal shocks.

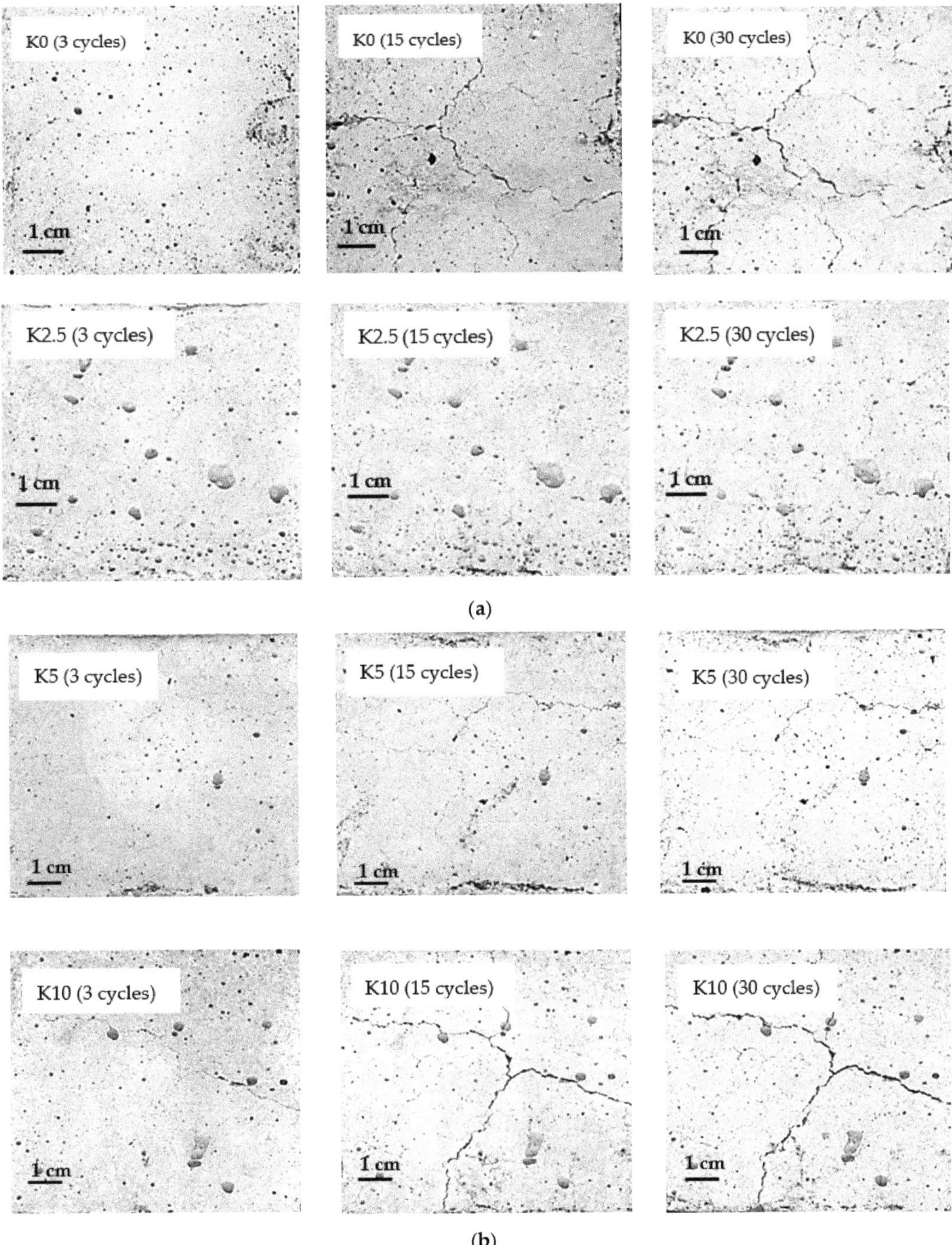

Figure 4. (**a**) Surfaces of castable specimens K0 and K2.5 after 3, 15, and 30 thermal cycles. (**b**) Surfaces of castable specimens K5 and K10 after 3, 15, and 30 thermal cycles.

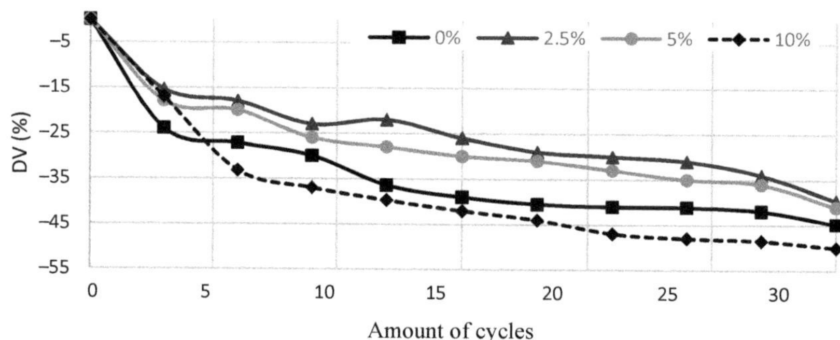

Figure 5. Variation of ultrasonic pulse velocity (DV) during thermal cycling of specimens with different HCM content.

A sharp decrease of Vup was observed in all specimens in the first stage of thermal cycling. This decrease is associated with the formation of the first microcracks in the material. Vup continued decreasing slowly and gradually with the increasing number of thermal cycles during further testing caused by gradual degradation of castable structure and accumulation of micro and macro cracks. The most significant decrease of Vup (~50%) after 30 cycles was observed for specimens K10, which cracked the most compared to the rest of the specimens (Figure 4). In specimens K2.5 and K5, the surface of which showed less prominent formation of macrocracks after 30 thermal cycles, Vup reduced by ~40%. The control specimens showed an intermediate value of Vup decrease, ~45%. Applying Equation (1), the strength degradation after 30 cycles for K0, K2.5, K5, and K10 reach 83.8 ± 1.1 MPa, 86.0 ± 0.9 MPa, 88.9 ± 1.2 MPa, and 69.9 ± 0.9 MPa respectively. Ultrasonic pulse velocity testing and strength degradation calculation results as well as visual analysis of specimens during thermal cycling (Figure 4) suggest that a small amount of HCM in refractory castable can reduce the formation and propagation of cracks and thus increase its thermal shock resistance.

The modulus of elasticity calculated from the Equation (2) for specimens K0, K2.5, K5, and K10 subjected to firing at 950 °C was 35.4 ± 0.2 GPa, 30.2 ± 0.3 GPa, 33.0 ± 0.7 GPa, and 33.9 ± 0.1 GPa, respectively. The results of the present work correlate with the data reported in the literature. Researchers [28] claim that thermal shock resistance of castables improves by lowering the modulus of elasticity.

Stress-to-elastic modulus ratio (σ/E), characterizes the ability of a material to resist crack nucleation, and strength damage factor, showing theoretical drop in material compressive strength after thermal cycling for K0, K2.5, K5, and K10 reach 3.2 ± 0.10, 3.6 ± 0.08, 3.5 ± 0.09, and 2.9 ± 0.06 respectively. These results show, that a stronger material (compositions K0, K2.5, and K5, Figure 3) with a lower modulus of elasticity (especially K2.5) have a higher thermal resistance.

The microstructure of castables was investigated by SEM. The HCM distribution in the structure of castable and the contact zone between the HCM and the binder were analyzed (Figure 6). A fairly uniform distribution of particles was observed in the structure of castable specimens containing up to 5% of HCM with each microsphere surrounded by the binder layer, tightly adjoining the particle surface (Figure 6a). As HCM are inert, they do not react with binder. That is evidenced by the clean, smooth, unchanged surface of HCM particles after hardening (Figure 6a,b) and drying of castable at 110 °C (Figure 6c,d), as well as the well-defined interface line between HCM surface and the binder (Figure 6d). The firing of castable at 950 °C resulted in an altered morphology of the particle surface (Figure 6e,f) and the formation of common phases in the HCM-binder contact zone (Figure 6g), which, in turn, ensure their bonding. A further change in HCM surface morphology and increased area of HCM adhesion with the binder (Figure 6h,i) was observed at the higher firing temperature of 1100 °C. At 1300 °C, the sintering of HCM with the matrix occurred all over the contact area (Figure 6j,k). After three thermal cycles at 950 °C, micro-cracks were detected on the surface of some HCM particles (Figure 6l). It can be assumed that the formation of cracks and fracture of HCM (presumably the thin-walled microspheres) occurs under the influence of compressive stresses caused by thermal deformation of castable while heating. Since part of the energy is spent on the fracturing of HCM, the overall stress level is reduced and thus significantly fewer micro-cracks are formed in hardened cement paste. It can also be assumed that the regular spherical shape of HCM particles has a positive effect. Stress concentration commonly occurs in the point of structure defects. A regular spherical shape has the lowest stress concentration factor compared to other shapes. Thus, evenly distributed in the matrix microspheres act not only as traps for cracks inhibiting crack propagation, but also reduce the level of stress concentration and more effectively suppress the propagation of microcracks at the binder and HCM interface.

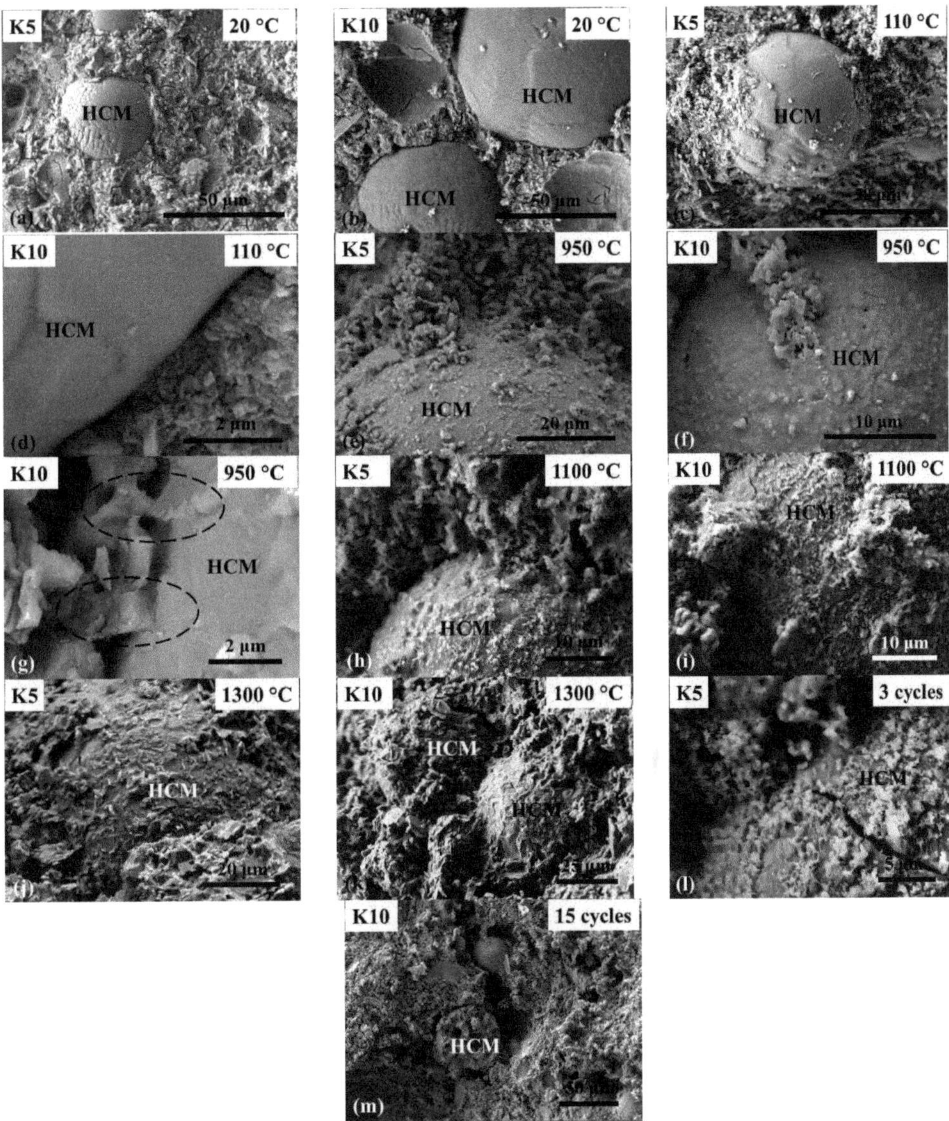

Figure 6. SEM images of the structure of castables containing 5% and 10% of HCM after hardening at 20 °C (**a,b**), drying at 110 °C (**c,d**), firing at 950 °C (**e–g**), 1100 °C (**h,i**), and 1300 °C (**j,k**), and after 3 and 15 thermal cycles (**l,m**).

A higher occurrence of microsphere clusters was observed in castables containing a higher amount of HCM (10%). In this case, the matrix does not completely cover the surface of the particle and spaces remain between the particles or a thin layer of matrix is formed between the particles, which breaks down when concrete is deformed during thermal loads, subsequently, forming cavities (Figure 6b). Thus, coarser void of irregular shape instead of regular spherical void is formed in the matrix in the location of microspheres cluster, diminishing the positive effect from HCM addition in such a way. It was determined also that during thermal cycling, macro-cracks passed through HCM clusters and joined the voids formed in the structure (Figure 6m). As is known, the presence of large, irregularly distributed voids across the material volume impairs the mechanical properties of castable. This impairment is illustrated by decreased CS and reduced thermal shock resistance of castable containing 10% of HCM.

4. Conclusions

The effect of hollow corundum microspheres (HCM) on both physical mechanical properties and thermal shock resistance behavior of refractory medium cement castable with bauxite aggregate was studied in the present work. The water quench test and nondestructive ultrasonic testing method along with calculated strength degradation and σ/E ratio were used to evaluate and compare thermal shock resistance of refractory castable. The following conclusions were drawn from the results of experimental research described in this paper:

A 2.5% and 5% addition of 0–100 μm size HCM improves the thermal resistance of alumina cement-based refractory castable without significantly affecting its density and compressive strength: as compared with reference samples, for these specimen series, the appearance of fewer outside cracks was established by visual control during water quench testing up to 30 cycles while less drop in ultrasonic pulse velocity revealed fewer inside damages. Stress-to-elastic modulus ratio (σ/E), characterizing the ability of a material to resist crack nucleation, and strength degradation, showing theoretical drop in material compressive strength after thermal cycling, were improved by approximately 12% and 10%, respectively. The most probable mechanism for increasing thermal resistance is the reduction of stress concentration in the material due to the regular spherical shape of the HCM.

The deterioration of thermal shock resistance parameters and compressive strength was observed with the increase of HCM mass fraction in castable up to 10%: the appearance of the outside cracks was comparable with that of reference samples while other parameters (Vup, σ/E, and strength degradation) were even worse. Such worsening effect may be associated with the clustering of HCM particles in the matrix, which was observed during microstructural analysis and, in terms of material matrix continuity, can be considered as a formation of coarser non-uniformly distributed cavities of irregular shape, that may cause the increase in stress concentration level.

Author Contributions: Conceptualization, R.S. and J.M.; methodology, J.Š. and V.A.; validation, R.S. and J.M.; formal analysis, J.M. and J.Š.; investigation, J.M. and J.Š.; resources, R.S. and V.A.; data curation, J.M. and J.Š.; writing—original draft preparation, R.S. and J.M.; writing—review and editing, J.Š. and V.A.; visualization, J.M. and J.Š.; supervision, V.A.; project administration, R.S. and V.A.; funding acquisition, V.A. All authors have read and agreed to the published version of the manuscript.

Funding: This research was funded by the Lithuanian Science Council (LMTLT), under contract No. S-MIP-19-41.

Institutional Review Board Statement: Not applicable.

Informed Consent Statement: Not applicable.

Data Availability Statement: Not applicable.

Acknowledgments: The work was supported by the Lithuanian Science Council (LMTLT), under contract No. S-MIP-19-41.

Conflicts of Interest: The authors declare no conflict of interest.

References

1. Ghassemi Kakroudi, M.; Yeugo-Fogaing, E.; Gault, C.; Huger, M.; Chotard, T. Effect of thermal treatment on damage mechanical behaviour of refractory castables: Comparison between bauxite and andalusite aggregates. *J. Eur. Ceram. Soc.* **2008**, *28*, 2471–2478. [CrossRef]
2. Auvray, J.M.; Gault, C.; Huger, M. Evolution of elastic properties and microstructural changes versus temperature in bonding phases of alumina and alumina–magnesia refractory castables. *J. Eur. Ceram. Soc* **2007**, *27*, 3489–3496. [CrossRef]
3. Kasheev, I.D.; Strelov, K.K.; Mamykin, P.S. *Chemical Technology of Refractory Materials*; Intermet Inziniring: Moscow, Russian, 2007; 752p. (In Russian)
4. Nishikawa, A. *Technology of Monolithic Refractories*; Japan by Toppan Printing Company: Tokyo Japan, 1984; 598p.
5. Ulbricht, J.; Dudczig, S.; Tomšu, F.; Palco, S. Technological measures to improve the thermal shock resistance of refractory materials. *Interceram Refract. Man.* **2012**, *2*, 103–106.
6. Bradt, R.C. Fracture of Refractories. In *Refractories Handbook*; Schacht, C.A., Ed.; CRC press: Boca Raton, FL, USA, 2004; Chapter 2, pp. 11–38.
7. Mori, J.; Maeda, E.; Sorimachi, K. *Effects of Coarse Grain Material on the Fracture Properties of Castable*; Proceedings of Unitecr; Technical Association of Refractories: Osaka, Japan, 2003; pp. 572–575.
8. Schnieder, J.; Traon, N.; Etzold, S.; Tonnesen, T.; Telle, R.; Tengstrand, A.; Malmgren, A.; Nylund, E. Fracture process zone in refractory castables after high-temperature thermal shock. *Refract. Worldforum* **2016**, *8*, 74–80.
9. Mráz, D.; Vlček, I.J. Evaluation of resistance of refractory concretes with heterogeneous structure to sudden thermal shocks. *Hutnické listy č.* **2017**, *2*, 33–38.
10. Strelov, K.K. *Structure and Properties of Refractories*, 2nd ed; Metallurgiya: Moscow, Russian, 1982; 216p. (In Russian)
11. Samadi, H.; Fard, F.G. *The Effect of Fiber Addition on Low Cement Castables*; Proceedings of Unitecr; Technical Association of Refractories: Osaka, Japan, 2003; pp. 268–271.
12. Antonovič, V.; Witek, J.; Mačiulaitis, R.; Boris, R.; Stonys, R. The effect of carbon and polypropylene fibers on thermal shock resistance of the refractory castable. *J. Civ. Eng. Manag.* **2017**, *23*, 672–678. [CrossRef]
13. Shen, L.; Liu, M.; Liu, X.; Li, B. Thermal shock resistance of the porous Al_2O_3/ZrO_2 ceramics prepared by gelcasting. *Mater. Res. Bull.* **2007**, *42*, 2048–2056. [CrossRef]
14. Wang, Y.; Li, X.; Chen, P.; Zhu, B. Matrix microstructure optimization of alumina-spinel castables and its effect on high temperature properties. *Ceram. Int.* **2018**, *44*, 857–868. [CrossRef]
15. Vlasov, A.S.; Polyak, B.I.; Postnikov, S.A. Heat-insulating corundum ceramics based on hollow microspheres. *Glass Ceram.* **1999**, *56*, 3–4. [CrossRef]
16. Krasnyi, B.L.; Tarasovskii, V.P.; Krasnyi, A.B.; Galganova, A.L.; Reznichenko, A.V. Heat-insulating refractory material based on hollow corundum microspheres. *Refract. Indust Ceram.* **2015**, *55*, 559–561. [CrossRef]
17. Li, M.; Li, Y.; Ouyang, D.; Wang, X.; Li, S.; Chen, R. Effects of alumina bubble addition on the properties of mullite castables. *J. Alloy. Comp.* **2018**, *735*, 327–337. [CrossRef]
18. Fu, L.; Gua, H.; Huang, A.; Zhang, M.; Hong, X.; Jin, L. Possible improvements of alumina–magnesia castable by lightweight microporous aggregates. *Ceram. Int.* **2015**, *41*, 1263–1270. [CrossRef]
19. Suvorov, S.A.; Kapustina, S.N.; Fishcher, V.N.; Tarnavskaya, I.A. Effect of composition and the amount of spherical-particle aluminous fillers on the density and strength of corundum refractories. *Therm. Eng* **1987**, *28*, 77–82. [CrossRef]
20. Canio, M.; Boccaccini, D.N.; Romagnoli, M. New Methods for the Assessment of Thermal Shock Resistance in Refractory Materials. In *Encyclopedia of Thermal Stresses*; Springer Netherlands: Heidelberg, Germany, 2014; pp. 3293–3307.
21. Martinovic, S.; Vlahovic, M.; Boljanac, T.; Majstorovic, J.; Volkov-Husovic, T. Influence of sintering temperature on thermal shock behavior of low cement high alumina refractory concrete. *Compos. Part. B* **2014**, *60*, 400–412. [CrossRef]
22. Kerienė, J.R.; Antonovič, V.; Stonys, R.; Boris, R. The influence of the ageing of calcium aluminate cement on the properties of mortar. *Constr Build. Mater.* **2019**, *205*, 387–397. [CrossRef]
23. Marenovica, S.; Dimitrijevic, M.; Volkov Husovic, T.; Matovic, B. Thermal shock damage characterization of refractory composites. *Ceram. Int.* **2008**, *34*, 1925–1929. [CrossRef]
24. Volkov Husovic, T.; Jangic-Heinemann, R.M.; Mitrakovic, D.; Acimovic-Pavlovic, Z.; Raic, K. The use of image analysis program for the determination of surface deterioriation of refractory specimen during thermal shock. *Mater. I Konstr.* **2006**, *49*, 60–63.
25. Niyogi, S.K.; Das, A.C. Prediction of thermal shock behaviour of castable refractories by sonic measurements. *Refractories* **1994**, *43*, 453–457.
26. Xu, J.; Wei, H. Ultrasonic Testing Analysis of Concrete Structure Based on S Transform. *Shock Vib.* **2019**, *2693141*. [CrossRef]
27. Boccaccini, D.N.; Romagnoli, M.; Kamseu, E.; Veronesi, P.; Leonelli, C.; Pellacani, G.C. Determination of thermal shock resistance in refractory materials by ultrasonic pulse velocity measurement. *J. Eur. Cer. Soc.* **2007**, *27*, 1859–1863. [CrossRef]
28. Szczerba, J.; Pedzicha, Z.; Nikiel, M.; Kapuscinska, D. Influence of raw materials morphology on properties of magnesia-spinel refractories. *J. Eur. Ceram. Soc.* **2007**, *27*, 1683–1689. [CrossRef]

materials

MDPI

Article

Investigating the Action Mechanism of Titanium in Alumina–Magnesia Castables by Adding Different Ti-Bearing Compounds

Hai Tang [1], Yuhao Zhou [1] and Wenjie Yuan [1,2,*]

[1] The State Key Laboratory of Refractories and Metallurgy, Wuhan University of Science and Technology, Wuhan 430081, China; tanghaiwust@sina.com (H.T.); heltlong@gmail.com (Y.Z.)
[2] National-Provincial Joint Engineering Research Center of High Temperature Materials and Lining Technology, Wuhan University of Science and Technology, Wuhan 430081, China
* Correspondence: yuanwenjie@wust.edu.cn

Abstract: To investigate the action mechanism of titanium, the effects of different Ti-bearing compounds, including $CaTiO_3$, $MgTiO_3$, and nano-TiO_2, on the properties of alumina–magnesia castables were studied. By analyzing the phase compositions, microstructures, and physical and mechanical properties of the castables, it was demonstrated that an intermediate product, $CaTiO_3$, was first generated. This was then consumed by solid-solution reactions, and titanium was involved in the liquid formation as the temperature increased. The solid-solution reaction of CA_6 ($CaAl_{12}O_{19}$) was more prominent due to the incorporation of more titanium in the crystal lattice of CA_6 instead of spinel ($MgAl_2O_4$). Moreover, the liquid formation was strongly promoted when more titanium accompanied the calcium, which finally accelerated the densification and improved the strengths of alumina–magnesia castables. On the whole, castables with $CaTiO_3$ addition presented higher bulk density and excellent strength after the heat treatment. Besides, the castables with 2 wt.% $CaTiO_3$ contents were estimated to possess greater thermal shock resistance.

Keywords: Ti-bearing compounds; CA_6; spinel; castables

Citation: Tang, H.; Zhou, Y.; Yuan, W. Investigating the Action Mechanism of Titanium in Alumina–Magnesia Castables by Adding Different Ti-Bearing Compounds. *Materials* **2022**, *15*, 793. https://doi.org/10.3390/ma15030793

Academic Editor: Jacek Szczerba

Received: 11 December 2021
Accepted: 18 January 2022
Published: 21 January 2022

Publisher's Note: MDPI stays neutral with regard to jurisdictional claims in published maps and institutional affiliations.

1. Introduction

Alumina-magnesia castables served at high temperatures mainly comprise alumina, spinel, and CA_6. The presence of spinel can improve the slag corrosion and erosion resistances of castables due to its high chemical resistance and mechanical properties [1–4]. Ko [5] reported that alumina-magnesia castables had better slag resistance than alumina-spinel castables, because the in-situ spinel possessed smaller grain sizes. Furthermore, CA_6 formed by the reaction between calcium aluminate cement and reactive alumina powders resulted in excellent mechanical strengths of the castables owing to the bridging and deflection mechanism of platelet CA_6 [6]. The high stability in the reducing atmosphere and low solubility of CA_6 in slag also allow it to be in contact with molten iron and steel [7]. However, the formation of CA_6 and spinel leads to volume expansions of 3.01 and 8%, respectively, which would result in the spalling of castables at elevated temperatures [8]. Thus, mineralizers, such as TiO_2, B_2O_3, ZrO_2, and rare earth oxides, have been introduced into this system to accelerate the densification of castables [9,10].

Among these additives, TiO_2 is regarded as one of the most effective multifunctional mineralizers, which can control the expansion behavior of alumina-magnesia castables and speed up the formation of CA_6 and spinel [11]. The densification of castables can be accelerated to a great extent due to the larger amount liquid phase during low viscosity forming when TiO_2 is incorporated [12]. In a previous study, it was found that the apparent activation energy of spinel formation varied with the content of TiO_2 in the same system [13]. Furthermore, the morphology of CA_6 could be influenced by the addition of nano-TiO_2 [14].

TiO_2 also functioned as a nucleation agent, promoting the recrystallization of secondary spinel from the liquid phase [15].

Although extensive investigations have found that TiO_2 has a significant influence on the phase evolution of alumina-magnesia castables, few studies have focused the differential impacts of titanium on the formation of spinel and CA_6. Therefore, it is necessary to further evaluate the contribution of titanium on the complex phase evolution of alumina-magnesia castables. Imbalanced distributions of titanium were designed using three kinds of Ti-bearing compounds ($CaTiO_3$, $MgTiO_3$, and nano-TiO_2) in this work. It was assumed that more titanium was located in the region where CA_6 and spinel formed in the samples with $CaTiO_3$ and $MgTiO_3$ addition, respectively. For the reference group (samples with nano-TiO_2 addition), titanium was considered to be evenly distributed in the starting raw materials of the castables. The local compositions were adjusted, and the different effects of titanium on the formation of spinel and CA_6 were investigated.

2. Materials and Methods

The formulations of the alumina-magnesia castables are listed in Table 1. Coarse tabular alumina (Almatis, Germany) was used as aggregates and fine particles served as the matrix of the castables. Reactive alumina (CL370, Almatis, Germany) and calcined magnesia (Dashiqiao, China) were designed to form in situ spinel at high temperatures. Calcium aluminate cement (Secar71, Kerneos, France) acted as the binder of the castables and a source for the CA_6 formation. The added mineralizers included nano-TiO_2 (Aladdin, Shanghai, China, 99.8 wt.% of purity), $CaTiO_3$ (Aladdin, Shanghai, China, 99.5 wt.% of purity), and $MgTiO_3$ (Aladdin, Shanghai, China, 99 wt.% of purity). Under the action of silica fume (951U, Elkem, Norway) and water reducer (FS60, BASF, Ludwigshafen, Germany), the amount of distilled water added was about 4.3 wt.%. The chemical compositions of raw materials are shown in Table 2. Samples with Ti-bearing compounds were divided into two groups, and the amounts of $CaTiO_3$ and $MgTiO_3$ were calculated to be 1 and 2 wt.% TiO_2, respectively. Moreover, the contents of calcined magnesia and calcium aluminate cement were adjusted to ensure that the proportions of calcium and magnesium in the same batch did not vary.

Table 1. Formulations of alumina-magnesia refractory castables.

Raw Materials	Content (wt.%)					
	NT1	NT2	CT1	CT2	MT1	MT2
Tabular alumina (d = 1–6 mm)	33.5	33.1	33.5	33.1	33.5	33.1
Tabular alumina (d = 0.2–1 mm)	27.5	26.9	27.5	26.9	27.5	16.9
Tabular alumina (d ≤ 0.074 mm)	15	15	15	15	15	15
Tabular alumina (d ≤ 0.045 mm)	3	3	4.6	6.3	3	3
Reactive alumina (CL370)	7	7	7	7	7	7
Calcined magnesia (d ≤ 88 μm)	6	6	6	6	5.5	5
Calcium aluminate cement (Secar71)	6	6	3.7	1.3	6	6
Silicon fume (951U)	1	1	1	1	1	1
Nano-TiO_2	1	2	-	-	-	-
$CaTiO_3$	-	-	1.7	3.4	-	-
$MgTiO_3$	-	-	-	-	1.5	3

Table 2. Chemical compositions of raw materials.

Raw Materials	Chemical Compositions (wt.%)					
	Al_2O_3	MgO	SiO_2	CaO	Na_2O	Fe_2O_3
Tabular alumina	99.5	-	≤0.02	-	0.4	-
Reactive alumina (CL370)	99.7	-	0.03	0.02	0.1	0.03
Calcined magnesia (d ≤ 88 μm)	0.52	94.6	1.6	1.41	0.32	1.31
Calcium aluminate cement (Secar71)	≥68.5	<0.5	<0.8	≤31.0	<0.5	<0.4
Silicon fume (951U)	0.4	0.6	96.1	0.3	0.2	0.1

The castables were mixed for 4 min in a rheometer with 4.3 wt.% distilled water. After mixing, the samples were cast in rectangle molds (150 × 25 × 25 mm) and then cured at 25 °C for 24 h in a climatic chamber with a relative humidity of 100%. After drying at 110 °C for 24 h, all the samples were calcined at 1150 °C, 1250 °C, 1350 °C and 1450 °C for 3 h, respectively. The permanent linear change (PLC) of the specimens was obtained by calculating the variation of the dimensions after cooling. The apparent porosity and bulk density of samples were measured by the Archimedes technique following the GB/T 2997-2000 standard [16]. The cold modulus of rupture was measured by a three-point bending test following GB/T 3001-2007 [17]. The elastic modulus of samples was measured by Elastic Modulus & Damping System (RFDA, HTVP1600, IMCE, Genk, Belgium). The phase compositions of the castables were characterized by X-ray diffraction (XRD, X'pert Pro MPD, Philips, Almelo, The Netherlands), and the relative results were analyzed based on the reference intensity ratio (RIR) method using the X'pert Highscore 2.0 Plus software. The microstructures of the castables were characterized by scanning electron microscopy (SEM, JEOL JSM-6610, JEOL Ltd., Tokyo, Japan). The chemical compositions and elemental distributions were analyzed using energy-dispersive X-ray spectroscopy (EDS, Bruker QUANTAX200-30, Karlsruhe, Germany).

3. Results

3.1. Phase Composition

XRD patterns of the castables containing different Ti-bearing compounds are presented in Figure 1. Spinel and CA_6 formed at 1150 and 1250 °C in all the specimens. As reported in a previous study, the formation temperature of CA_6 was up to 1400 °C in alumina-magnesia castables without mineralizer addition [12]. This demonstrated that the formation of CA_6 was accelerated to a great extent by the introduction of Ti-bearing compounds. In addition, other minor phases, including β-Al_2O_3 ($NaAl_{11}O_{17}$), CA_2 ($CaAl_4O_7$), MgO, and $CaTiO_3$, as well as multicomponent products, such as nepheline ($NaAlSiO_4$), anorthite ($CaAl_2Si_2O_8$), and $Ca_3Ti_8Al_{12}O_{37}$, were also detected in the samples calcined at 1150 °C. It is well known that β-Al_2O_3 is derived from commercial alumina, and the liquid phase is usually first generated between β-Al_2O_3 and silica [18]. More β-Al_2O_3 participated in the formation of the liquid as the temperature increased, which resulted in a gradual decrease in the intensity of the diffraction peak. Calcium existed in the form of CA_2, $CaTiO_3$, and anorthite at a relatively low temperature (1150 °C) and then gradually transformed to CA_6 as the temperature increased. MgO was gradually consumed simultaneously, and more spinel was generated with the increase in temperature.

$CaTiO_3$ was detected instead of $MgTiO_3$ (trace $Ca_3Ti_8Al_{12}O_{37}$ was also found for samples MT1 and MT2) in all the samples calcined at 1150 °C. This indicated that nano-TiO_2 and $MgTiO_3$ reacted with calcium aluminate to form $CaTiO_3$ at this temperature. As the temperature increased further, Ti-bearing compounds gradually took part in the complex phase evolution of the matrix. Titanium could be incorporated into the crystal lattice of spinel and CA_6, forming a solid solution [13]. Additionally, Ti-bearing compounds could participate in the formation of liquid [12]. These factors could account for the disappearance of Ti-bearing compounds, especially for $CaTiO_3$ with calcination temperatures above 1150 °C.

Because the formation of a solid solution would cause a change of the lattice parameters, the variations of the diffraction peak positions for spinel and CA_6 were compared to evaluate the doping level of titanium, as presented in Figure 2. The (104) plane of corundum was set as the base plane, and the selected crystal planes of spinel and CA_6 were (311) and (114), respectively. Basically, the diffraction peaks of spinel gradually shifted to higher angles as the temperature increased, as presented in Figure 2a. This could be explained by the formation of Al-rich spinel caused by the aluminum substitution on the magnesium sites [19]. Because the radius of Al (0.143 nm) was less than that of Mg (0.160 nm) [20], the substitution of magnesium with aluminum could reduce the lattice parameter of spinel followed by an increase in the diffraction angle (2θ).

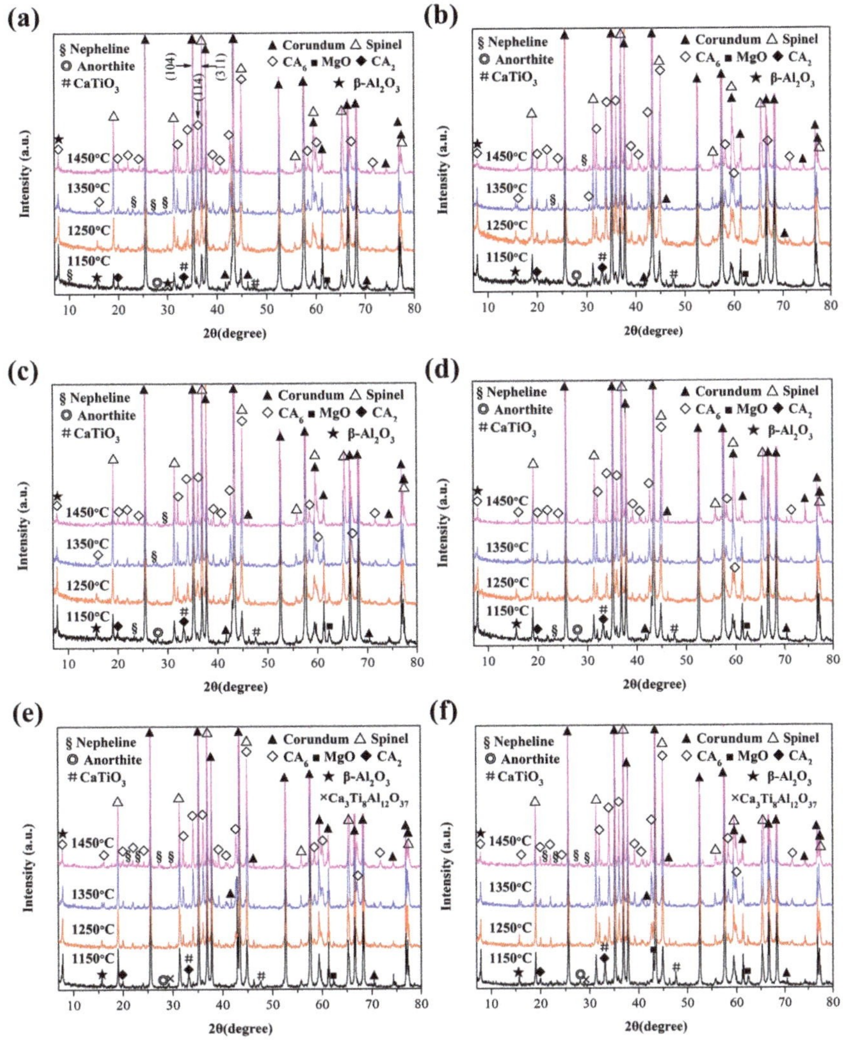

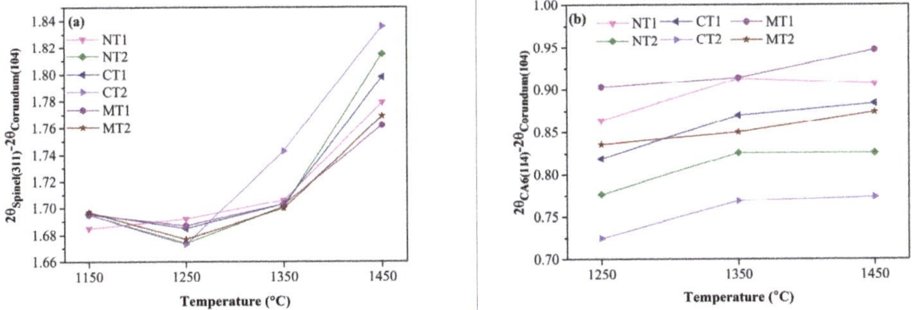

Figure 1. XRD patterns of the castables with the addition of different Ti-bearing compounds after calcination at different temperatures: (**a**) NT1, (**b**) NT2, (**c**) CT1, (**d**) CT2, (**e**) MT1, and (**f**) MT2.

Figure 2. The shift of diffraction angle of spinel and CA_6 in samples calcined at different temperatures: (**a**) spinel and (**b**) CA_6.

Additionally, the shift of the diffraction peak for spinel and CA_6 depended on different Ti-bearing compounds. It has been demonstrated that the doping of titanium into the spinel and CA_6 crystal lattices could cause a slight increase in the lattice constants [12,21]. The diffraction peaks of spinel in the samples with $MgTiO_3$ addition calcined at 1450 °C were located at somewhat lower angles than the other peaks. Similarly, the diffraction angle of the CA_6 peak in specimens with $CaTiO_3$ addition was smaller than those of the other peaks. Moreover, the differences of the diffraction peaks of CA_6 between the specimens were more significant compared with the differences of the peaks of the spinel phase. This demonstrated that the solid-solution reaction of CA_6 strongly depended on Ti-bearing compounds.

CA_6 and spinel contents in the specimens were calculated by the RIR method, as presented in Figure 3. The formation of spinel in the samples with $MgTiO_3$ addition (MT1 and MT2) was accelerated at 1150 and 1350 °C, as shown in Figure 3a. When the calcination temperature reached 1350 °C, the differences of spinel and CA_6 contents between the samples were limited. Nevertheless, $MgTiO_3$ addition favored the formation of spinel and CA_6 at 1450 °C. In contrast, the amounts of spinel and CA_6 in the samples with $CaTiO_3$ addition were significantly lower than those of the other samples calcined at this temperature, which indicated that the formation of liquid consumed MgO, Al_2O_3, and CaO in the castables [18]. Thus, the decline of CA_6 and spinel contents was likely related to greater liquid formation, which is discussed in the following section.

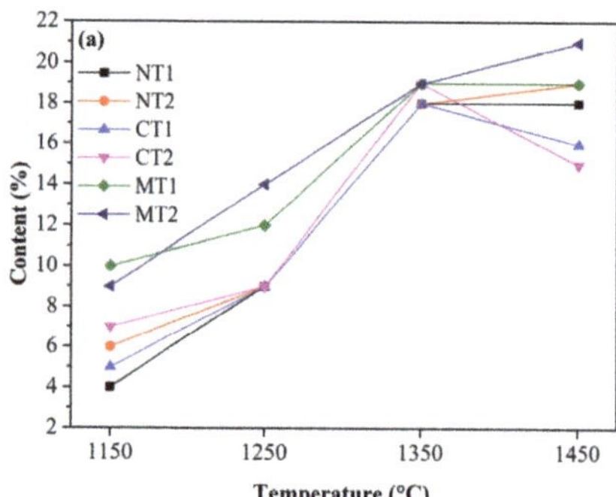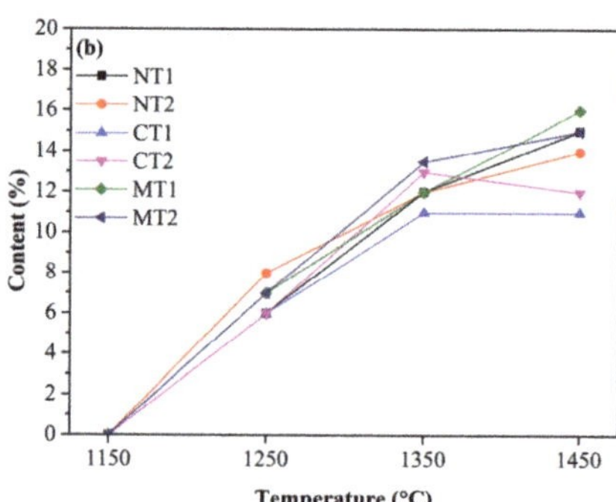

Figure 3. Spinel (**a**) and CA_6 (**b**) contents in alumina-magnesia refractory castables containing Ti-bearing compounds as a function of the calcination temperature.

3.2. Microstructure

As illustrated above, Ti-bearing compounds participated in the complex reactions at higher temperatures, followed by the disappearance of diffraction peaks, as shown in Figure 1. The distributions of titanium in samples CT2 and MT2 calcined at 1450 °C were characterized by SEM/EDS mapping analysis (Figure 4). Titanium was mainly identified in the region where calcium was located rather than magnesium when $MgTiO_3$ or $CaTiO_3$ was added to the castables. Calcium and magnesium mainly appeared in the form of CA_6 and spinel in the samples calcined at 1450 °C, as shown by the XRD patterns. Therefore, the incorporation of titanium in the CA_6 crystal lattice was more prominent. It was reported that large amounts of titanium can dissolve into the structure of CA_6, and the TiO_2 concentration in hibonite can be as high as 10 wt.% [22]. In contrast, the occupation of Ti^{4+} at Al^{3+} sites in the spinel crystal lattice requires to overcome a higher energy barrier [23]. Thus, it was easier for titanium to form a solid solution with CA_6 rather than spinel. Additionally, the transformation of $MgTiO_3$ to $CaTiO_3$ at a low temperature of 1150 °C (as discussed at the phase compositions) also resulted in the preferential distribution of titanium with calcium. These factors explained why the titanium was almost always detected in the area where calcium existed in the CA_6 region.

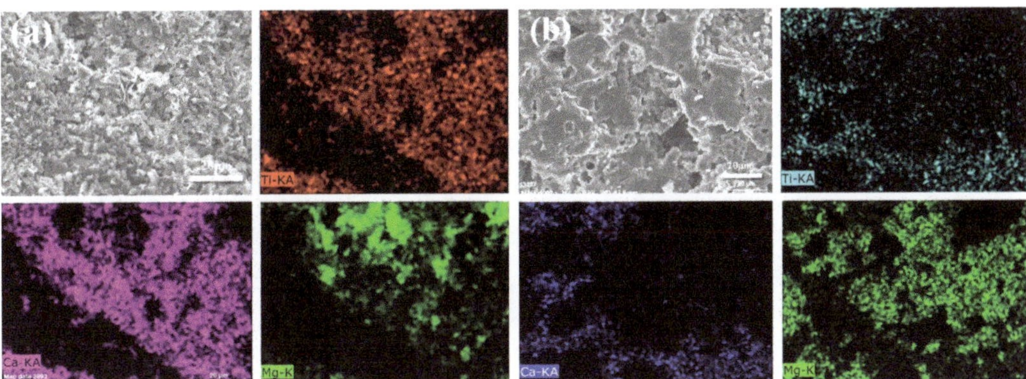

Figure 4. Titanium, calcium, and magnesium distribution in specimens after calcination at 1450 °C for 3 h: (**a**) MT2 and (**b**) CT2.

The microstructures of CA_6 and spinel observed by SEM are shown in Figure 5. Thinner, flaky CA_6 was stacked in the matrix of sample MT2 (shown in Figure 5a). CA_6 in samples NT2 and CT2 exhibited thicker geometric characteristics (Figure 5b,c). The grain sizes of spinel in samples NT2 and CT2 were significantly larger than those in samples MT2. To reveal the mechanism that caused the morphological changes of CA_6 and spinel, the chemical compositions of CA_6 and spinel characterized by EDS analysis are listed in Table 3 (the EDS location is marked with a white point in Figure 5). The proportions of titanium in CA_6 of samples NT2 and CT2 were significantly higher than that of sample MT2. Some studies found that the substitution of aluminum by titanium can greatly promote the mass transfer and grain growth along vertical axes (c-axes) [24]. This indicates that more titanium replaced aluminum in the CA_6 structure, and the thicker, flaky CA_6 was generated in the matrix of the castables. In addition, the substitution of aluminum by titanium can lead to numerous vacancies, which might promote the solid solution formation in spinel and CA_6. Therefore, the doping amount of magnesium increased with more titanium doping into the CA_6 crystal lattice. Only small amounts of titanium and calcium were incorporated into the crystal lattice of spinel (Table 3), which was consistent with the element distribution obtained by the EDS mapping (Figure 4). The liquid phase was an important factor for the morphology of the phase. In general, the formation of liquid likely promoted the crystallization and growth of grains due to the faster diffusion rate [19]. Thus, the formation of granular spinel was most likely related to a lower amount of liquid phase formation in sample MT2.

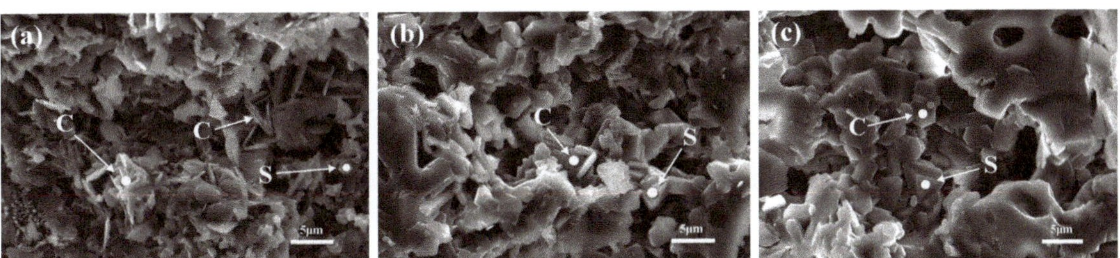

Figure 5. Microstructure of CA_6 (C) and spinel (S) in samples calcined at 1450 °C for 3 h: (**a**) MT2, (**b**) NT2, and (**c**) CT2.

Table 3. EDS analysis of CA$_6$ and spinel in different samples calcined at 1450 °C for 5 h (at. %).

Phase	Samples	Ca	Ti	Mg	Al	O
CA$_6$	MT2	3.9	2.1	1.8	22.4	69.8
	NT2	3.9	3.0	2.2	31.5	59.4
	CT2	3.5	3.3	3.0	33.6	56.6
Spinel	MT2	0.4	0.2	11.8	26.9	60.7
	NT2	0.2	0.3	12.1	38.3	49.1
	CT2	0.3	0.3	12.8	29.3	57.3

3.3. Physical and Mechanical Properties

PLC is an important parameter that represents the volume stability of refractories. As presented in Figure 6, the formation of spinel and CA$_6$ caused volume expansions of 8 and 3% [10], respectively, which resulted in the continuous increase in PLC for all samples until 1350 °C, except for sample CT2. With the temperature further increasing, the shrinkage caused by sintering effects and phase evolution (liquid formation) dominated the calcination process [6], which resulted in a dramatic decrease in PLC for all samples above 1350 °C. Martinez [25] and Sako [18] stated that a liquid phase was first generated in the CaO-Al$_2$O$_3$-SiO$_2$ ternary system, and its chemical composition was similar to that of gehlenite (C$_2$AS), containing a small amount of Na$_2$O in the matrix of alumina–magnesia castables. In addition, titania could assist the formation of liquid in the CaO-Al$_2$O$_3$-SiO$_2$ ternary system [13]. It was demonstrated that liquid was generated in the local area where CaO, Al$_2$O$_3$, SiO$_2$ and Na$_2$O co-existed. As more titanium followed the calcium in the system described above, more liquid was generated. As mentioned above, the uneven distribution of titanium was designed by the incorporation of three different Ti-bearing compounds. It was assumed that titanium tended to distribute into calcium and magnesium in samples with CaTiO$_3$ and MgTiO$_3$ addition, respectively, and titanium was uniformly distributed in castables with nano-TiO$_2$ addition. Thus, more liquid formed, and less volume expansion occurred for the samples with CaTiO$_3$ and nano-TiO$_2$ addition after calcination at 1350 and 1450 °C. In contrast, samples with MgTiO$_3$ addition contained less liquid phase and exhibited greater volume expansion.

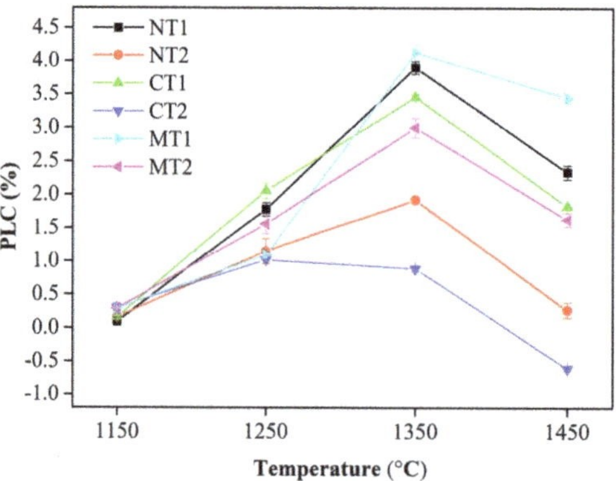

Figure 6. Permanent linear change (PLC) of specimens calcined at different temperatures.

The densification of castables was closely related to the thermal shock and slag resistances of the materials. The variations of the apparent porosity and bulk density of the samples as the temperature increased are presented in Figure 7. It was found that the variations of the apparent porosity of castables were consistent with those of the PLC of samples and were opposite to the variations of the bulk density. The apparent porosity of alumina-magnesia castables without mineralizer addition reached 23% after calcination at 1450 °C for 5 h [13]. For comparison, the apparent porosities of samples MT1 and MT2 were about 24% and 20%, respectively, which indicated that $MgTiO_3$ had little effect on the densification of alumina-magnesia castables. However, a denser body of castables was obtained by the incorporation of two other compounds, especially $CaTiO_3$.

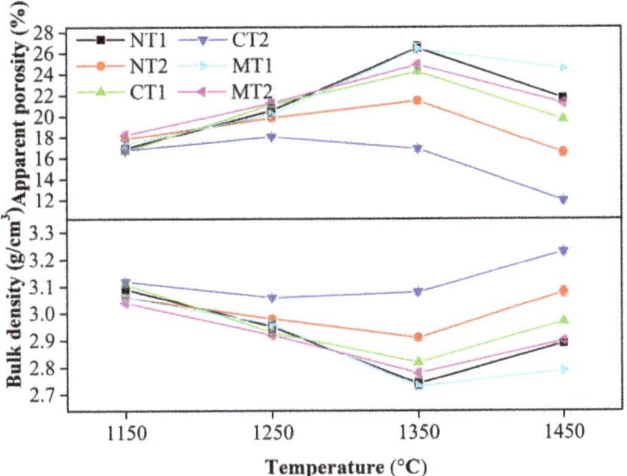

Figure 7. Apparent porosity and bulk density of samples after calcination at different temperatures.

Figure 8 presents the cold modulus of rupture (CMOR) values of the castables calcined at different temperatures. The strengths of the samples had the lowest values after calcination at 1350 °C, except for sample CT2, which underwent the largest volume expansion. In general, a higher bulk density produces a positive influence on the strengths of materials. Nevertheless, the strengths of samples MT2, NT2, and CT2 were similar after calcination at 1450 °C, while their bulk densities and apparent porosities were significantly different, as presented in Figure 7. When CA_6 was interwoven and stacked, the strengths of the samples were enhanced [6,26]. The amounts of CA_6 in samples CT2 and NT2 were less than that in sample MT2, as shown in Figure 3. Thinner, flaky CA_6 was stacked in the matrix of sample MT2. However, thicker plates of CA_6 were scattered in samples NT2 and CT2, as shown in Figure 5. The decrease in the CA_6 content and its morphological changes hindered the further improvement of the strength.

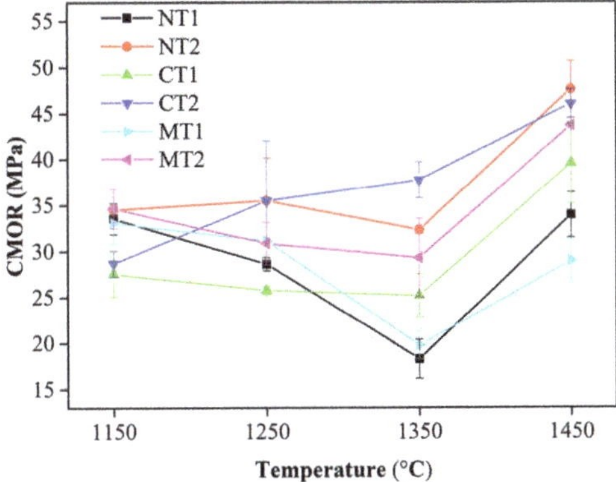

Figure 8. Variation of cold modulus of rupture for the castables with calcination temperatures.

The evolution of the phase, texture, and apparent porosity have a major impact on the elastic modulus (E) of refractories [27,28]. The elastic modulus of the samples calcined at different temperatures were measured and are presented in Figure 9. The variations of the elastic modulus of the samples were the same as those of the bulk densities, which indicated the elastic modulus of alumina-magnesia castables strongly depended on the densification degree.

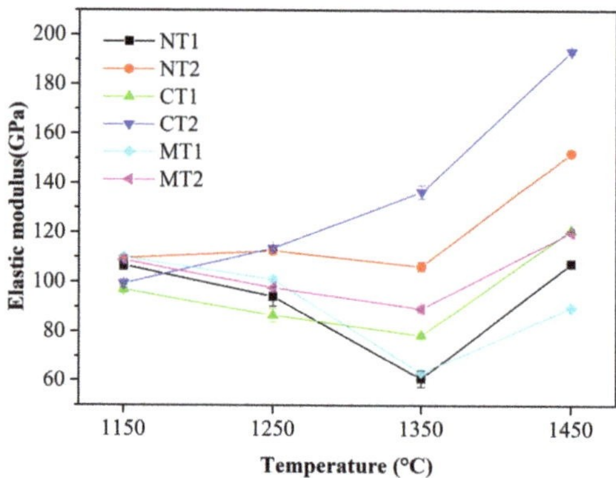

Figure 9. Variation of elastic modulus for the castables with calcination temperatures.

For practical applications, refractories need to undergo rapid temperature changes. High-strength refractories are much more susceptible to thermal shock damage [29]. The stored elastic strain energy at the stress level of σ can be quantified as σ^2/E [30], which plays a critical role on the crack extension. A lower value of σ^2/E corresponds to better thermal shock resistance. The values of σ^2/E and the strengths and elastic modulus for samples calcined at 1450 °C for 3 h are listed in Table 4, where parameters σ and E represented CMOR and elastic modulus in this work, respectively. Although the σ^2/E value of sample MT1 was lower than the others, this sample could not bear thermal stress, as it had the lowest strength. In general, there were no significant differences in the strengths of samples NT2, CT2, and MT2. The elastic modulus of sample CT2 was significantly higher than those of NT2 and MT2 because of the lower porosity. The resistance to thermal shock damage was inversely proportional to the elastic strain energy [31]. Thus, sample CT2, with a relative lower value of σ^2/E, was estimated to have greater thermal shock resistance.

Table 4. Results of strength, elastic modulus, and σ^2/E of samples calcined at 1450 °C for 5 h.

Parameters	NT1	NT2	CT1	CT2	MT1	MT2
CMOR (MPa)	33.9	47.6	39.5	46.0	28.9	43.7
Elastic modulus (GPa)	107.4	152.3	121	193.7	89.9	120.1
σ^2/E (kPa)	10.70	14.85	12.90	10.92	9.32	15.86

4. Conclusions

After comparing the physical properties of samples containing different Ti-bearing compounds (nano-TiO_2, $MgTiO_3$, and $CaTiO_3$), the mechanisms through which titanium affected the phase and microstructural evolution were revealed. $CaTiO_3$ formed first in alumina-magnesia castables containing TiO_2 and $MgTiO_3$ calcined at 1150 °C. As the calcination temperature increased, Ti gradually dissolved in CA_6 and spinel. The influence of titanium on the solid-solution reaction of CA_6 was more prominent than that of spinel. The thicknesses of the CA_6 platelets increased as more titanium was incorporated into the crystal lattice. The enrichment of titanium with calcium promoted the formation of liquid. Therefore, the sintering of castables and the growth of spinel were greatly accelerated due to the faster transfer rate in liquid. Lower apparent porosities and greater strengths of alumina-magnesia castables were achieved by the addition of $CaTiO_3$. Although castables with $CaTiO_3$ and TiO_2 additions had higher bulk densities than those of the castables with $MgTiO_3$, their strengths were similar due to the decreasing contents and different morphologies of CA_6. Based on the lower values of σ^2/E, the castables containing 3.4 wt.% $CaTiO_3$ (equivalent weight of 2 wt.% TiO_2) were estimated to exhibit better thermal shock resistance.

Author Contributions: Conceptualization, W.Y.; methodology, W.Y.; data curation, H.T. and Y.Z.; formal analysis, H.T.; investigation, H.T. and Y.Z.; writing—original draft Preparation, H.T.; writing—review and editing, W.Y.; supervision, W.Y. All authors have read and agreed to the published version of the manuscript.

Funding: This research received no external funding.

Institutional Review Board Statement: Not applicable.

Informed Consent Statement: Not applicable.

Data Availability Statement: The data are not publicly available.

Conflicts of Interest: The authors declare no conflict of interest.

References

1. Fuhrer, M.; Hey, A.; Lee, W. Microstructural evolution in self-forming spinel/calcium aluminate-bonded castable refractories. *J. Eur. Ceram. Soc.* **1998**, *18*, 813–820. [CrossRef]
2. Li, X.; Li, J.; Zhang, L. Enhancing mechanism of improved slag resistance of Al_2O_3-spinel castables added with pre-synthesized $(Al,Cr)_2O_3$ micro-powder. *Ceram. Int.* **2021**, *47*, 33322–33329. [CrossRef]
3. Tang, H.; Li, C.; Gao, J.; Touzo, B.; Liu, C.; Yuan, W. Optimization of Properties for Alumina-Spinel Refractory Castables by CMA (CaO-MgO-Al$_2$O$_3$) Aggregates. *Materials* **2021**, *14*, 3050. [CrossRef] [PubMed]
4. Hong, L.; Chen, W.; Hou, D. Kinetic analysis of spinel formation from powder compaction of magnesia and alumina. *Ceram. Int.* **2019**, *46*, 2853–2861. [CrossRef]
5. Ko, Y.-C. Influence of the Characteristics of Spinels on the Slag Resistance of Al_2O_3-MgO and Al_2O_3-Spinel Castables. *J. Am. Ceram. Soc.* **2004**, *83*, 2333–2335. [CrossRef]
6. An, L.; Chan, H.M. R-Curve Behavior of In-Situ-Toughened Al_2O_3:$CaAl_{12}O_{19}$ Ceramic Composites. *J. Am. Ceram. Soc.* **1996**, *79*, 3142–3148. [CrossRef]
7. Domínguez, C.; Chevalier, J.; Torrecillas, R.; Fantozzi, G. Microstructure development in calcium hexaluminate. *J. Eur. Ceram. Soc.* **2001**, *21*, 381–387. [CrossRef]
8. Braulio, M.; Rigaud, M.; Buhr, A.; Parr, C.; Pandolfelli, V. Spinel-containing alumina-based refractory castables. *Ceram. Int.* **2011**, *37*, 1705–1724. [CrossRef]
9. Braulio, M.A.L.; Pandolfelli, V.C. Tailoring the Microstructure of Cement-Bonded Alumina-Magnesia Refractory Castables. *J. Am. Ceram. Soc.* **2010**, *93*, 2981–2985. [CrossRef]
10. Yuan, W.; Zhu, Q.; Deng, C.; Zhu, H. Effects of rare earth oxides additions on microstructure and properties of alumina-magnesia refractory castables. *Ceram. Int.* **2017**, *43*, 6746–6750. [CrossRef]
11. Yuan, W.; Tang, H.; Zhou, Y.; Zhang, D. Effects of fine reactive alumina powders on properties of alumina-magnesia castables with TiO_2 additions. *Ceram. Int.* **2018**, *44*, 5032–5036. [CrossRef]
12. Yuan, W.; Deng, C.; Zhu, H. Effects of TiO_2 addition on the expansion behavior of alumina–magnesia refractory castables. *Mater. Chem. Phys.* **2015**, *162*, 724–733. [CrossRef]
13. Yuan, W.; Tang, H.; Shang, H.; Li, J.; Deng, C.; Zhu, H. Effects of TiO_2 Addition on Kinetics of In Situ Spinel Formation and Properties of Alumina-Magnesia Refractory Castables. *J. Ceram. Sci. Technol.* **2017**, *8*, 121–128. [CrossRef]
14. Badiee, S.H.; Otroj, S. The effect of Nano-titania addition on the properties of high-alumina low-cement self-flowing refractory castables. *Ceram-Silikaty* **2011**, *55*, 319–325.
15. Maitra, S.; Das, S.; Sen, A. The role of TiO_2 in the densification of low cement Al_2O_3–MgO spinel castable. *Ceram. Int.* **2007**, *33*, 239–243. [CrossRef]
16. Refractories standard compilation. In *Test Method for Bulk Density, Apparent Porosity and True Porosity of Dense Shaped Refractory Products*, 4th ed.; Standards Press of China: Beijing, China, 2010; Volume 2, pp. 3–8.
17. Refractories standard compilation. In *Refractory Products-Determination of Modulus of Rupture at Ambient Temperature*, 4th ed.; Standards Press of China: Beijing, China, 2010; Volume 2, pp. 35–42.
18. Sako, E.Y.; Braulio, M.A.L.; Zinngrebe, E.; van der Laan, S.R.; Pandolfelli, V.C. In-Depth Microstructural Evolution Analyses of Cement-Bonded Spinel Refractory Castables: Novel Insights Regarding Spinel and CA_6 Formation. *J. Am. Ceram. Soc.* **2012**, *95*, 1732–1740. [CrossRef]
19. Okuyama, Y.; Kurita, N.; Fukatsu, N. Defect structure of alumina-rich nonstoichiometric magnesium aluminate spinel. *Solid State Ionics* **2006**, *177*, 59–64. [CrossRef]
20. Lange, N.A.; Speight, J.G. *Lange's Handbook of Chemistry*, 16th ed.; McGraw-Hill: New York, NY, USA, 2005. [CrossRef]
21. Jouini, A.; Sato, H.; Yoshikawa, A.; Fukuda, T.; Boulon, G.; Kato, K.; Hanamura, E. Crystal growth and optical absorption of pure and Ti, Mn-doped $MgAl_2O_4$ spinel. *J. Cryst. Growth* **2006**, *287*, 313–317. [CrossRef]

22. Armstrong, J.T.; Meeker, G.; Huneke, J.; Wasserburg, G. The Blue Angel: I. The mineralogy and petrogenesis of a hibonite inclusion from the Murchison meteorite. *Geochim. Cosmochim. Acta* **1982**, *46*, 575–595. [CrossRef]
23. Xu, L.; Chen, M.; Yin, X.-L.; Wang, N.; Liu, L. Effect of TiO_2 addition on the sintering densification and mechanical properties of $MgAl_2O_4$–$CaAl_4O_7$–$CaAl_{12}O_{19}$ composite. *Ceram. Int.* **2016**, *42*, 9844–9850. [CrossRef]
24. Rahaman, M.N. *Ceramic Processing Gand Sintering*, 2nd ed.; Marcel Dekker Inc.: New York, NY, USA, 2003.
25. Martinez, A.T.; Luz, A.; Braulio, M.; Sako, E.; Pandolfelli, V. Revisiting CA_6 formation in cement-bonded alumina-spinel refractory castables. *J. Eur. Ceram. Soc.* **2017**, *37*, 5023–5034. [CrossRef]
26. Simonin, F.; Olagnon, C.; Maximilien, S.; Fantozzi, G.; Diaz, L.A.; Torrecillas, R. Thermomechanical Behavior of High-Alumina Refractory Castables with Synthetic Spinel Additions. *J. Am. Ceram. Soc.* **2004**, *83*, 2481–2490. [CrossRef]
27. Boccaccini, D.; Romagnoli, M.; Kamseu, E.; Veronesi, P.; Leonelli, C.; Pellacani, G. Determination of thermal shock resistance in refractory materials by ultrasonic pulse velocity measurement. *J. Eur. Ceram. Soc.* **2007**, *27*, 1859–1863. [CrossRef]
28. Rendtorff, N.M.; Aglietti, E.F. Thermal Shock Resistance (TSR) and Thermal Fatigue Resistance (TFR) of Refractory Materials. Evaluation Method Based on the Dynamic Elastic Modulus. In *Encyclopedia of Thermal Stresses*; Hetnarski, R.B., Ed.; Springer: Dordrecht, The Netherlands, 2014; pp. 5119–5128. [CrossRef]
29. Harmuth, H.; Bradt, R.C. Investigation of refractory brittleness by fracture mechanical and fractographic methods. *Interceram Int. Ceram. Rev.* **2010**, *62*, 6–10.
30. Schacht, C.A. *Refractories Handbook*; Marcel Dekker, Inc.: New York, NY, USA, 2004.
31. Hasselman, D.P.H. Elastic Energy at Fracture and Surface Energy as Design Criteria for Thermal Shock. *J. Am. Ceram. Soc.* **1963**, *46*, 535–540. [CrossRef]

Article

Development and Testing of Castables with Low Content of Calcium Oxide

David Zemánek * and Lenka Nevřivová

Faculty of Civil Engineering, Brno University of Technology, Veveří 331/95, 602 00 Brno, Czech Republic
* Correspondence: zemanek.d@fce.vutbr.cz; Tel.: +420-604341018

Abstract: Colloidal silica is used in many kinds of industry. It is an aqueous dispersion of SiO_2 nanoparticles. SiO_2 colloidal solutions are commercially available in different concentrations, with different particle sizes and are stabilized with different ions. Colloidal SiO_2 was used in this study as a cement replacement in refractory castable. The present study, in its first stage, offers an assessment of five different SiO_2 colloidal solutions. The particle size of the solutions was 15 nm, the particle concentration was 30% and 40% and the colloidal solutions were stabilized with Na^+, OH^- and Cl^- ions. The effect of the colloidal solutions on selected characteristics of the refractory pastes and on their mineralogical composition after firing at 1000 °C and 1500 °C was described. The most suitable SiO_2 colloidal solution from the first stage was subsequently used for the refractory castable test samples' preparation in the second stage. Refractory castables, unlike paste, contain a coarse aggregate (grog) up to a grain size of 6 mm. Four types of coarse refractory grog were evaluated. Their effect on selected characteristics of the refractory castable and on its mineralogical composition after firing at 1000 °C and 1500 °C was described. The selected characteristics, within the scope of this study, include bulk density, apparent porosity, cold modulus of rupture and linear changes after firing. Finally, the study describes the effect of the sol particle concentration and the effect of pore size distribution on corrosion resistance and on the internal structure of the material. Mineral and chemical compositions and microstructures of both the raw materials and designed aggregates were thoroughly investigated by the means of X-ray fluorescence spectroscopy, powder X-ray diffraction and scanning electron microscopy. An analysis of the transition zone between corrosive media (K_2CO_3) and tested castables showed better corrosion resistance for a sol-gel castable than an ultra-low cement castable.

Keywords: sol-gel; refractory; castables; corrosion

Citation: Zemánek, D.; Nevřivová, L. Development and Testing of Castables with Low Content of Calcium Oxide. *Materials* **2022**, *15*, 5918. https://doi.org/10.3390/ma15175918

Academic Editor: Haijun Zhang

Received: 25 July 2022
Accepted: 24 August 2022
Published: 26 August 2022

Publisher's Note: MDPI stays neutral with regard to jurisdictional claims in published maps and institutional affiliations.

1. Introduction

There has been worldwide growth of castable usage, mainly attributed to the fact that more and more brick-laid linings have been and can further be replaced by castables that possess a variety of advantages over bricks, in terms of production cost, installation efficiency, safety, material consumption, etc. Monolithic ceramics have evolved over the years into a widely used class of refractories. The progress of monolithic refractories is closely related to improvements in the quality of raw materials, new binders, ultra-fine powders, and efficient additives [1]. Refractory castables can be classified based on different aspects including the calcium oxide content, binder source, chemical composition, bulk density, application method and others [2]. A medium cement castable (MCC) is characterized by a calcium oxide content higher than 2.5%, a low cement castable (LCC)'s CaO content is 1.0–2.5%, an ultra-low cement castable (ULCC)'s CaO content is 0.2–1.0% and a no cement castable (NCC) CaO content until 0.2% [3]. Different sorts of binding systems have been developed throughout the years, starting with hydraulic bonding, in which higher amounts of calcium aluminate (CA) cement were used, towards coagulating binders such as colloidal silica or alumina [4].

Hydraulic binders were used at the very beginning of castable production. Conventional cement-bonded castables were produced since the early 1920s until the 1960s when CA cement started to be used. The purity of CA cement was rising over the time and in the 1990s, the next step of evolution was done. Hydratable alumina (HA) has been used as a castable binder. It is a transition alumina, having specific crystallinity containing over 90% of alumina. It forms $Al_2O_3 \cdot 3H_2O$ (AH$_3$) and $Al_2O_3 \cdot H_2O$ (AH) gel in combination with water which, upon heating, dehydrate and eventually form ceramic bonds [5]. Lower porosities are achieved in the samples as a consequence of the ultrafine pore structure resulting from hydratable alumina. The HA-bonded products thus have an improved corrosion resistance to liquids and gases. Compared to the cement-bonded system, HA requires a longer mixing time and the requirement for water is also high due to the high specific surface area of the binder [6,7]. Thus, they have a specific application to areas where the cement removal is beneficial and where drying out can be controlled tightly, such as in precast [8]. The very new type of hydraulic binders is spinel-containing CA cement and calcium magnesia alumina (CMA) cement [9]. The production of chemically bonded castables started in the 1950s with the usage of phosphates and water glass. Sulfate- and chloride-bonded castables appeared a bit later. Chemical binders promote bonding by the formation of new products or by a polymerization reaction between the binder and the refractory material (oxide aggregates) or between the binder and the hardener [4,10].

Coagulation bonding began to be used in the 1980s. The binders which work by coagulation binding are fine clay powder, ultrafine oxide powder, silica sol, alumina sol, etc. Nanoparticles of the colloidal suspension are attracted by Van der Waal forces, if the equilibrium in the sol is disturbed, it leads to the coagulation of these particles. The addition of a coagulant increases the coagulation activity. When colloidal silica (silica sol) is mixed with a refractory micro filler, the equilibrium of the sol is disturbed and a gel forms around the microparticles which holds the particles together. The gel is responsible for the manipulation strength in the green state [10].

Among the available colloidal systems, only silica sol is commercially exploited as a binder for refractory systems. The high-stability, high-solid content, as well as the possibility of mullite formation at a low temperature in high alumina systems, favors its wide industrial application. However, the presence of free silica in the system can promote the appearance of a vitreous phase and thus decrease refractoriness, which is especially critical in alkali-containing systems, restricting its use for high-temperature applications [11]. Carbon bonding (MgO-C) and nano-engineered castables started to be used in the beginning of the 21st century. Nano engineered castables use colloidal binders and sintering additives at one time [12].

The advancements in nanotechnology in the last two decades could benefit the refractory industry if explored properly [13,14]. The literature shows that the utilization of nanopowders and colloidal suspensions in refractory castables has increased in recent years, mainly to improve the bonding and densification in castables at lower sintering temperatures [13,15,16]. The agglomeration of nanopowder particles (due to its high reactivity owing to a high surface area) results in poor dispersion in the matrix and thus possesses a challenge to use these as additives [17,18]. If the agglomeration problem is kept under control, the use of a nanoparticle addition could improve the castable properties. Another limitation which hinders the use of nanopowders as additives is the availability and the cost of available nanopowders. Nanoparticles containing colloidal suspensions (colloidal binders) are a better alternative and are preferred [18].

A colloid or sol is a stable dispersion of particles in water, wherein particles are too small for gravity to make them settle. Particle size typically ranges from 1 to 1000 nanometers [19]. The name sol-gel derives from the fact that microparticles or molecules in a solution (sols) agglomerate which, under controlled conditions, eventually link together to form a coherent network (gel) [20]. The various advantages associated with sol-gel-bonded castables are less mixing time, cohesive and self-flowing nature, easier installation as there are no vibration works, shorter drying time and reduced drying effects due to a lower moisture content, better corrosion resistance (absence of CaO), better oxidation resistance, better refractoriness, longer lining life and larger self-life as no hydratable phase is present [21]. The bonding action of colloidal binders is based on the gelation (coagulation) of the colloidal particles, which leads to a high permeable and porous material structure. Technically this can be advantageous as the composition can be safely and quickly dried; reducing the risk of explosive spalling and thus the overall processing time is reduced [13].

Various colloidal systems are available for refractory castables production, e.g., silica sol, alumina sol, mullite sol and spinel sol. The use of alumina sol as a binder for refractory castables is mentioned by a few authors only [15,18,22–26]. The advantage of alumina sol is its purity. The use of mullite sol as a binder for refractory castables is mentioned as well, for example, in combination with CA cement, A cement or micro silica [21,27–29]. Use of spinel sol as a binder has also been reported by some authors in ultra-low cement compositions [21,27]. However, silica sol is the most common for the refractory castable production. Recently, several studies were focused on the "cement free" refractory castables [30,31].

That is why silica sol was chosen for the production of high-alumina no-cement castable samples in the present study. The study evaluates the properties of no-cement castable samples bonded by a sol-gel method together with an ultra-low cement castable sample, for comparison. Other possible alternative bonds, which can be suitable for comparison as a clay bond, kaolin bond or hydratable alumina bond, were not assumed. Possible applications of developed castables may be in the glass industry (mechanically stressed parts), primary aluminum industry (anode coke oven) or stressed parts of boilers.

The corrosion of refractories can be defined by the loss of thickness and mass from the exposed face of the refractory as a consequence of a chemical attack by a corroding fluid in a process in which the refractory and the corroding fluid react, approaching the chemical equilibrium in the zone of contact [32,33]. The influence of the used slag and influence of the used refractory grog on the microstructure was observed. There are a few tests used for the corrosion resistance determination of refractory material and they are either static or dynamic, such as hot stage microscopy, the crucible test, finger test, dynamic finger test, induction furnace test or rotary slag test [34]. The crucible test was applied in our study. The corrosion resistance of the sol-gel-bonded castables is assumed to be better than the resistance of the ultra-low cement castable. Spinel and silica sols-bonded castables were found to achieve the best results [35].

2. Experimental Procedure

2.1. Methods

Chemical composition analysis of the raw materials was performed by wavelength-dispersive X-ray spectroscopy (WDXRF) using SPECTROSCAN MAKC-GV (Spectron Company, St. Petersburg, Russia) instrument equipped with QUANTITATIVE ANALYSIS software (version 4.0, Spectron Company, St. Petersburg, Russia). The samples were analyzed in forms of fused beads. Powder X-ray diffraction analysis of the raw materials and engineered aggregates was conducted on Panalytical Empyrean diffractometer (Panalytical B.V., Almelo, The Netherlands) equipped with Cu-anode, 1-D position-sensitive detector at convention Bragg–Brentano reflection geometry. The setting was: step size—0.013° 2θ, time per step—188 s, and angular range 5–80° 2θ. Quantitative phase analysis was done via the Rietveld method using Panalytical High Score 3 plus software (version 4.8, Panalytical B.V., Almelo, The Netherlands).

Scanning electron microscopy with X-ray microanalysis (SEM/EDS) was conducted on gold-coated mechanically broken specimens (for morphological analyses) and on polished carbon-coated thin sections (for chemical microanalyses) using TESCAN MIRA 3 instrument (Tescan Orsay Holding a.s., Brno, Czech Republic) with the accelerating voltage of 30 kV.

Cold crushing strength (CCS) according to standard EN 993-5:2018 (using machine MEGA 11-600 D-S, Brio Hranice s.r.o., Hranice, Czech Republic) and cold modulus of rupture (CMOR) according to standard EN 993-6:2018 (using machine Testometric M350-20CT, Testometric Co. Ltd., Rochdale, UK) were carried out after drying and after firing. The apparent porosity, water absorption and bulk density were determined by a vacuum water absorption method with subsequent hydrostatic weighing (standard EN 993-1:2018). Determination of permanent change in dimensions on heating was tested according to standard EN 993-10:2018.

Abrasion test was carried out under conditions of standard ASTM C 704. Abrasion test used SiC material with particle size 250 μm with feed rate 50 g·min^{-1} and impact angle 90°.

Corrosion testing was carried out according to CEN/TS 15418:2006. Prepared castables were casted into silicone mold with required parameters of the testing samples—outer dimensions of the samples were $100 \times 100 \times 100$ mm, inner hollow cylindrical shaped crucible was 55 mm in diameter and 55 mm in depth. Used corrosion medium was K_2CO_3, specifically 20 g was precisely weighed in each tested crucible before testing. Corrosion cup test or crucible test, as it can be entitled, was carried out at 950 °C to achieve melting point of K_2CO_3 in this study. Standard CEN/TS 15418:2006 specifies the heating rate for the corrosion cup test to 5 °K/min and soaking time to five hours at the maximum temperature.

Apparent porosity and pore size distribution were determined by mercury intrusion porosimetry using Thermo Finnigan POROTEC Pascal 140–240 instruments (ThermoFisher Scientific, Waltham, MA, USA) with SOL.I.D software (version 3.0.1, ThermoFisher Scientific, Waltham, MA, USA).

2.2. Raw Materials and Mixtures
2.2.1. First Stage—Fine Pastes

Raw materials used in this study were obtained from their producers—andalusite (A) from Imerys (Glomel, France), tabular alumina (TA) from Almatis (Ludwigshafen, Germany), reactive alumina (RA) from Nabaltec (Schwandorf, Germany), ground alumina (GA) from Nabaltec (Schwandorf, Germany) and silica fume (SF) from RW Silicium (Pocking, Germany). Colloidal silicas used in this study were chosen from SChem (Ústí nad Labem, Czech Republic). For comparison with traditional hydraulic bond used for ULCC, calcium aluminate cement (CAC) from Almatis (Ludwigshafen, Germany) was chosen, Chemical and phase compositions of the used raw materials are given in Table 1. As the gelling agent, material responsible for transformation from colloidal sol to gel, ammonium chloride (NH_4Cl, 1 M) from PENTA (Prague, Czech Republic), was used.

Table 1. Chemical and phase composition of fine matrix compounds in wt. %.

Compound/ Composition	Fine Matrix Compounds				
	A	TA	RA	GA	SF
Chemical composition					
SiO_2	37.01	0.09	0.16	0.80	98.46
Al_2O_3	60.49	99.55	99.45	98.35	0.04
TiO_2	0.19	–	–	0.03	0.01
Fe_2O_3	0.95	0.01	–	0.09	0.07
CaO	0.20	–	0.01	0.22	0.62
MgO	0.18	–	0.14	0.23	0.04
K_2O	0.29	–	0.04	0.03	0.55
Na_2O	0.29	0.45	0.20	0.25	0.01
Phase composition					
Quartz	1.37	–	–	–	0.26
Corundum	–	94.98	100	98.90	–
Andalusite	90.12	–	–	–	–
Diaoyudaoite	–	5.02	–	1.10	–
Amorphous phase	8.51	–	–	–	99.74

In the first stage, fine matrix was prepared from dry raw materials, which were precisely weighed (Table 2), mixed and well homogenized in a laboratory homogenizer for 1 h. After homogenization process, mixtures were mixed for 15 min with designed amount of colloidal silica 8.5% to achieve fine paste. Then, addition of 0.1% of gelling agent (NH_4Cl) was carried out with subsequent mixing for 5 min.

Table 2. Composition of fine matrix in % wt.

Component	Proportion in the Fine Matrix [% wt.]
Andalusite <400 μm	16.75
Tabular alumina <44 μm	18.25
Reactive alumina 0.5–3.0 μm	36.00
Ground alumina 0.5–3.0 μm	14.75
Silica fume <0.15 μm	14.25

Fine pastes designation and the type of colloidal silicas stabilization is shown in Section 3.1. Type of stabilization is declared by its producer. Colloidal silica designation is typically, e.g., K1530—where K is a commercial name, 15 represents average particle size diameter d_{50} and 30 is a concentration of solid particles.

2.2.2. Second Stage—Castables

Four types of refractory grog with different amounts of Al_2O_3 were used—fused mullite (M) from MOTIM Electrocorundum (Mosonmagyaróvár, Hungary), fireclay grog (F) from P-D Refractories (Velké Opatovice, Czech Republic), high alumina grog (H) from P-D Refractories (Velké Opatovice, Czech Republic) and andalusite (A) from Imerys (Glomel, France). Their chemical and phase compositions are shown in Table 3.

Table 3. Chemical and phase composition of coarse grog in wt.%.

Compound/ Composition	Coarse Grog			
	M	F	H	A
Chemical composition				
SiO_2	22.92	56.48	47.38	36.52
Al_2O_3	76.57	39.66	50.33	59.74
TiO_2	0.00	1.40	0.97	0.24
Fe_2O_3	0.05	1.00	0.50	0.58
CaO	0.14	0.50	0.07	0.26
MgO	0.02	0.11	0.13	0.11
K_2O	0.10	0.30	0.30	0.05
Na_2O	0.20	0.55	0.13	2.43
Phase composition	M	F	H	A
Quartz	0.1	2.5	0.1	1.4
Mullite	80.5	35.7	78.6	–
Corundum	0.2	–	1.5	–
Cristobalite	–	7.4	2.1	–
Andalusite	–	–	–	90.1
Amorphous phase	19.2	54.4	17.7	8.5

Castables were designed in accordance with Andreasen's particle distribution model with packing coefficient q = 0.24. Deflocculants or plasticizers were not used in this study. Designed composition of castables based on mentioned coarse grogs is presented in Table 4. Castables are labeled, e.g., NA—where the first letter indicates the type of castables (N—no-cement castable, U—ultra-low cement castable), the second letter represents the type of used refractory grog (A—andalusite, H—high-alumina grog, F—fireclay, M—fused mullite).

Table 4. Designed composition of castables in % wt.

Designation	NM	NF	NH	NA	UA
Component	**[%]**	**[%]**	**[%]**	**[%]**	**[%]**
Castable type	NCC	NCC	NCC	NCC	ULCC
Coarse grog	Mullite	Fireclay	High alumina	Andalusite	Andalusite
Coarse grog content	78.0	78.0	78.0	78.0	78.0
Fine matrix	22.0	22.0	22.0	22.0	22.0
Water	2.2	5.5	3.6	3.1	5.5
L1540	4.75	4.75	4.75	4.75	-
CAC	-	-	-	-	2.0

Legend: N—no cement castable, U—ultra-low cement castable.

2.3. Samples Preparation

2.3.1. First Stage—Fine Pastes

Fine pastes were casted into molds with dimensions $20 \times 20 \times 100$ mm and covered by plastic foil. Samples were demolded after 24 h. Green bodies were then dried in the laboratory dryer at 110 °C for 24 h. All the test samples were then fired in an electrical laboratory furnace with air atmosphere at 1000 °C and 1500 °C (heating rate of 4 °K/min and soaking time of 300 min at the maximum temperature).

In the second stage, castables were prepared from dry raw materials. After dry homogenization process that lasted for 1 h, mixtures were mixed for 5 min with designed amount of water, colloidal silica and calcium aluminate cement (CAC) for UA mixture. For no-cement castables (NCC), gelling agent was added to the mixture and second wet mixing lasted 2 min.

2.3.2. First Stage—Castables

Mixtures were casted into molds with dimensions $40 \times 40 \times 160$ mm and covered by plastic foil. Samples were demolded after 24 h. Green bodies were then dried in the laboratory dryer at 110 °C for 24 h. All the test samples were then fired in an electrical laboratory furnace with air atmosphere at 1000 °C and 1500 °C (heating rate of 4 °K/min and soaking time of 300 min at the maximum temperature).

3. Results and Discussion

3.1. First Stage—Fine Pastes

The objective of the first stage of this study was to describe the effect of colloidal silica on the properties of the fine pastes after drying and after firing. The raw material used for the preparation of the fine matrix is shown in Table 1. The formula for the preparation of the fine matrix was designed based on previous experience and research [36] (see Table 2). Ammonium chloride (NH_4Cl, 1 M) was used as the gelling agent.

The main output of this stage is the determination of the most suitable colloidal SiO_2 solution for the preparation of the no-cement refractory castable.

Table 5 presents the results of the experiments that utilized the vacuum water absorption method, together with hydrostatic weighing to determine the apparent porosity (AP) and bulk density (BD), as well as the results of the Cold Modulus of Rupture (CMOR) and linear change after firing (LCf) for all five tested fine pastes.

When fired at 1000 °C, the lowest porosity was achieved using colloidal silica K1530AD. When fired at 1500 °C, the apparent porosity decreased by up to 5.5%. When using a colloid with a higher SiO_2 solids content (40%), the fine paste has a higher porosity after firing at 1500 °C than when using a colloidal silica with a lower solids content (30%). The lowest decrease in porosity was observed when using L1540 (see Figure 1).

Table 5. Five types of colloidal silica and their influence on final properties of fine pastes after firing [% wt.].

Sample	Colloidal Silica	Stabilization	T [°C]	BD [kg·m⁻³]	AP [%]	CMOR [MPa]	LCf [%]
A	L1530	Na$^+$	1000	2490	25.21	6.2	−0.2
			1500	2599	20.54	6.4	−1.9
B	L1540	Na$^+$	1000	2510	25.07	6.2	−0.1
			1500	2565	22.04	6.2	−1.0
C	K1530	Na$^+$	1000	2493	26.21	6.1	−0.4
			1500	2590	20.66	6.4	−2.5
D	K1530AD	OH$^-$	1000	2517	24.46	6.6	−0.3
			1500	2608	20.74	6.6	−2.2
E	K1530K	Cl$^-$	1000	2512	25.08	6.1	−0.2
			1500	2600	20.88	6.6	−1.9

T—temperature, BD—bulk density, AP—apparent porosity, CMOR—cold modulus of rupture, LCf—linear change after firing (1000 °C).

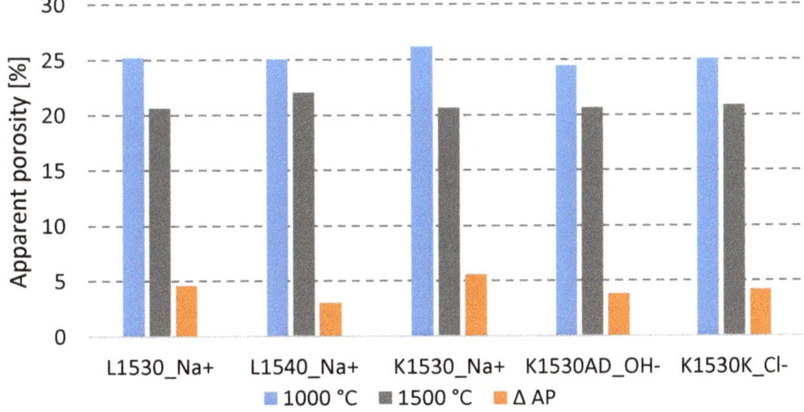

Figure 1. Comparison of different colloidal silica and its influence on apparent porosity after firing at 1000 °C and 1500 °C.

For sample B, which is also stabilized with Na$^+$ ions but contains 10% more of silica nanoparticles, the apparent porosity after firing at 1500 °C is 2% higher. This also may be observed in terms of linear shrinkage after firing, where the higher content of silica nanoparticles in sample B leads to lower linear shrinkage at both firing temperatures. Using colloidal silica with a higher content of solids determined the lowest decrease in apparent porosity and a low increase in bulk density in comparison to the concentration of solids at 30%. The highest bulk density of 2608 kg·m⁻³ was measured for sample D after firing at 1500 °C, where the second highest linear shrinkage of −2.2% was measured and the highest CMOR of 6.6 MPa was obtained. Other results at both firing temperatures are comparable and in the case of physical and mechanical properties, the four tested silicas with the same particles size and concentration provided similar results.

The XRD diffractograms of tested pastes after firing at 1000 °C and 1500 °C are presented in Figure 2.

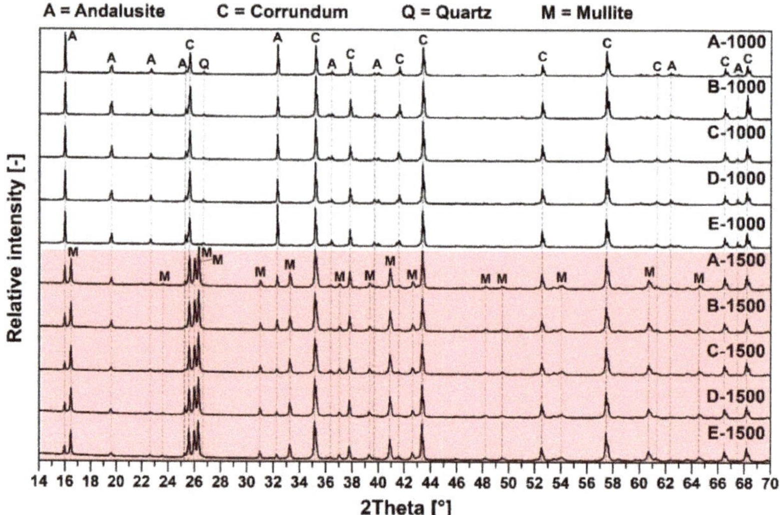

Figure 2. Mineralogical composition of the fine matrix fired at 1000 °C (white background) and 1500 °C (red background).

The major crystalline phases for samples fired at 1000 °C are andalusite ($Al_2(SiO_4)O$), corundum (Al_2O_3) and quartz (SiO_2). The major crystalline phases for samples fired at 1500 °C are andalusite, corundum and mullite ($3Al_2O_3 \cdot 2SiO_2$). Major differences in the samples' mineralogical composition are shown in the following Figure 3. Residual quartz is contained only in samples after firing at 1000 °C. The mullite presence was detected at the main mullite peaks with an intensity of 3.39(1), 3.428(0.95) and 2.206(0.6) only after firing at 1500 °C. At that temperature, all quartz was transformed to mullite and the glassy phase, which is also presented in Table 4.

(a)   **(b)**

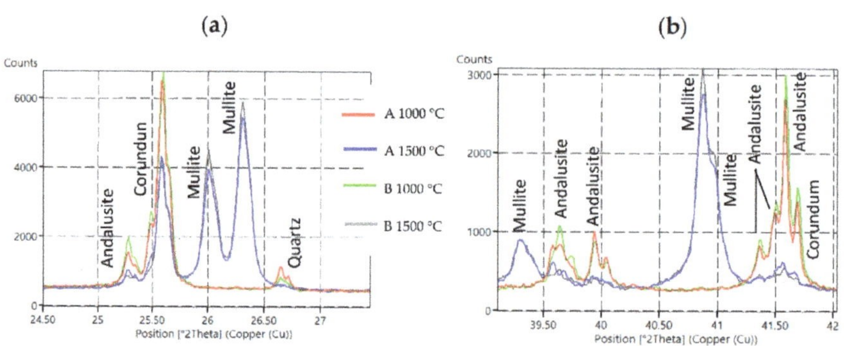

Figure 3. XRD diffractograms for Sample A and B fired at 1000 °C and 1500 °C with focus on specific sections—(**a**) 24.5°–27.5° 2θ section, (**b**) 39.0°–27.5° 2θ section.

The mineral andalusite was identified after firing at 1000 °C on all major diffraction lines, 5.54(1), 2.77(0.9) and 4.53(0.9). Corundum 2.085(1), 2.552(0.9) and 1.601(0.8) and quartz 3.342(1) were also identified after firing at 1000 °C. After firing at 1500°C, the content of andalusite and corundum was reduced at the expense of mullite and the glassy phase. The decrease in the andalusite content is documented on diffraction lines 25.3, 39.6, 39.9, 41.4, 41.5 and 41.6° 2θ. A decrease in the corundum content is documented on diffraction lines 25.6 and 41.7° 2θ (see Figure 3 and also Table 6).

Table 6. Phase composition of samples fired at 1000 °C and 1500 °C [%].

Firing Temperature [°C]	1000 °C					1500 °C				
Samples Formula	A	B	C	D	E	A	B	C	D	E
Quartz	0.5	0.6	0.6	0.5	0.5	0.2	0.1	0.2	-	0.1
Corundum	68.8	67.9	67.0	66.7	67.9	28.7	28.6	28.9	28.9	30.1
Andalusite	15.1	15.7	16.3	16.2	15.1	-	-	-	-	-
Mullite	-	-	-	-	-	71.1	71.3	70.9	71.1	69.8
Amorphous phase	15.6	15.8	16.1	16.7	16.7	-	-	-	-	-

Based on the discussed results, silica sol with the designation L1540, sample formula B, which contains 40% of solid SiO_2 particles, was selected. The minimum LSf values (-0.1% at 1000 °C, -1.0% at 1500 °C) were recorded when using this sol and good physical and mechanical properties were obtained. The highest mullite content was measured as 71.3% for Sample B due to the highest content of solid particles (40%). The higher mullite content in the fine paste when using L1540 (B) is also documented in Figure 3 on the diffraction lines 26.0, 26.3, 39.3 and 40.9° 2θ. In terms of used colloidal silicas with 30% of particles, the highest content was measured as 71.1% for Sample A. The higher mullite content in the prepared fine pastes when colloidal silica with a higher solid-phase content was used is a positive signal and precondition to achieve good physical, mechanical and refractory castable properties.

3.2. Second Stage—Castables

Based on the results of the first stage, silica sol L1540 was selected and used in further testing. The second stage was focused on the design of the refractory castable with coarse aggregate (grog) fractions of 0–1, 1–3 and 3–6 mm. The NCC bond was realized by the sol-gel method and was tested with refractory grog with different Al_2O_3 contents. The effect of the NCC and ULCC bond on the basic physical and mechanical properties of refractory castable was also compared in this stage.

Five types of castables were prepared in the terms mentioned above. Table 7 shows the chemical composition of the tested mixtures after firing, all tested mixtures met the condition of standard ASTM C401-12:2018 for the calcium oxide content for ultra-low cement castables (ULCC) and no-cement castables (NCC).

Table 7. Calculated chemical compositions of castables based on WDXRF analysis in % wt.

Mixture	SiO_2	Al_2O_3	TiO_2	Fe_2O_3	CaO	MgO	K_2O	Na_2O
NM	24.27	74.91	0.05	0.25	0.03	0.08	0.03	0.18
NF	42.13	54.79	0.95	0.79	0.13	0.30	0.54	0.17
NH	39.45	58.49	0.82	0.56	0.06	0.08	0.20	0.15
NA	33.05	65.49	0.11	0.68	0.08	0.12	0.11	0.15
UA	32.32	65.45	0.16	0.88	0.79	0.04	0.01	0.13

3.2.1. Physical and Mechanical Properties after Firing

Table 8 presents castables' physical properties after drying and firing at selected temperatures.

The apparent porosity (shown in Figure 4 and further discussed in Section 3.2.3.) and water absorption decreases with the increasing firing temperature. The less Al_2O_3 contained in the castable, the higher the decrease of WA and AP was measured to be. The porosity decreasing is also related to the further sintering of the coarse grog with a lower alumina (F, H) content.

Table 8. Physical properties of tested castables after drying at 110 °C and after firing at 1000 °C and 1500 °C.

Mixture	T [°C]	BD [kg·m^{-3}]	WA [%]	AP [%]	LCf [%]
NM	110	2690	6.1	16.3	-
	1000	2700	5.8	15.6	−0.52
	1500	2700	5.7	15.2	−0.82
NF	110	2410	6.7	16.2	-
	1000	2430	6.4	15.6	0.00
	1500	2370	5.7	13.5	0.63
NH	110	2300	9.8	22.5	-
	1000	2320	9.4	21.8	0.02
	1500	2315	9.1	21.1	0.11
NA	110	2730	5.5	15.0	-
	1000	2715	5.6	15.2	0.00
	1500	2630	5.4	14.2	1.16
UA	110	2740	5.3	14.5	-
	1000	2700	5.8	15.7	0.10
	1500	2640	5.5	14.4	0.83

Legend: T—temperature, BD—bulk density, WA—water absorption, AP—apparent porosity, LCf—linear change after firing.

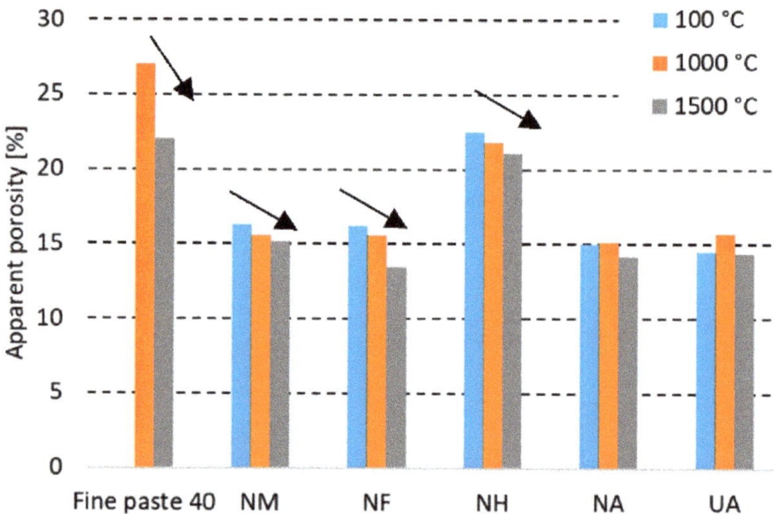

Figure 4. Apparent porosity of tested castables at elevated temperatures.

A comparison of the values for the dried state with the maximum firing temperature shows the highest decrease for sample NF, where the water absorption declined by 1.0% and the apparent porosity declined by 2.7%. Differences in the bond type for UA and NA in terms of apparent porosity and water absorption are similar. The used bond affected the apparent porosity and water absorption marginally. The porosity and mineralogical composition of the used coarse grog is crucial for the final porosity of the refractory castable.

Figure 5 represents linear changes that occurred during the firing of tested castables. After firing at 1000 °C, the sample NA (sol-gel bonded) performed a permanent linear change of 0.00% in comparison to UA (CAC bonded), where an expansion of 0.10% was measured.

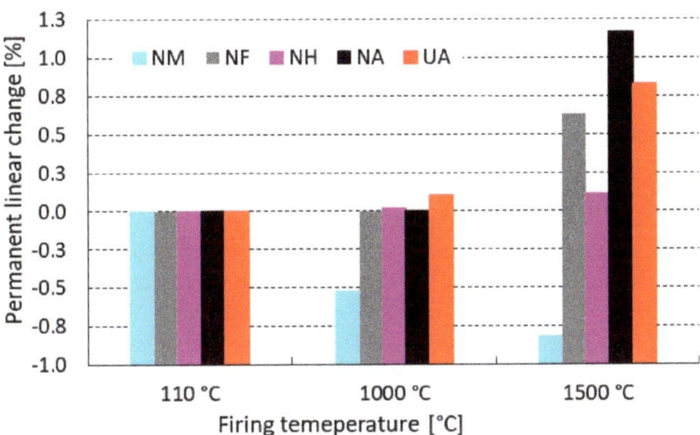

Figure 5. Influence of firing temperature on LCf of tested castables.

3.2.2. Abrasion Test

Andalusite (A) is well-known for its expansion behavior during firing and this phenomenon is obvious by the highest expansion rates, as shown in Figure 5. The opposite mechanism was measured for mullite (M), where the highest shrinkage occurred after firing at both tested temperatures. The most dimensionally stable grog appears to be high alumina grog (HA) with miniscule expansion 0.11% at 1500 °C. To achieve a similar LCf value using andalusite, a combination with mullite is necessary to compensate their expansion and shrinkage behavior after firing.

An abrasion test was carried out under conditions of standard ASTM C 704. Figure 6 represents the loss of abrasion tested at room temperature for tested castables after pre-firing at 1500 °C. The abrasion test used SiC material with a particle size 250 μm with a feed rate of 50 g·min^{-1} and impact angle of 90°. Comparing the NCC refractory castable, the use of a fireclay grog F (abrasion = 5.3 cm^3) appears to be the most suitable and a mullite grog M (abrasion = 6.8 cm^3) is the least suitable. NH and NA castables performed the same value of abrasion loss at 5.8 cm^3. For the refractory castable with andalusite grog, the effect of the used bond on the abrasion resistance was investigated. When the sol-gel bond is used (NA), the abrasion resistance is 17% lower than when using the cement bond (UA). It was shown that abrasion was 17% lower using sol-gel bond (NA) than the calcium cement bond (UA), despite the fact that the apparent porosity and average pore diameter of the UA and NA refractory castables were very similar.

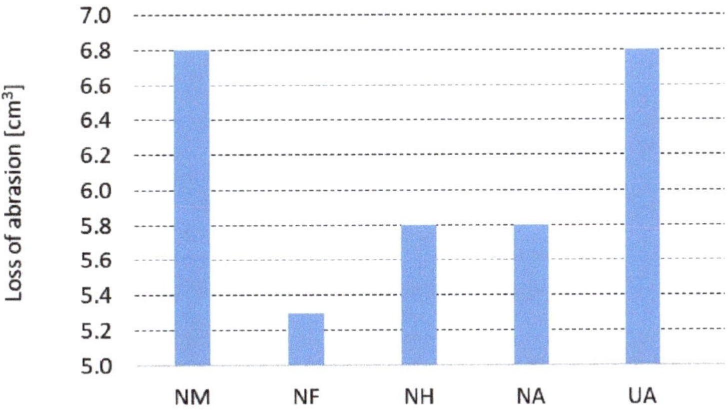

Figure 6. Loss of abrasion determined according to standard ASTM C 704.

3.2.3. Pore Structure

The effect of the used refractory grog on the pore structure of the refractory castables after firing at 1500 °C is shown in Figure 7. The pore size distribution of the refractory castable with mullite (M) and castable with high alumina (H) grog is very similar. The average pore diameter is 17.48 and 13.1 μm, respectively. Ninety percent of the pores are 1–40 μm in diameter. When andalusite (A) is used, the apparent porosity is lower, as is the average pore diameter, which is 1.24 μm and 2.42 μm, respectively. Seventy-six percent of the pores are within the 1–40 μm interval. For the refractory castable with andalusite grog, we can also compare the effect of the bond used in the structure of the refractory castable. When the sol-gel (NA) bond is used, there are more fine pores in the material structure than when the hydraulic (UA) bond is used, with almost the same apparent porosity (14.9% and 14.2%, respectively). The average pore diameter when sol-gel bonding is used is 50% smaller than when using the hydraulic bond.

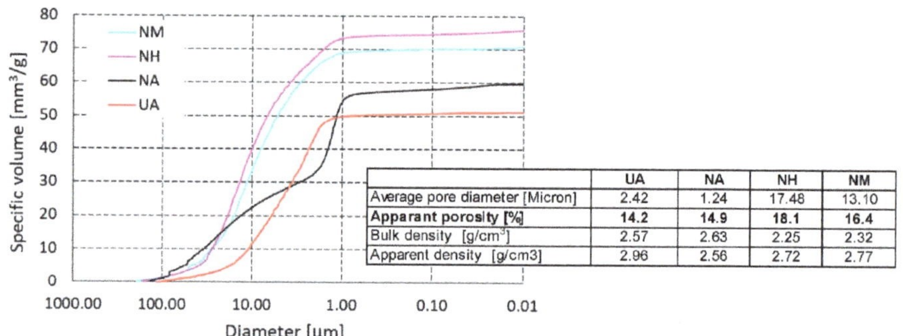

	UA	NA	NH	NM
Average pore diameter [Micron]	2.42	1.24	17.48	13.10
Apparant porosity [%]	14.2	14.9	18.1	16.4
Bulk density [g/cm³]	2.57	2.63	2.25	2.32
Apparent density [g/cm3]	2.96	2.56	2.72	2.77

Figure 7. Total cumulative pore volume of laboratory-prepared samples and reference material.

3.2.4. Corrosion Crucible Test and EDS Element Mapping

The corrosion crucible test was carried out according to the standard CEN/TS 15418:2006. Crucibles were pre-fired at 1500 °C with a soaking time of 5 h at the maximum temperature, with a heating rate of 1.5 °K/min. Crucibles were filled with a corrosive medium (K_2CO_3) and subsequently fired at 950 °C with a soaking time of 5 h at the maximum temperature, with a heating rate 1.7 °K/min.

Based on the general state of the specimen after the corrosion test, the standard [37] defines four stages of corrosion. The decisive parameter for assessing the degree of corrosion is the size of the corroded area and the maximum depth of penetration of the sample by the corrosive agent on a cross section of the test specimen (U: unaffected, for samples with no visible degradation; LA: lightly attacked, for samples with a minor attack; A: attacked, for samples with clearly visible degradation; C: corroded, for completely corroded samples).

The next section is focused on specifically UA and NA castables to compare differences between the sol-gel bond and the hydraulic bond from CAC. A visual evaluation performed on cross-cut crucibles and its results are presented in Appendix A, Figure A1 for the NA castable and in Appendix B, Figure A3 for the UA castable. According to the standard, examined castables may be labeled as unaffected (U) for both samples, as shown in Table 9. Sample visual observation shows a darker color for the UA castable in comparison to the NA castable. The maximal penetration depth of 15.24 mm was measured for the UA castable, whereas the NA castable performed better results and the value of 14.87 mm was obtained. In addition, the total corroded area A was higher by 3.9% for the UA castable than the NA castable, although the apparent porosity of the NA castable is higher than the UA castable (+0.2% after firing at 1500 °C).

Table 9. Visual evaluation of crucible test for tested castables.

Castable	UA	NA
Corrosion category	U	U
d_{max} [mm]	15.24	14.87
A [mm^2]	1376	1326 (-3.9%)

Corroded crucibles were also examined using SEM analysis, element mapping and phase analysis. Cross-cut sections of corroded crucibles were analyzed using SEM and its outputs are presented in Appendix A—Figure A1, and Appendix B—Figure A3. Corroded and non-corroded areas were sectioned-cut under water, samples were then dried, their surface was goldened and the microstructure was observed using SEM. Comparing the microstructure of the corroded and non-corroded samples, the corroded sample shows a higher amount of glassy phase and newly created minerals caused by corrosion. This finding was also confirmed by quantitative phase analysis (Appendix A—Figure A2 for NA, Appendix B—Figure A4 for UA), where an increase in the glassy phase of 4.4% for NA was measured and 2.7% for UA, respectively. The new mineral detected in the phase composition of the corroded samples is kalsilite $K(AlSi_2O_6)$ and its amount is 0.9% for NA and 0.7% for UA. During the corrosion process, the content of mullite decreased by 5.6% for NA and 0.5% for UA. Despite these findings, the no-cement castable resisted corrosion better, as shown in Table 8.

Samples for EDS element mapping were prepared by cutting 20×20 mm cubic samples, subsequently polished under water. Outputs are presented in Appendix A—Figure A2, and in Appendix B—Figure A4. The EDS probe localized the presence of the K elements in both sections—the optically visual transition zone is indicated at a magnification of $100\times$ by the grey dashed line. When analyzing the NA castable, the peak intensity for the K element was measured two times higher for the corroded section in comparison to the non-corroded section. Analysis of the UA castable shows a four-times-higher peak intensity for the K element in the corroded section than in the non-corroded section. When comparing the bond type, the UA castable visually appears to be more infiltrated by K_2CO_3 based on the presence of K^+ ions. This is caused by a higher average pore size although the apparent porosity is slightly higher for the NA castable. The castable containing pores with a higher diameter allows the corrosive media to infiltrate the castable easier than the castable containing pores with a lower diameter. Additionally, calcium-containing mixtures or castables usually perform with a lower refractoriness than calcium-free mixtures [10].

4. Conclusions

- Fine pastes prepared from the fine matrix and five types of colloidal silica performed comparable properties with minimal differences (especially for BD and CMOR). Higher differences were obtained in AP. To achieve a high density paste, K1530K colloidal silica is the most suitable.
- The phase composition of fine pastes after firing can be slightly modified using a different type of colloidal silica. Increasing the content of SiO2 solids by 10% in colloidal silica results in a 1.3% increase of the mullite content (at 1500 °C).
- The performance and properties of castables can be controlled by using different types of grog. The most stable properties after firing were obtained using fireclay and mullite grog. To achieve specific properties at specific conditions, the castables mixture may be tailored for its purpose—e.g., using 70% of A and 30% of M to achieve better abrasion resistance and a permanent linear change close to the zero.
- The pore structure mainly influences the corrosion resistance. A higher average pore diameter leads to deteriorating the corrosion resistance. Mercury intrusion porosimetry (MIP) can be used to evaluate the proper castable design in terms of the pore structure, especially for castables used in corrosive environments.

- The sol-gel-bonded castable, compared to the more traditional calcium aluminate-bonded castable, performed successfully at all tested aspects. Key property values such as permanent linear changes, the apparent porosity and bulk density were almost at the same level regardless of the used bond. The sol-gel-bonded castable performed better for corrosion resistance.

Author Contributions: Conceptualization, D.Z. and L.N.; methodology, D.Z.; investigation, L.N.; resources, L.N., data curation, D.Z. and L.N.; writing, D.Z.; review and editing, L.N.; visualization, D.Z. All authors have read and agreed to the published version of the manuscript.

Funding: The research was funded by the Internal Grant Agency of Brno University of Technology, specific junior research No. FAST-J-22-7943, with project name: Study of microstructure of castables prepared by the sol-gel method.

Institutional Review Board Statement: Not applicable.

Informed Consent Statement: Not applicable.

Data Availability Statement: The data presented in this paper are available upon request from the corresponding author.

Conflicts of Interest: The authors declare no conflict of interest.

Appendix A

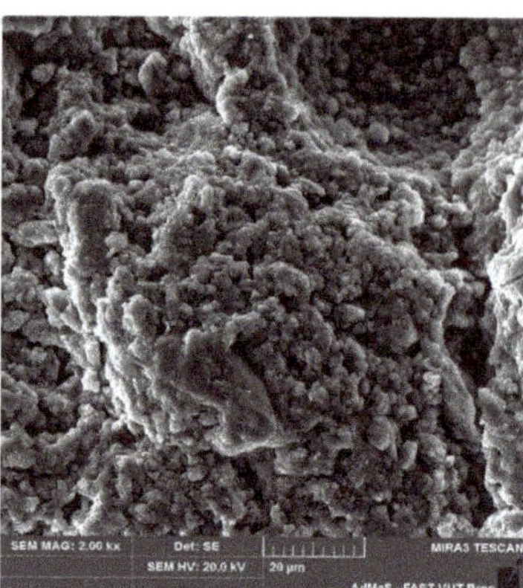

Figure A1. SEM microstructure of the NA castable after corrosion crucible test with magnification of 2000×, with specifically non-corroded area (**left**) and corroded area (**right**).

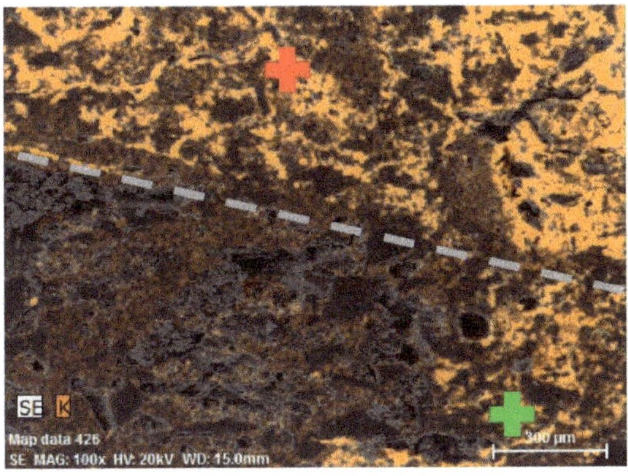

	NA-N	NA-K
mullite	86.0	80.4
corundum	6.5	7.3
cristobalite low	0.5	0.4
amorphous	6.9	11.2
kalsilite	–	0.7

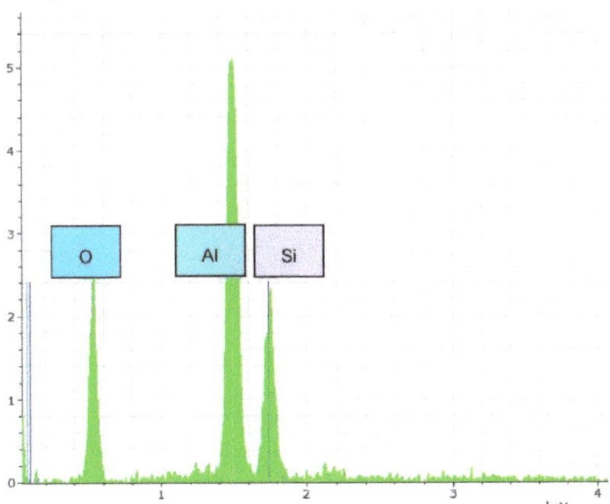

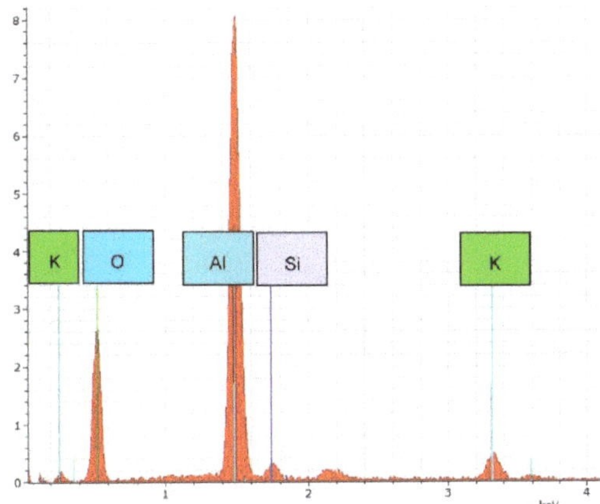

Figure A2. Element mapping of the transition zone between corroded (red point, NA-K) and non-corroded (green point, NA-N) area with magnification of $100\times$, both sections divided by grey line, phase composition and EDS spectrums.

Appendix B

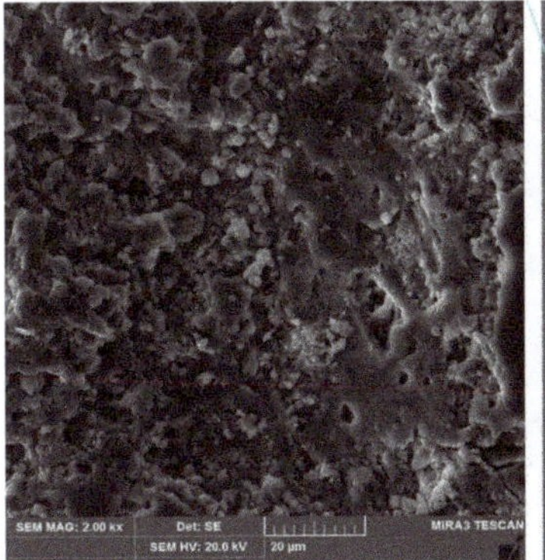

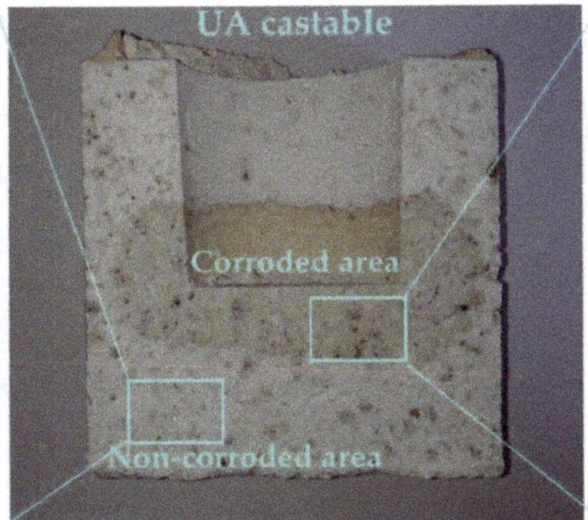

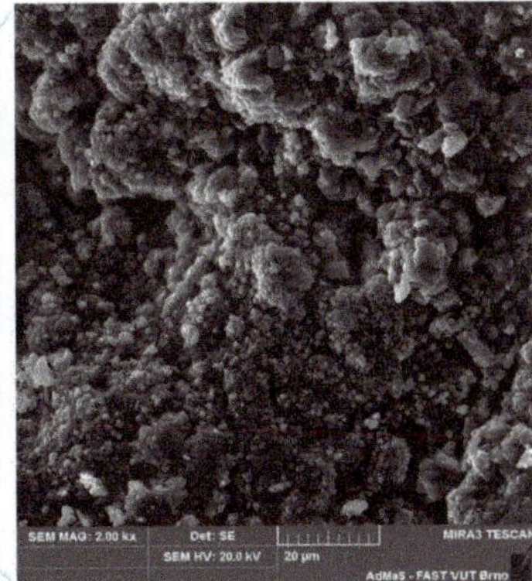

Figure A3. SEM microstructure of the UA castable after corrosion crucible test with magnification of $2000\times$, with specifically non-corroded area (**left**) and corroded area (**right**).

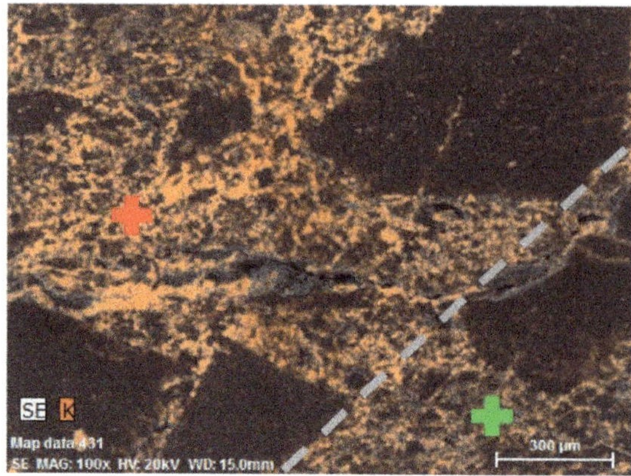

	UA-N	UA-K
mullite	83.3	82.8
corundum	9.0	6.4
cristobalite low	0.9	0.4
amorphous	6.8	9.5
kalsilite	-	0.9

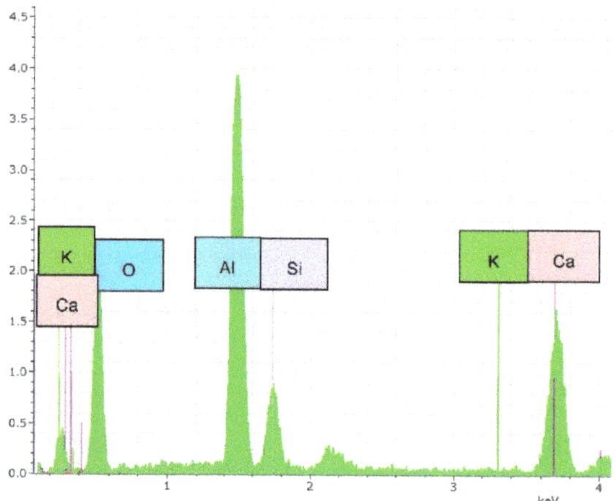

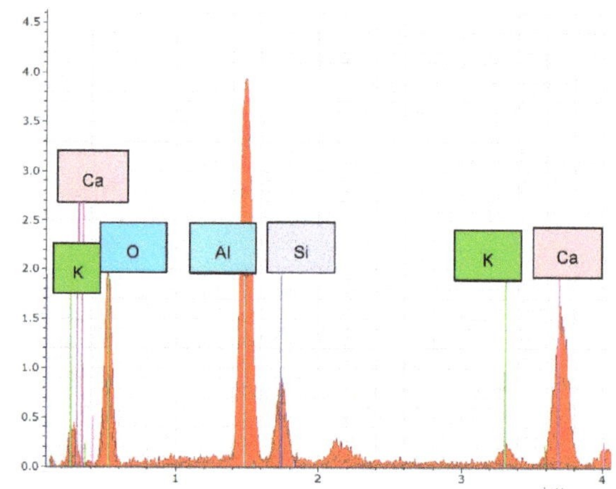

Figure A4. Element mapping of the transition zone between corroded (red point, UA-K) and non-corroded (green point, UA-N) area with magnification of 100×, both sections divided by grey line, phase composition and EDS spectrums.

References

1. Banerjee, S. *Monolithic Refractories-a Comprehensive Handbook*; World Scientific/The American Society: Singapore, 1998.
2. Nishikawa, A. *Technology of Monolithic Refractories*; Plibrico Japan Company Limited: Tokyo, Japan, 1984; pp. 67–71.
3. *EN ISO 1927-1*; Monolithic (unshaped) refractory products–Part 1: Introduction and Classification. ISO: Geneva, Switzerland, 2006.
4. Krietz, L. Refractory castables. In *Refractories Handbook*; Marcel Dekker: New York, NY, USA, 2004.
5. Ismael, M.R.; Salomao, R.; Pandolfelli, V.C. Refractory Castables Based on Colloidal Silica and Hydratable Alumina. *Am. Ceram. Soc. Bull.* **2007**, *86*, 58–61.
6. Lorenz, R.; Buchel, G.; Buhr, A.; Aronni, J.; Racher, R. Improved Workability Of Calcia Free Alumina Binder Alphabond For Non-Cement Castables. In Proceedings of the 47th International Colloquium on Refractories, Aachen, Germany, 13–14 October 2004; pp. 67–71.
7. New Almatis Alphabond 300 Global Product Data Sheet. Gp-Rcp/015/R04/1207/ Msds 834. 2004. Available online: https://www.almatis.com/media/c5vhaad0/gp-rcp_015_alphabond_300_0812.pdf (accessed on 6 May 2022).
8. Parr, C.; Wohrmeyer, C. The advantages of calcium aluminate cement as a castable bonding system. In Proceedings of the St. Louis Section Meeting of American Ceramic Society, St. Louis, MO, USA, 29 March–2 April 2006.
9. Wohrmeyer, C.; Parr, C.; Assis, G.; Fryda, H.; Auvray, J.M.; Touzo, B.; Guichard, S. New spinel containing calcium aluminate cement for corrosion resistant castables. In Proceedings of the UNITECR-11, Kyoto, Japan, 30 October–2 November 2011; pp. 476–479.
10. da Luz, A.P.; Braulio, M.A.L.; Pandolfelli, V.C. *Refractory Castable Engineering*; Göller Verlag: Baden, Germany, 2015.
11. Pilate, P.; Tirlocq, J.; Cambier, F. Refractory Castables: An Overview. *Ceram. Forum Int.* **2007**, *84*, 43–49.

12. Nouri-Khezrabad, M.; Braulio, M.A.L.; Pandolfelli, V.C.; Golestani-Fard, F.; Rezaie, H.R. Nano bonded refractory castables. *Ceram. Int.* **2013**, *39*, 3479–3497. [CrossRef]
13. Ismael, M.R.; Anjos, R.D.; Salomao, R.; Pandolfelli, V.C. Colloidal Silica as A Nanostructured Binder For Refractory Castables. *Refract. Appl. News* **2015**, *11*, 16–20.
14. Antonovic, V.; Pundiene, I.; Stonys, R.; Cesniene, J.; Keriene, J. A Review of The Possible Applications Of Nanotechnology In Refractory Concrete. *J. Civ. Eng. Manag.* **2010**, *16*, 595–602. [CrossRef]
15. Braulio, M.A.L.; Tontrup, C.; Medeiros, J.; Pandolfelli, V.C. Colloidal Alumina as A Novel Castable Bonding System. *Refract. Worldforum* **2011**, *3*, 135–141.
16. Anderson, M. Better Refractories through Nanotechnology. *Ceram. Ind. Mag.* **2005**, *155*, 29–35.
17. Lipinski, T.R.; Drygalska, E.; Tontrup, C. The Influence of Additions Of Nanostructured Al_2O_3 Powder on The High Temperature Strength of High Alumina Refractories. In Proceedings of the UNITECR-09, Salvador, Brazil, 13–16 October 2009; pp. 3–10.
18. Lipinski, T.R.; Tontrup, C. The Use of Nano-Sca led Alumina In Alumina-Based Refractory Materials. In Proceedings of the UNITECR-07, Dresden, Germany, 18–21 September 2007; pp. 391–393.
19. Bergna, H.E.; Roberts, W.O. *Colloidal Silica: Fundamentals and Applications*; CRC Press, Taylor and Francis: Boca Raton, FL, USA, 2005.
20. Ghosh, S.; Majumdar, R.; Sinhamahapatra, B.K.; Nandy, R.N.; Mukherjee, M.; Mukhopadhyay, S. Micro Structures Of Refractory Castables Prepared With Sol Gel Additives. *Ceram. Int.* **2003**, *29*, 671–677. [CrossRef]
21. Banerjee, S. Recent Developments in Monolithic Refractories. *Ceram. Bull.* **1998**, *77*, 59–63.
22. Mukhopadhyay, S.; Dutta, S.; Majumdar, M.; Kundu, A.; Das, S.K. Synthesis and Characterization Of Alumina Bearing Sol For Application In Refractory Castables. *Ind. Ceram.* **2000**, *20*, 88–92.
23. Mukhopadhyaya, S.; Das, S.K. Role of Alumina Sol on the Physical Properties of Low Cement Castables. *Trans. Indian Ceram. Soc.* **2000**, *59*, 68–74. [CrossRef]
24. Mukhopadhyay, S.; Mahapatra, S.; Mukherjee, P.; Dasgupta, T.; Das, S.K. Effect of Alumina Sol In No-Cement Refractory Castables. *Trans. Indian Ceram. Soc.* **2001**, *60*, 63–67. [CrossRef]
25. Mukhopadhyay, S.; Mahapatra, S.; Mondal, P.K.; Das, S.K. Alumina and Silica Sols as Binders In A Typical ULC Castable. *Trans. Indian Ceram. Soc.* **2003**, *62*, 106–111. [CrossRef]
26. Han, K.R.; Park, S.W.; Kim, C.S.; Yang, J.O. Alumina Bonded Unshaped Refractory and Manufacturing Method Thereof. International Patent WO 2011/115353, 30 November 2010.
27. Mukhopadhyay, S.; Ghosh, S.; Mahapatra, M.K.; Mazumder, R.; Barick, P.; Gupta, S.; Chakraborty, S. Easy-To-Use Mullite and Spinel Sols as Bonding Agents in a High-Alumina Based Ultra Low Cement Castable. *Ceram. Int.* **2002**, *28*, 719–729. [CrossRef]
28. Cao, F.; Long, S.; Wu, X.; Telle, R. Properties of Sol–Gel Bonding Castables. *Key Eng. Mater.* **2007**, *336–338*, 1484–1487. [CrossRef]
29. Borges, O.H.; Santos, T.; Salvini, V.R.; Pandolfelli, V.C. CA6-based macroporous refractory thermal insulators containing mineralizing agents. *J. Eur. Ceram. Soc.* **2020**, *40*, 6141–6148. [CrossRef]
30. Voronkov, M.E.; Brykov, A.S.; Nekrasova, O.K.; Pavlov, S.S. Effect of Pyrocatechin on the Properties of Cement-Free Refractory Concrete Mixtures Based on Silica-Containing Colloidal Binders. *Refract. Ind. Ceram.* **2019**, *59*, 545–548. [CrossRef]
31. Klárová, M.; Vlček, J.; Topinková, M.; Burda, J.; Martaus, A.; Priesol, I.; Szczerba, J. Cement Substitution in High-Temperature Concrete. *Minerals* **2021**, *11*, 1161. [CrossRef]
32. Schacht, C.A. *Refractories Handbook*; Taylor & Francis Inc.: Boca Raton, FL, USA, 2004.
33. Kannabiran, S.; Zhang, M. Effects of Grain Size and Temperature on the Abrasion Resistance of Low Cement Castables. In Proceedings of the UNITECR 2017, Santiago, Chile, 27–29 September 2017; pp. 180–183, ISBN 978-3-9815813-3-1.
34. Szczerba, J.; Sniezek, E.; Reynaert, C. Corrosion Tests for Refractory Materials Intended for the Steel Industry–A Review. *Ceram. Silikáty* **2020**, *64*, 278–288.
35. Singh, A.K.; Sarkar, R. High alumina castables: A comparison among various sol-gel bonding systems. *J. Aust. Ceram. Soc.* **2017**, *53*, 553–567. [CrossRef]
36. Zemánek, D.; Nevřivová, L. Use of the sol-gel method for the production of no cement castables. *AIP Conf. Proc.* **2021**, *2322*, 020004.
37. *DIN CEN/TS 15418:2006-09*; Methods of Test for Dense Refractory Products-Guidelines for Testing the Corrosion of Refractories Caused by Liquids. Deutsches Institut für Normung: Berlin, Germany, 2006.

 *materials*

Article

Enhancing the Oxidation Resistance of Al₂O₃-SiC-C Castables via Introducing Micronized Andalusite

Xiaoyu Wang [1], Saixin Wang [1], Yuandong Mu [1,*], Ruijie Zhao [2], Qingfeng Wang [3], Chris Parr [4] and Guotian Ye [1,*]

[1] Henan Key Laboratory of High Temperature Functional Ceramics, School of Materials Science and Engineering, Zhengzhou University, Zhengzhou 450001, China; wxysop13@163.com (X.W.); 202012192013587@gs.zzu.edu.cn (S.W.)
[2] Gongyi Shunxiang Refractories Co., Ltd., Zhengzhou 450001, China; 15538387111@163.com
[3] School of Materials Science and Engineering, Luoyang Institute of Science and Technology, Luoyang 471023, China; qfwang@lit.edu.cn
[4] Imerys Aluminates, 92800 Paris, France; chris.parr@imerys.com
* Correspondence: myd1228@zzu.edu.cn (Y.M.); gtye@zzu.edu.cn (G.Y.)

Abstract: Additions of andalusite aggregates (19 wt%) were shown in previous literature to enhance the antioxidation of Al₂O₃-SiC-C (ASC) castables. This work aims to investigate whether micronized andalusite has a greater influence on antioxidation improvement than andalusite aggregates. Various low contents (5 wt% and below) of micronized andalusite ($\leq$5 μm) were introduced as a substitute for brown fused alumina in the matrix of ASC castables. The antioxidation of castable specimens was estimated by the oxidized area ratio on the fracture surface after a thermal shock test. The microstructure and phases of micronized andalusite and the castable specimens were characterized by scanning electron microscopy (SEM) and X-ray diffraction (XRD), respectively. The results suggest that the antioxidation effects of ASC castables with a low addition of micronized andalusite are effectively enhanced. The heat-induced transformation of andalusite produces SiO₂-rich glass, favoring the sintering of the castable matrix and impeding oxygen diffusion into the castable's interior. Therefore, the castable antioxidation is enhanced without deteriorating the hot modulus of rupture.

Keywords: micronized andalusite; antioxidation; Al₂O₃-SiC-C castables; hot modulus of rupture

Citation: Wang, X.; Wang, S.; Mu, Y.; Zhao, R.; Wang, Q.; Parr, C.; Ye, G. Enhancing the Oxidation Resistance of Al₂O₃-SiC-C Castables via Introducing Micronized Andalusite. *Materials* **2021**, *14*, 4775. https://doi.org/10.3390/ma14174775

Academic Editors: Jacek Szczerba and Ilona Jastrzębska

Received: 15 July 2021
Accepted: 21 August 2021
Published: 24 August 2021

Publisher's Note: MDPI stays neutral with regard to jurisdictional claims in published maps and institutional affiliations.

1. Introduction

Al₂O₃-SiC-C (ASC) castables are widely employed in blast furnace runners for their excellent thermal shock resistance and erosion resistance [1,2]. These favorable properties are mainly obtained from SiC and carbon/graphite (C) additions, which have a low wettability to molten iron and slags and provide thermal shock resistance [3,4]. However, C is vulnerable to oxidation at high temperatures, and the volatiles generated during heat treatment leads to a decrease in the compactness and strength of the castable [3,4]. Simultaneously, the decarburized layer formed after oxidation increases the castable corrosion by the slag [2]. Antioxidants are generally added to inhibit the oxidation of carbonaceous materials in castables. In this regard, the effect of various antioxidants (such as Si, B₄C, Si₂BC₃N, etc.) on the oxidation resistance of ASC castables has been widely studied [5–7]. Although the above-mentioned antioxidants could effectively increase the antioxidation of ASC castables, they are expensive and have a high energy consumption [6,7]. However, we recently found that the partial substitution of brown fused alumina by andalusite aggregates (8–5 mm/5–3 mm/3–1 mm) could also enhance the antioxidation of ASC castables [8–11]. Andalusite is a natural mineral, which is commonly applied in refractory materials [12–14]. Employing andalusite could improve the volume stability and thermal shock resistance of the refractory, due to the mullitization of andalusite accompanied by a 3–5% volume expansion [8].

Based on our previous study, adopting 19 wt% andalusite aggregates could significantly promote the antioxidation of ASC castables [8–11], as the high-temperature heating prompts the transformation of andalusite into mullite and silica-rich glass. This silica-rich glass obstructs the pore channels and impedes oxygen diffusion into the castable interior [15–17]. Furthermore, the exuded SiO_2-rich glass from andalusite particles reacts with fine alumina to produce secondary mullite and raise the castable densification, thereby reducing the rate of oxygen diffusion into the castable interior and enhancing the oxidation resistance [8–11].

It is suggested that the addition of micronized andalusite into the matrix may provide the enhanced oxidation resistance of the ASC castable through two mechanisms; The transformation of andalusite particles is a reaction that proceeds from the exterior to the interior, and there are more defects on the finer particle surface. Andalusite with a smaller particle size would achieve a higher conversion extent (mullitization rate) at the same heating temperature [8]. According to previous research, only 20% of the silica-rich glass phase produced by mullitization is released to the surface, and the remaining 80% remains inside [10]. Hence, reducing the particle size of andalusite aggregate is conducive to increasing the release of silica-rich glass and enhancing castable antioxidation. Conversely, most castable pore channels are distributed in the matrix, so the silica-rich glass phase produced by andalusite aggregates cannot uniformly fill the pores in the castable matrix. Moreover, there is a higher probability that the exuded silica-rich glass in the matrix will combine with the nearby alumina powder to form secondary mullite. In summary, introducing micronized andalusite into the matrix, compared with aggregates, is speculated to be a more effective route to enhancing castable oxidation resistance.

Moreover, the substitution of high-density and low-porosity brown fused alumina would deteriorate the slag corrosion resistance of the castables [18,19]. Maintaining a low content of andalusite to enhance the oxidation resistance is important for corrosion resistance. Few previous studies have researched the performance of castables containing micronized andalusite in the matrix. The current work aims to examine the influence of micronized andalusite on the antioxidation of ASC castables. The amount of micronized andalusite introduced was limited to 5 wt% or below, which was much lower than the reported andalusite aggregate content (19 wt%). Thermal shock tests coupled with the castable oxidation index were applied to evaluate the antioxidation impact [8,9]. The antioxidation mechanism was further illuminated by the phase assemblage, hot modulus of rupture (HMOR), apparent porosity (AP), residual strength ratio (RSR), cold modulus of rupture (CMOR), and castable microstructure analysis.

2. Experimental Procedures

2.1. Materials

The raw materials for the ASC castables include brown fused alumina ($Al_2O_3 \geq 95$ wt%, ≤ 8 mm; Hengjia, Shandong, China), silicon carbide ($SiC \geq 97$ wt%, ≤ 1 mm; Hengjia, Shandong, China), ball pitch ($C \geq 56$ wt%, 1~0.2 mm; Hengjia, Shandong, China), Si powder ($Si \geq 98$ wt%; Hengjia, Shandong, China), reactive alumina powder (CL370, $Al_2O_3 \geq 99.56$ wt%, $d_{50} = 2.14$ µm; Almatis na (Qingdao) Co., Ltd., Qingdao, China), calcium aluminate cement (CAC, Secar 71, $d_{50} = 13.6$ µm; Imerys, Tianjin, China), aluminum powder ($Al \geq 99$ wt%; Hengjia, Shandong, China), and silica fume (951U, Elkem Materials, Oslo, Norway). As presented in Table 1, the brown fused alumina was partially substituted with micronized andalusite (Durandal 59, IMERYS Co., Paris, France) in this experiment. The particle size distribution of micronized andalusite is shown in Figure 1.

Table 1. Mass fractions (wt%) and particle sizes (mm) of raw materials in Al$_2$O$_3$-SiC-C (ASC) castables.

Raw Material	Particle Size (mm)	wt (%)				
		M	F1	F2	F3	F4
Brown fused alumina	8–5	19	19	19	19	19
	5–3	19	19	19	19	19
	3–1	19	19	19	19	19
	1–0	7.5	7.5	7.5	7.5	7.5
	≤0.074	5	4	3	2	0
Micronized andalusite	≤0.005	0	1	2	3	5
Pitch	1–0.2	3	3	3	3	3
SiC	1–0	8	8	8	8	8
	≤0.045	8	8	8	8	8
Silica fume		2	2	2	2	2
Secar 71		3	3	3	3	3
α-Al$_2$O$_3$ micro powder		5	5	5	5	5
Si powder		1.5	1.5	1.5	1.5	1.5
Al powder		+0.1	+0.1	+0.1	+0.1	+0.1

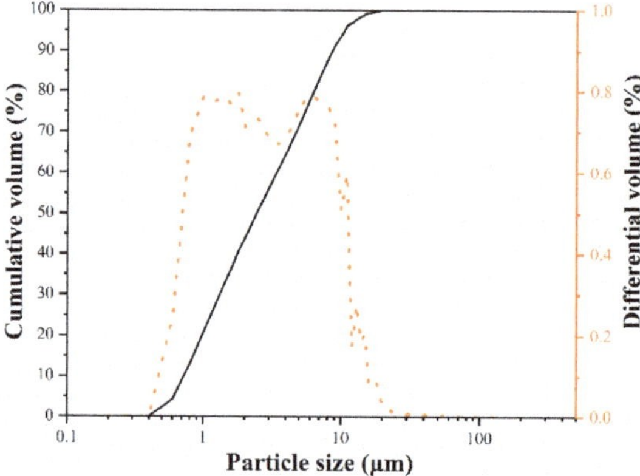

Figure 1. Particle size distribution of micronized andalusite.

2.2. Sample Preparation and Characterization

The ASC castable specimens were prepared according to the same procedures in the standards reported in previous work [8–10]. The castable specimens containing different micronized andalusite levels are denoted as M, F1, F2, F3, and F4, as listed in Table 1. The air quenching method (ΔT = 950 °C, 10 cycles) was utilized to test the fired castable thermal shock resistance. The castable specimens were held for 0.5 h in a 950 °C furnace, followed by 5 min of air cooling with a fan (air speed: 8870 m^3/h) at room temperature. Finally, the CMOR of the fired castable bars before and after thermal shock tests were tested, and the RSR was calculated to assess the castable thermal shock resistance. Afterward, the castable specimens were cut to observe the cross-sectional oxidized layer to analyze the oxidation level after the thermal shock processes. The pixel method was used to calculate the oxidation index, which refers to the ratio of the oxidized area to the total area of the specimen section, and is used to evaluate the castable oxidation degree.

The CMOR and AP of the fired castables were obtained with a loading rate of 0.15 MPa/s, based on the Chinese Standards GB/T 3001-2007 and GB/T 2997-2015 [8–10]. The HMOR of the fired castables at elevated temperatures was attained via a high temperature strength tester with castables firing at 1450 °C for 3 h, according to GB/T 3002-2004 [9]. A laser particle size analyzer (HELOS & RODOS, SYMPATEC, Clausthal, Germany) was applied to detect the particle size distribution of micronized andalusite, based on the Chinese Standard GB/T 19077-2016. Scanning electron microscopy (SEM; Sigma HD, Zeiss, Oberkochen, Germany) was used to characterize the microstructures of the polished castable specimens after heat treatment at 1450 °C and of the micronized andalusite before and after heat treatment at 1450 °C for 3 h. X-ray diffraction (XRD, D8 advance, Bruker, Karlsruhe, Germany) was applied to detect the micronized andalusite phase compositions after firing at temperatures ranging from 1200 °C to 1450 °C for 3 h.

3. Results and Discussion

The antioxidation effect of ASC castables is evaluated through photographs and the oxidation index (Figures 2 and 3) of the calcined castable cross-sections after thermal shock tests. As shown in Figure 2, the dark regions represent the unreacted primary core, and the gray areas indicate the oxidized layer where the amount of C and SiC declines. Figure 2 also shows that the specimen M has a noticeably thicker oxidized layer than the specimen F1, and the oxidized layer thickness declines with the elevated micronized andalusite content. The oxidation index in Figure 3 quantitatively shows that the M samples have a significantly higher oxidation index than F1 samples, and the castable oxidation index gradually decreases with the increase of the micronized andalusite content. The above results denote that the antioxidation of ASC castables is noticeably improved with 1 wt% micronized andalusite and that a higher level (up to 5 wt%) of micronized andalusite leads to higher castable antioxidation.

Figure 2. Photographs of the cross-section of the fired castables without micronized andalusite (M) and with 1 wt% (F1), 2 wt% (F2), 3 wt% (F3), and 5 wt% (F4) micronized andalusite after thermal shock tests.

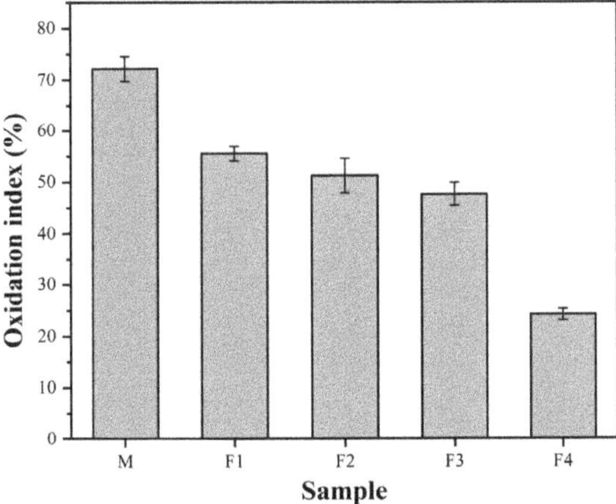

Figure 3. Oxidation index of the castable specimens after heat treatment at 1450 °C for 3 h.

The previous study proved that the oxidation index of the castables was not below 30% when 19 wt% andalusite aggregate (3–1 mm) was added [9]. In comparison, when only 5 wt% of micronized andalusite was applied, the castables' oxidation index was 24.1%, illustrating that castables with a small amount of micronized andalusite achieve a comparable antioxidation effect to andalusite aggregates.

Compared with the previous study [8–10], since the content of the micronized andalusite in the castables of this study is low (below 5 wt%), it is hard to detect the appearance of andalusite in the XRD patterns and SEM images of the castables. Therefore, XRD patterns of pure micronized andalusite subjected to different heating procedures are illustrated in Figure 4, which visually exhibits the phase evolution of micronized andalusite during heating at high temperatures. The residual andalusite peaks gradually drop when the temperature rises from 1200 °C to 1450 °C. Meanwhile, the mullite peaks appear in specimens after calcining at 1250 °C, indicating that a portion of andalusite transforms to mullite when fired at 1250 °C. Figure 4 also signifies that the mullite peaks noticeably increase after firing at 1400 °C and that the andalusite peaks disappear after calcining at 1450 °C, demonstrating that the heat-induced decomposition of micronized andalusite is drastic at 1400 °C and completed at 1450 °C. Compared with previous research results stating that andalusite aggregates generally begin the mullitization process gradually at 1400 °C–1500 °C [10], the conversion of micronized andalusite can be completed at a lower temperature than coarse andalusite. Therefore, more silica-rich glass phases could be generated to obstruct the pore channel and increase the castable densification, thereby impeding oxygen diffusion into the castable interior. This also proves why adding low-content (5 wt%) micronized andalusite can achieve similar antioxidation properties to adding 19 wt% andalusite aggregate [9], as shown in Figure 3.

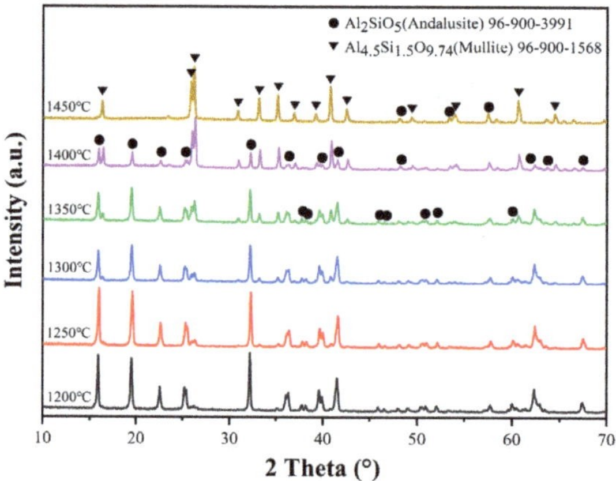

Figure 4. XRD patterns of micronized andalusite ($\leq$5 µm) after firing in the temperature range of 1200 °C to 1450 °C for 3 h.

Figure 5a,b shows the morphology of micronized andalusite before and after heat treatment at 1450 °C for 3 h, respectively. As shown in Figure 5a,b, the degree of conversion of thermal-treated micronized andalusite is already very high when compared with uncalcined micronized andalusite. EDS spot analyses on micronized andalusite after heat treatment at 1450 °C, as shown in Figure 5c,d, detect that the particles still maintain the morphology of the original parent phase, but most of them should have transformed into mullite and SiO$_2$-rich glass phase. The XRD of the andalusite after heat treatment at 1450 °C (Figure 4) also corroborates this point. The mullite produced by micronized andalusite has a short columnar shape due to its own size limitation, which is different to the mullite resulting from andalusite aggregates [15,20]. The previous study showed that a small amount of silica-rich glass phase extruded onto the surface of andalusite aggregate [9]. In contrast, as the conversion of micronized andalusite is complete after heat treatment at 1450 °C (Figure 4), the larger specific surface area and smaller particle size (Figure 1) of micronized andalusite make it easier for the silica-rich glass phase to be released to the surface, effectively bonding the surrounding particles. This further explains why adding 5 wt% micronized andalusite can achieve a similar antioxidant effect to adding 19 wt% andalusite aggregate [10]. As a result, the pore channels can be filled to a higher degree by the more extruded SiO_2-rich glass from transformed micronized andalusite, which strongly prevents oxygen diffusion and promotes antioxidation (Figure 2).

Figure 5. Secondary electron images of uncalcined micronized andalusite (**a**) before and (**b**) after calcining for 3 h at 1450 °C, and (**c,d**) the energy disperse spectroscopy (EDS) spot analyses of selected points in Figure 4b.

Figure 6 displays the backscattered images of the castable specimen's oxidized layers after heat treatment at 1450 °C for 3 h. It is apparent from Figure 6a that many pores exist in the matrix, and gaps are located between the matrix and the aggregate in the M specimen, suggesting that the aggregate is not closely combined with the matrix. In comparison, the boundary between the aggregate and matrix in the F4 sample is more blurred (Figure 6b). Figure 6a,b clearly indicates that the castables containing andalusite have a much stronger aggregate-matrix interfacial transition zone. This is credited to the heat-induced transformation of andalusite which produces silica-rich glass (Figure 5) and introduces impurities, and the generated silica can partially react with alumina in the castable to produce secondary mullite during heating. Therefore, the generation and reaction of silica-rich glass creates such a strong interfacial bond. The denser connection generated by silica-rich glass and secondary mullite between the aggregate and the matrix would also block oxygen diffusion [21], thereby promoting castable antioxidation.

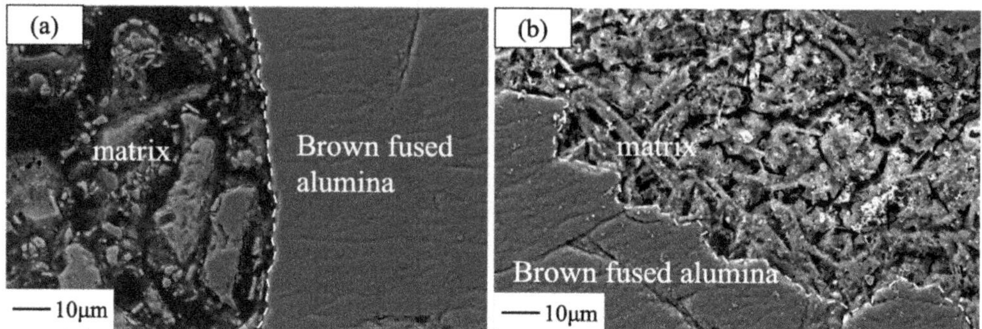

Figure 6. Backscattered images of the polished castable specimens (**a**) M and (**b**) F4 after calcining for 3 h at 1450 °C.

Figure 7 implies that the AP of the castable specimens F1–F4 is slightly lower than that of the castable sample M. It should be noted that, theoretically, the formation of mullite and secondary mullite would produce a volume expansion of approximately 3% and 8–10%, respectively [2,8,15–17]. Due to these volume expansions, the AP in the castable should increase. In addition, the oxidation of C would also increase the porosity of samples. However, silica-rich glass (Figure 5) and the larger amount of liquid generated from the conversion of andalusite in the matrix would block the pores of castables and accelerate the sintering, consequently offsetting the volume expansion. Moreover, the improvement of oxidation resistance results in less C being oxidized in the castable, thereby reducing the porosity caused by the oxidation of C. This explains why the AP declines with an elevated micronized andalusite content.

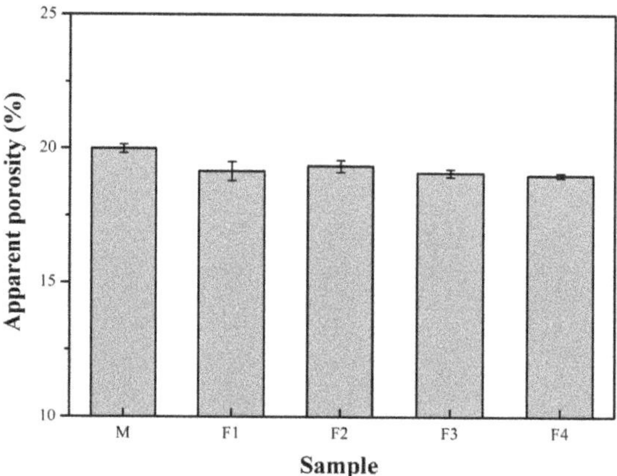

Figure 7. AP of the castable specimens after heat treatment at 1450 °C for 3 h.

The higher amount of high-temperature liquid not only benefits the decline of AP but also enhances the CMOR of the castables. Figure 8 illustrates that the CMOR of castable specimens F1–F4 is higher than specimen M, although due to the strong covalent bonding C and SiC have a low diffusivity and inhibit sintering in the oxide compositions [3,4]. The silica-rich glass and larger amount of liquid improve the castable sintering, promote the castable densification (Figure 6b), and raise the castable strength after calcining at 1450 °C.

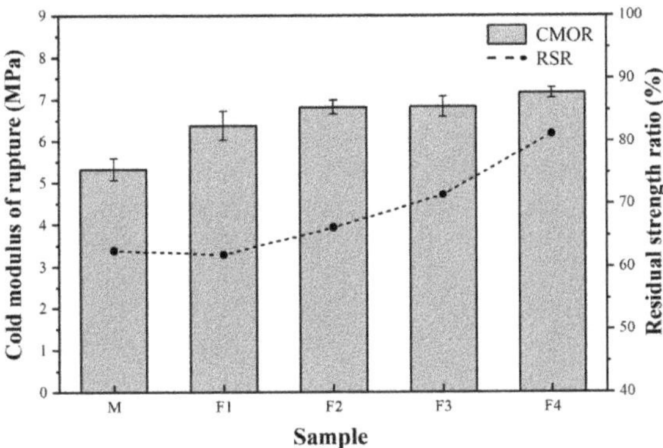

Figure 8. CMOR and RSR of the castable specimens after heat treatment at 1450 °C for 3 h.

Figure 8 also shows that the RSR of the castable specimen M is much lower than that of the castable samples F1–F4, indicating that the addition of andalusite can remarkably enhance the thermal shock resistance of the castables. The improved thermal shock resistance of castables containing micronized andalusite should mainly benefit from the silica-rich glass produced from the heat-induced transformation of andalusite, which could absorb the thermal shock stress [21,22]. As mentioned above, the improved castable antioxidation is caused by the replacement of brown fused alumina with micronized andalusite, consequently declining the oxidation degree of SiC and C. The enhanced thermal shock resistance could also be ascribed to more SiC and C remaining in the castable specimens containing micronized andalusite [3,4]. Therefore, the integration of micronized andalusite could improve the material antioxidation and enhance the castable thermal shock resistance.

As shown in Figure 9, the HMOR of the specimens F1–F4 is higher than that of the specimens without andalusite. The higher amount of secondary mullite produced from silica-rich glass (Figure 5) and alumina powder is possibly beneficial to improving the high-temperature performance of the castables containing micronized andalusite, as the secondary mullite neutralizes the negative influence of the high amount of high-temperature liquid. Moreover, the addition of micronized andalusite in excess of 5 wt% may further increase the oxidation resistance of the ASC castables, but it can be imagined that adding more micronized andalusite would produce more silica-rich glass phase, which may not benefit the HMOR of the ASC castables [23]. Further studies would be necessary to confirm the optimum addition level of micronized andalusite to ensure oxidation resistance without a reduction in HMOR and while maintaining corrosion resistance.

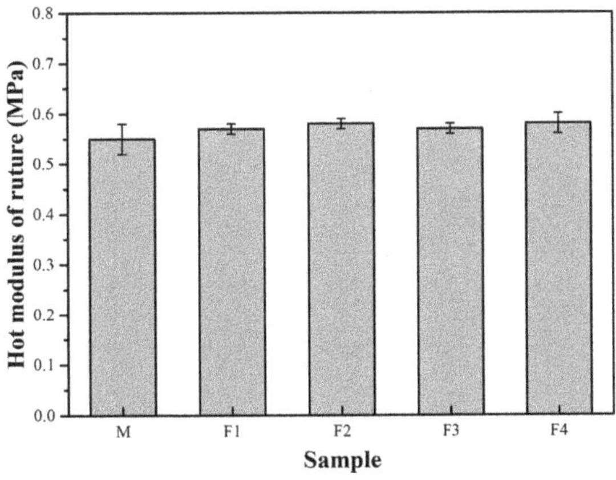

Figure 9. HMOR of the castable specimens after heat treatment at 1450 °C for 3 h.

4. Conclusions

In this study, the antioxidation property of ASC castables with different micronized andalusite contents was investigated. The results suggest that the antioxidation effects of ASC castables with a low addition (5 wt% and below) of micronized andalusite are effectively enhanced. The main conclusions are drawn as follows:

The conversion ratio of micronized andalusite is higher than that of andalusite aggregates at the same heating temperature. The morphology of the fired micronized andalusite indicates that much more silica-rich glass was generated and released to the surface, which is beneficial to blocking the oxygen transition channels and hindering oxidation. This explains why a small amount of micronized andalusite addition achieves a similar antioxidation effect that a high content of andalusite aggregate does.

Additionally, the generated glassy silica phase from micronized andalusite and the subsequent formation of secondary mullite could strengthen the bonding between the matrix and aggregate, resulting in an improved cold and hot modulus of rapture. Additionally, as more SiC and C are protected from oxidation, the thermal shock resistance of the castables is also enhanced.

Author Contributions: Conceptualization, G.Y.; methodology, G.Y.; software, Q.W.; validation, G.Y., Y.M. and C.P.; formal analysis, X.W.; investigation, X.W. and R.Z.; resources, G.Y.; data curation, X.W.; writing—original draft preparation, X.W.; writing—review and editing, Y.M. and X.W.; visualization, X.W.; supervision, S.W.; project administration, Y.M.; funding acquisition, G.Y. and C.P. All authors have read and agreed to the published version of the manuscript.

Funding: This research was funded by the National Natural Science Foundation of China (51572244 and U1604252) and Imerys Aluminates.

Institutional Review Board Statement: Not applicable.

Informed Consent Statement: Not applicable.

Data Availability Statement: All the data is available within the manuscript.

Acknowledgments: The authors acknowledge the National Natural Science Foundation of China (51572244 and U1604252) for the financial support. This work was also financed by Imerys Aluminates.

Conflicts of Interest: The authors declare no conflict of interest.

References

1. Prompt, N.; Ouedraogo, E. High temperature mechanical characterization of an alumina refractory concrete for Blast Furnace main trough: Part I. General context. *J. Eur. Ceram. Soc.* **2008**, *28*, 2859–2865. [CrossRef]
2. Zhang, S.; Lee, W.E. Carbon containing castables: Current status and future prospects. *Br. Ceram. Trans.* **2002**, *101*, 1–8. [CrossRef]
3. Gallet-Doncieux, A.; Bahloul, O.; Gault, C.; Huger, M.; Chotard, T. Investigations of SiC aggregates oxidation: Influence on SiC castables refractories life time at high temperature. *J. Eur. Ceram. Soc.* **2012**, *32*, 737–743. [CrossRef]
4. Zhao, Y.; Zhao, S.; Nie, J. Effect of ball pitch on properties of Al_2O_3-SiC-C castables for iron trough. *Refractories* **2010**, *44*, 126–128.
5. Wu, J.; Bu, N.; Li, H.; Zhen, Q. Effect of B_4C on the properties and microstructure of Al_2O_3-SiC-C based trough castable refractories. *Ceram. Int.* **2017**, *43*, 1402–1409. [CrossRef]
6. Liu, G.; Jin, X.; Qiu, W.; Ruan, G.; Li, Y. The effect of microsilica on the oxidation resistance of Al_2O_3-SiC-SiO_2-C castables with Si and B_4C additives. *Ceram. Int.* **2016**, *42*, 251–262. [CrossRef]
7. Bi, Y.; Zhang, H.; Wang, H.; Duan, S.; Jia, Q.; Ge, S.; Zhang, S. Catalytic Fabrication of SiC/SiO_2 coated graphite and its behaviour in Al_2O_3–C castable systems. *Ceram. Int.* **2019**, *45*, 16180–16187. [CrossRef]
8. Ding, D.; Tian, X.; Ye, G. Effect of andalusite aggregates on oxidation resistance of Al_2O_3-SiC-C. castables. *Ceram. Int.* **2019**, *45*, 19237–19241. [CrossRef]
9. Chen, A.; Wang, X.; Zhou, W.; Mu, Y.; Parr, C.; Ye, G. Oxidation resistance of Al_2O_3-SiC-C castables with different grades of andalusite. *J. Alloys Compd.* **2020**, *851*, 156836. [CrossRef]
10. Tian, X.; Liu, M.; Luan, J.; Zhu, L.; Zhao, J.; Ye, G. Effect of andalusite aggregate pre-fired at different temperatures on volume stability and oxidation resistance of Al_2O_3-SiC-C. castables. *Ceram. Int.* **2020**, *46*, 22745–22751. [CrossRef]
11. Tian, X.; Ding, D.; Zhang, P.; Liao, G.; Ahouanto, F.; Ye, G. Effect of andalusite with different particle sizes on properties of Al_2O_3-SiC-C castables. *Refractories* **2018**, *52*, 422–425.
12. Prigent, P.; Bouchetou, M.L.; Poirier, J. Andalusite: An amazing refractory raw material with excellent corrosion resistance to sodium vapours. *Ceram. Int.* **2011**, *37*, 2287–2296. [CrossRef]

13. Rebouillat, L.; Rigaud, M. Andalusite-based high-alumina castables. *J. Am. Ceram. Soc.* **2002**, *85*, 373–378. [CrossRef]
14. Dubreuil, P.; Filari, E.; Sobolev, V.M. Use of andalusite refractories in ferrous metallurgy. *Refract. Ind. Ceram.* **1999**, *40*, 252–259. [CrossRef]
15. Hülsmans, A.; Schmücker, M.; Mader, W.; Schneider, H. The transformation of andalusite to mullite and silica: Part I. Transformation mechanism in [1] A direction. *Am. Mineral.* **2000**, *85*, 980–986. [CrossRef]
16. Hülsmans, A.; Schmücker, M.; Mader, W.; Schneider, H. The transformation of andalusite to mullite and silica: Part II. Transformation mechanism in [100] and [10] A direction. *Am. Mineral.* **2000**, *85*, 987–992. [CrossRef]
17. Ding, D.; Ye, G.; Li, N.; Liao, G.; Tian, X.; Chen, L. Andalusite transformation and properties of andalusite-bearing refractories fired in different atmospheres. *Ceram. Int.* **2019**, *45*, 3186–3191. [CrossRef]
18. Ouedraogo, E.; Prompt, N. High-temperature mechanical characterization of an alumina refractory concrete for Blast Furnace main trough: Part II. Material behaviour. *J. Eur. Ceram. Soc.* **2008**, *28*, 2867–2875. [CrossRef]
19. Liao, G.; He, K.; Lil, L.; Jiang, M. Study on application of alumina in high-purity andalusite based refractory. *J. Miner. Mater. Charact. Eng.* **2004**, *3*, 81–89. [CrossRef]
20. Kakroudi, M.G.; Huger, M.; Gault, C.; Chotard, T. Anisotropic behaviour of andalusite particles used as aggregates on refractory castables. *J. Eur. Ceram. Soc.* **2009**, *29*, 571–579. [CrossRef]
21. Li, C.; Bian, C.; Han, Y.; Wang, C.A.; An, L. Mullite whisker reinforced porous anorthite cermics with low thermal conductivity and high strength. *J. Eur. Ceram. Soc.* **2016**, *36*, 761–765. [CrossRef]
22. Ribeiro, G.C.; Resende, W.S.; Rodrigues, J.A.; Ribeiro, S. Thermal shock resistance of a refractory castable containing andalusite aggregate. *Ceram. Int.* **2016**, *42*, 19167–19171. [CrossRef]
23. Bouchetou, M.L.; Ildefonse, J.P.; Poirier, J.; Daniellou, P. Mullite grown from fired andalusite grains: Role of impurities and of the high temperature liquid phase on the kinetics of mullitization and consequences on thermal shocks resistance. *Ceram. Int.* **2005**, *31*, 999–1005. [CrossRef]

MDPI

Communication

Structure–Property Functions of Inorganic Chemical Binders for Refractories

Vanessa Hopp *, Ali Masoudi Alavi, Dominik Hahn and Peter Quirmbach

Faculty for Natural Science, Campus Koblenz, University of Koblenz-Landau, 56070 Koblenz, Germany; masoudi@uni-koblenz.de (A.M.A.); dominikhahn@uni-koblenz.de (D.H.); pquirmbach@uni-koblenz.de (P.Q.)
* Correspondence: vanmueller@uni-koblenz.de

Abstract: For refractory application, amongst others, inorganic chemical binders are used to shape and process loose, unpacked materials. The binder influences the chemical composition within the ceramic body during setting, aging and firing and thus the finally reached properties of the refractory material. For an effective design of tailored materials with required properties, the mode of action of the binder systems should carefully be investigated. A combination of both structure analysis techniques and macroscopic property investigations proved to be a powerful tool for a detailed description of structure–property correlations. This is shown on the basis of X-ray powder diffraction and nuclear magnetic resonance spectroscopy analyses combined with observation of (thermo)mechanical and chemical investigations.

Keywords: inorganic chemical binders; refractories; phosphates; water glasses

Citation: Hopp, V.; Masoudi Alavi, A.; Hahn, D.; Quirmbach, P. Structure–Property Functions of Inorganic Chemical Binders for Refractories. *Materials* **2021**, *14*, 4636. https://doi.org/10.3390/ma14164636

Academic Editors: Jacek Szczerba and Ilona Jastrzebska

Received: 30 June 2021
Accepted: 13 August 2021
Published: 17 August 2021

Publisher's Note: MDPI stays neutral with regard to jurisdictional claims in published maps and institutional affiliations.

1. Introduction

Ceramics obtain their typical properties such as strength and hardness specifically through sintering of ceramic particles by heat treatment above their sintering temperature [1]. In practice, therefore, there is usually a need to increase the green strength of the ceramic body to enable components to be shaped and processed. For this purpose, binder additives are used which lead to the bonding of loose, unpacked materials such as monolithic refractory ceramics [2].

In principle, binders act through adhesive and cohesive forces. Cohesive forces take effect especially within the binder system itself, while adhesion results from bonding forces at the interface between the binder and the connected material, e.g., ceramic material [3]. However, depending on the type of binder used, their modes of action and effective operating temperatures differ significantly. For example, temporary (organic) binders or permanent (hydraulic or chemical) binders are applied both within the refractory industry production and other ceramic materials sectors.

Besides the major technically advances of binders, considerations must be made in respect to binders being able to decisively influence the chemical composition within the ceramic body composition during setting, aging and firing. Binders may actively enter into interactions and chemical reactions with ceramic components which, consequently, has a great influence on materials properties, e.g., strength, thermal properties or resistance to chemical attack (erosion, wear, corrosion). In particular, it must be considered that especially the bonding phase, as the interphase between the grains of the refractory material, shows high sensitivity to physical-chemical and mechanical impact. For refractory manufacturers, selecting a suitable binder additive is therefore of decisive importance to increase production efficiency and to set the desired product quality.

Therefore, the evaluation of influences of the binder additive (1) on the property development of the ceramic (e.g., mechanical, chemical and thermal stability) is essential and, furthermore, (2) mineralogical and chemical structural changes within the ceramic composition caused by the binder interactions must be observed analytically. Simultaneous monitoring and correlating of changes in materials properties and structures leads to a fundamental understanding of bonding mechanism and a targeted use of binders.

2. The Principle of Binders

Bonding systems for refractory application can be distinguished into cold bonding systems and hot bonding types. Hot bonding describes mainly the development of a bonding phase due to sintering processes of the ceramic particles at high temperatures. This so-called ceramic bonding can replace other types of binders in the case of temperature treatment of the material. Cold bonding systems always require the addition of a binder and are classified as organic, hydraulic and chemical binders (see overview in Table 1) [4].

Table 1. Bonding systems and their effectivity in respect to the temperature range according to [4].

Bonding System	Effectivity in Respect to the Temperature Range	
	Beginning Approx. °C	End Approx. °C
Ceramic bonding	1000	1200 to 1500
Organic bonding	50	250
Hydraulic bonding	20	500 to 600
Inorganic chemical bonding	20 to 250	1000 to 1450

The group of organic binders (e.g., resins, polymers) are mainly applied as temporary binder systems. After the firing process at high temperatures, no organic residues remain in the refractory material due to the decomposition and burning of the organic species.

The bonding effect of hydraulic binders is based on the formation of interlocking hydrate phases. In the refractory sector, mainly calcium aluminate cement phases are applied [4,5]. In special cases the application of ρ-alumina as a source for hydratable alumina are present as cold bonding agents [6].

Another class of binders for industrial applications are the inorganic chemical binders. This group includes all the types, whose bonding effect is based on chemical reactions either with ceramic particles or within the binder itself [4,7]. Examples of chemical binders are colloidal sols like silica, alumina and sols like water glass and phosphate binders. Phosphate binders and water glasses can be applied both as liquid and solid-state basic materials. In the case of phosphates as chemical binders, mainly phosphoric acid (H_3PO_4) or metal phosphate salts, e.g., mono aluminum phosphate (MALP) are applied [8]. In the group of colloidal sols, mainly water glasses, for refractories especially sodium water glasses, are present.

As a requirement for the bonding ability of inorganic chemical binders in refractory applications only substances, that are able to generate a three-dimensional network structure either by chemical activation or due to a drying process are suitable. Therefore, glass-forming structures like silicates and phosphates are preferred as bonding agents.

Silicon is situated in the fourth group of the periodic table of the elements which causes a high variety of bonding possibilities. The oxides of silicon, so-called silicates, are specified as network forming structures by linkage of SiO_4 tetrahedrons as fundamental units. The oxygen can perform as bridging oxygen and is therefore essential for the development of a three-dimensional network. In similar way, the oxides of phosphorus are network formers, as well. Phosphorus is situated in the fifth group of the periodic table of the elements and thus shows another connectivity compared to silicon; however, due to the radius of phosphorus cations, which is very similar to silicon, the P^{5+} cation forms tetrahedral units. In this case a double bond to one of the oxygen atoms is formed, causing a disconnecting point.

3. Setting and Bonding Mechanism

In the group of inorganic chemical binders, the use of phosphate and water glass binders brings advantages for industrial refractory applications. Phosphate binders generate great green strength of shaped and unshaped refractories leading to user-friendly handling for example while demolding and transportation. Water glass binders are based on a sol-gel transformation that allows running the setting process at room temperature.

3.1. Phosphates

The phosphate bonding mechanism generally consists of three reaction steps summarized in Figure 1 [9–11]. Bonding in a first step is initiated by an acid–base reaction between the phosphates and the ceramic oxides during setting in the presence of process-related water. This includes dissolution processes of both the ceramic oxides (formation of aquasols, M-OH) and the phosphates (phosphoric acidic phase). Therefore, initiation strongly depends on the water-solubility of the phosphates as well as pH-values. The acid–base reactions result in the formation of network-forming phosphates (e.g., aluminum phosphates, crystalline and amorphous), that lead to adhesive and cohesive bonding forces. In a second reaction step the P-crosslinking of the bonding phase and thus the bonding effect is increased by temperature treatment below T < 800 °C. This network formation is mainly due to condensation and polymerization reactions of the phosphates. In a third reaction step (T > 800 °C) high-temperature reactions result in the formation of sintered phosphates with higher crystallinity, mostly ortho phosphates. In addition, further reactions between the phosphate species and the ceramic oxides occur with frequently cation exchange reactions. Simultaneously, sintering of ceramic oxides takes places causing the phosphate bonding to merge into ceramic bond, thus achieving permanent bonding.

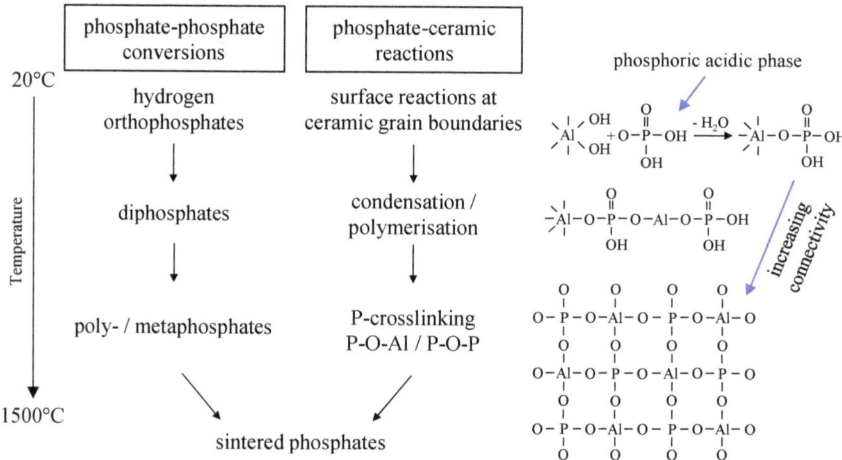

Figure 1. Overview of the bonding mechanism of inorganic phosphate binders in an alumina-based ceramic application as a function of temperature.

Additionally, it should be pointed out that only certain phosphate structures (network-forming, e.g., aluminum phosphates) can generate bonding effects. The formation of bonding networks (as a consequence of reactions with ceramic oxides) is crucial for achieving cohesion and adhesion in the bonding phase/matrix. Accordingly, bonding capacities cannot be attributed to inert phosphates that do not react or interact with the ceramic oxides at low temperatures (e.g., calcium or magnesium phosphates). The conversion schemes of these phosphates (hydrogen orthophosphates—diphosphates—poly- or metaphosphates) do not generate bonding networks.

3.2. Silicates

As silicate systems of inorganic binders mainly water glasses like sodium water glasses are applied either as a colloidal solution or as a dry powder. The setting of the water glass binders can be initiated by different reaction routes depending on the desired properties.

The simplest setting is activated by heat. Due to the thermal dehydration of the water glass systems polycondensation occurs forming an amorphous three-dimensional binder framework (see overview in Figure 2).

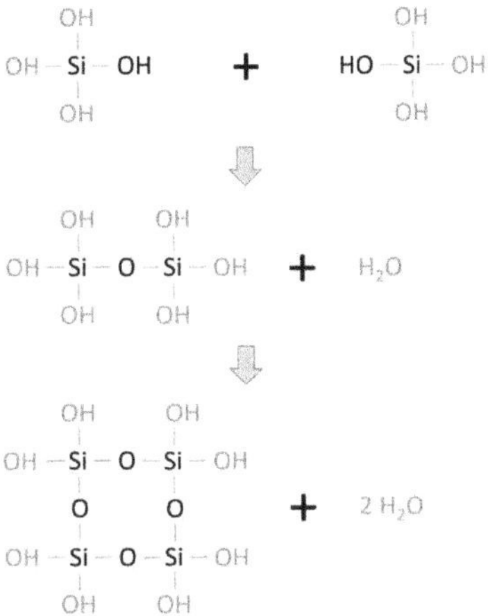

Figure 2. Overview of the network formation in silicate systems due to the polycondensation of silanol groups to siloxanes.

To reach an increased green strength and an improved thermal stability of the refractory system, the setting mechanism is activated chemically, for example due the use of inorganic phosphates, mainly aluminum phosphates as hardening agents. In this case, the setting mechanism occurs in a multi-step reaction including (a) an ion-exchange reaction between the alkali ions of the water glass and the metal ions of the phosphate hardener, (b) the formation of crystalline alkali phosphate structures and (c) an acid–base reaction. The drop of pH value lead to a reduction of the electrochemical stabilization of the sol, thus causing the silanol groups to condensate to siloxanes. In Figure 3 the influence on the particle size of the silicate structures and the network formation is illustrated in terms of the pH value. With an increase in pH value the particle size of colloidal silicates is increased forming a sol state, which is mainly stabilized due to electrostatic forces by the negative surface charges based on the DVLO theory [12,13]. With lowering pH value, the stabilization decreases and the polycondensation of the cross-linking of the silicate particles occurs forming amorphous three-dimensional frameworks to a gel state. In the case of water glasses the alkali modulus defines the pH value due to the alkaline nature of the M_2O content [14].

In this multi-reaction setting all of the single steps are in competition to each other and have a massive impact on the properties (e.g., change in mechanical strength determined by cold bending strength, Young's modulus) of the refractory system. Therefore, a good insight into the setting mechanism is the key step to the adjustment of tailored sample properties based on the binder performance.

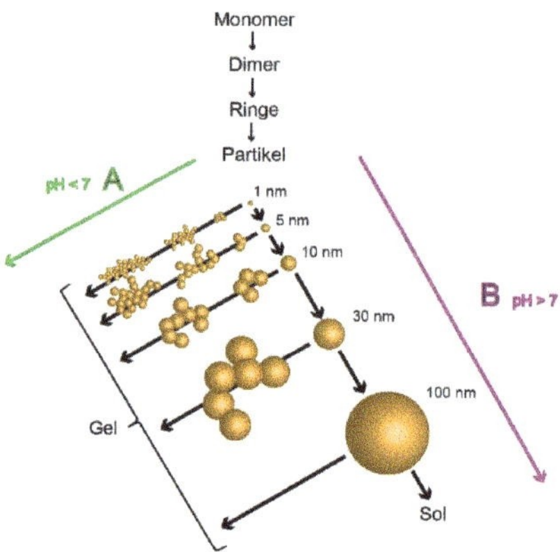

Figure 3. Influence of pH value on the particle size and network formation of silicates. Adapted from [14].

4. Measurements of Structure and Property Information

Due to the enormous impact of the bonding matrix structure on the properties of the refractory material, the chemical and mineralogical structure investigations of the binders itself as well as the binder–ceramic compound have a high relevance in the process of material design. For determining the chemical and/or mineralogical structure of materials a wide range of different analytical methods are quite well known in the field of material science. In the case of the three-dimensional phosphate or silicon-based binder matrices, the appropriate combination of different measuring methods is essential for an overarching description. This is mainly due to the existence of both amorphous and crystalline phases. A strong base for the structure determination of phosphate or silicon-based binder matrices is therefore the combination of the complementary X-ray powder diffraction and nuclear magnetic resonance spectroscopy.

X-ray powder diffraction (PXRD) is a standard and widely used method for the investigation of crystalline phases. Crystalline phases can be identified by comparing to indexed diffraction patterns. The use of an internal standard allows a quantification of the identified phases and in addition of the amorphous phase content. Though amorphous structures cannot be described sufficiently due to the missing requirement of long-range order X-ray.

A powerful technique for the description of network-forming structures is the solid-state nuclear magnetic resonance spectroscopy (NMR) [15]. It can be considered as a complementary method to the PXRD since crystalline and amorphous structures can be assigned. The signal's full width at half maximum (FWHM) is correlated with the phase crystallinity.

The main restriction in the NMR technique is the requirement of NMR active isotopes of the nuclei of interest. Since common nuclei in materials for refractory applications like ^{1}H, ^{27}Al, ^{29}Si, and ^{31}P are accessible, the NMR technique is a powerful tool for phase assignment. The position of the NMR signal, namely the chemical shift, is influenced by the electronic environment of the observed nucleus. With this information direct conclusions of the structures can be generated. The signal width in the solid-state NMR spectra is mainly affected by the magnetic field strength, the nucleus spin, the natural isotope abundance, and the gyromagnetic ratio. Several pulse sequences were developed to increase the resolution of the NMR spectra.

In contrast to high resolution liquid NMR spectroscopy, in solid-state NMR the signal widths are significantly broader due to the impact of the anisotropy of the chemical shift, homonuclear and heteronuclear correlation, and quadrupole interactions. The approach of magic angle spinning at high spinning frequencies and the use of special pulse sequences allows the reduction of signal width increasing the resolution.

For a correct phase assignment signal comparison with databases or literature is required. Since the availability of records of inorganic compounds in databases is limited the measurements need to be extended by other spectroscopic methods that generate information of the existence of functional groups, like FT-IR and Raman spectroscopy.

The combination of PXRD and NMR techniques can be complemented by FTIR and Raman spectroscopy and scanning electron microscopy (SEM). FTIR and Raman spectroscopy are great tools for the identification of individual phosphates and in addition provides information about phase transitions due to temperature-depending measurements. The use of SEM is typical for the optical analysis of the microstructure, combined with energy dispersive X-ray spectroscopy (EDX) the element distribution in the matrix can be investigated.

For a precise and systematic design of a material with tailored properties it is necessary to understand the structure–property relationship. Therefore, complementing the before mentioned structure measurements with additional investigations on macroscopic properties (mechanical, chemical and thermal) is reasonable. This includes for example the investigation of the strength development in dependence of temperature, time or composition via dynamic mechanical analysis (DMA) or cold bending strength measurements.

5. Correlation

5.1. Strength Development of Phosphate-Bonded Al_2O_3-$MgAl_2O_4$ High-Temperature Ceramics

Generating sufficient early stage strength, that allows ceramic components to be shaped and processed, is of utmost importance in ceramic manufacturing. By adding phosphate binders to the ceramic mix, the bonding strength of ceramics can be improved. Through extensive investigation of structural changes of the bonding phase, differences and changes in the bonding strengths can be explained analytically. For example, structure–property correlations can be achieved by comparing the cold modulus of rupture of phosphate bonded Al_2O_3-$MgAl_2O_4$-ceramics with structural changes in the bonding phase after a defined thermal treatment monitored, for instance, by ^{31}P MAS NMR (Figure 4).

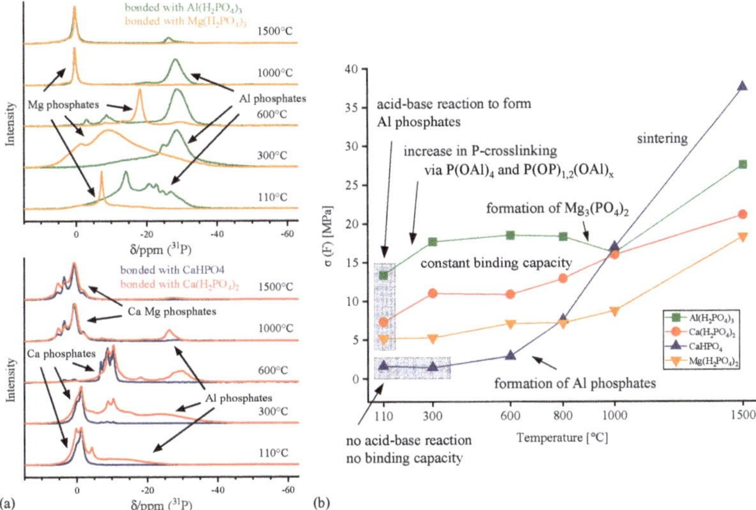

Figure 4. Correlation of the cold modulus of rupture and structural changes in the bonding phase of phosphate bonded $Al_2O_3MgAl_2O_4$ ceramics (composition: 85 wt.% Al_2O_3, 15 wt.% $MgAl_2O_4$; binder concentration: 3 wt.% P_2O_5): (**a**) ^{31}P MAS NMR spectra of the phosphate bonded ceramics after thermal treatment (normalized) and (**b**) cold modulus of rupture tests at appropriate temperatures. The strength development of ceramic bodies can be directly correlated with the evolution of an aluminum phosphate bonding network [16].

The strength development of the ceramic components varies depending on the phosphate used. Early stage strength is achieved by applying water soluble phosphates as binders (water solubility at 25 °C: $Al(H_2PO_4)_3$: >250 g/L; $Mg(H_2PO_4)_2$: 70 g/L; $Ca(H_2PO_4)_2$: 20 g/L; $CaHPO_4$ < 1 g/L). Further heating generally leads to an increased strength of the components, but the strength developments as well as the end values after thermal treatment at T = 1500 °C vary significantly.

The strength development of the phosphate-bonded ceramics can be explained by evaluating the structure of phosphatic bonding phases. Investigating the phosphate connectivity by solid-state ^{31}P NMR allows conclusions to be drawn about the bonding capacity of the developing phosphate structures. It can be stated that the formation of a P-crosslinked aluminum phosphate bonding network leads to an increasing strength of the ceramic components. Aluminum is incorporated in the glass structure and stabilizes crosslinked phosphate bonding networks. Within the refractory body, the binder creates a branched three-dimensional network because of chemical reactions, polymerization and polycondensation. With increasing temperature, the degree of P-crosslinking continuously increases via multiple crosslinking reactions (P–O–Al and P–O–P bonds) within the initialization and network-forming stages The bonding effect is based on adhesive and cohesive forces of the amorphous aluminum phosphate network thus leading to the strength development seen in Figure 4. The development of aluminum phosphates has been verified by solid-state ^{27}Al NMR and PXRD investigations.

When using water-insoluble phosphates as binders, no early stage strength is achieved since no acid–base reactions are initiated. However, bonding strength can be generated by heating (T > 600 °C) resulting in a thermally-induced development of aluminum phosphates, thus, bonding strength.

Besides aluminum phosphates, other phosphate species (such as calcium or magnesium phosphates) do not generate bonding effects, since they do not react with ceramic oxides at T < 600 °C and generally are not capable to form broad phosphate networks because of non-bridging-oxygens [10,11,16].

At temperatures T $\geq$ 600 °C the phosphate phases react with $MgAl_2O_4$ to form $Mg_3(PO_4)_2$, $Ca_xMg_y(PO_4)_6$ and other crystalline ortho phosphates. The formation of these low-melting high-temperature phases leads to a partial break in the bonding network. Since high temperature processes include the sintering of ceramic particles, in total, temperature treatment at T > 1000 °C leads to a significant increase in bonding strength.

In conclusion, phosphate binders actively influence the chemistry of the ceramic components (e.g., cation exchange, spinel decomposition). By monitoring structural changes in the bonding phase, property developments can be explained. Ultimately, this correlation leads to a better understanding of phosphate bonding mechanisms and, consequently, the mode of action of phosphate binders.

5.2. Insights in the Setting Process of Sodium Water Glasses Using a Combination of Dynamic Mechanical Analysis, NMR and PXRD

For the kinetic description of the setting process itself, the dynamic mechanical analysis (DMA) can obtain valuable information by measuring the viscoelastic properties of a material as a function of temperature, frequency and time. This allows time-dependent tracking of the strength development during setting. When combined with the structure information obtained by the above-mentioned methods, insights in the mechanism of the setting process are gained. During the measurement stress is put onto the material by an oscillating movement of a feeler stamp, the so-called exciter signal. This leads to a response signal of the sample, that is material-specific and depends on the viscoelastic behavior (Figure 5). DMA describes the changes in the materials rigidity by the complex modulus of elasticity (MOE), which is composed of the storage modulus E^I representing the elastic part of energy and the loss modulus E^{II} standing for the energy that is lost due to plastic deformation. The ratio between E^{II} and E^I is defined by the loss factor $\tan\delta$ [17]. Specifically for the investigations of sol-gel-based binder systems, DMA offers the possibility to determine the gel and the glass point during the setting mechanism [18].

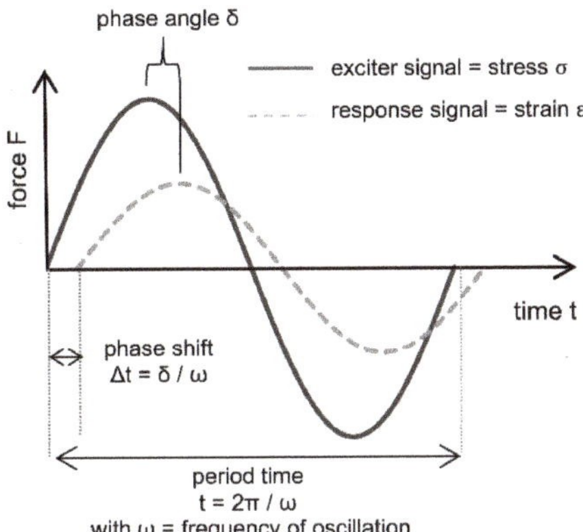

Figure 5. Wave function of exciter and response signal in DMA according to [18].

Dynamic Mechanical Analysis investigations of a sol-gel-based pure binder system consisting of a sodium water glass hardened by either aluminum or boron orthophosphate were carried out with the aim to determine the influence of the attendant cation thus leading to a better understanding of the setting process of the alkali silicate systems [19]. The results of the DMA as shown in Figure 6 exhibit the overall setting of the water glass to be much faster when using boron orthophosphate as hardening agent instead of aluminum orthophosphate. Three different phases of the setting process can be discriminated. In an initial phase the storage modulus increases only very moderate until reaching the gel point, which is very similar for both hardeners. This point is followed by a much more intense increase of the storage modulus. This phase now differs highly using the two phosphates. The use of boron phosphate instead of aluminum phosphate leads to a significant faster increase and achievement of the final plateau that can be seen as the third stage of setting.

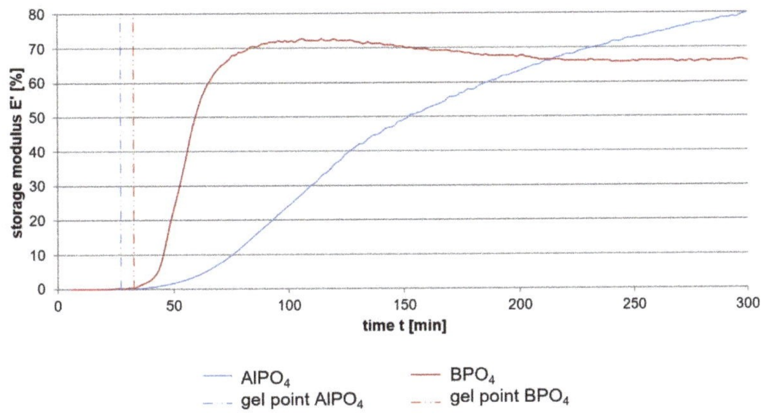

Figure 6. DMA measurements of a sodium water glass hardened with either $AlPO_4$ or BPO_4 [19].

The increase in the storage modulus in the initial phase is quite similar for both hardeners. Investigations with different hardener portions (10, 20, 30 mass-%) and at different starting temperatures (25 °C, 35 °C, 45 °C) show that this phase of setting is more determined by the temperature and the solid to liquid ratio then by the chemistry of the hardener [20]. Due to the very low solid to liquid ratio in the initial phase, the distances between the engaged species are very high leading to a very low influence of the network building processes of the different phosphate hardeners. Instead, condensation reactions due to the temperature induced release of water and the electrochemical destabilization caused by the change of both the pH value and the surface charge of the silicate particles, determine the setting progress.

When reaching the gel point, the distances between the different species in the sol are small enough to cause the differences in the hardening mechanisms of aluminum and boron phosphate becoming noticeable: The samples hardened with aluminum phosphate show a significant slower increase in the storage modulus and reach a later but clearly higher final level. NMR of silicon ^{29}Si, aluminum ^{27}Al and boron ^{11}B nuclei explain the reasons for this in the structure of the three-dimensional network, specifically in the coordination of the boron and aluminum compounds integrated in the silicon network. Whereas Al is represented by high coordinated Al species (4,5,6) in the silicon network, boron compounds show a lower grade coordination (3,4) (Figure 7). This is congruent with the observation of higher n-values for Si in the water glass network when using aluminum phosphate instead of boron orthophosphate, indicating a higher linking degree [20]. Hardening with boron orthophosphate leads thus to a fast formed but less interconnected and wide-meshed network. The longer setting of the water glass gained by the use of aluminum orthophosphate results in a higher connected and denser network that leads to the observed higher values of E^I and thereby in a higher strength.

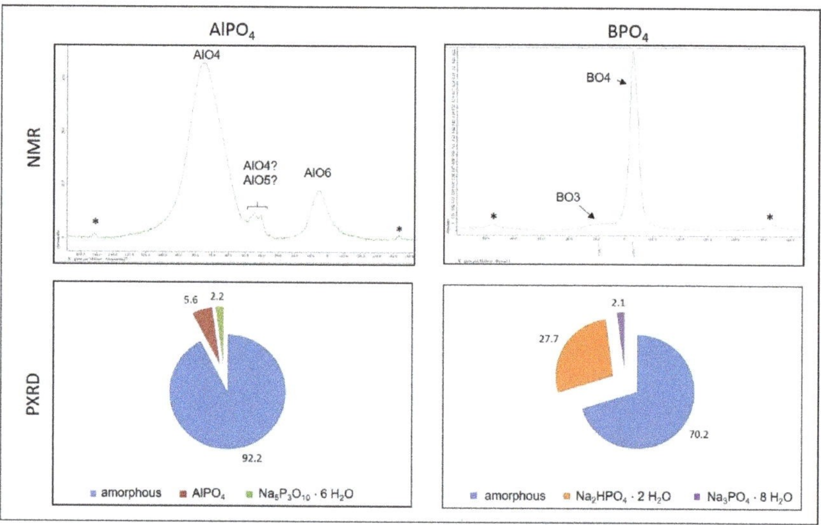

Figure 7. Results of ^{27}Al and ^{11}B MAS NMR (**1st row**) and PXRD (**2nd row**) measurements of sodium water glass hardened either with AlPO$_4$ (**1st column**) or BPO$_4$ (**2nd column**) * spinning sidebands [20].

PXRD measurements of fully hardened samples (Figure 7, 2nd row) indicate an additional aspect with regard to the different speed of the storage modulus' increase. Besides the amorphous silicate network both mechanisms lead to the formation of crystalline structures. The use of boron orthophosphate causes the formation of short-chained hydrogen phosphates whereas the addition of the aluminum phosphate lead to the development of water-free long-chained phosphates. In combination with extensive STA measurements, it could be seen, that the two mechanisms cause differences in the release of water. In the case of hardening using AlPO$_4$, the samples show a higher weight loss due to the loss of water. However, the results indicated that the hardening mechanism of AlPO$_4$ do not generate more free water than the mechanism of BPO$_4$, but the mechanism of BPO$_4$ to cause a stronger prevention of water what results in the formation of the hydrogen phosphates. The reason for this different retention of water is seen in the inner structure of the originated silicate network with regard to the arrangement of the linking points. The essential aspect here is not the connectivity itself but the location in the network where the linking points are formed. It is known that different energy barriers of the condensation reactions lead to different locations of preferred interaction. This leads to the hypothesis that the network-forming reactions using boron orthophosphate as hardening agent exhibit lower energy barriers and thus induce condensation reactions in the periphery of the originating network. The result is a very fast formation of a far-reaching network that on the one hand hinders free water from release and one the other hand causes a fast hardening and increase in storage modulus. Instead, aluminum phosphate shows a higher energy barrier and thus needs more attempt for a successful condensation. This is even more likely in network areas with a higher density. This results in smaller, but higher linked domains which are surrounded by a certain content of water that thus can easily leave the samples. These structures result in the lower increase in storage modulus of the aluminum phosphate samples.

In an additional test series based on thermally induced linear change, melting behavior indicate a strong impact of the mechanism and the structure on the material properties. Therefore, it is important to identify the determining factors on the setting mechanism for the development of tailor-made material solutions. The outcome of this work points out the combination of the DMA with the methods for structure analysis and the thermo analysis as a powerful tool to connect the setting mechanism itself with the resulting structures and thus with the properties.

5.3. Inorganic Phosphate Hardeners in Chemical Setting of Potassium Water Glass

In the case of water glass as silicate binders the setting procedure can be conditioned by the admission of inorganic aluminum phosphates as hardening agents. Introducing cyclic aluminum metaphosphate modifications effects significantly the phase formation and the bonding properties, which can be determined by different measurement and spectroscopy techniques like PXRD, solid-state NMR, FT-IR, and Raman spectroscopy. As a monitor of the polycondensation progress of water glass systems the common Q^n notation is applied. The Q^n distribution in pure sodium water glasses can be considered in Figure 8 left in terms of ^{29}Si liquid NMR spectra. The signals for the Q^0 to Q^4 can be clearly distinguished and assigned. The spectra in Figure 8 on the right-hand side illustrates the ^{29}Si MAS NMR spectra of chemically hardened potassium water glass with aluminum tetrametaphosphate. Due to the nature of solid-state NMR technique and especially of the ^{29}Si experiments very broad signals are detected with high overlap. The deconvolution of the signal region resolves four signals, where two can be assigned to Q^3 and Q^4 of pure water glass systems. The other two signals are formed by the introduction of aluminum ions into the silicate framework creating alumo-silicate phases. Therefore, it can be demonstrated that the polycondensation shifts the Q^n distribution to higher coordinated silicate specimen, since no Q^0 to Q^2 signals can be detected.

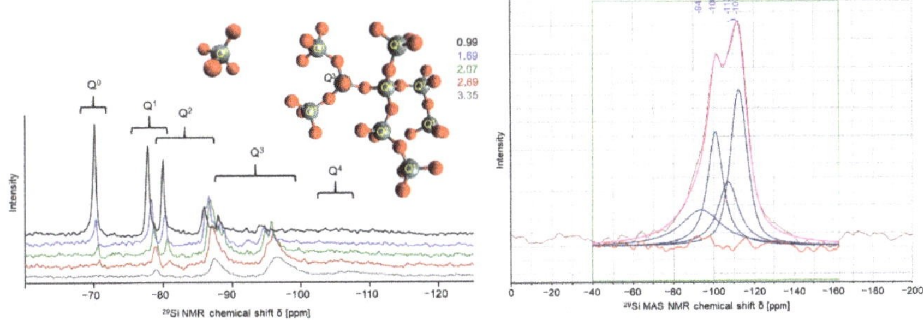

Figure 8. left: Q^n distribution in liquid ^{29}Si NMR with RIDE pulse sequence of pure sodium water glasses, adapted from [21] and **right**: deconvoluted ^{29}Si MAS NMR spectra of a potassium water glass with aluminum tetrametaphosphate [22].

The impact of the aluminum in the formation of alumo-silicate phases in respect to the alkali modulus (molar ratio of SiO_2 in respect to M_2O) of the potassium water glass and the modification of the aluminum metaphosphate on the ^{29}Si NMR signals is shown in Table 2. The existence of newly formed Si-O-Al connections is additionally verified spectroscopically by FT-IR and Raman measurements.

Table 2. Relative intensities of Q^n and $Q^n(Al^m)$ signals in the ^{29}Si MAS NMR spectra of aluminum metaphosphates hardened potassium water glass binders after deconvolution, where Q^n stands for the coordination of the silicate group and the m stands for the substituted O-bridged Si-heteroatom connections. Samples with potassium water glasses K1 and K2 (alkali modulus K1 < K2), and 30 wt% of aluminum metaphosphates A = aluminum tetrametaphosphate, $Al_4(P_4O_{12})_3$ and B = aluminum hexametaphosphate, $Al_2P_6O_{18}$. [construct. [22], assignment of the $Q^n(Al)^m$ signals based on studies of [23].

Sample	Relative Intensities of Deconvoluted ^{29}Si NMR Signals			
	$Q^4(Al^3)/Q^3(Al^1)$ at −95 ppm	$Q^3/Q^4(Al^2)$ at −108 ppm	$Q^4(Al^1)$ at −108 ppm	Q^4 at −112 ppm
K1A30	0.23	0.37	0.13	0.28
K2A30	0.23	0.23	0.00	0.55
K2B30	0.20	0.08	0.44	0.29

Since the mechanical resistance towards deformation is mainly reasoned in the condensation degree of the silicate framework the formation of alumo-silicate sections has an impact on the mechanical properties.

The quantification of the assigned $Q^n(Al^m)$ groups by deconvolution of the solid-state ^{29}Si NMR spectra characterizes the degree of alumo-silicate content inside the sample. It can be demonstrated that in case of aluminum hexametaphosphate as hardening agent an increase in alumo-silicate content is detected. Therefore, comparing the alumo-silicate content to the mechanical properties like the cold bending strength and the Young's modulus by resonance frequency damping analysis (RFDA) a decrease in mechanical stability is obtained (Figure 9). Additionally, the water glass' alkali modulus influences the mechanical properties, since the samples with a higher K_2O content show the highest cold bending strength values with the same hardener.

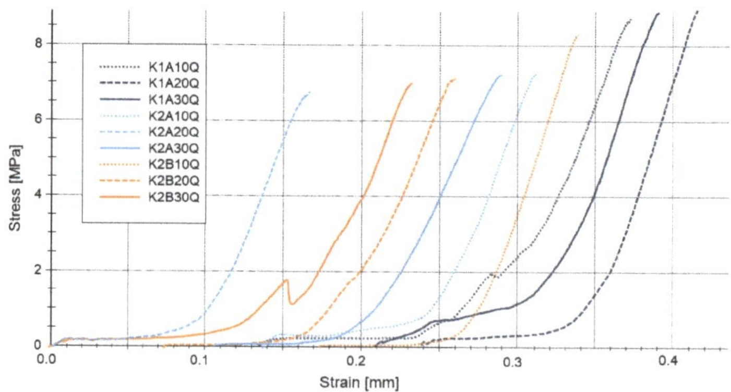

Figure 9. Cold bending strength analysis of potassium water glasses with aluminum metaphosphates (leaned on testing standard DIN EN 853-2-2007). Samples analogous to Table 2 [24].

As a result, samples with potassium water glass with a lower alkali modulus and the application of aluminum tetrametaphosphate as hardening agent show the highest values in the cold bending strength tests. This can be correlated to the increase in polycondensation of the silicate framework and a lower degree in formed alumo-silicate phases due to the stronger interception of the aluminum ions by ion-exchange reaction.

Additionally, the solubility of aluminum ions in the alkaline water glass environment and existing ion-exchange reaction of aluminum-ions from the aluminum metaphosphate and the alkali ions of the water glass are mainly responsible for the formation of alumo-silicate phases. Dissolved aluminum ions are the main source of the alumo-silicate formation. The ion exchange reaction reduces the available aluminum ions and therefore reduces its impact on integrating into the silicate framework.

For the correct interpretation of the phase formation while setting procedure ^{31}P MAS NMR experiments give insights into the depolymerization mechanism of the hardener. In Figure 10 all samples show crystalline potassium hydrogen phosphate as a result of the depolymerizing metaphosphate ring structure. The incrementally decomposition can be found in the of amorphous di- and oligophosphates ranging the Q^1 area. Crystalline potassium tetrametaphosphate dihydrate was determined as a set of four signals as Q^2 which proves the existence of an ion-exchange reaction between the aluminum ions from the metaphosphate and the potassium ions from the water glass. As a result, in the case of aluminum tetrametaphosphate as chemical hardener an ion-exchange reaction affects the amount of aluminum ions for a potential formation of alumo-silicate frameworks. Due to the stepwise depolymerization of the cyclic phosphate structure potassium ions are consumed by the formation of alkali phosphate and alkali hydrogen phosphate structures affecting the condensation progress due the lowering of the alkali concentration. As a consequence, the polycondensation of the silicate framework is forced towards highly coordinated networks.

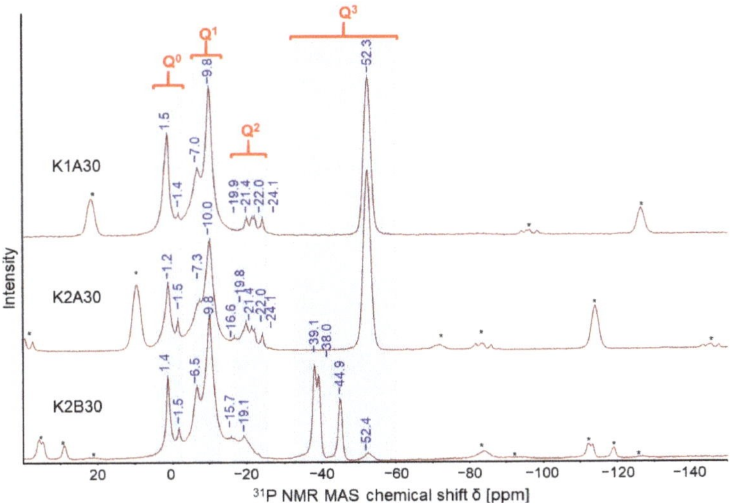

Figure 10. ^{31}P MAS NMR spectra of two potassium water glasses with different alkali modulus K1 = 3.09, K2 = 3.53 with 30 wt% aluminum tetrametaphosphate KXA30 and aluminum hexametaphosphate K2B30 [22].

The chemical resistance of water glasses towards acidic attack (see Figure 11) is a function of the alkali concentration of the silicate and alumo-silicate structures, since it is the region of attack. Therefore, the influence of the alkali modulus of the water glass has the strongest impact on the chemical resistance of the water glass system.

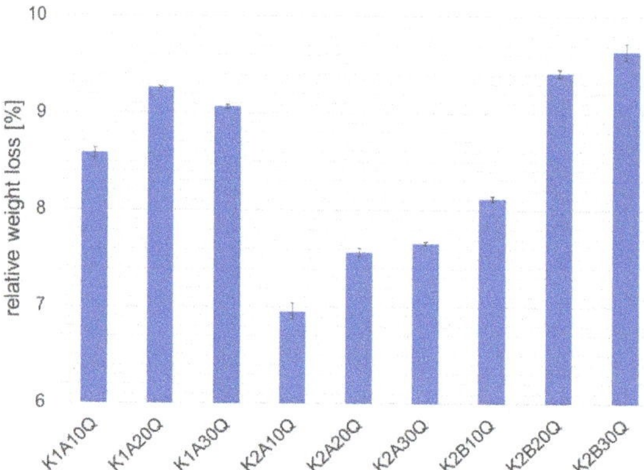

Figure 11. Relative weight loss of potassium water glasses with aluminum metaphosphate hardeners after treatment in hot sulfuric acid according to testing norm DIN 51-102 Part 1. Samples with 30 wt% of potassium water glasses K1 and K2 (alkali modulus K1 < K2), and 10–30 wt% of aluminum metaphosphates A = aluminum tetrametaphosphate, $Al_4(P_4O_{12})_3$ and B = aluminum hexametaphosphate, $Al_2P_6O_{18}$, and 70 wt% quartz as filler [22].

Using aluminum metaphosphates as hardening agents for alkali silicate binders increases the chemical stability towards acidic attack, since the available alkali ions are incorporated into alkali phosphate and alkali hydrogen phosphate structures as a result of the incrementally depolymerization of the cyclic metaphosphate structure.

Therefore, aluminum metaphosphate, especially aluminum tetrametaphosphate as hardening agent increases the chemical resistance towards acidic attack in the setting procedure of water glass binder system.

As a summary the chemical setting of potassium water glass binders with aluminum metaphosphates as hardening agents proceeds in a multi-step reaction mechanism involving the incrementally depolymerization of the cyclic metaphosphate structures, the polycondensation of the amorphous three-dimensional silicate network, an ion-exchange reaction, and the incorporation of alkali ions into the silicate framework. All reaction steps mutually influence each other and define the ultimate material properties.

By the structural investigation of the samples in terms of spectroscopic and diffraction techniques, information about the different phase contents are generated while chemical setting. Combining the results of the reaction mechanism with macroscopic properties like cold-bending strength, young's modulus and chemical resistance towards acidic attack structure–property correlations can be derived. This can be used as a powerful tool in predicting sample properties based on the starting compounds and brings the ability in optimizing the sample matrix for required functionalities.

6. Conclusions

Inorganic phosphate binders and phosphate hardened silicates play an important role as refractory binders due to their ability in forming thermally stable phases ranging from room temperature up to $T \leq 1600\ °C$. Based on the tetrahedral nature of PO_4 and SiO_4 units three-dimensional network structures are generated that result in an increased mechanical stability of both the green body as well as the ceramic body after sintering. Analytically, it could be demonstrated that monitoring the multi-step reactions while setting and bonding by the specific combination of complementary structure analysis techniques (PXRD; NMR, Raman Spectroscopy) and macroscopic property investigations (mechanical: DMA cold bending strength analysis), chemical: acid resistance) are the key to the design of tailored material with required properties in respect to the field of application. This correlation allows extracting of the crucial steps for a selective and systematical range of raw material–

binder combinations to adjust materials with tailored properties. In contrast to the usual trial and error method this leads to a more efficient approach in material design.

Author Contributions: Conceptualization, V.H., A.M.A. and D.H.; methodology, V.H., A.M.A. and D.H.; formal analysis, V.H., A.M.A. and D.H.; investigation, V.H., A.M.A. and D.H.; writing—original draft preparation, V.H., A.M.A. and D.H.; writing—review and editing, V.H., A.M.A., D.H. and P.Q.; supervision, P.Q.; funding acquisition, P.Q. All authors have read and agreed to the published version of the manuscript.

Funding: This research was financially supported by the European Union (European Regional Development Fund, project: ReSpaZ), by the funding line "Cooperative Research and Training Group Rhineland Palatinate" (Max-von-Laue Institute of Advanced Ceramic Material Properties Studies, CerMaProS) of the Rhineland-Palatinate Ministry for Science, Further Education and Culture and by the Alexander Tutsek-Foundation, Munich. We also thank the German Research Foundation (INST 366/6-1) for the purchase of the NMR spectrometer.

Institutional Review Board Statement: Not applicable.

Informed Consent Statement: Not applicable.

Data Availability Statement: The study does not include publicly archived datasets.

Conflicts of Interest: The authors declare no conflict of interest.

References

1. Bordia, R.K.; Kang, S.J.L.; Olevsky, E.A. Current understanding and future research directions at the onset of the next century of sintering science and technology. *J. Am. Ceram. Soc.* **2017**, *100*, 1252–2314. [CrossRef]
2. Parr, C.; Auvray, J.M.; Szepizdyn, M.; Wöhrmeyer, C.; Zetterstrom, C. A review of bond systems for monolithic castable refractories. *Refract. World Forum* **2015**, *7*, 63–70.
3. Girifalco, L.A. *Statistical Mechanics of Solids*; Oxford University Press: New York, NY, USA, 2003.
4. Gelsdorf, G. Gliederung, Klassifikation und Prüfung von Feuerbetonen. *Keram. Z.* **1982**, *34*, 348–481.
5. Hewlett, P.C. *Lea's Chemistry of Cement and Concrete*, 4th ed.; Elsevier: Burlington, NC, USA, 1998.
6. Viadya, S.D.; Thakkar, N.V. Study of phase transformations during hydration of rho alumina by combined loss of ignition and X-ray diffraction technique, *J. Phys. Chem. Solids* **2001**, *62*, 977–986. [CrossRef]
7. Kalyoncu, R.S. *Chemically Bonded Refractories—A Review of the State of the Art*; U.S. Department of the Interior: Washington, DC, USA, 1982.
8. Morris, J.H.; Perkins, P.G.; Rose, A.E.A.; Smith, W.E. The chemistry and binding properties of aluminium phosphates. *Chem. Soc. Rev.* **1977**, *6*, 173–194. [CrossRef]
9. Kingery, W.D. Fundamental study of phosphate bonding in refractories. *J. Am. Ceram. Soc.* **1950**, *33*, 239–250. [CrossRef]
10. Hahn, D.; Alavi, A.M.; Quirmbach, P. Powder XRD and ^{31}P and ^{27}Al solid state MAS NMR investigations of phase transformations in aluminium phosphate bonded Al_2O_3-$MgAl_2O_4$ refractories. *Mater. Chem. Phys.* **2021**, *267*, 124663. [CrossRef]
11. Hahn, D.; Alavi, A.M.; Hopp, V.; Quirmbach, P. Phase development of phosphate bonded Al2O3-MgAl2O4 high-temperature ceramics: XRD and solid-state NMR investigations. *J. Am. Ceram. Soc.* **2021**, *00*, 1–18. [CrossRef]
12. Derjaguin, B.; Landau, L. Theory of the stability of strongly charged lyophobic sols and of the adhesion of strongly charged particles in solutions of electrolytes. *Prog. Surf. Sci.* **1993**, *43*, 30–59. [CrossRef]
13. Verwey, E.J.W.; Overbeek, J.T.G.; van Nes, K. *Theory of the Stability of Lyophobic Colloids: The Interaction of Sol Particles Having an Electric Double Layer*; Elsevier: Amsterdam, The Netherlands, 1948.
14. Iler, R.K. *The Chemistry of Silica: Solubility, Polymerization, Colloid and Surface Properties and Biochemistry of Silica*; Wiley: New York, NY, USA, 1979.
15. Brow, R.K.; Kirkpatrick, R.J.; Turner, G.L. Local structure of $xAl_2O_3 \cdot (1-x)NaPO_3$ glasses: An NMR and XPS study. *J. Am. Ceram. Soc.* **1990**, *73*, 2293–2300. [CrossRef]
16. Hahn, D. Strukturelle Analyse Phosphatischer Bindephasen in Al_2O_3-$MgAl_2O_4$-Hochtemperaturkeramiken. Ph.D. Thesis, University of Koblenz-Landau, Mainz, Germany, June 2021.
17. Menard, K.P. *Dynamic Mechanical Analysis—A Practical Introduction*, 2nd ed.; Taylor and Francis Group: Boca Raton, FL, USA, 2008.
18. Hopp, V.; Sax, A.; Helmus, D.; Quirmbach, P. Adaptation of the dynamic mechanical analysis to determine the gel point during the setting of liquid alkali silicates. *Int. J. Appl. Ceram. Technol.* **2019**, *16*, 2331–2341. [CrossRef]
19. Hopp, V.; Alavi, A.M.; Sax, A.; Quirmbach, P. Influence of Aluminium and Boron Orthophosphate on the Setting and the Resulting Structure of Alkali Silicate Binders for Refractory Application. *Ceramics* **2020**, *3*, 1. [CrossRef]
20. Hopp, V. Einfluss von Aluminium- und Bororthophosphat auf die Chemische Härtung von Natron-Wasserglas-Gebundenen Feuerfesten Rieselmassen. Ph.D. Thesis, University of Koblenz-Landau, Mainz, Germany, 12 September 2019.

21. Jansson, H.; Bernin, D.; Ramser, K. Silicate species of water glass and insights for alkali-activated green cement. *AIP Adv.* **2015**, *5*, 067167. [CrossRef]
22. Alavi, A.M. Einfluss der Struktur von Aluminium-Metaphosphaten auf die Chemische Härtung von Kalium-Wasserglas-Bindern. Ph.D. Thesis, University of Koblenz-Landau, Mainz, Germany, 26 April 2019.
23. Thomas, J.M.; Klinowsky, J.; Ramadas, S.; Anderson, M.W.; Fyfe, C.A.; Gobbi, G.C. New Approaches to the Structural Characterization of Zeolites: Magic-Angle Spinning NMR (MASNMR). In *Intrazeolite Chemistry*; American Chemical Society: Washington, DC, USA, 1983; pp. 159–180. [CrossRef]
24. Alavi, A.M.; Breitzke, H.; Hemberger, Y.; Sax, A.; Buntkowsky, G.; Quirmbach, P. Insights into the mechanism of the chemically initiated setting of potassium silicate solutions with aluminum metaphosphates. Correlation of the structural and macroscopic parameters on the performance of the binding properties. *Constr. Build. Mater.* **2021**, *277*, 120060. [CrossRef]

Article

Corrosive Effect of Wood Ash Produced by Biomass Combustion on Refractory Materials in a Binary Al–Si System

Hana Ovčačíková *, Marek Velička, Jozef Vlček, Michaela Topinková, Miroslava Klárová and Jiří Burda

Department of Thermal Engineering, Faculty of Materials Science and Technology, VSB-Technical University of Ostrava, 17. listopadu 2172/15, 708 00 Ostrava, Czech Republic
* Correspondence: hana.ovcacikova@vsb.cz; Tel.: +4-205-9732-1523

Abstract: In terms of its chemical composition, biomass is a very complex type of fuel. Its combustion leads to the formation of materials such as alkaline ash and gases, and there is evidence of the corrosive effect this process has on refractory linings, thus shortening the service life of the combustion unit. This frequently encountered process is known as "alkaline oxidative bursting". Corrosion is very complex, and it has not been completely described yet. Alkaline corrosion is the most common cause of furnace-lining degradation in aggregates that burn biomass. This article deals with an experiment investigating the corrosion resistance of 2 types of refractory materials in the Al_2O_3-SiO_2 binary system, for the following compositions: I. (53 wt.% SiO_2/42 wt.% Al_2O_3) and II. (28 wt.% SiO_2/ 46 wt.% Al_2O_3/12 wt.% SiC). These were exposed to seven types of ash obtained from one biomass combustion company in the Czech Republic. The chemical composition of the ash is a good indicator of the problematic nature of a type of biomass. The ashes were analyzed by X-ray diffraction and X-ray fluorescence. Analysis confirmed that ash composition varies. The experiment also included the calculation of the so-called "slagging/fouling index" (I/C, TA, Sr, B/A, Fu, etc.), which can be used to estimate the probability of slag formation in combustion units. The corrosive effect on refractory materials was evaluated according to the norm ČSN P CEN/TS 15418, and a static corrosion test was used to investigate sample corrosion.

Keywords: corrosion; refractory; biomass; thermal processing; wood ash

Citation: Ovčačíková, H.; Velička, M.; Vlček, J.; Topinková, M.; Klárová, M.; Burda, J. Corrosive Effect of Wood Ash Produced by Biomass Combustion on Refractory Materials in a Binary Al–Si System. *Materials* **2022**, *15*, 5796. https://doi.org/ 10.3390/ma15165796

Academic Editor: Panos Tsakiropoulos

Received: 31 July 2022
Accepted: 18 August 2022
Published: 22 August 2022

Publisher's Note: MDPI stays neutral with regard to jurisdictional claims in published maps and institutional affiliations.

1. Introduction

Worldwide, 80% of electricity is produced using fossil fuels. According to the International Energy Agency (IEA), electricity production reached approximately 25.8 T-kWh in 2020, and an increase to 36.5 T-kWh is expected by 2040 [1]. As stated by the World Bioenergy Association, 59.2 NPP/year, i.e., 10.3% of the global energy supply, comes from biomass [2]. Biomass is becoming a popular source of energy which can be used in various ways. Electricity produced from biomass currently corresponds to 493 TWh, which is approximately 2% of the world's electricity production [2]. Using biomass as a raw material for powerplants is certainly interesting and useful; however, this technology also has certain disadvantages. The use of biomass in powerplants leads to the formation of residue, called biomass ash. It is estimated that around 480 million tons of ash are produced every year by biomass powerplants worldwide. This is similar to coal ash, with 780 million tons per year [2].

The most frequently burned material is wood (64%), followed by cereals and plant residue from agricultural production. In general, it can be said that the average percentage of ash produced by burning biomass ranges between 1 and 6%; for wood, it is 0.6–1.6%; for bark, it rarely exceeds 3%; straw produces an ash content of around 5%, while grass produces 7%. At the other end of the spectrum, the ash content produced by black coal is significantly higher, reaching 20–30%, and from brown coal, this amount can be even greater [3]. Ash represents a variable composition of mineral and inorganic components.

During the combustion process, ash continuously changes its physical and chemical properties, the final product being a molten mixture of original minerals, various eutectics, and elements. Ash causes various problems, especially corrosion, erosion, stickers, etc. [4] If the melting temperature of ash during combustion is $t_{ash} < t_{flame}$, then the grate of the hearth can become clogged. Ash layers on the walls of the furnace diffuse into the lining, which then peels off in thin layers. The combustion chamber of the boiler must therefore be structurally adjusted such that the flame temperature drops below the ash melting temperature, i.e., the temperature on the grate should be lower than the melting temperature of the biomass ash [5].

The major problem from a chemical point of view is corrosion, which comes from the interaction between a refractory and a corrosive medium: gas, molten metals, molten glass, molten salts, or slag. It results in a loss of mass and thickness and in the degradation of the material properties [6]. The corrosion of refractory materials is a combination of external and internal physical and chemical influences.

The process is basically a chemical reaction between the refractory material and the slag or metal. Reactants are transported to the interface of the refractory material, and, in turn, the product reacts and is transported to the liquid phase. The dissolution of refractory materials in the melt is controlled by diffusion. Three types of corrosion have been defined: surface, dimple, and undersurface corrosion [7].

Alkaline corrosion, or "alkaline oxidative bursting", is extremely common, effective, and particularly harmful to alumina–silicon (Al–Si) lining systems, and it is usually observed in the temperature range of aggregates of 800–1000 °C. During biomass combustion, damage to the refractory lining is observed (Figure 1) as the peeling of surface layers, cracking, the bending of individual parts of the lining, the bulging of entire walls, and eventually their collapse [8,9].

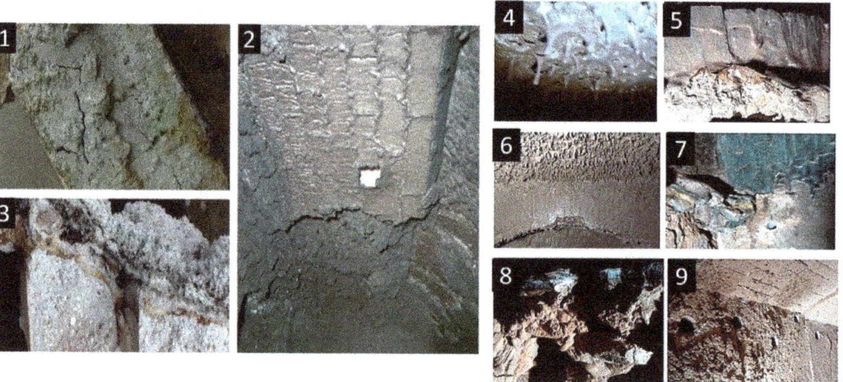

Figure 1. Degradation of refractory materials in boilers after combustion of different types of biomasses [8,9]: (**1**) the damage to refractory materials after 1 year of the combustion of wood chips; (**2**) the furnace vault after 1.5 years of the combustion of chipboard; (**3**) the corroded part of refractory samples after 2 years of combustion of plant biomass, and (**4–9**) the presentation of the defects of the refractory lining after the combustion of biomass for 10 years of operation.

The main difference between coal ash and biomass ash is that coal ash contains higher amounts of SiO_2 and Al_2O_3, but it contains lower amounts of K_2O and Na_2O. The eutectic of Al–Si forming fly ash lies above 1200 °C, while the eutectic of plant fly ash is much lower. Eutectic temperatures for mixtures of alkali metals together with silica or phosphorus have a low melting point: $Na_2O.2SiO_2$ (874 °C), $K_2O.4SiO_2$ (770 °C), and $2CaO.3P_2O_5$ (774 °C) [10].

Aluminosilicate refractories are based on the SiO_2-Al_2O_3 system. The equilibrium diagram of this system is given in Figure 2, marking various refractories. The main phase in the Al–Si binary diagram is mullite ($3Al_2O_3.2SiO_2$) [11], which increases the resistance of the refractory material against the corrosive effects of ash [11].

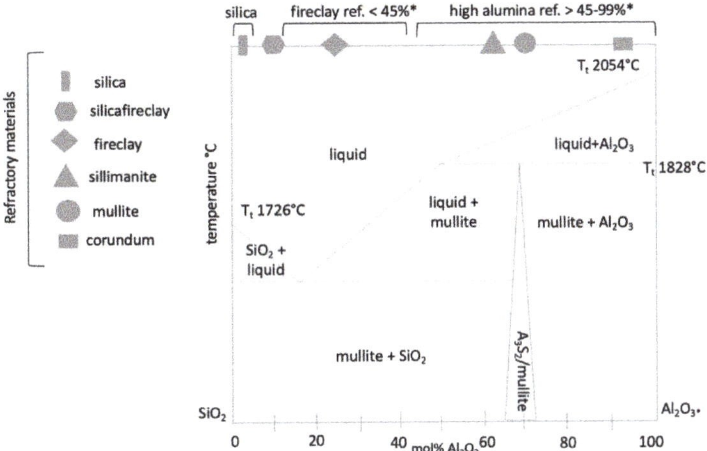

Figure 2. Categorization of basic refractory materials in binary diagram of SiO_2-Al_2O_3. Note: * the amount of Al_2O_3.

In the AL–SI system, new phases are often formed as a result of different chemical reactions, gradually degrading the system. The newly formed products have a larger volume than the original material, with expansion being reported between 7 and 30%. This creates compounds in the lining or on its surface that have chemical compositions and physical parameters different from the lining itself [12]. Ternary diagrams of the types Na_2O-Al_2O_3-SiO_2 and K_2O-Al_2O_3-SiO_2 also describe the formation of individual phases in the given system (see Figure 3).

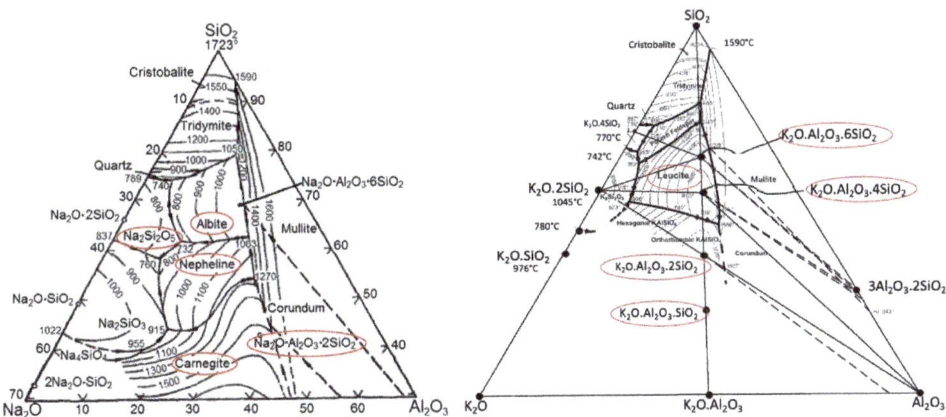

Figure 3. Ternary diagram of Na_2O-SiO_2-Al_2O_3 [13] and K_2O-Al_2O_3-SiO_2 [14], marking the individual phases formed during alkaline corrosion.

The corrosion mechanism in the Na–Al–Si system includes the formation of albite ($NaAlSi_3O_8$), nosean ($Na_8Al_6Si_6O_{28}S$) [15,16], and natrosilite ($Na_2Si_2O_5$) by Equation (1), which further reacts with mullite ($Al_6Si_2O_{13}$) to form albite ($NaAlSi_3O_8$) and aluminum oxide according to Equation (2). Nepheline ($NaAlSiO_4$) can also be formed according to Equation (3). Nosean is rarely reported in the literature as a corrosion product. However, due to its structural similarity to nepheline, it can also be expected to produce swelling. The reaction can be described by Equation (4) [17]:

$$Na_2SO_4 + 2SiO_2 = Na_2Si_2O_5 + SO_2 + 1/2O_2 \tag{1}$$

$$Na_2Si_2O_5 + 2Al_6Si_2O_{13} = 2NaAlSi_3O_8 + 5Al_2O_3 \tag{2}$$

$$2NaAlSi_3O_8 + Al_6Si_2O_{13} + 3Na_2SO_4 = 8NaAlSiO_4 + 3SO_2 + 3/2O_2 \tag{3}$$

$$4Na_2SO_4 + 3Al_6Si_2O_{13} = Na_8Al_6Si_6O_{28}S + 6Al_2O_3 + 3SO_2 + 3/2O_2 \tag{4}$$

In the case of high-alumina refractories (>45% Al_2O_3) containing mullite (A_3S_2) and cristobalite (SiO_2), reaction with NaO_2 above 1000 °C forms nepheline (NaS_2) and α-Al_2O_3 according to Equation (5). As can be deduced from the ternary diagram K_2O-Al_2O_3-SiO_2, at a lower content of $Al_2O_3 < 30\%$, orthoclase KAS_6 is formed, and at a content of $Al_2O_3 > 30\%$, new phases of leucite (KAS_4) are formed according to Equation (6):

$$3Al_2O_3 \cdot 2SiO_2 + Na_2O \rightarrow Na_2O \cdot Al_2O_3 \cdot 2SiO_2 + 2Al_2O_3 \tag{5}$$

$$K_2O \cdot Al_2O_3 \cdot 6SiO_2 \rightarrow K_2O \cdot Al_2O_3 \cdot 4SiO_2 + 2SiO_2 \tag{6}$$

Since the composition of biomass ash encourages the formation of eutectic melts, it is advisable to use high-alumina refractory materials with an Al_2O_3 content > 80% or to add silicon carbide for these linings. The compound, aluminosilicate-based materials mainly include products containing oxide-less constituents—graphite and silicon carbide

Materials in Al_2O_3-SiO_2-SiC systems combine the high thermal conductivity and chemical inertness of silicon carbide with the chemical and thermal stability of aluminosilicate and corundum. The products are therefore highly resistant to corrosion by liquid metals, as well as to sudden changes in temperature.

SiC oxidizes according to Equations (7) and (8) and creates an amorphous SiO_2 film on the surface [18,19]:

$$SiC + 1.5O_2 \rightarrow SiO_2 + CO \tag{7}$$

$$SiC + 2O_2 \rightarrow SiO_2 + CO_2 \tag{8}$$

To prevent graphite oxidation, firing is carried out without any contact between the fired product and oxygen. The firing temperature is chosen to create a ceramic bond in the products. At present, the process of quick firing is used, ensuring a reducing atmosphere in the kilns at higher temperatures and while cooling the products.

K_2O and Na_2O, in the form of alkaline vapors, are capable of diffusing into the refractory matrix, and then they react with Al_2O_3 and SiO_2 components to form K-aluminosilicate and Na-aluminosilicate phases [20]. In the AL–SI binary system, potassium pairs react according to Equations (9)–(11). The most harmful is the presence of free SiO_2 and Na_2O, which increase the reaction rate at high temperatures and support the formation of reactive glassy phases, according to Equation (12) [20]:

$$K_2O + SiO_2 \rightarrow K_2O.SiO_2 \tag{9}$$

$$3 (K_2O.2SiO_2) + 3Al_2O_3.2SiO_2 \rightarrow 3 (K_2O\ Al_2O_3.2SiO_2) + 2SiO_2 \tag{10}$$

$$K_2O.Al_2O_3.2SiO_2 + 2SiO_2 \rightarrow K_2O.Al_2O_3.4SiO_2 \tag{11}$$

$$2SiO_2 + Na_2O \rightarrow Na_2O \cdot 2SiO_2 \tag{12}$$

The so-called slagging/fouling index can be used to estimate the probability of slag formation in combustion units during biomass combustion. Slagging/fouling means the formation of layers (sticky, melted, or soft) of ash particles on heat exchange surfaces. A summary of slagging and fouling indices and their calculation are presented in Table 1.

Table 1. Ash characterization indices [10,21–24].

Index	Equation	Tendency Slagging/Fouling			
		Low	Middle	High	Ex. High
SiO_2 (%)	-	<20	20–25	>25	
Cl (%)	-	<0.2	0.2–0.3	0.3–0.5	>0.5
B/A	$\frac{B}{A} = \frac{Fe_2O_3 + CaO + MgO + Na_2O + K_2O}{SiO_2 + Al_2O_3 + TiO_2}$	<0.5	0.5–1	1–1.75	>1.75
S/A	$S/A = \frac{SiO_2}{Al_2O_3}$	<0.31	-	0.3–3	-
I/C	$\frac{I}{C} = \frac{Fe_2O_3}{CaO}$	<0.31	0.3–3	>3	-
Fu	$Fu = \frac{B}{A} \cdot (Na_2O + K_2O)$	<0.6	0.6–40	>40	-
TA	$TA = Na_2O + K_2O$	<0.3	$0.3 < TA < 0.4$	>0.4	-
Sr	$Sr = \frac{SiO_2}{SiO_2 + Fe_2O_3 + CaO + MgO} \cdot 100$	>72	65–72	<65	

The SiO_2 *index* is often the predominant element in biomass samples and causes the formation of melt, or "stickers", therefore giving it the characteristic of being slag-forming.

The *chlorine index Cl* acts as an accelerator of the reaction between K and SiO_2, which leads to the formation of fused glass deposits and the formation of slag at boiler operating temperatures of 800–900 °C [23].

Ash-deposition potential may be evaluated in terms of base-to-acid (B/A). The *basicity index B/A* (base/acid ratio) is based on the general rule that basic oxide compounds lower the melting point, and acidic compounds raise it. The B/A ratio is an indication of the fusion and slagging potential of ash. *I/C* (iron/calcium ratio) stands for Fe_2O_3/CaO, e.g., ash with a ratio of $Fe_2O_3/CaO = 0.3/3.0$ containing eutectics that increase slag formation.

The *Fouling index Fu* (fouling index) is the B/A ratio, also taking into account the alkali content ($Na_2O + K_2O$). Fouling refers to the dry deposition of ash particles or the condensation of volatile inorganic components on heat transfer surfaces. The normal percentage of alkali in biomass ash is between 25 and 35%, and it forms a eutectic in combination with silica.

Ash has a high viscosity (Sr) value, *slag viscosity index Sr* [24], so it will have a low tendency to slag. The *TA (total alkali) index* assesses the fuel's ability to form ash layers. Values of individual ash samples, defined based on the above-mentioned indices, are summarized in Section 3.2.

The chemical composition of ash is a good indicator of the problematic nature of biomass. For biomass fuels, massive slagging of heat exchange surfaces of boilers occurs during combustion. Ash composition and atmosphere in a combustion chamber influence the ash-melting temperature [10]. Indicators tell us of the characteristics of ash in terms of their influence on the formation of the glassy phase, and thus their tendency to slag and clog linings, heat exchange surfaces, and gas flow routes. These indices are based on chemical composition of biomass and its combustion. The equations are mainly based on fuel evaluation. However, since there is no specific index for biomass, it is possible to apply these indices to this type of fuel as well.

2. Materials and Methods

2.1. Ashes from Wood Biomass Combustion

Seven different types of ash from different types of wood biomass were used for the experimental portion of our study. All of these were obtained from the Czech Republic, mainly from the Moravian–Silesian Region, but one was from the Central Bohemian Region. Ashes utilized during the experimental portion were used in the original form for the crucible test. the granulometry was not adjusted. More information about the ash samples is presented in Table 2.

Table 2. Characterization of wood ash used for experiment.

Type of Wood Biomass	Disposal Method	Labeled
Spruce pellets	combustion	P_{019}
Woodchips	combustion	P_{020}
Woodchips	combustion	P_{031}
Woodchips, woodbark, sawdust, pellets, scraps	combustion	P_{032}
Woodchips, woodbark, sawdust, pellets, scraps	combustion	P_{033}
Woodchips	gassification	P_{059}, P_{060}

P_{019} P_{020} P_{031} P_{032} P_{033} P_{059} P_{060}

2.2. Refractory Materials

Tested refractory materials were manufactured by one of the largest producers and suppliers of refractory products and raw materials in the Czech Republic. Two types of shaped refractory materials, belonging to the silica–aluminum group, were selected for the corrosion experiment.

The first type was quality labeled as STV. It is a shaped refractory material classified as standard fire clay. The second type was quality labeled as ARS60N and is classified high alumina. The parameters of the mentioned tested materials with their properties are shown in Table 3.

Table 3. Chemical composition and properties of refractory materials.

Oxides wt.%	STV	ARS60N
SiO_2	53.5	28.40
Al_2O_3	40.5	46.60
TiO_2	2.1	-
Fe_2O_3	2.1	0.88
CaO	0.3	0.2
MgO	0.3	0.27
$K_2O + Na_2O$	0.8 + 0.2	0.5
SiC	-	13.2
Bulk density (kg/m^3)	2150	2700
Apparent porosity (%)	18.0	15
Cold crushing strength (MPa)	30	70
Refractory qualities under load (RUL) T0.5 (°C)	1360	>1500

2.3. Corrosion Crucible Test and Evaluation Method

The crucible test gives only approximate results. The refractory cube was filled with corrodent and heated to the testing temperature for a specified period. The testing conditions (temperature and corrodent composition) may reflect the expected service conditions, but in some situations, a more aggressive corrodent and/or high temperature may be used to speed up the attack to determine the resistance of the refractory to the corrosive liquid in a relatively short time. The crucible test is described step by step in Figure 4. The refractory cuboid sample with a cylindrical hole in the central portion was filled with corrosive, medium/powdered ash with a heating temperature of 1200 °C for 2 h. After cooling, the tested sample was cut through along the vertical axis, and the corroded portion was measured.

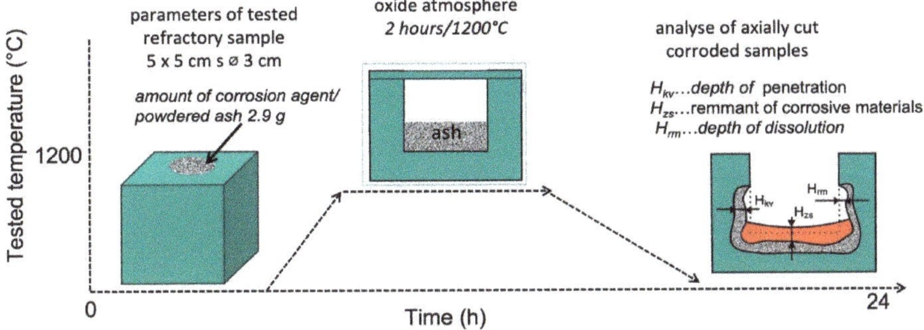

Figure 4. Schematic diagram of the crucible test of refractory materials.

After the corrosion test, samples were visually checked for compactness, potential cracks, and holes in the sample and walls. The ČSN P CEN/TS 15418 method [25] and the internal regulation method of P-D Refractories CZ a.s. [26] were used for test evaluation.

The classification used for reporting the condition of the crucible with defined parameters [25] U: unaffected/no visible attack; LA: lightly attacked/minor attack; A: attacked/clearly attacked and C: corroded/completely corroded. In addition to the above-mentioned evaluation regulations, another internal regulation method of P-D Refractories CZ was also used [26].

Table 4 shows the parameters of the classification after the corrosion test. Two evaluation methods may sometimes be requested by a customer or company testing laboratory, and the parameters can be used for comparison.

Table 4. Alkali test classification after internal regulation of P-D Refractories CZ [26].

Class	Classification		
	Corrosion	Infiltration	Cracks
A	not attacked	no corrosion and/or infiltration	No
B	slight attack	<6 mm corrosion and/or infiltration	No
C	distinctive attack	>7 mm corrosion and/or infiltration	Slight
D	severe attack	>9 mm corrosion and/or infiltration	large, clearly visible cracks

2.4. Characterization Methods

The chemical composition (XRF) of the ash was determined by energy-dispersive X-ray fluorescence spectroscopy (ED-XRF) on the SPECTRO XEPOS (Spectro Analytical Instruments, Kleve, Germany). Powdered samples were shaped/pressed into tablets for XRD measurement.

The mineralogical composition (XRPD) of the samples was evaluated using X-ray diffraction analysis on the X-ray diffractometer MiniFlex 600 (Rigaku, Tokyo, Japan) equipped with a 0Co tube and a D/teX Ultra 250 detector. XRD patterns were recorded in a 5–90° 2θ range with a scanning rate of 5° min^{-1}.

3. Results and Discussion

3.1. Ash Characterization

Chemical analysis is a good indicator for determining the problematic nature of biomass. The chemical composition of all of the ash types is presented in Figure 5. Biomass ash almost always contains carbonates, especially calcite, and very often portlandite, as well as a proportion of organic carbon.

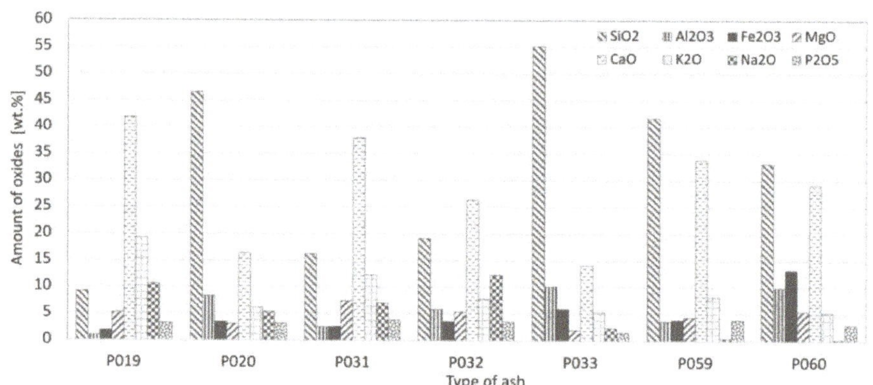

Figure 5. Concentrations of major elements in ash after wood biomass combustion.

Oxides in biomass ash can be divided into acidic (SiO_2, Al_2O_3, TiO_2, etc.) and basic (K_2O, CaO, MgO, Na_2O, Fe_2O_3, P_2O_5, etc.). Acidic oxides increase the melting point of ash. The higher the content of acidic oxides, the higher the melting point. On the other hand, basic oxides lower the melting point of the ash.

The predominant oxides are SiO_2 and CaO. A high level of CaO is typical for wood. The higher the content of basic oxides, the lower the melting point. SiO_2 plays an important role as a glass-forming oxide, while CaO and K_2O reduce the viscosity of the resulting glass-forming melt. The nature of the oxides and their representation determines the formation of other compounds and the behavior of the refractory material in contact with the corrosive agent. Ash was analyzed by XRDF, and this showed variable sample composition. The percentage of single oxides is as follows: SiO_2 9.13–55.17 wt.%, CaO 16.33–41.79 wt.%, Al_2O_3 0.98–10.14 wt.%, Fe_2O_3 1.80–13.16 wt.%. For alkali oxides it is Na_2O 0.38–12.23 wt.% and K_2O 6.11–19.17 wt.%. The amount of Cl is around 0.6 wt.%.

In terms of chemical composition, ash resembles low-melting glass. The variability of chemical composition complicates accurate representation in a ternary diagram. An approximate composition based on the largest content of wt.% of oxides is shown in the diagram. Four ash types, labeled P_{020}, P_{033}, P_{059} and P_{060}, are marked in the CaO-Al_2O_3-SiO_2 ternary diagram, and two types, labeled P_{031} and P_{019} are marked in the K_2O-SiO_2-CaO system, as presented in Figure 6.

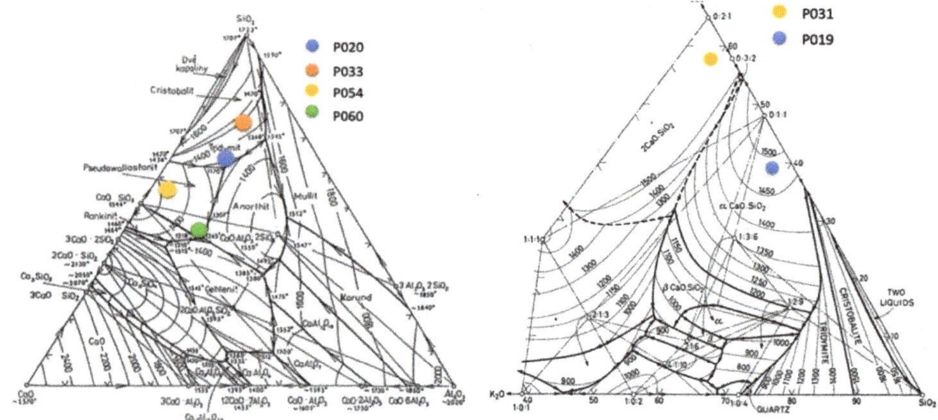

Figure 6. Approximate position of ash types P_{020}, P_{033}, P_{059}, P_{060}, in ternary diagram CaO-Al_2O_3-SiO_2 [27] and P_{031} and P_{019} in ternary diagram K_2O-Al_2O_3-SiO_2 [28].

The next method of ash characterization was X-ray powder diffraction phase analysis (XRPD). The samples were compared to the reference diffractogram database published by ICDD (PDF-2) in the range of 5–90° 2theta. The results for the analyzed samples are presented in Table 5, where there is an overview of the phases in the samples.

Table 5. Phase composition of analyzed biomass ash samples.

Phase Composition	Labeled of Sample						
	P_{019}	P_{020}	P_{031}	P_{032}	P_{033}	P_{059}	P_{060}
quartz (SiO_2)	x	x		x	x	x	
calcite ($CaCO_3$)	x	x	X	x	x	x	X
graphite C	x						
CaO	x	x	X	x			
magnesite ($MgCO_3$)			X				
MgO			X	x			
anorthite ($CaAl_2Si_2O_8$)		x					
microcline ($KAlSi_3O_8$)		x			x	x	x
arcanite (K_2SO_4)			X				
anhydrite ($CaSO_4$)				x			
anorthoclase					x		
leucite ($KAlSi_2O_6$)						x	x
orthoclase ($KAlSi_3O_8$)						x	
sylvite (KCl)							x
portlandite $Ca(OH)_2$						x	x
hematite (Fe_2O_3)						x	x
mullite ($Al_{4.59}Si_{1.41}O_{0.97}$)							x
analcime ($NaAlSi_2O_6$)							x

As confirmed by the analysis, the most frequently recurring phases are quartz, anorthite, calcium silicate, hematite, anhydrite, and microcline. In ash samples P_{059} and P_{060}, there were seven phases identified as portlandite; microcline, leucite, and portlandite occur in both. Samples P_{019} and P_{032}, were especially rich in the glass phase.

3.2. Calculation of the Slagging and Fouling Indices

To predict slagging/fouling in a combustion furnace, it is possible to use indices for the SiO_2, basic/acid ratio, silica/alumina ratio, fouling, iron/calcium ratio, and total alkalis, as summarized in Table 6. A special index only for biomass does not exist, but many authors have calculated these indices with regard to the probability of slag forming in combustion units.

Table 6. Calculation of slagging and fouling indices for individual ash types.

Ash	Index							
	SiO_2 (%)	Cl (%)	B/A	S/A	I/C	Fu	TA	Sr
P_{019}	9.1 [l]	0.21 [s]	8.1 [ex]	9.3 [h]	0.0 [l]	240.1 [h]	29.7 [h]	15.8 [h]
P_{020}	46.5 [h]	0.41 [h]	0.7 [m]	5.6 [h]	0.2 [l]	7.7 [m]	11.5 [h]	67.1 [m]
P_{031}	16.2 [l]	0.16 [l]	3.7 [ex]	6.4 [h]	0.1 [l]	72.1 [h]	19.3 [h]	25.3 [h]
P_{032}	19.1 [l]	1.74 [ex]	2.3 [ex]	3.3 [h]	0.1 [l]	46.4 [h]	20.0 [h]	35.2 [h]
P_{033}	55.1 [h]	0.10 [l]	0.5 [l]	5.4 [h]	0.4 [m]	3.6 [m]	7.6 [h]	71.7 [m]
P_{059}	41.7 [h]	-	1.2 [h]	11.6 [h]	0.1 [l]	10.2 [m]	8.0 5 [h]	49.8 [h]
P_{060}	33.1 [h]	-	1.3 [h]	3.4 [h]	0.5 [m]	6.9 [m]	5.3 [h]	41.0 [h]

Note: X [l]: low value; X [m]: middle value; X [h]: high value; X [ex]: extreme value.

In the case of SiO_2 content in P_{020}, P_{033}, P_{059} and P_{060}, they have a high inclination towards slagging. The high levels of silica in wood biomass ashes may have been caused by contamination with different elements (clay, sand, etc.); also, each part of the wood plant may contain different amounts of oxides. According to chloride content, extremely high fouling inclinations were observed in samples P_{032} = 1.74 and P_{020} = 0.47, while a low fouling inclination with a value > 0.2 was calculated for P_{033} = 0.1.

Index B/A confirmed that all wood biomass ash samples included in this study showed high to extremely high values of slagging (>1), as shown in Table 6 and Figure 7. The highest value B/A = 8.1 was for ash P_{059}, which will have a tendency toward slag formation. The low B/A index for P_{033} ash can be attributed to its high SiO_2 content, which implies an increased presence of acidic compounds. No ash samples had a B/A value lower than 0.5. Samples which had the highest B/A index and a high Na_2O content showed an inclination towards fouling.

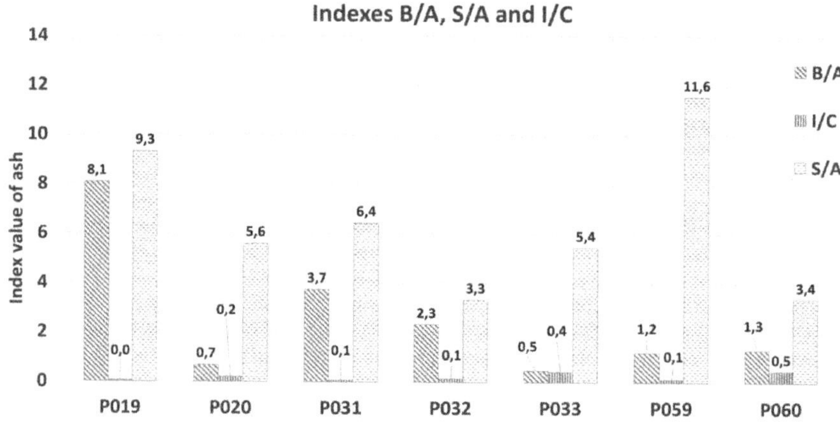

Figure 7. Indices B/A, S/A, and I/C for individual ash samples.

The I/C value for the tested samples was >0.3 and represents a small risk for the degradation of refractory materials. S/A index values were higher for all tested ash samples. the slagging tendency based on the silica/alumina ratio (S/A) was high for all the ash samples because no value was lower than 3 (Figure 7).

As a final result, higher Fu values correspond to higher fouling tendencies. The extreme value of the fouling index (Fu) was observed in sample P_{019}, with Fu = 240.1, and P_{031} Fu = 72.1, as presented in Figure 8. Fouling index values over 40 indicate a high tendency to fouling. The fouling tendency based on total alkalis was high for the types of ash mentioned below.

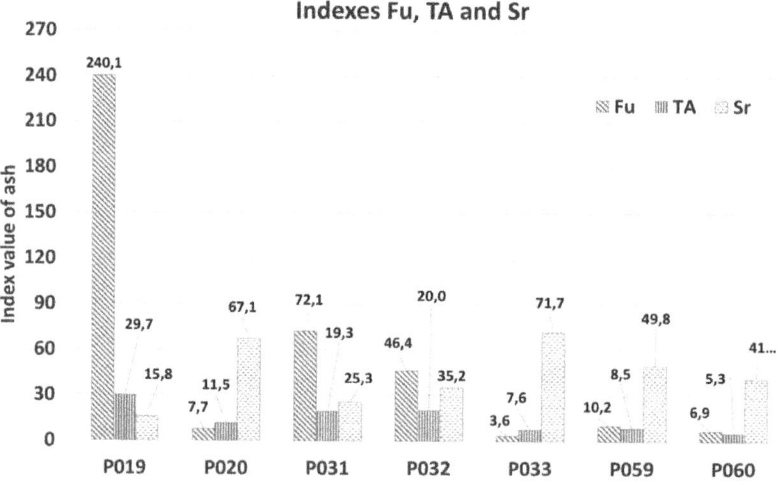

Figure 8. Indices for Fu, TA, and Sr for individual ash samples.

The value of total alkali content (TA) was never less than 0.4. (Figure 8). The lowest TA index was observed in P_{060} TA = 5.3, and it was 6 times higher in P_{019} = 29.7. The maximum value was 0.4, and there was a higher tendency towards the fouling of refractory materials during biomass combustion. Slagging probability was high for ash samples based on their Sr ratios, which were lower than 65. In this study, the calculated index for Sr corresponded to five ash types. Samples P_{020} and P_{033} had the middle value between 65 and 72. Based on calculations, the ash will tend to form slag and deposits.

3.3. Corrosion of Refractory Materials

Two types of refractory materials were selected for this portion of the corrosion experiment, STV fire clay (containing 53 wt.% SiO_2, 42 wt.% Al_2O_3), shown in Figure 9, and ARS60N high alumina (containing 28 wt.% SiO_2, 46 wt.% Al_2O_3, and 13 wt.% SiC), shown in Figure 10. Refractory materials underwent a crucible test under the following conditions: 2.9 g of ash as corrosion agent/2 h on maximum temperature at 1200 °C. All ash types melted at the suggested testing temperature.

Figure 9. Corrosion of STV refractory materials using different types of wood ash and defined parameters of crucible test: 2.9 g ash/2 h/1200 °C.

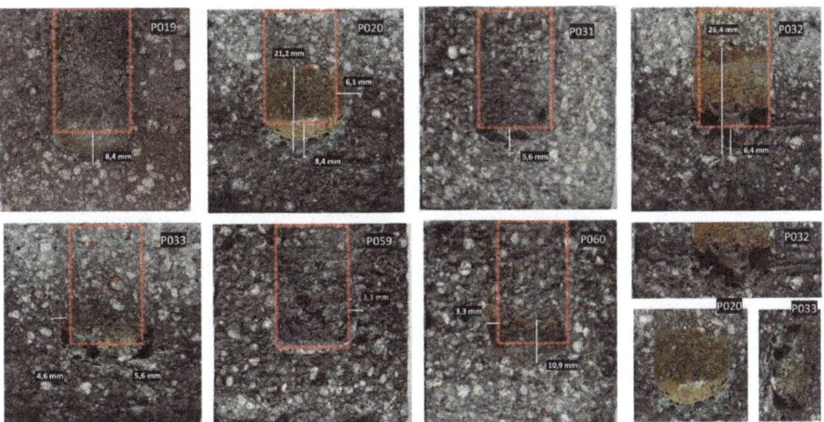

Figure 10. Corrosion of ARS60N quality refractory material with different types of wood ash with defined parameters of crucible test; 2.9 g ash/2 h/1200 °C.

Resistance to eutectic melts was determined by the refractory quality of STV at 1200 °C. As shown in Figure 9, infiltration was relatively low in samples P_{019} and P_{060} with refractory materials, whereas the attack of ashes P_{032} a P_{031} achieved a high intensity with STV refractory materials.

The evaluation of the corrosion attack by two methods is presented in Table 7. The corroding interaction was most intensive between P_{032} and STV refractory surface. P_{032} had the highest proportion, 1.77 wt.% Cl, a high content of alkalis ($Na_2O + K_2O$), namely 16 wt.%, a high CaO content of 26 wt.%, and 19 wt.% SiO_2. B/A = 2.3 ^{ex} was extreme, and the ash can be defined as basic. The slag viscosity index was Sr = 35.2 ^{h} and index Fu = 46.6 ^{h} was high, too. The ash showed a tendency towards slagging, and CaO and K_2O reduced the viscosity of the slag. Basic oxides lowered the ash melting point.

Table 7. Evaluation of STV refractory material corrosion at 1200 °C.

Quality	STV						
Ash	P_{019}	P_{020}	P_{031}	P_{032}	P_{033}	P_{059}	P_{060}
Cracks	no	no	no	no	no	no	no
ČSN P CEN/TS 15418 [25]	U	U	A	A	LA	LA	U
PD Refractories CZ [26]	B	B	B	C	B	B	B

Note: Evaluation of crucible test according to [25] U: unaffected/no visible attack: −; LA: lightly attacked/minor attack: +; A attacked/clearly attacked: ++; C: corroded/completely corroded: +++. Evaluation of crucible test according to [26] A: not attacked; B: slight attack/<6 mm corrosion and/or infiltration/no cracks: +; C: distinctive attack/>7 mm corrosion and/or infiltration/slight cracks: ++; D: severe attack corrosion/>9 mm corrosion and/or infiltration/large cracks, visible disruption: +++.

Cl acts as an accelerator of reactions between K_2O and SiO_2, which leads to the formation of fused glass deposits and the formation of slag at the boiler operating temperature of 800–900 °C [23]. According to Table 7 and normative regulation [25], the reaction between the STV quality refractory and ash P_{032} was evaluated as attacked/clearly attacked, and according to [26], the distinctive attack is >7 mm corrosion and/or infiltration/slight cracks.

P_{031} attack had an effect in contact with refractory materials that was very similar to that of P_{032}. All index values were categorized as high, but the amount of Cl was only 0.16 wt.%, unlike that of P_{031}. The crucible test of STV corrosion by P_{031} was evaluated according to [25] as attacked/clearly attacked and [26] slight attack/<6 mm corrosion and/or infiltration/no cracks.

From the visual point of view (see Figure 9), the sample tested by P_{033} formed melt, which represents a low-melt eutectic compound, because K_2SiO_3 and Na_2SiO_3 present low melting temperatures. P_{033} contained 55.1 wt.% SiO_2, alkali ($Na_2O + K_2O$) 7 wt.%, and only 0.1 wt.% Cl. Index B/A was only 0.5, but a high index Sr = 71.7, which had a negative effect to refractory materials. A similar effect was observed on samples attacked by P_{020}. STV refractory materials, after the application of P_{019} and P_{060}, may appear to be inert to ash, but a little infiltration was noticeable upon closer examination. Nevertheless, the evaluation according to the methodology was uniform.

An increase in temperature can be a significant indicator of changes in refractory materials. A problem may occur if temperature fluctuations occur–in this case, the combustion temperature could be 200 °C higher than the normal operating temperature [29], and the corrosion of refractory materials can be very intensive.

The formation of slag in the furnace depends on the chemical and mineralogical composition of the ash and on the conditions in the furnace (temperature, reduction, or oxidation zone, etc.). When fuel is burnt, combustion produces ash. In this case, residue will always form on the walls of the hearth and the heat exchange surfaces [18]. Slag deposits can be further divided into so-called "sintered deposits". (They are formed by dust particles trapped in a liquid or plastic state and stick to the wall.) The second type is hard deposits having a layered structure. The viscosity of the slag depends on the composition of the ash [13,14,30].

The second tested sample was high-alumina refractory ARS60N with a content of 46 wt.% Al_2O_3, 28 wt.% SiO_2, and 13 wt.% SiC (Figure 10). From the visual point of view, the refractory material was resistant to ash attack at 1200 °C. None of the samples showed total corrosion leading to disintegration. P_{032} showed negative effects on the refractory material ARS60N, just as it did with the STV quality. P_{032} caused slight changes on the surface structure with very fine cracks. The infiltration of ash was 6.4 mm, and in the structure of the tested refractory materials, it formed spaces without grains, which changed to compact slag in the middle of the sample. However, the integrity of the samples was preserved. A detailed photo of the changing structure of P_{032} is presented under the main pictures. As already mentioned, this ash has a tendency towards slagging. A very similar effect was observed in the middle of the refractory material, which formed after the attack of ash P_{033}. According to Table 8 and the literature [25], the corrosion effect between refractory quality ARSN60 and P_{032} described as attacked/clearly attacked, and according to Plibrico [26], it presented as a distinctive attack />7 mm corrosion and/or infiltration/slight cracks.

Table 8. Evaluation corrosion of refractory materials ARS60N at temperature 1200 °C.

| Quality | ARS60N | | | | | | |
Ash	P_{019}	P_{020}	P_{031}	P_{032}	P_{033}	P_{059}	P_{060}
Cracks	No	no	no	small	no	no	no
ČSN P CEN/TS 15418 [25]	LA	A	U	A	LA	U	LA
PD Refractories CZ [26]	B	C	B	C	B	B	B

Note: Evaluation of crucible test according to [25] U: unaffected/no visible attack: −; LA: light attacked/minor attack: +; A: attacked/clearly attacked: ++; C: corroded/completely corroded: +++. Evaluation of crucible test according to [26] A: not attacked; B: slight attack/<6 mm corrosion and/or infiltration/no cracks: +; C: distinctive attack/>7 mm corrosion and/or infiltration/slight cracks: ++; D: severe attack corrosion/>9 mm corrosion and/or infiltration/large cracks, visible disruption: +++.

The viscosity index of P_{032} is Sr = 35.2 [h], and that of P_{031} Sr = 25.3 [h]. According to [24], ash with a value (<65) viscosity Sr will have a high tendency towards slagging. On the other hand, P_{020}, with index Sr = 67 [m] and P_{033} Sr = 71.7 [m], will display the opposite trend. The effect of P_{020} is visually comparable for both STV and ARS60N refractory materials. The interaction between P_{020} and ARS60N samples formed a very slight layer of glassy character with an infiltration of about 6–8 mm. In terms of chemical composition, P_{020} contained 46 wt.% SiO_2, 16 wt.% CaO, and 11 wt.% alkali oxides ($Na_2O + K_2O$). This ash is acidic, as the ratio is B/A = 0.7 [m]. TA = 11.5 [m] and Fu = 7.7 [m] [24], so this ash should have a low tendency towards slagging. Other tested ash types showed invisible or minimum attack with refractory materials. ARS60N presented a higher level of corrosion resistance against the tested ash types than to STV. This portion of the experiment showed that the suggested temperature of 1200 °C during the corrosion crucible test should not be exceeded for tested STV and ARS60N refractory materials with the above-mentioned tested wood ash types.

4. Conclusions

STV quality refractory materials contain 53 wt.% SiO_2 and 42 wt.% Al_2O_3), while in ARS60N, the quantity of oxides is 46 wt.% Al_2O_3, 28 wt.% SiO_2, and 13 wt.% SiC). These materials belong to the alumina–silica system and were tested for their corrosion resistance. The results obtained within this research can be summarized as follow:

- Seven types of wood ash were used as corrosive media in their original forms. All types of ash were obtained by biomass combustion. The testing method was a crucible test at 1200 °C.
- The intensity of the corrosive effect was determined by two regulations. Neither a corrosion test nor visual evaluation revealed a greater penetration or destruction of the samples and that was not recorded, but only light/middle corrosion. From all tested ash samples, the one labeled P_{032} achieved the highest corrosive effect on both tested alumina–silica refractory materials.

- A less-intensive corrosion attack was observed in P_{031} with STV and ARS60N with P_{020}. Visible signs of corrosion included melts, fine cracks, infiltration, and in some cases, structural changes. The above-mentioned results suggest that the poorer the material is in silicon dioxide, the better it resists corrosion.
- The chemical stability of the phases increases as follows: amorphous silica (SiO_2) < cristobalite (SiO_2) < quartz (SiO_2) < andalusite (Al_2SiO_5) < mullite ($Al_6Si_2O_{13}$) < corundum (Al_2O_3).
- Upon a visual check, fire clay materials with $SiO_2 > Al_2O_3$ content presented a higher infiltration of the corrosion medium into the sample with an apparent porosity similar to that of sillimanite refractory materials.
- This may lead us to the conclusion that a higher level of Al_2O_3 content can increase corrosion resistance, as is often reported. An effective solution is also a partial replacement of the oxide with silicon carbide SiC, which has shown a favorable effect on the resistance of refractory materials. A high refractory-material-testing temperature can be a good indicator of a possible corrosive effect on the tested refractory material if there is undesirable temperature fluctuation during combustion.
- The nature and behavior of biomass ash can be defined by calculating slagging/fouling indices (B/A, Sr, Fu, TA, etc.), which are applied to the calculations of conventional fuels.
- Based on the chemical composition and calculated indices for wood biomass ash samples, it was confirmed that many of the values are above the limits and should tend to form slag and fouling. However, not all samples confirmed this prediction.

Author Contributions: Conceptualization, H.O.; Methodology, M.K.; Software, J.B.; Validation, M.K.; Formal analysis, M.T.; Investigation, J.V.; Resources, H.O.; Data curation, J.V.; Writing—original draft preparation, H.O.; Writing—review and editing, H.O.; Visualization, H.O.; Supervision, J.V.; Project administration, M.V.; Funding acquisition, M.V. All authors have read and agreed to the published version of the manuscript.

Funding: This work was supported by Project No. (SP2022/13), Project No. (SP2022/68), and Project No. (CZ.02.1.01/0.0/0.0/17_049/0008426).

Informed Consent Statement: Informed consent was obtained from all subjects involved in the study.

Data Availability Statement: The data presented in this study are available on request from the corresponding author.

Acknowledgments: Luděk Gryžbon, Smolo a.s nam. Svobody, 527, Lyžbice, 73961, Třinec, Czech Republic cooperated during experiment in case provision of ashes and their partly characterizations.

Conflicts of Interest: The authors declare no conflict of interest.

References

1. Munawar, M.A.; Khoja, A.H.; Naqvi, S.R.; Mehran, M.T. Challenges and opportunities in biomass ash management and its utilization in novel applications. *Renew. Sustain. Energy Rev.* **2021**, *150*, 111451. [CrossRef]
2. Bioenergy, I.E.A. *Options for Increased Use of Ash from Biomass Combustion and Co-Firing*; IEA: Paris, France, 2018.
3. Pastorek, Z.; Kára, J.; Jevič, P. *Biomasa: Obnovitelný Zdroj Energie*; FCC PUBLIC: Praha, Czech Republic, 2004; p. 288. ISBN 80-86534-06-5.
4. Ochodek, T. *Charakteristika Paliv. Výzkumné Energetické Centrum*; VŠB-TU: Ostrava, Czech Republic, 2018.
5. Blahůšková, V.; Vlček, J.; Jančar, D. Study connective capabilities of solid residues from the waste incineration. *J. Environ. Manag.* **2019**, *231*, 1048–1055. [CrossRef] [PubMed]
6. Reynaert, C.; Sniezek, E.; Szczerba, J. Corrosion tests for refractory materials intended for the steel industry—A review. *Ceramics-Silikáty* **2020**, *64*, 227–288. [CrossRef]
7. Kutzendörfer, J. *Koroze Žárovzdorných Materiálů*; Silikátová Společnost České Republiky: Praha, Czech Republic, 1998; ISBN 80-02-01204-6.
8. Plešingerová, B.; Derin, B.; Vadász, P.; Medveď, D. Analysis of deposits from combustion chamber of boiler for dendromass. *Fuel* **2020**, *266*, 117069. [CrossRef]
9. Kovář, P.; Lang, K.; Vlček, J.; Ovčačíková, H.; Klárová, M.; Burda, J.; Velička, M.; Topinková, M. *Inovované Žáromateriály pro Vyzdívky Pecních Agregátů Spalující Biomasu*; Tanger: Ostrava, Czech Republic, 2019; pp. 95–101. ISBN 978-80-87294-93-2.
10. Horák, J.; Kuboňová, L.; Dej, M.; Laciok, V.; Tomšejová, Š. Effect of type biomass and ashing temperature on the properties of solid fuel ashes. *Pol. J. Chem. Technol.* **2019**, *21*, 43–51. [CrossRef]
11. Routschka, G. *Refractory Materials*, 2nd ed.; Vulkan-Verlag: Essen, Germany, 2004; ISBN 3-8027-3154-920.

12. Schlegel, E.; Aneziris, C.G.; Fischer, U. Illustration of alkali corrosion mechanisms in high temperature thermal insulation materials. In Proceedings of the Conference: UNITECR'07—Technical conference on refractories, Dresden, Germany, 18–21 September 2007; pp. 117–120, ISBN 978-3-00-021528-5.
13. Kazmina, O.V.; Tokareva, A.Y.; Vereshchagin, V.I. Using quartzofeldspathic waste to obtain foamed glass material. *Resour.-Effic. Technol.* **2016**, *2*, 23–29. [CrossRef]
14. Levin, E.M.; Robbins, C.R.; McMurdie, H.F. *Phase Diagrams for Ceramics*; The American Ceramics Society Inc.: Westerville, OH, USA, 1964.
15. Tomšů, F. *Principy Zvyšování Odolnosti Proti Náhlým Změnám Teploty Žáruvzdorných Materiálů a Možná Aplikace na Žáromonolity*; Průmyslová keramika: Rájec-Jestřebí, Czech Republic, 2008.
16. Jacobson, N.S.; Lee, K.N.; Yoshio, T. Corrosion of Mullite of molten Slag. *J. Am. Ceram. Soc.* **1996**, *79*, 2161–2167. [CrossRef]
17. Weinberg, A.V.; Varona, C.; Chaucherie, X.; Goeuriot, D.; Poirier, J. Corrosion of Al_2O_3 SiO_2 refractories by sodium and sulfur vapors: A case study on hazardous waste incinerators. *Ceram Int.* **2017**, *43*, 5743–5750. [CrossRef]
18. Weinberg, A.V. Understanding the Failure and Development of Innovative Refractory Materials for Hazardous Waste Incineration. Ph.D. Thesis, University of Lion, Lyon, France, 2017; p. 449.
19. Jacobson, N.; Myers, D.; Opila, E.; Copland, E. Interactions of water vapor with oxides at elevated temperatures. *J. Phys. Chem. Solids.* **2005**, *66*, 471–478. [CrossRef]
20. Ren, B.; Li, Y.; Jin, S.; Sang, S. Correlation between chemical composition and alkali attack resistance of bauxite-SiC refractories in cement rotary kiln. *Cem. Int.* **2017**, *43*, 14161–14167. [CrossRef]
21. Vamvuka, D.; Kakaras, E. Ash properties and enviromental impact of various biomass and coal fuels and their blends. *Fuel Process Technol.* **2011**, *92*, 570–581. [CrossRef]
22. Park, S.W.; Jang, C.H. Characteristics of carbonized sludge for co-combustion in pulverized coal power plants. *Waste Manag.* **2011**, *31*, 523–529. [CrossRef] [PubMed]
23. Garcia-Maraver, A.; Mata-Sanchez, J.; Carpio, M.; Perez-Jimenez, J.A. Critical review of predictive coefficients for biomass ash deposition tendency. *J. Energy Inst.* **2017**, *90*, 214–228. [CrossRef]
24. Pronobis, M.; Kalisz, S.; Polok, M. The impact of coal characteristics on the fouling of stoker-fired boiler convection surfaces. *Fuel* **2013**, *112*, 473–482. [CrossRef]
25. *ČSN P CEN/TS 15418 (726022)*; Method of Test for Dense Refractory Products—Guidelines for Testing the Corrosion of Refractories Caused by Liquid. Czech Standarts Institute: Praha, Czech Republic, 2007.
26. P-D Refractories CZ a.s. *Internary Regulation for Corrosion Testing of Refractory*; P-D Refractories CZ: Velké Opatovice, Czech Republic, 2006.
27. Hlaváč, J. *Základy Technologie Silikátů*; SNTL: Praha, Czech Republic, 1988; pp. 1–516.
28. Roedder, E. Silicate melt systems. *Phys. Chem. Earth* **1959**, *3*, 224–297. [CrossRef]
29. Vlček, J.; Ovčačíková, H.; Klárová Topinková, M.; Burda, J.; VeličkA, M.; Kovař, P.; Lang, K. Refractory materials for biomass combustion. *AIP Conf. Proc.* **2019**, *2170*, 020024. [CrossRef]
30. Werther, J.; Saenger, M.; Hartge, E.-U.; Ogada, T.; Siagi, Z. Combustion of agricultural residues. *Prog. Energy Combust. Sci.* **2000**, *26*, 1–27. [CrossRef]

Article

Corrosion Resistance of Novel Fly Ash-Based Forsterite-Spinel Refractory Ceramics

Martin Nguyen * and Radomír Sokolář

Faculty of Civil Engineering, Institute of Technology of Building Materials and Components, Brno University of Technology, Veveri 331/95, 602 00 Brno, Czech Republic; sokolar.r@fce.vutbr.cz
* Correspondence: nguyen.m@fce.vutbr.cz

Abstract: This article aims to investigate the corrosion resistance of novel fly ash–based forsterite–spinel (Mg_2SiO_4-$MgAl_2O_4$) refractory ceramics to various corrosive media in comparison with reactive alumina–based ceramics. Because fly ash is produced in enormous quantities as a byproduct of coal-burning power stations, it could be utilized as an affordable source of aluminum oxide and silicon oxide. Corrosion resistance to iron, clinker, alumina, and copper was observed by scanning electron microscope with an elemental probe. The influence on the properties after firing was also investigated. Fly ash–based and reactive alumina–based mixtures were designed to contain 10%, 15% and 20% of spinel after firing. Raw material mixtures were sintered at 1550 °C for two hours. X-ray diffraction analysis and scanning electron microscopy were used to analyze sintered samples. The apparent porosity, bulk density, modulus of rupture, and refractory and thermo–mechanical properties were also investigated. The experimental results disclosed that the modulus of rupture, thermal shock resistance and microstructure were improved with increasing amounts of spinel in the fired samples. An analysis of the transition zones between corrosive media and ceramics revealed that all mixtures have good resistance against corrosion to iron, clinker, aluminum and copper.

Keywords: forsterite; spinel; fly ash; corrosion resistance; refractory ceramics

Citation: Nguyen, M.; Sokolář, R. Corrosion Resistance of Novel Fly Ash-Based Forsterite-Spinel Refractory Ceramics. *Materials* **2022**, *15*, 1363. https://doi.org/10.3390/ma15041363

Academic Editors: Jacek Szczerba, Ilona Jastrzębska and Andres Sotelo

Received: 27 December 2021
Accepted: 10 February 2022
Published: 12 February 2022

Publisher's Note: MDPI stays neutral with regard to jurisdictional claims in published maps and institutional affiliations.

1. Introduction

Refractory forsterite ceramics have played a significant role since the development of modern steelmaking technology. Due to the high melting point of forsterite refractory ceramics and their non-reactivity with iron at high temperatures, they have been predominantly implemented as a refractory lining of furnaces and regenerators in the metallurgical industrial sector. They have also been utilized in the cement and lime production industries as refractory lining for rotary kilns [1,2]. In the past decades, forsterite ceramics have also been utilized in electrotechnical engineering for ceramic-metal joints. Forsterite ceramics have relatively high coefficients of thermal expansion, which is comparable to the coefficient of metals used for joining [3].

In recent years, new ways of utilizing forsterite ceramics have emerged. Several researchers have investigated the sintering process of forsterite nanofibers with low thermal conductivity for their potential application as thermal insulation [4–7]. Other researchers are exploring the feasibility of a potential application of forsterite as a biomaterial in biomedicine for bone transplants due to its good compatibility with live tissue and high fracture toughness [8–11]. Researchers have also focused on the utilization of forsterite nanocrystals in the optical industry as a medium for optical lasers due to their great optical and mechanical properties [12,13].

Forsterite ceramics have a low thermal shock resistance due to their comparatively high coefficient of thermal expansion. However, this adverse effect of forsterite ceramics can be reduced by the incorporation of magnesium alumina spinel ($MgO·Al_2O_3$) directly into the raw material mixtures or indirectly by the addition of aluminum oxide as a raw material and its subsequent synthesis with magnesium oxide for the creation of spinel. Previous authors' studies have explored the feasibility of synthesis of spinel from the addition of raw materials containing aluminum oxide into forsterite mixtures [14,15]. Refractory spinel ceramics are also widely adopted in various fields, predominantly as linings of various industrial kilns and furnaces due to their numerous advantages, i.e., a very high melting point of 2135 °C, a low coefficient of thermal expansion compared to coefficient of forsterite, better thermal shock resistance, and chemical and corrosion resistance [16–20].

Previous studies have proven that spinel incorporation into forsterite ceramics has been shown to improve physico-mechanical properties such as microstructure, mechanical properties and thermal shock resistance due to the embedment of spinel crystals that are located predominantly near the grain boundaries of larger forsterite crystals or bound in the amorphous matrix and filling the empty sections of cavities and pores in between [14,15,21].

Spinel refractory ceramics of industrial grade are commonly synthesized from alumina or bauxite combined with magnesium oxide [18,20,22]. However, fly ash, which is generated in vast amounts as a byproduct from coal-burning power stations all over the world, could be used as an affordable source of aluminum oxide and silicon oxide. Many researchers have focused on the implementation of fly ash into the mixture as a raw material for sintering predominantly aluminosilicate refractory ceramics [23,24]. Despite the increased attention on the implementation of fly ash in the synthesis of aluminosilicate refractories, fly ash has rarely been looked at as the research object in the synthesis of other ceramic refractories that include silicon oxide and/or aluminum oxide, for example, forsterite–spinel refractory ceramics.

Corrosion resistance of the refractory ceramics is an important characteristic of every refractory product made. It is the ability of the tested ceramic material to withstand deterioration against the corrosive substance. Therefore, corrosion resistance to various corrosive media is a key characteristic of every refractory ceramic. It is known from previous research and the literature that different refractories have different corrosion resistance based on the chemical composition of the refractory and the corrosive materials. Acidic refractories such as silica, zirconia and aluminosilicate refractories are generally highly resistant to acidic materials but have low corrosion resistance to basic materials with MgO or CaO in their chemical composition [1,2,25,26].

On the contrary, basic refractories are highly resistant to basic materials such as clinker, lime, basic slags and alkaline materials with lower corrosion resistance to acidic materials. Moreover, they are stable, with high corrosion resistance to various metals and their slags. Many researchers have focused on testing the corrosion resistance of refractory ceramics to different corrosive media on magnesia or spinel refractory ceramics with positive results [25–28].

The main objective of this research is to investigate the influence of corrosion resistance by various corrosive media on fly ash–based forsterite–spinel ceramics in comparison with reactive alumina–based ceramics. The corrosion resistance of all designed forsterite–spinel mixtures was evaluated by examining the transition zone between corrosive media and ceramics using a scanning electron microscope with an elemental probe. In addition, the phase composition of fired forsterite–spinel ceramics was analyzed through X-ray diffraction analysis. The microstructure of fired test samples was further observed using scanning electron microscopy, and the physico–mechanical, refractory and thermo–mechanical properties of forsterite–spinel ceramics were also evaluated.

2. Materials and Methods

2.1. Raw Materials

Calcined caustic magnesite (CCM) was obtained from Magnesite Works (Jelsava, Slovakia), olivine from Norway, talc from Fichema (Brno, Czech Republic), coal fly ash from Mělník power plant (Mělník, Czech Republic), reactive alumina from Almatis (Ludwigshafen, Germany) and kaolin from Sedlecký kaolin (Božičany, Czech Republic). The chemical composition of the involved raw materials is presented in Table 1. Chemical analysis and X-ray fluorescence was used to determine the chemical composition.

Table 1. The chemical composition of all used raw materials.

Raw Materials	MgO	SiO$_2$	Al$_2$O$_3$	CaO	Fe$_2$O$_3$	K$_2$O+Na$_2$O	TiO$_2$	LOI [1]
CCM	85.6	0.5	0.1	5.2	7.4	0.1	0.1	1.0
Olivine	24.1	64.7	1.0	0.7	8.8	0.5	0.1	0.1
Talc	31.5	59.1	1.0	1.0	0.7	0.2	0.0	6.5
Fly ash	1.4	57.4	29.3	2.2	5.1	1.7	1.7	1.2
Alumina	0.0	0.0	99.3	0.0	0.1	0.3	0.0	0.3
Kaolin	0.5	46.8	36.6	0.7	0.9	1.2	0.1	13.2

[1] Loss on ignition.

The coarser raw materials were pre-treated in the vibration mill to achieve a particle size range of 1–64 μm (where d$_{50}$ = 10–30 μm). Particle size distribution was determined by laser granulometer (Malvern Mastersizer, Malvern Panalytical, Malvern, UK). The raw materials were then subjected to determination of mineralogical composition by means of X-ray diffraction analysis (XRD; Panalytical Empyrean, Panalytical B.V., Almelo, Netherlands). with CuKα as a radiation source, an accelerating voltage of 45 kV and a beam current of 40 mA. Olivine, fly ash and alumina were also subjected to determination of morphology in a scanning electron microscope (Tescan MIRA3, Tescan Orsay Holding a.s., Brno, Czech Republic).

Figure 1 represents the mineralogical composition of the raw materials. The major crystalline phase in CCM was periclase (MgO), with trace amounts of iron compounds. Forsterite (2MgO·SiO$_2$) was the major crystalline phase in olivine, with minor crystalline phases of fayalite (2FeO·SiO$_2$), serpentinite (3MgO·2SiO$_2$·2H$_2$O) and quartz (SiO$_2$). The only crystalline phase in talc powder was talc (3MgO·4SiO$_2$·H$_2$O). The crystalline phases in coal fly ash were mullite (3Al$_2$O$_3$·2SiO$_2$) and quartz (SiO$_2$). The only crystalline phase in reactive alumina was corundum (Al$_2$O$_3$). Kaolin was primarily composed of kaolinite (Al$_2$O$_3$·2SiO$_2$·2H$_2$O) and traces of biotite (K(Mg,Fe)$_3$AlSi$_3$O$_{10}$(OH)$_2$). The existence of an amorphous glass phase is indicated by the background curvature in olivine and fly ash diffractograms, and the background scattering (noise) is caused by the presence of iron compounds due to the interference with a CuKα radiation source.

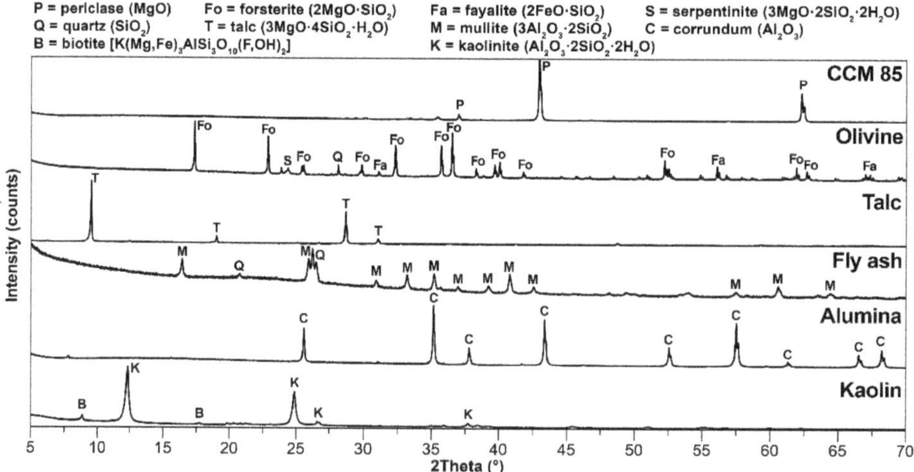

Figure 1. X-ray diffraction analysis of raw materials.

The scanning electron microscope (SEM) microphotographs of all untreated raw materials are presented in Figure 2a–f. The particles in CCM agglomerated together, with a particle size in a range of 5–50 μm. Untreated olivine has an apparent fibrous microstructure. Particles in talc have a foliated or fibrous microstructure, with different size flakes. The reactive alumina is composed of fine particles with d_{50} = 1.9 μm and that tend to cluster together. The untreated coal fly ash is mainly composed of spherical particles ranging in size, with a diameter of 0.4–90 μm (d_{50} = 14 μm). Kaolin platelets are clustered together in a general sheet appearance.

Figure 2. Scanning electron microscope images of (**a**) CCM; (**b**) olivine; (**c**) talc; (**d**) fly ash; (**e**) alumina; (**f**) kaolin with 1000× magnification.

2.2. Sample Preparation

In this work, six different mixtures were designed. Coal fly ash was the source of Al_2O_3 in mixtures FA-S10, FA-S15 and FA-S20, and reactive alumina was the source of Al_2O_3 in mixtures RA-S10, RA-S15 and RA-S20. The different sources of aluminum oxide were selected for spinel synthesis and comparison of properties of the final forsterite–spinel ceramics. The designations S10, S15 and S20 correspond to the theoretical amount of spinel in the mixture after synthesis. The mixture composition from raw materials is presented in Table 2.

Table 2. Composition of all designed mixtures from raw materials.

Raw Materials	FA-S10	FA-S15	FA-S20	RA-S10	RA-S15	RA-S20
CCM [wt.%]	43.5	45.9	48.2	40.0	39.6	39.2
Olivine [wt.%]	24.8	15.9	7.1	34.0	32.1	30.2
Talc [wt.%]	12.4	8.0	3.5	17.0	16.0	15.1
Fly ash [wt.%]	14.3	25.2	36.2	-	-	-
Alumina [wt.%]	-	-	-	4.1	7.3	10.5
Kaolin [wt.%]	5.0	5.0	5.0	5.0	5.0	5.0

First, the raw materials were accurately weighed, mixed and homogenized in a container by means of a rotary mechanical homogenizer for 24 h. After homogenization, mixtures were then mixed with a varying amount of water, utilizing a Pfefferkorn deformation apparatus (standard ČSN 72 1074) to achieve the optimal plasticity. The optimal plasticity (P_{opt}) was achieved when the ratio of sample height after deformation (h_{def}) to sample height before deformation (h_0) was equal to 0.6, as defined in Equation (1).

$$P_{opt} = h_{def}/h_0. \tag{1}$$

Test samples from all mixtures for all tests were obtained by molding plastic paste into the brass molds. The green samples were then dried until a constant weight in a laboratory drier at 105 °C. The dried samples were fired at 1550 °C with a heating rate of 4 K/min in a laboratory furnace with an air atmosphere. The soaking time was two hours at maximum temperature.

2.3. Characterization

The dimensions of the test samples were $20 \times 25 \times 100$ mm^3 for the measurement of a change in dimension during firing (standard EN 993-10:1997) and modulus of rupture (MOR; Testometric M350-20CT, Testometric Co. Ltd., Rochdale, UK), according to the standard EN 993-6:1995. A vacuum water absorption method with subsequent hydrostatic weighing (standard EN 993-1:1995) was used to determine apparent porosity, water absorption and bulk density on the same test samples. Refractoriness (standard EN 993-12:1997) of the mixtures was tested on a set of three pyrometric cones that were prepared according to the standard EN 993-13:1995. The refractoriness was performed in a small laboratory furnace with an observation port equipped with a digital camera that allowed real-time observation of the furnace and of the pyrometric cones.

On cylindrical test samples with a height of 50 mm and 50 mm in diameter, refractoriness under load (standard ISO 1893:2007) was investigated, and the temperature at 0.5% deformation ($T_{0.5}$) was evaluated. Thermal shock resistance (standard EN 993-11:2007; method B) was determined by a parameter *residual MOR (MOR_{res})* in percent, which is a ratio between the MOR of thermally cycled samples (MOR_{cyc}) and samples at a laboratory temperature (MOR), according to Equation (2). A residual MOR parameter enables a quantitative approach for measuring the thermal shock resistance. Test samples for thermal shock resistance were prisms with dimensions of $230 \times 64 \times 54$ mm^3.

$$MOR_{res} = (MOR_{cyc}/MOR) \times 100 \tag{2}$$

X-ray diffraction analysis was also performed on test samples to determine their mineralogical composition. Fluorite (CaF_2) was added to the samples as an inert standard for the quantitative analysis of all samples.

2.4. Testing of Corrosion Resistance of Forsterite–Spinel Ceramic Samples

Corrosion resistance (standard CEN/TS 15418:2006) was tested using the crucible method, with prism-shaped crucibles made and sintered from the same plastic paste for all six designed mixtures. The crucibles were made with utilization of a two-part brass mold with a prism base of $110 \times 110 \times 84$ mm^3 and the top cylindrical part with a diameter of 60 mm and height of 60 mm. The dimensions of the crucibles after firing were due to the firing shrinkage of approximately $100 \times 100 \times 76$ mm^3, with hollow cylindrical centers with a 55 mm diameter and 55 mm depth, which is in conformity with the dimensions specified in the standard CEN/TS 15418:2006.

The corrosive media used were iron, clinker, copper and aluminum to test the endurance of the designed forsterite–spinel mixtures against the corrosion of the molten materials. According to the literature [1,2], the corrosion resistance of the industrially produced forsterite ceramics is good against all used corrosive media mentioned above. The firing temperature for the corrosion resistance test was set at the melting point of the individual corrosive media—1538 °C for iron, 1450 °C for clinker, 1085 °C for copper and 660 °C for aluminum—to simulate the exposure of the refractory lining inside the kiln or furnace. The heating rate for the corrosion resistance test was 5 K/min, and soaking time was five hours at the maximum temperature, as specified by the standard CEN/TS 15418:2006.

The results of the corrosion resistance of all six mixtures were evaluated by SEM analysis with a secondary electron (SE) detector and a back-scattered electron (BSE) detector to analyze the microstructure and phase transition zones at the area of contact between the ceramics and corrosive media. An energy dispersive X-ray spectroscopy (EDX) probe was also utilized to determine the elemental analysis of the areas near the transition zone.

3. Results and Discussion

3.1. Mineralogical Composition and Microstructure

The XRD diffractograms of all mixtures are presented in Figure 3. The major crystalline phase for all mixtures is forsterite ($2MgO \cdot SiO_2$), with minor crystalline phases of spinel ($MgO \cdot Al_2O_3$), periclase (MgO) and monticellite ($CaO \cdot MgO \cdot SiO_2$) minerals. Fluorite ($CaF_2$) was added to the samples for the quantitative phase determination. The curved background of the XRD diffractograms signifies the presence of an amorphous glass phase, and the background scattering indicates the presence of iron compounds due to the use of CuKα as a radiation source. The quantitative analysis of the phase composition is presented in Table 3.

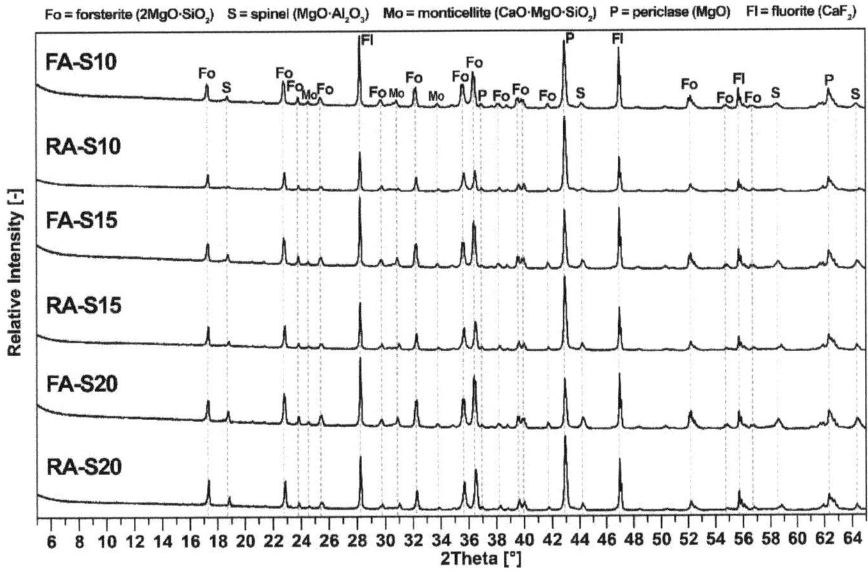

Figure 3. X-ray diffraction analysis of all fly ash–based and reactive alumina–based mixtures.

Table 3. Quantitative phase composition of all mixtures.

Phase	FA-S10	FA-S15	FA-S20	RA-S10	RA-S15	RA-S20
Forsterite	57.8	48.5	42.4	72.2	66.5	58.3
Spinel	11.1	14.2	19.7	9.8	14.7	19.8
Periclase	13.4	14.8	11.9	8.3	10.1	10.6
Monticellite	2.7	2.2	2.4	1.6	2.0	1.9
Amorphous phase	14.9	20.4	23.6	8.1	6.7	9.4

As can be seen in Figure 3, unreacted Al_2O_3 was not detected in the fired samples, and at the same time, no traces of mullite were detected. Therefore, the mullite in fly ash (Figure 1) had completely decomposed and recrystallized with magnesium oxide into spinel. It can therefore be concluded that all aluminum oxide reacted and formed spinel. Unreacted periclase was observed in the XRD diffractogram due to the presence of an amorphous glass phase, which inhibited periclase's reaction in forming forsterite. However, forsterite, spinel and periclase have excellent refractory properties. The amount of amorphous phase was higher in samples with fly ash due to its higher content in fly ash and due to the presence of flux oxides. All samples also contained a minor amount of monticellite, which formed due to the presence of calcium oxide in the raw materials. The theoretical value of formed spinel in all mixtures correlated with the results of quantitative analysis.

Figure 4a–f represents the SEM microphotographs of the morphology and microstructure of the samples with fly ash (FA-S10, FA-S15, FA-S20) fired at 1550 °C for two hours. It can be observed in Figure 4d–f that the tetragonal dipyramidal spinel crystals formed in clusters with diameters of 2–4 µm. They were located predominantly at the grain boundaries of larger forsterite crystals or bound in the amorphous matrix, filling some sections of the cavities and pores in between (see Figure 4a–c).

(a) (b) (c)

(d) (e) (f)

Figure 4. SEM images of fired forsterite–spinel ceramics mixtures with fly ash (FA-S10, S15, S20) with a magnification of 2000× (**a–c**) and 20,000× (**d–f**).

Spinel crystals that were synthesized from the mixtures with fly ash (Figure 4d–f) were less uniform and more irregular, with indications of polymorphism and crystal deformations. The crystal deformations can be observed in Figure 4d,e, and the polymorphic crystallization can be observed in the bottom of Figure 4f. This can be explained by the fact that spinel crystals formed indirectly from the decomposition of mullite and the subsequent reaction with magnesium oxide in mixtures with fly ash.

Figure 5a–f represents the SEM microphotographs of the morphology and microstructure of the samples with reactive alumina (RA-S10, RA-S15, RA-S20) fired at 1550 °C for two hours. It can be observed in Figure 5d–f that the tetragonal dipyramidal spinel crystals formed in clusters with diameters of 2–6 µm. With increasing content of reactive alumina in the mixture, the spinel crystal size was also increased. They were located primarily at the grain boundaries of larger forsterite crystals or bound in the amorphous matrix and filling some sections of the cavities and pores in between (Figure 5a–c).

Figure 5. SEM images of fired forsterite–spinel ceramics mixtures with reactive alumina (RA-S10, S15, S20) with a magnification of 2000× (**a–c**) and 20,000× (**d–f**).

The spinel crystals that were synthesized from the mixtures with reactive alumina were more uniform, and the crystallization was complete, with smooth surfaces and without any indication of polymorphism or deformations. This could be attributed to spinel crystals that crystallized directly from reactive alumina and magnesium oxide without any intermediate phase.

3.2. Physico-Mechanical Properties

Table 4 presents the results of the experiments that utilized the vacuum water absorption method, together with hydrostatic weighing to determine the apparent porosity, water absorption and bulk density, as well as the results of the MOR of fired samples from all six designed mixtures.

Table 4. Results of experiments to determine physico-mechanical properties.

Mixture	Apparent Porosity (%)	Water Absorption (%)	Bulk Density (kg·m^{-3})	Modulus of Rupture (MPa)
FA-S10	24.2	14.7	2365	15.5
RA-S10	17.5	4.5	2745	16.0
FA-S15	21.8	11.3	2460	18.4
RA-S15	15.6	4.2	2735	22.6
FA-S20	16.3	8.6	2510	17.3
RA-S20	14.4	2.9	2750	19.1

Apparent porosity and water absorption decreased with increased amounts of formed spinel in the structure in both fly ash–based mixtures and reactive alumina–based mixtures. This can be attributed to the higher firing shrinkage with increasing amounts of fly ash and reactive alumina in the mixtures for the subsequent spinel synthesis. The higher firing shrinkage resulted in higher densification of samples, which led to a decrease in apparent porosity and water absorption. The higher decrease in porosity and water absorption in fly ash–based mixtures is due to the creation of a more amorphous glass phase, resulting from increased amounts of flux oxides due to the increased content of fly ash in mixtures FA-S15 and FA-S20 for the subsequent spinel synthesis. Apparent porosity was higher in fly ash–based mixtures than alumina–based mixtures due to the expansion stage at 1250 °C, which was caused by the reaction of flux oxides and amorphous phase. This phenomenon is described in more detail in [15].

A higher bulk density of reactive alumina mixtures was caused by the denser structures of these samples due to the lower porosity and utilization of larger quantities of raw materials with higher bulk densities. The highest MOR in fly ash mixtures was achieved in mixture FA-S15, with a MOR value of 18.4 MPa. Similarly, the highest MOR in reactive alumina mixtures was achieved by the RA-15 mixture, with an MOR value of 22.6 MPa. The decrease of MOR in fly ash mixture FA-S20 is attributed to the large quantity of flux oxides from fly ash. Comparably, the decrease of MOR in reactive alumina mixture RA-S20 was caused by the increased amount of amorphous phase.

When the quantity of synthesized spinel increased in the mixture, the apparent porosity and water absorption decreased, while the bulk density and MOR increased. It can be concluded that increasing the quantity of synthesized spinel in forsterite ceramics leads to improved physico-mechanical properties.

3.3. Refractory and Thermo-Mechanical Parameters

Table 5 contains the results of firing shrinkage, refractoriness, refractoriness under load and residual MOR, which is a parameter for the determination of thermal shock resistance for all six designed mixtures.

Table 5. Results of experiments to determine refractory and thermo-mechanical parameters.

Mixture	Firing Shrinkage (%)	Refractoriness (°C)	Refractoriness under Load $T_{0.5}$ (°C)	Residual MOR (%)
FA-S10	5.9	1694	1593	21.3
RA-S10	6.2	1737	1664	23.0
FA-S15	8.5	1676	1561	28.6
RA-S15	6.9	1742	1678	35.3
FA-S20	11.3	1655	1532	22.5
RA-S20	7.4	1731	1645	26.7

As can be seen in Table 5, the firing shrinkage increased in mixtures with fly ash due to the elevated volume of flux oxides, which promote sintering. Firing shrinkage was higher in fly ash–based mixtures as opposed to reactive alumina–based mixtures, in which the spinel crystallized from mullite and magnesium oxide. Consequently, spinel has higher density than mullite, which also promoted shrinkage during firing in fly ash–based mixtures.

Larger quantities of flux oxides in fly ash mixtures also caused lower refractoriness and refractoriness under load of these mixtures, as opposed to mixtures with reactive alumina. However, the maximum impairment caused only a 5% decrease in the refractoriness of S20 mixtures and a 7% decrease in refractoriness under load of the S15 mixtures. Residual MOR is a parameter of thermal shock resistance. The highest values of residual MOR of both fly ash–based and reactive alumina–based mixtures had samples with 15% spinel (S15) after firing.

The mixtures with 20% spinel (S20) contained larger quantities of the amorphous phase, which caused marginally lower values of refractoriness, refractoriness under load and residual MOR. In general, the addition of a small quantity of spinel into the forsterite ceramics leads to better MOR and thermal shock resistance due to the improvement in microstructure. In addition, spinel ceramics have higher thermal shock resistance due to the lower value of coefficient of thermal expansion compared to forsterite [3,19].

3.4. Corrosion Resistance of Forsterite–Spinel Ceramics

Sections 3.4.1–3.4.4. contain the results of the corrosion resistance of forsterite–spinel ceramics, prepared from all six designed mixtures with utilization of the SEM microphotographs of the transition zones between the corrosive media and the ceramics. The resulting resistance against the corrosion by the corrosive media is evaluated by the overall microstructure near the transition zone between the ceramics and corrosive media and the depth of penetration of the corrosive media into the ceramics. The corrosion resistance of forsterite–spinel ceramics was tested by molten iron, copper, aluminum and clinker.

3.4.1. Corrosion Resistance of Ceramics to Iron

The microphotographs from SEM of the transition zone between iron and forsterite–spinel ceramics of all designed mixtures are presented in Figures 6 and 7. Due to the heavier atomic number of iron, the BSE detector can be utilized to detect various levels of signal and differentiate between iron (light grey) and ceramics (dark grey).

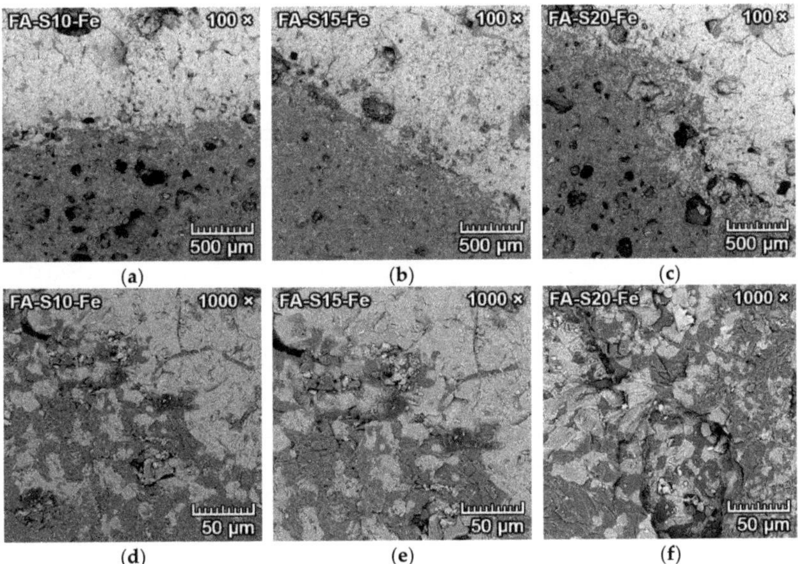

Figure 6. SEM microstructure of the transition zone between iron (light grey) and forsterite–spinel ceramics (dark grey) of fly ash–based mixtures FA-S10, FA-S15 and FA-S20, with a magnification of 100× (**a–c**) and a magnification of 1000× (**d–f**).

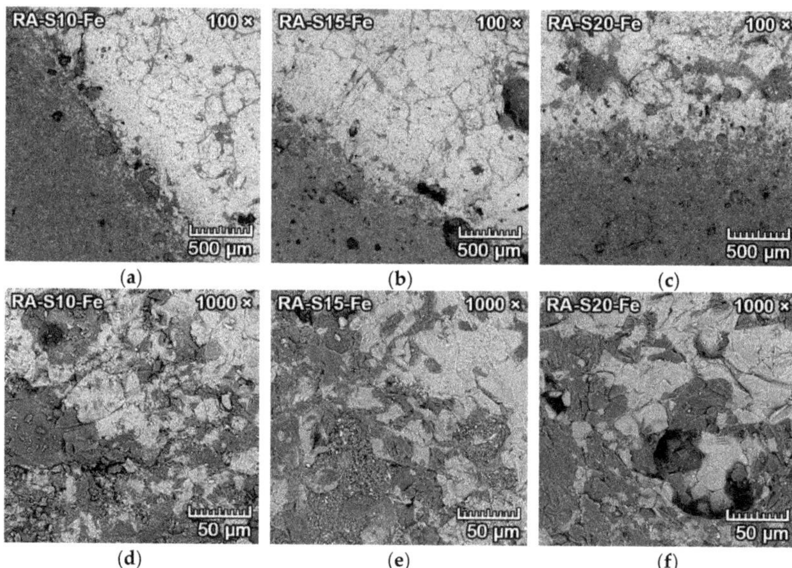

Figure 7. SEM microstructure of the transition zone between iron (light grey) and forsterite–spinel ceramics (dark grey) of reactive alumina–based mixtures RA-S10, RA-S15 and RA-S20, with a magnification of 100× (**a–c**) and a magnification of 1000× (**d–f**).

Figure 6a–f represents the corrosion resistance to iron of samples of fly ash–based mixtures FA-S10, FA-S15 and FA-S20. These samples had an increased porosity in the proximity of the transition zone between iron and ceramics, which was caused by the creation and formation of fayalite ($2FeO \cdot SiO_2$) from iron oxide and an equimolar amount of amorphous silica, according to Equation (3).

$$2FeO + SiO_2 \rightarrow 2FeO \cdot SiO_2. \tag{3}$$

The increase in porosity, which can be seen in Figure 6a–c, was stronger in fly ash–based mixtures, as the amount of the amorphous phase was higher than that of reactive alumina–based mixtures. The higher ratio of flux oxides and mullite decomposition also led to the formation of additional amorphous phase. Due to the higher porosity in the proximity of the transition zone, the EDX probe also detected olivine with a higher concentration of iron oxide, which diffused more easily into the pore structure. Olivine is a solid solution between forsterite ($2MgO \cdot SiO_2$) and fayalite ($2FeO \cdot SiO_2$), with a general formula of (Mg^{2+}, Fe^{2+})$_2SiO_4$. The olivine found in the sample contained 5–10% of fayalite, with up to 30% in the proximity of the transition zone. The depth of penetration of iron into the fly ash–based mixtures was 1–2 mm, with lower values for mixtures with increased content of spinel (FA-S15, FA-S20).

Figure 7a–f represents the corrosion resistance to iron of samples of reactive alumina–based mixtures RA-S10, RA-S15 and RA-S20. These samples also had weakly increased porosity (Figure 7a–c) in the proximity of the transition zone between iron and ceramics, which was caused by the creation and formation of fayalite. However, the increase in porosity was lower than that of fly ash–based mixtures. This also allowed the diffusion of iron into the porous ceramic structure, which is clearly visible in Figure 7d–f.

However, due to the low solubility of spinel to iron oxide [29], the increased amount of spinel in the forsterite ceramics (mixtures S15, S20) led to more distinct transition zones between iron and ceramics with larger grains of forsterite–spinel matrix (dark grey) and iron oxide-fayalite (light grey). As a result, the corrosion resistance of forsterite–spinel ceramics to molten iron was negligible, with a depth of penetration of iron into the ceramics of 0–2 mm in reactive alumina–based mixtures. The depth of penetration was lower with increased content of spinel in the mixtures.

3.4.2. Corrosion Resistance of Ceramics to Clinker

The microphotographs from SEM of the transition zone between clinker and forsterite–spinel ceramics of all prepared mixtures are presented in Figures 8 and 9. Due to the similar atomic numbers of clinker compounds and ceramics, the BSE detector could not be utilized to differentiate between clinker and ceramics. Therefore, a secondary electron (SE) detector with an EDX probe and larger magnification was used. Figures 8a–c and 9a–c illustrate the microstructure near the transition zone between the clinker and ceramics. Figures 8d–f and 9d–f illustrate the microstructure with a larger magnification.

Figure 8a–f represents the corrosion resistance to clinker of fly ash–based mixtures FA-S10, FA-S15 and FA-S20. The clinker reacted with ceramics at the transition zone and caused additional formation of monticellite ($CaO \cdot MgO \cdot SiO_2$) from forsterite and dicalcium silicate, according to Equation (4), and the creation of merwinite ($3CaO \cdot MgO \cdot 2SiO_2$) from one mole of forsterite and three moles of dicalcium silicate, according to Equation (5). Both minerals were identified by the EDX probe.

$$2MgO \cdot SiO_2 + 2CaO \cdot SiO_2 \rightarrow 2(CaO \cdot MgO \cdot SiO_2), \tag{4}$$

$$2MgO \cdot SiO_2 + 3(2CaO \cdot SiO_2) \rightarrow 2(3CaO \cdot MgO \cdot 2SiO_2). \tag{5}$$

Due to the slow cooling of the samples, tricalcium silicate dissolved into dicalcium silicate and lime (CaO), according to Equation (6) [30]. Subsequently, lime reacted with spinel in the proximity of the transition zone to form tricalcium aluminate, with simultaneous precipitation of periclase (MgO), according to Equation (7). Isometric hexoctahedral periclase crystals are visible in Figure 8d–e, as small cube-shaped crystals with diameters of 1–2 µm were scattered on the dicalcium aluminate/amorphous silica melt.

$$3CaO \cdot SiO_2 \rightarrow 2CaO \cdot SiO_2 + CaO, \tag{6}$$

$$3CaO + MgO \cdot Al_2O_3 \rightarrow 3CaO \cdot Al_2O_3 + MgO. \tag{7}$$

Figure 8f represents the spinel crystals in the amorphous phase located in the proximity of the transition zone. The corrosion resistance of fly ash–based mixtures to clinker was minimal, with the depth of penetration of 2–3 mm. With the increased content of spinel in the mixtures (FA-S15, FA-S20), the depth of penetration was lower than in mixture FA-S10.

Figure 9a–f represents the corrosion resistance to clinker of reactive alumina–based mixtures RA-S10, RA-S15 and RA-S20. Figure 9a–c represents the overall microstructure in the proximity of the transition zone, and Figure 9d–f represents the microstructure with a higher magnification of 5000×. Periclase crystals are clearly visible in Figure 9d–f as small white cube-shaped crystals with diameters of 1–2 µm scattered on the dicalcium aluminate/amorphous silica melt (darker grey). The corrosion resistance of reactive alumina–based mixtures to clinker was minimal, with better results than fly ash–based mixtures. The depth of penetration of clinker into the reactive alumina–based mixtures was 1–2.5 mm. With the increased content of spinel in the mixtures (RA-S15, RA-S20), the depth of penetration of clinker was lower than in mixture RA-S10.

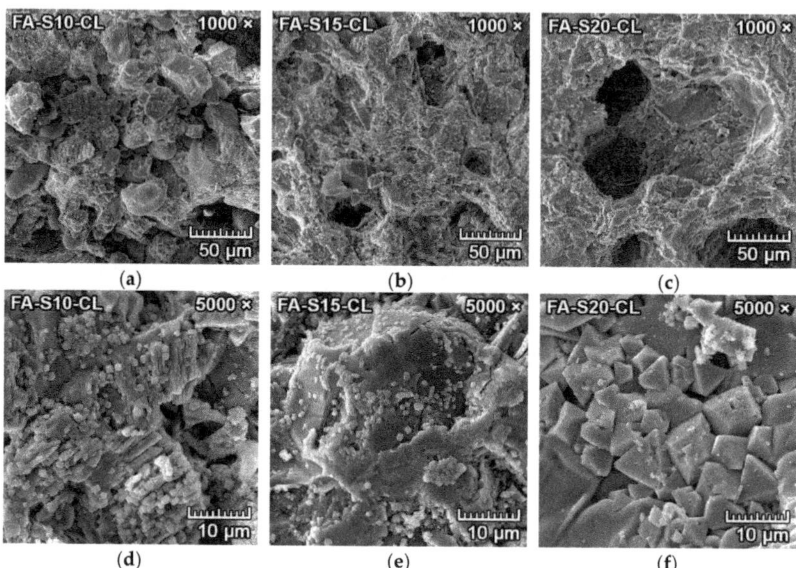

Figure 8. SEM microstructure of the transition zone between clinker and forsterite–spinel ceramics of fly ash–based mixtures FA-S10, FA-S15 and FA-S20, with a magnification of 1000× (**a–c**) and a magnification of 5000× (**d–f**).

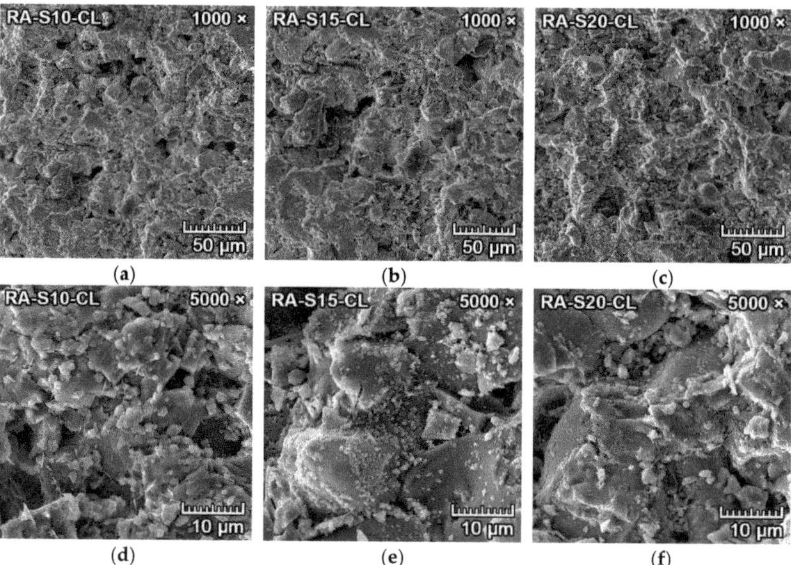

Figure 9. SEM microstructure of the transition zone between clinker and forsterite–spinel ceramics of reactive alumina–based mixtures RA-S10, RA-S15 and RA-S20, with a magnification of 1000× (**a–c**) and a magnification of 5000× (**d–f**).

3.4.3. Corrosion Resistance of Ceramics to Aluminum

The microphotographs from SEM of the transition zone between aluminum and forsterite–spinel ceramics synthesized from all mixtures are presented in Figures 10 and 11. The SE detector could clearly distinguish between the aluminum metal (darker and smooth) and ceramics (lighter and porous). The transition zone is clearly visible in the center of

Figures 10a–c and 11a–c. Between the aluminum metal and the ceramics is a darker smooth layer, visible in the bottom of Figures 10d–f and 11d–f. The EDX probe detected that this darker layer was composed mainly of melted aluminum, which filled the outer pores of the ceramics, and aluminum oxide (Al_2O_3).

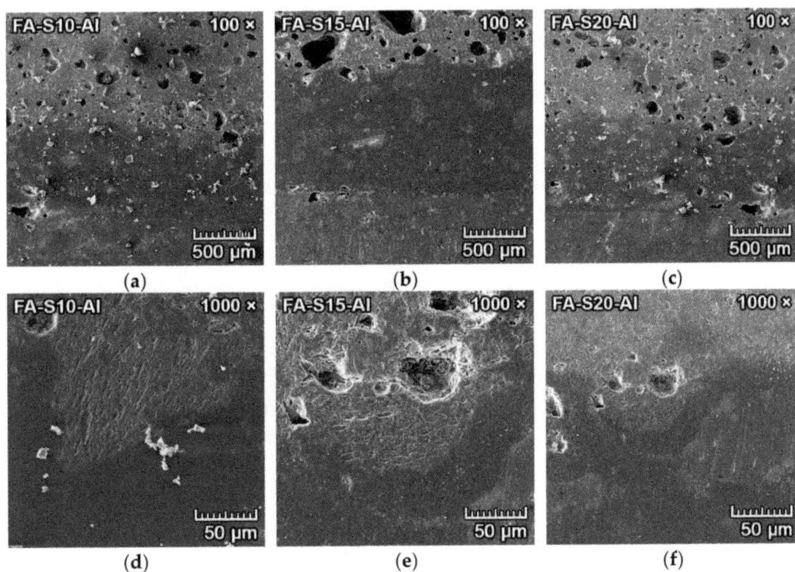

Figure 10. SEM microstructure of the transition zone between aluminum (dark grey and smooth) and forsterite–spinel ceramics (lighter and porous) of fly ash–based mixtures FA-S10, FA-S15 and FA-S20, with a magnification of $100\times$ (**a–c**) and a magnification of $1000\times$ (**d–f**).

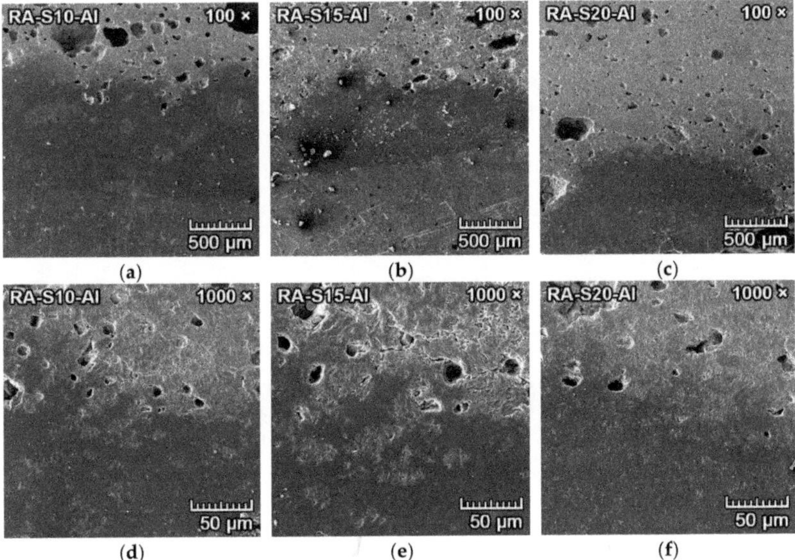

Figure 11. SEM microstructure of the transition zone between aluminum (dark grey and smooth) and forsterite–spinel ceramics (lighter and porous) of reactive alumina–based mixtures RA-S10, RA-S15 and RA-S20, with a magnification of $100\times$ (**a–c**) and a magnification of $1000\times$ (**d–f**).

Figure 10a–f illustrates the corrosion resistance to aluminum of fly ash–based mixtures FA-S10, FA-S15 and FA-S20. The corrosion resistance of fly ash–based mixtures to aluminum metal was very good, with the depth of penetration of aluminum around 1 mm for fly ash–based mixtures. The aluminum oxide layer (dark grey) was thinner in mixture FA-S20 due to the higher content of spinel, which also led to better corrosion resistance of this mixture.

Figure 11a–f illustrates the corrosion resistance to aluminum of reactive alumina–based mixtures RA-S10, RA-S15 and RA-S20. The darker grey layer in the middle of Figure 11a,b and in the bottom of 11c represents the layer of oxidized aluminum. The corrosion resistance of reactive alumina–based mixtures to aluminum metal was very good, with a depth of penetration of aluminum of 0.2–0.6 mm for reactive alumina–based mixtures. The layer of oxidized aluminum (dark grey) was very thin in mixtures RA-S15 and RA-S20 due to the higher content of spinel, which also led to better corrosion resistance of these mixtures.

3.4.4. Corrosion Resistance of Ceramics to Copper

The microphotographs from SEM of the transition zone between copper and forsterite–spinel ceramics are presented in Figures 12 and 13. The SE detector could clearly distinguish between the copper metal (lighter and smooth) and ceramics (darker and porous). The transition zone between copper and ceramics is clearly visible in the middle of Figures 12a–c and 13a–c. Between the copper metal and the ceramics is a very thin layer of a few micrometers that is darker than the copper metal and lighter than ceramics (middle of Figure 12d–f). The EDX probe detected that this thin layer was composed primarily of melted copper, which filled the pores of the ceramics, and part of the copper that oxidized into copper oxide (CuO). The bottom halves of Figure 12a–c contain almost pure copper, with small quantities of copper oxide.

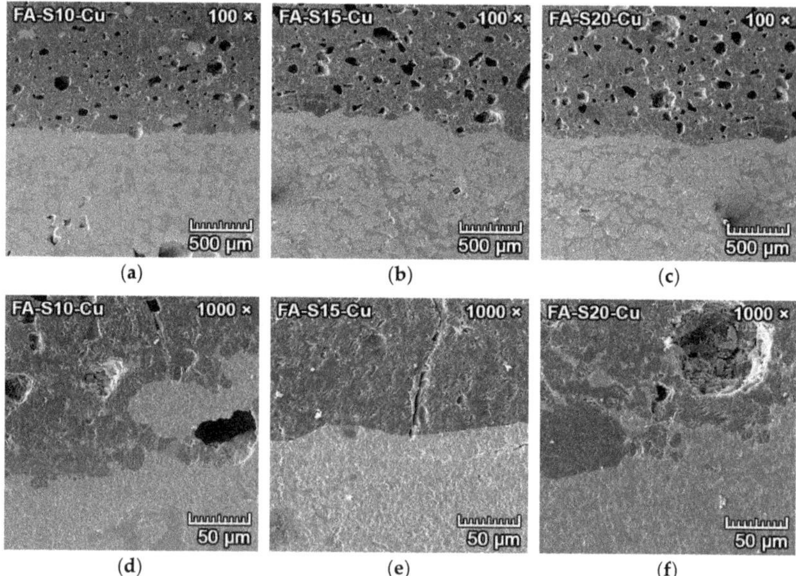

Figure 12. SEM microstructure of the transition zone between copper (light grey and smooth) and forsterite–spinel ceramics (darker and porous) of fly ash–based mixtures FA-S10, FA-S15 and FA-S20, with a magnification of 100× (**a–c**) and a magnification of 1000× (**d–f**).

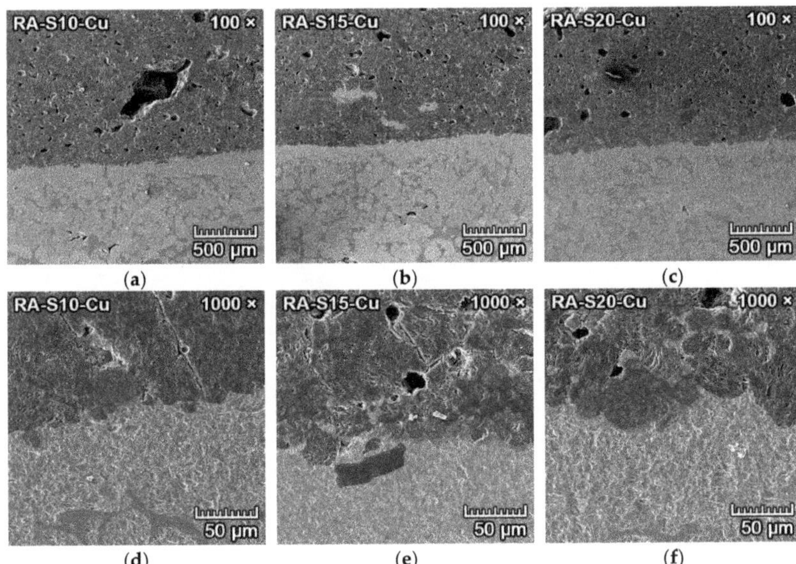

Figure 13. SEM microstructure of the transition zone between copper (light grey and smooth) and forsterite–spinel ceramics (darker and porous) of reactive alumina–based mixtures FA-S10, FA-S15 and FA-S20, with a magnification of 100× (**a–c**) and a magnification of 1000× (**d–f**).

The corrosion resistance of fly ash–based mixtures FA-S10, FA-S15 and FA-S20 to copper metal was excellent, with a depth of penetration of 0.01–0.013 mm. All three fly ash–based mixtures have very high corrosion resistance to molten copper.

The transition zone between copper and forsterite–spinel ceramics is clearly visible in the middle of Figure 13a–c. Between the copper metal and the ceramics is a very thin layer of a few micrometers that is darker than the copper metal and lighter than the ceramics, visible in the middle of Figure 13d–f. The corrosion resistance of reactive alumina–based mixtures RA-S10, RA-S15 and RA-S20 to copper metal was excellent, with a depth of penetration of 0.003–0.008 mm. All three reactive alumina–based mixtures have very high corrosion resistance to molten copper.

3.4.5. Discussion of Corrosion Resistance Results

According to the literature, the corrosion resistance of forsterite ceramics is very good in response to the corrosive effect of molten iron or iron slags. Similarly, clinker does not react with forsterite up to 1500 °C; therefore, the corrosion resistance of forsterite to clinker is also very high [1,2,31,32]. This is also consistent with the results of corrosion resistance to iron and clinker in this paper. The corrosion resistance to iron and clinker of both fly ash–based and reactive alumina–based mixtures was very high. With increased content of spinel in the mixtures (FA-S15, RA-S15, FA-S20 and RA-S20), the corrosion resistance was even better. This was due to the increased content of Al_2O_3, which has a low solubility to iron oxide [26,28,29,31].

The high content of MgO (in forsterite and periclase) in all mixtures also results in very good corrosion resistance to non-ferrous metals such as aluminum and copper due to the excellent oxidation resistance of MgO through solid solution phase formation [29,31]. The aluminum and copper metals that are in contact with MgO-based refractories such as forsterite form a protective oxide layer on the surface of the ceramics (transition zone), which inhibits any further corrosion of the ceramics. This is also in agreement with the results of the corrosion resistance of forsterite–spinel ceramics to aluminum and copper in this paper. The corrosion resistance was slightly improved in mixtures with reactive alumina (RA-S10, RA-S15, RA-S20) as opposed to fly ash–based mixtures (FA-S10, FA-S15, FA-S20) due to the lower content of amorphous phase and flux oxides.

4. Conclusions

Refractory forsterite–spinel ceramics were successfully sintered from fly ash–based and reactive alumina–based raw materials to compare the resulting properties after the firing of fly ash–based mixtures (FA-S10, FA-S15, FA-S20) compared to reactive alumina–based mixtures (RA-S10, RA-S15, RA-S20). The corrosion resistance of all six ceramic mixtures to iron, clinker, aluminum and copper was also tested. The increased content of spinel in the forsterite–spinel ceramics led to improved physico–mechanical properties such as MOR and thermal shock resistance, especially in mixtures FA-S15 and RA-S15 with 15% spinel and mixtures FA-S20 and RA-S20 with 20% spinel. The spinel crystals in reactive alumina–based mixtures that formed from alumina and magnesium oxide were more uniform and without cracks, and the resulting properties, such as MOR and thermal shock resistance, improved with increased alumina (RA-S15, RA-S20) content in the mixture, without impairing refractory properties. The spinel crystals in fly ash–based mixtures that formed from mullite decomposition in the presence of magnesium oxide in mixtures FA-S10, FA-S15 and FA-S20 were less uniform and had cracks, but the resulting properties (MOR, thermal shock resistance) were improved in mixture FA-S15 compared to mixtures FA-S10 and FA-S20, with minor impairments to the refractory properties in comparison with alumina–based mixtures. Mixtures FA-S15 and RA-S15, containing 15% spinel after firing, had the best resulting properties of all designed mixtures.

A microstructural analysis by SEM of the transition zones between the corrosive media and forsterite–spinel ceramics revealed that all mixtures had good resistance against corrosion from iron, clinker, aluminum and copper. The highest corrosion resistance to all tested corrosive media was for mixtures FA-S15, RA-S15 and RA-S20, with 15% and 20% spinel. The depth of penetration of iron was 0–2 mm in all mixtures. In corrosion from clinker, the depth of penetration was 1–2 mm in alumina–based mixtures (RA-S10, RA-S15 and RA-S20) and 2–3 mm in fly ash–based mixtures (FA-S10, FA-S15 and FA-S20). In corrosion from aluminum metal, the depth of corrosion was 0.5–1 mm and only 0.005–0.01 mm in corrosion by copper metal in all tested mixtures.

In conclusion, mixtures FA-S15 and RA-S15, with 15% spinel in forsterite ceramics, improved the microstructure, MOR and thermal shock resistance while retaining good refractory properties. Corrosion resistance to all tested corrosive media was also very promising.

Author Contributions: Conceptualization, M.N. and R.S.; methodology, M.N.; software, R.S.; validation, M.N.; formal analysis, M.N.; investigation, M.N.; resources, R.S.; data curation, M.N.; writing—original draft preparation, M.N.; writing—review and editing, R.S.; visualization, M.N.; supervision, R.S.; project administration, M.N.; funding acquisition, M.N. All authors have read and agreed to the published version of the manuscript.

Funding: This research was funded by the Internal Grant Agency of Brno University of Technology, specific junior research No. FAST-J-21-7279, with project name: Influence of microstructure of refractory forsterite–spinel ceramics on its high-temperature behavior and physico-mechanical properties, and by Czech Science Foundation GAČR, grant number 18-02815S, with project name: Elimination of sulphur oxide emission during the firing of ceramic bodies based on fly ashes of class C.

Institutional Review Board Statement: Not applicable.

Informed Consent Statement: Not applicable.

Data Availability Statement: The data presented in this paper are available upon request from the corresponding author.

References

1. Kingery, W. *Introduction to Ceramics*, 2nd ed.; Wiley: New York, NY, USA, 1960; pp. 261–284, ISBN 9780471478607.
2. Budnikov, P.P. *Technology of Ceramics and Refractories*; MIT Press Ltd.: Cambridge, MA, USA, 1964; pp. 270–283, ISBN 9780262523776.
3. Bouhifd, M.; Andrault, D.; Fiquet, G.; Richet, P. Thermal expansion of forsterite up to the melting point. *Geoph. Res. Lett.* **1996**, *23*, 1143–1146. [CrossRef]
4. Ji, W.; Wei, H.; Cui, Y. Facile synthesis of porous forsterite nanofibres by direct electrospinning method based on the Kirkendall effect. *Mater. Lett.* **2018**, *211*, 319–322. [CrossRef]
5. Zhao, F.; Zhang, L.; Ren, Z.; Gao, J.; Chen, X.; Liu, X.; Ge, T. A novel and green preparation of porous forsterite ceramics with excellent thermal isolation properties. *Ceram. Int.* **2019**, *45*, 2953–2961. [CrossRef]
6. Tsai, M.T. Synthesis of nanocrystalline forsterite fiber via a chemical route. *Mater. Res. Bull.* **2002**, *37*, 2213–2226. [CrossRef]
7. Chen, D.; Gu, H.; Huang, A.; Zhang, M.; Zhou, F.; Wang, C. Mechanical Strength and Thermal Conductivity of Modified Expanded Vermiculite/Forsterite Composite Materials. *J. Mater. Eng. Perform.* **2016**, *25*, 15–19. [CrossRef]
8. Kharaziha, M.; Fathi, M.H. Synthesis and characterization of bioactive forsterite nanopowder. *Ceram. Int.* **2009**, *35*, 2449–2454. [CrossRef]
9. Ni, S.; Chou, L.; Chang, J. Preparation and characterization of forsterite (Mg_2SiO_4) bioceramics. *Ceram. Int.* **2007**, *33*, 83–88. [CrossRef]
10. Ramesh, S.; Yaghoubi, A.; Sara Lee, K.Y.; Christopher Chin, K.M.; Purbolaksono, J.; Hamdi, M.; Hassan, M.A. Nanocrystalline forsterite for biomedical applications: Synthesis, microstructure and mechanical properties. *J. Mech. Behav. Biomed.* **2013**, *25*, 63–69. [CrossRef]
11. Saqaei, M.; Fathi, M.; Edris, H.; Mortazavi, V.; Hosseini, N. Effects of adding forsterite bioceramic on in vitro activity and antibacterial properties of bioactive glass-forsterite nanocomposite powders. *Adv. Powder. Technol.* **2016**, *27*, 1922–1932. [CrossRef]
12. Zampiva, R.; Acauan, L.; Dos Santos, L.; De Castro, R.; Alves, A.; Bergmann, C. Nanoscale synthesis of single-phase forsterite by reverse strike co-precipitation and its high optical and mechanical properties. *Ceram. Int.* **2017**, *43*, 16225–16231. [CrossRef]
13. Sun, H.T.; Fujii, M.; Nitta, N.; Mizuhata, M.; Yasuda, H.; Deki, S.; Hayashi, S. Molten-Salt Synthesis and Characterization of Nickel-Doped Forsterite Nanocrystals. *J. Amer. Ceram. Soc.* **2009**, *92*, 962–966. [CrossRef]
14. Nguyen, M.; Sokolář, R. Formation and Influence of Magnesium-Alumina Spinel on Properties of Refractory Forsterite-Spinel Ceramics. *Mater. Tehnol.* **2020**, *54*, 135–141. [CrossRef]
15. Nguyen, M.; Sokolář, R. Impact of Fly Ash as a Raw Material on the Properties of Refractory Forsterite–Spinel Ceramics. *Minerals* **2020**, *10*, 835. [CrossRef]
16. Ewais, E.M.M.; El-Amir, A.A.M.; Besisa, D.H.A.; Esmat, M.; El-Anadouli, B.E.H. Synthesis of nanocrystalline $MgO/MgAl_2O_4$ spinel powders from industrial wastes. *J. All. Com.* **2017**, *691*, 822–833. [CrossRef]
17. Mustafa, E.; Khalil, N.; Gamal, A. Sintering and microstructure of spinel–forsterite bodies. *Ceram. Int.* **2002**, *28*, 663–667. [CrossRef]
18. Braulio, M.A.L.; Rigaud, M.; Buhr, A.; Parr, C.; Pandolfelli, V.C. Spinel-containing alumina-based refractory castables. *Ceram. Int.* **2011**, *37*, 1705–1724. [CrossRef]
19. Ganesh, I. A Review on Magnesium Aluminate ($MgAl_2O_4$) Spinel: Synthesis, Processing and Applications. *Int. Mater. Rev.* **2013**, *58*, 63–112. [CrossRef]
20. Graf, R.B.; Wahl, F.M.; Grim, R.E. Phase transformations in silica-alumina-magnesia mixtures as examined by continuous X-ray diffraction: II. Spinel-silica compositions. *Am. Mineral.* **1963**, *48*, 150–158.
21. Michel, R.; Ammar, M.R.; Simon, P.; de Bilbao, E.; Poirier, J. Behaviour of olivine refractories at high temperature: Agglomeration in a fluidized bed reactor. *Refract. Worldforum* **2014**, *6*, 95–98.
22. Orosco, P.; Barbosa, L.; Carmen Ruiz, M. Synthesis of magnesium aluminate spinel by periclase and alumina chlorination. *Mater. Res. Bull.* **2014**, *59*, 337–340. [CrossRef]
23. Cao, J.; Dong, X.; Li, L.; Dong, Y.; Hampshire, S. Recycling of waste fly ash for production of porous mullite ceramic membrane supports with increased porosity. *J. Eur. Ceram. Soc.* **2014**, *34*, 3181–3194. [CrossRef]
24. Dong, Y.; Hampshire, S.; Zhou, J.; Ji, Z.; Wang, J.; Meng, G. Sintering and characterization of fly ash-based mullite with MgO addition. *J. Eur. Ceram. Soc.* **2011**, *31*, 687–695. [CrossRef]
25. Gengfu, L.; Yawei, L.; Tianbin, Z.; Yibiao, X.; Jun, L.; Shaobai, S.; Quanyou, L.; Yuanjin, L. Influence of the atmosphere on the mechanical properties and slag resistance of magnesia-chrome bricks. *Ceram. Int.* **2020**, *46*, 11225–11231. [CrossRef]
26. Sako, E.Y.; Braulio, M.A.L.; Pandolfelli, V.C. The corrosion resistance of microsilica-containing Al_2O_3–MgO and Al_2O_3–spinel castables. *Ceram. Int.* **2012**, *38*, 4783–4789. [CrossRef]
27. Mukhopadhyay, S.; Pal, T.K.; DasPoddar, P.K. Improvement of corrosion resistance of spinel-bonded castables to converter slag. *Ceram. Int.* **2009**, *35*, 373–380. [CrossRef]
28. Jeon, J.; Kang, Y.; Park, J.H.; Chung, Y. Corrosion-erosion behavior of $MgAl_2O_4$ spinel refractory in contact with high MnO slag. *Ceram. Int.* **2017**, *43*, 15074–15079. [CrossRef]
29. Park, H.S.; Kim, Y.; Kim, S.; Yoon, T.; Kim, Y.; Chung, Y. A study on the wetting behavior of liquid iron on forsterite, mullite, spinel and quasi-corundum substrates. *Ceram. Int.* **2018**, *44*, 17585–17591. [CrossRef]
30. Guo, Z.; Ding, Q.; Liu, L.; Zhang, X.; Luo, X.; Duan, F. Microstructural characteristics of refractory magnesia produced from macrocrystalline magnesite in China. *Ceram. Int.* **2021**, *47*, 22701–22708. [CrossRef]

31. Rutman, D.S.; Shchetnikova, I.L.; Kelareva, E.I.; Zhobolova, L.S.; Perepelitsyn, V.A. Effect of metallurgical slag on ceramics processed from pure oxides, spinels, and forsterite. *Refractories* **1973**, *14*, 701–705. [CrossRef]
32. Braulio, M.A.L.; Tomba Martinez, A.G.; Luz, A.P.; Liebske, C.; Pandolfelli, V.C. Basic slag attack of spinel-containing refractory castables. *Ceram. Int.* **2011**, *37*, 1935–1945. [CrossRef]

materials

Article

Corrosion of Alumina-Spinel Refractory by Secondary Metallurgical Slag Using Coating Corrosion Test

Sina Darban [1,2,*,†], Camille Reynaert [1,†], Maciej Ludwig [1], Ryszard Prorok [1], Ilona Jastrzębska [1] and Jacek Szczerba [1]

[1] Department of Ceramics and Refractories, Faculty of Materials Science and Ceramics, AGH University of Science and Technology, al. A. Mickiewicza 30, 30-059 Krakow, Poland; camille.reynaert@vesuvius.com (C.R.); ludwig@agh.edu.pl (M.L.); rprorok@agh.edu.pl (R.P.); ijastrz@agh.edu.pl (I.J.); jszczerb@agh.edu.pl (J.S.)

[2] Laboratoire Matériaux et Durabilité des Constructions, INSA/UPS Génie Civil, Université de Toulouse, CEDEX 04, 31077 Toulouse, France

* Correspondence: darban@insa-toulouse.fr

† Early Stage Researcher of the European Commission, Marie Skłodowska-Curie Actions Innovative Training Networks in the Frame of the Project ATHOR—Advanced Thermomechanical Multiscale Modelling of Refractory Linings 764987 Grant.

Abstract: In this paper, the corrosion mechanism of commercial alumina-spinel refractory was investigated at 1350 and 1450 °C. Disc samples were coated with shells of two different slags containing 4 and 10 wt.% SiO_2. The after-corrosion refractory was investigated in view of changes in its microstructure and phase composition by SEM/EDS and XRD techniques, respectively. At 1350 °C slags slightly infiltrated the microstructure, whereas at 1450 °C slags infiltrated the alumina-spinel refractory causing its significant corrosion. As a result of corrosion, new phases were formed, including calcium dialuminate ($Ca_2Al_4O_7$), calcium hexaluminate ($CaAl_{12}O_{19}$), and gehlenite ($Ca_2AlSi_2O_7$). Formation of calcium aluminate layers in the microstructure of the refractory inhibited further dissolution of alumina aggregates; however, expansive behavior of $CaAl_{12}O_{19}$ raised the microstructure porosity. The additional SiO_2 in the slag doubled the amount of low melting gehlenite in the matrix, accelerating the corrosion process of alumina-spinel brick at high temperatures.

Keywords: corrosion; alumina-spinel; refractories; metallurgical slag; XRD; SEM/EDS

Citation: Darban, S.; Reynaert, C.; Ludwig, M.; Prorok, R.; Jastrzębska, I.; Szczerba, J. Corrosion of Alumina-Spinel Refractory by Secondary Metallurgical Slag Using Coating Corrosion Test. *Materials* **2022**, *15*, 3425. https://doi.org/10.3390/ma15103425

Academic Editor: Andrea Piccolroaz

Received: 24 March 2022
Accepted: 5 May 2022
Published: 10 May 2022

Publisher's Note: MDPI stays neutral with regard to jurisdictional claims in published maps and institutional affiliations.

1. Introduction

The primary consumption of refractory materials—around 70%—is attributed to the steel industry. These high-temperature materials should have excellent chemical inertness, high thermal shock resistance, and adequate mechanical properties to show acceptable performance in aggressive and high-temperature environments in the steel industry, especially in a steel ladle.

Alumina-spinel bricks are refractory materials used in the sidewalls and bottom of steel ladles in secondary steel production. These materials directly contact the aggressive slag during the steel ladle operation. Previous studies have revealed that spinel is advantageous for corrosion resistance, and it prevents the dissolution of sidewalls of refractory materials in the steel ladle [1–6].

Chemical reactions and thermal gradients are two phenomena that take place—simultaneously—during work, and severely influence the corrosion process. Several zones can be distinguished within the refractory microstructure after the corrosion, including slag zone, reaction zone, impregnated zone, and unaffected zone, as shown in Figure 1. The thermal gradient affects the liquid viscosity, and it results in the crystallization of different phases in various temperature zones of the refractory. In most cases, the first stage in the corrosion of the refractory by liquid slag is its infiltration to the refractory through open pores. The open pores in the original refractory material facilitate infiltration of the liquid medium under the influence of capillary forces, and ease the diffusion process due to the larger surface area exposed to corrosion reactions. This infiltration often proceeds with the dissolution of grains starting from grain boundaries. The composition of the liquid phase, located mostly along the grain boundaries, is altered during the impregnation process.

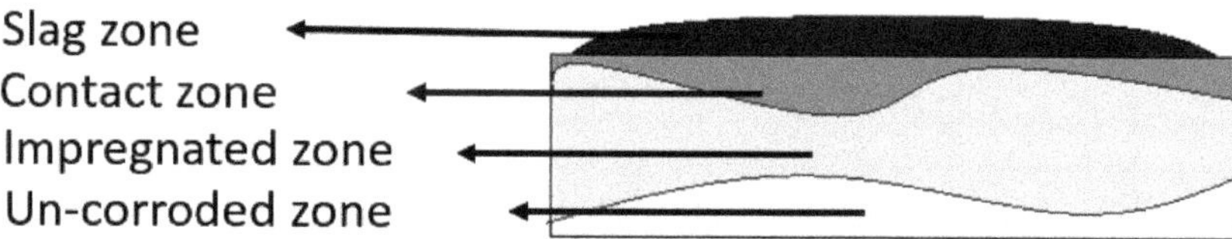

Figure 1. The schema of different corrosion zones in a contact corrosion test.

Refractories are designed to maintain their original texture and chemical composition. However, due to liquid slag infiltrating the refractory, both chemical composition and texture can be changed. Figure 1 shows zones that typically occur in corroded materials. The impregnated zone extends between the un-corroded and the corroded side where the dissolution of refractory components is followed by precipitation of secondary phases.

The impregnation of a refractory by liquid slag can be limited by reduction of pore size in the refractory, or increasing the slag viscosity. However, applying the specific solution to preserve the refractory lining against infiltration must respect factors required by the metallurgical process [3,7–9].

Lee and Zhang [10] studied the corrosion mechanism of oxide and oxide-carbon refractories through static corrosion methods, i.e., sessile drop, dipping, and crucible test, as well as dynamic techniques, like the rotating finger or slag rotary test [7]. Their studies showed that two main mechanisms govern the corrosion of refractory materials. The first one is direct dissolution, in which the solid-phase from the refractory dissolves into a liquid slag with no intermediate phases, while the second is the indirect dissolution mechanism, in which new solid phases are excluded from the aggressive slag, which derives lesser risk of refractory spalling. Song et al. [11] studied the corrosion behavior of alumina-based refractories by sessile drop test in contact with SiO_2-MgO-FeO-CaO-Al_2O_3-based slag obtained from smelting nickel laterite in a rotary heart furnace (RHF). They concluded corrosion in the system by three stages: melting and wetting, dissolution and diffusion, and final crystallization. The 0.43 and 0.63 basicity slags showed low apparent contact angles of 25° and 30°, respectively. They concluded that—for both types of slags—with extended corrosion time, two different covering layers formed between the alumina substrate and slag, preventing the further dissolution of Al_2O_3. The first layer was created by diffusion of Mg^{2+} and Si^{4+} from the slag towards Al_2O_3, while the second was formed by dissolution of Al_2O_3 and subsequent diffusion of Al^{3+} into the slag.

Braulio et al. [12] proposed a corrosion mechanism of spinel-containing castable in contact with slag of basicity 0.36. The pore size distribution was indicated as a crucial factor responsible for penetration of the refractory by slag and it was found that fumes of CaO and SiO_2 reduce crack resistance by increasing slag infiltration. Martinez et al. [13] suggested that the formation of CA_6 layers at alumina grain boundaries could potentially be helpful for improving corrosion resistance by inhibiting the refractory infiltration on the layer. From their perspective, the formation of CA_6 and CA_2 can be optimized by controlling the quantity of SiO_2 fumes and using a proper binder, other than calcium aluminate cement.

Corrosion of refractory depends on several factors, such as the composition as well as the viscosity of the slag, and work temperature, specifically, thermal gradient. So far, the majority of studies have concerned the slag corrosion at the working temperature of a steel ladle (1550–1650 °C); however, the thermal gradient in refractory is a determinant factor in the corrosion mechanism. The present study investigates the corrosion phenomena in alumina-spinel brick used in steel ladles at 1350 and 1450 °C; i.e., temperatures that actually occur at a cross-section of refractory during work. The alumina-spinel refractory was corroded by metallurgical slags, with one of them having increased acidity.

2. Materials and Methods

2.1. Preparation of Slag

Two slags of different compositions were used in this work, as presented in Table 1. The first slag was a steel ladle slag, delivered by the industrial partner in project ETN-ITN-ATHOR, designated as "R" slag. The second slag was prepared based on the composition of R slag by adding additional SiO_2 (Sigma Aldrich 797863-5MG, St. Louis, MO, USA) to obtain 10% SiO_2 in the slag—the slag is designated as "S" slag. The chemical compositions of R and S slags were measured by a Malvern PANalytical WDXRF Axios mAX spectrometer, Malvern, UK. A sample for XRF analysis was prepared by grinding a 5 kg batch of each slag, followed by its homogenization for 5 h. The main components of R slag were CaO (49%) and Al_2O_3 (41%), while SiO_2 constituted 4%. The main components of S slag were CaO (46%) and Al_2O_3 (38%), while SiO_2 constituted 10%. Thus, the amount of SiO_2 in S slag was about 6% higher than in R slag.

Table 1. XRF chemical composition of slags used in the study.

Component (wt.%)	R Slag	S Slag
CaO	48.5	45.6
Al_2O_3	40.9	38.4
SiO_2	4.20	10.0
Fe_2O_3	2.20	2.10
TiO_2	1.90	1.80
MgO	0.60	0.60
P_2O_5	0.47	0.42
SO_3	0.40	0.38
K_2O	0.33	0.29
Cr_2O_3	0.05	0.05
MnO	0.08	0.05
SrO	0.05	0.04
ZrO_2	0.09	0.08
Na_2O	0.21	0.18
Cl	0.02	0.01

2.2. Alumina-Spinel Brick

Table 2 presents the chemical composition and physicochemical properties of test commercial alumina-spinel refractory. The main component of the test brick was alumina, which reached about 88%, the contents of MgO and SiO_2 were at a similar level of 5%, while other impurities constituted about 2%. The refractory was characterized by 18.4% open porosity and 3.1 g/cm^3 apparent density.

Table 2. Characteristics of alumina-spinel.

Element	Chemical Composition by XRF (wt.%)
Al_2O_3	87.7
MgO	5.8
SiO_2	4.8
Na_2O	0.5
Fe_2O_3	0.3
CaO	0.2
Others	0.8
Bulk density (g/cm^3)	3.08
Apparent porosity (%)	18.42

2.3. Corrosion Test

Figure 2 shows the schema of sample preparation for corrosion testing (left side) and the investigated refractory sample prepared accordingly (right side). Firstly, the alumina-spinel bricks were cut and drilled into disk-shaped samples (d = 25 mm, h = 25 mm) and dried for 24 h at 150 °C. Then, 20 g of each finely powdered slag (<63 µm; R and S) was pressed as a shell around the alumina-spinel disk samples, using a hydraulic pressing machine with a pressure of 50 kN. This methodology for corrosion testing was proposed by the author to maximize the contact between the slag and refractory sample, thus enhancing the impact of increased SiO_2 content in the slag to the refractory. Another benefit of the applied method was ensuring that the liquid slag directly contacted the refractory at high temperatures. The prepared samples were heated in graphite crucibles (h = 55 mm, d = 35 mm) as shown in Figure 3. The slag did not react with carbon from the crucible because of the low wetting angle between slag and carbon. The corrosion tests were conducted in an air atmosphere, in an electrical furnace at 1350 °C and 1450 °C for 10 h with a heating rate of 5 °C/min. Table 3 presents the heat treatment conditions and designations of the corroded samples.

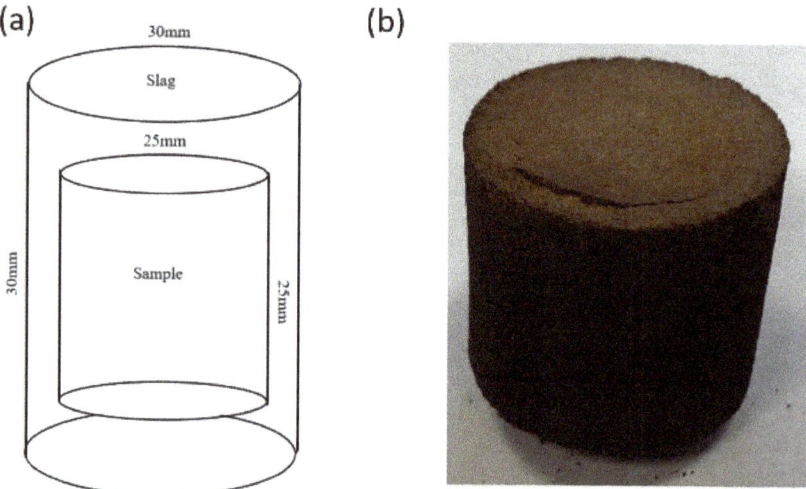

Figure 2. The samples for corrosion testing by the slag-coating method: (**a**) scheme of the corrosion system, (**b**) the commercial refractory sample covered with the shell of slag.

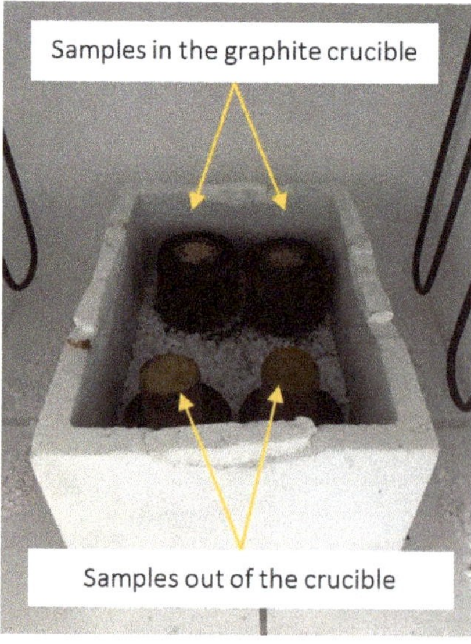

Figure 3. Samples prepared for corrosion testing, presented in two views: in graphite crucibles and out of graphite crucibles.

Table 3. Program of corrosion testing.

Sample Name	Slag Type	Temperature (°C)	Dwell Time (h)	Atmosphere
Al-1-1	R	1350	10	
Al-1-2	R	1450	10	Oxidizing
Al-2-1	S	1350	10	(air)
Al-2-2	S	1450	10	

2.4. Analysis of Material before and after Corrosion

The phase compositions of starting alumina-spinel refractory, slags, and after-corrosion materials were identified on powder samples using X-ray powder diffractometry (XRD). X-ray patterns were collected with an automated PanAnalytical X-Pert diffractometer (Almerow, The Netherlands) with Cu K_α radiation (λ = 1.5418 Å), in the 2θ range from 5° to 90°, in steps of 0.008°. The obtained diffractograms were analysed and matched with a database produced by the International Centre for Diffraction Data-ICDD PDF2, using X-Pert High Score Plus software v.3.0.5 produced by PANalytical B.V (Almelo, The Netherlands).

The microstructures of the refractory, before and after corrosion, were investigated by scanning electron microscopy NOVA NANO SEM 200 (FEI, Hillsboro, OR, USA), equipped with an EDS system (EDAX). The samples for microstructure analysis were prepared by a standard ceramographic technique; resin-embedded specimens were polished and coated with a carbon layer. For delivered alumina-spinel brick, both polished cross-section and fractured section were observed.

Hot-stage microscopy was employed to investigate in situ dimensional changes of slag R and S as a function of temperature. For this purpose, the homogenized powder slag was shaped into a cube of 3 mm dimensions. The measurement was conducted using a Carl Zeiss MH01 microscope (München, Germany) up to 1450 °C with a heating rate of 10 °C/min. The dimensions of the samples (height, width) were collected every 10 °C during the experiment and were used to prepare curves of in situ changes in cross-section dimensions of the test sample. The relative change in sample height (δ_h) was calculated according to Equation (1). The sintering temperature was established as a dimensional change corresponding to 2% shrinkage of the sample compared to its original dimensions at 25 °C. The melting and flow temperatures were referenced to 2/3 and 1/3 of initial sample dimensions, respectively.

$$\delta_h(T) = \frac{h_{(T)} - h_0}{h_0} \times 100 \tag{1}$$

h_0—the initial height of the sample;
$h_{(T)}$—the height of the sample at a specific temperature.

3. Results

3.1. Reference Slag

According to the XRD analysis of R slag (Figure 4), mayenite ($Ca_{12}Al_{14}O_{33}$, denoted as $C_{12}A_7$), was identified as the only phase in the commercial slag.

In this study various calcium aluminates formed. The diagram CaO-Al_2O_3 consists of five binary phases: C_3A, $C_{12}A_7$ (mayenite), CA, CA_2 (grossite), and CA_6 (hibonite), where C is CaO and A is Al_2O_3. Therefore, the stoichiometric aluminates can be characterized by different Ca/Al ratios, which are C_3A-1.5, $C_{12}A_7$-0.9, CA-0.5, CA_2-0.3, and CA_6-0.1. Figure 5 shows the SEM micrograph of R slag. In the SEM images, three types of micro-areas of different contrast can be distinguished. According to the EDS chemical analysis in points 1 and 2, Ca/Al has the same value of 0.6, despite different contrast. They represent the mayenite phase, $C_{12}A_7$, which was also identified as the main phase by XRD; however, this phase is non-stoichiometric (Ca/Al$_{theor,}$$C_{12}A_7$ = 0.9; point 1: 0.6, Table 4). The phase surrounding the mayenite is non-homogenous and contains a great number of small light inclusions. The atomic ratio in this area of Ca/Al equals 1.1, which indicates that it constitutes the non-stoichiometric phase from the system C_3A-$C_{12}A_7$, with increased content of SiO_2.

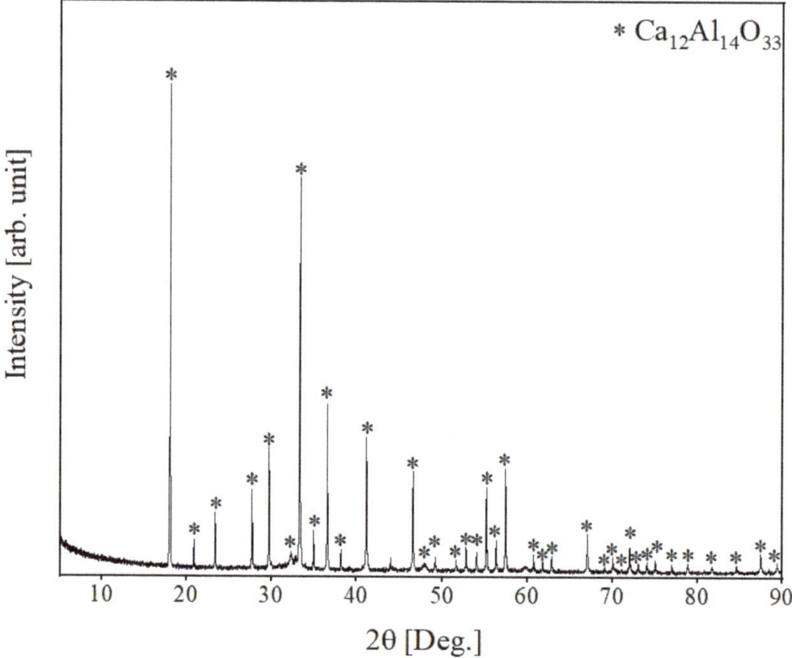

Figure 4. XRD pattern of R slag.

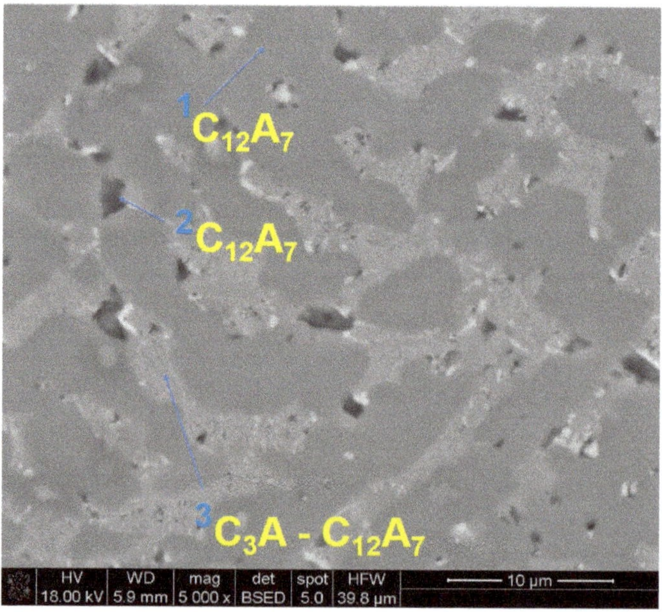

Figure 5. SEM micrograph with EDS in selected micro-areas for reference slag R.

Table 4. EDS analysis of micro-areas indicated in Figure 5.

Point	Chemical Composition by EDS, at.% *						Ratio Ca/Al **	Phase
	Ca	Al	Si	Mg	Ti	Fe		
1	21.6	38.4	0.3	0.1	-	-	0.6	$C_{12}A_7$
2	22.6	35.4	2.3	0.3	0.9	-	0.6	$C_{12}A_7$
3	25.1	22.9	7.2	0.8	2.8	1.5	1.1	C_3A-$C_{12}A_7$

* the rest to 100% is oxygen; ** theoretical Ca/Al ratio in aluminates: C_3A-1.5, $C_{12}A_7$-0.9, CA-0.5, CA_2-0.3, CA_6-0.1.

3.2. Hot-Stage Microscopy Analysis

Hot-stage microscopy is a widely implemented method in many research studies to identify the in situ high-temperature geometrical response of refractory material to slag, such as expansion or shrinkage during heating [14]. The heating microscopy permits the identification of sintering, melting, and flow temperatures to design the heating conditions of static corrosion testing. Figure 6a,b show the sintering (T_s), melting (T_M), and flow (T_F) temperatures of R and S slags during the hot-stage microscopy test.

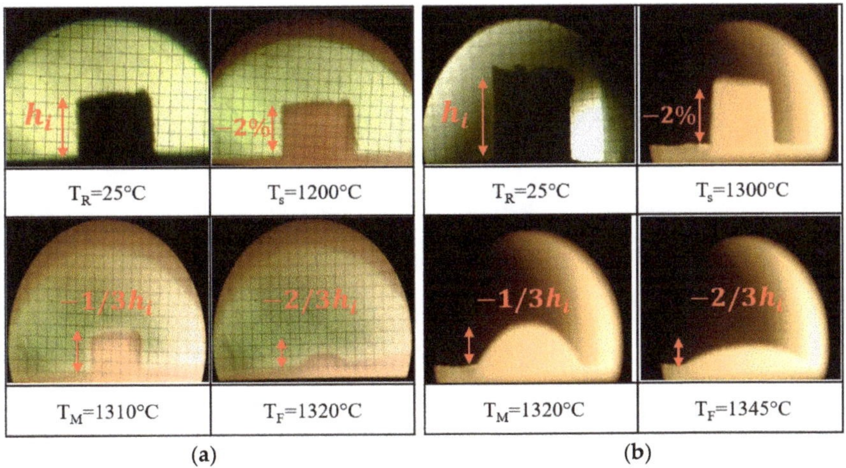

Figure 6. Hot-stage microscopy images of (**a**) R slag, and (**b**) S slag at selected temperatures (−2% h_i—sintering, −1/3 h_i—melting point, −2/3 h_i—flow point).

Based on the obtained images, the R slag flowed around 1320 °C, while S slag flowed at a temperature 25 °C higher than R slag. According to these results, it was determined that the selected temperatures for the corrosion testing must be at least 1320 °C and 1345 °C for R and S slag, respectively.

3.3. Alumina-Spinel Refractory

According to the XRD results shown in Figure 7, corundum (αAl_2O_3) and non-stoichiometric spinel ($Mg_{7.5}Al_{16}O_{32}$) were the main phases, along with beta-alumina ($NaAl_{13}O_{17}$) and silica (SiO_2) as minor phases in the alumina-spinel brick. According to Jing et al. [15], the difference between spinel and alumina-rich spinel in the XRD pattern is a slight shift in the reflex position between 44° and 45° of 2θ. In our work, the reflex characteristic for spinel was shifted towards higher 2θ angles and equaled 44.95°, which confirms the presence of alumina-rich spinel in the test refractory.

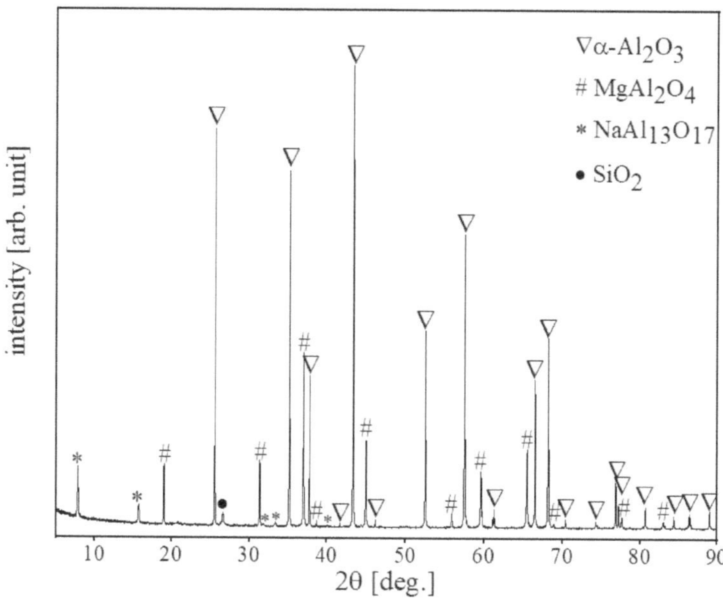

Figure 7. XRD pattern of alumina-spinel refractory.

SEM micrograph of alumina-spinel brick in fracture and cross-section views are presented in Figure 8a,b, respectively. EDS measurements conducted at polished cross-sections were employed to obtain elemental analysis on a definite grain. As indicated in Table 5, point 1 represents the area of the alumina grain, while the composition in point 2 represents the non-stoichiometric Al-rich spinel distributed within the matrix, characterized by Al/Mg of 3.3 (Al/Mg is 2 in MgAl$_2$O$_4$), which confirms the XRD results.

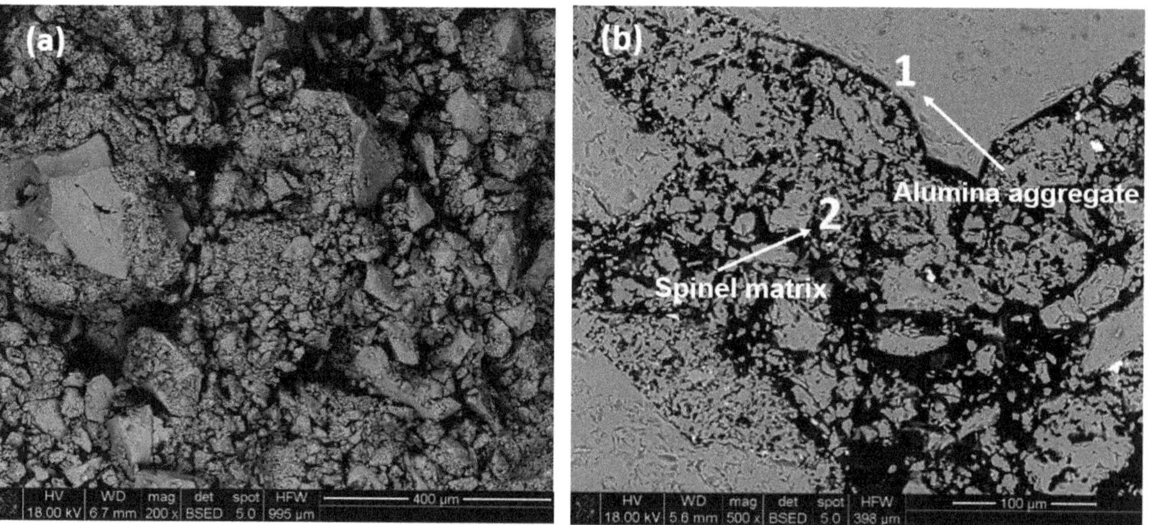

Figure 8. SEM micrographs with EDS analysis in selected areas of alumina-spinel brick: (**a**) fracture surface, (**b**) polished cross-section (point 1: alumina, and point 2: spinel).

Table 5. EDS analysis of micro-areas indicated in Figure 8.

Point	Element (at.%) *	
	Mg	Al
1	-	53
2	13.8	45.5

* the rest to 100% is oxygen.

3.4. Alumina-Spinel Refractory after Corrosion Test

Figure 9a–d show macro-photographs of all the samples after corrosion testing at 1350 °C and 1450 °C for 10 h. For corrosion at a lower temperature, both slags remained on the samples and no significant changes in the sample dimensions were registered. Even after cutting, the internal parts of the sample Al-1-1 (corroded by R slag) and Al-2-1 (corroded by S slag) remained unaltered, which shows minor infiltration of slag towards the refractory at 1350 °C.

Figure 9. The alumina-spinel refractory after corrosion by slag R (**a**) Al-1-1 (1350 °C) and (**b**) Al-2-1 (1450 °C), and slag S (**c**) Al-2-1 (1350 °C) and (**d**) Al-2-2 (1450 °C).

However, at an increased temperature of 1450 °C a low amount of slags remained on the external parts of both samples Al-1-2 and Al-2-2 due to intense infiltration of slag to refractory, and the sample expansion registered high at 25%.

3.4.1. X-ray Diffractometry

Due to the minor infiltration of slag into samples at 1350 °C, two different areas in the samples were selected for XRD examination, namely the edge (external) area and core (internal) area. The first sample was collected from the interaction zone where the slag was in direct contact with the surface of the refractory, while the second sample was gathered from the interior area, intact by slag as observed with naked-eye observation. Thanks to this separation, the difference in phases between the interaction zone and the un-reacted zone can be distinguished and investigated.

XRD patterns of corroded samples taken from the edge and core parts are presented in Figure 10a,b for R and S samples, respectively. For corrosion at 1350 °C by S slag, gehlenite ($Ca_2Al_2SiO_7$, C_2AS) was identified in the core and edge parts of sample Al-2-1, confirmed by the presence of a characteristic reflex at 22° of 2θ. This is attributed to the increased content of SiO_2 in the composition of slag S, which reacted with components of refractory, causing its corrosion. Based on the SiO_2-CaO-Al_2O_3 ternary phase diagram [16,17], gehlenite is the low melting phase (T_m = 1593 °C) located in the eutectic zone, and it can generate the liquid phase during the high-temperature corrosion process. Apart from C_2AS in Al-2-1, phase composition in the core parts of both samples was similar.

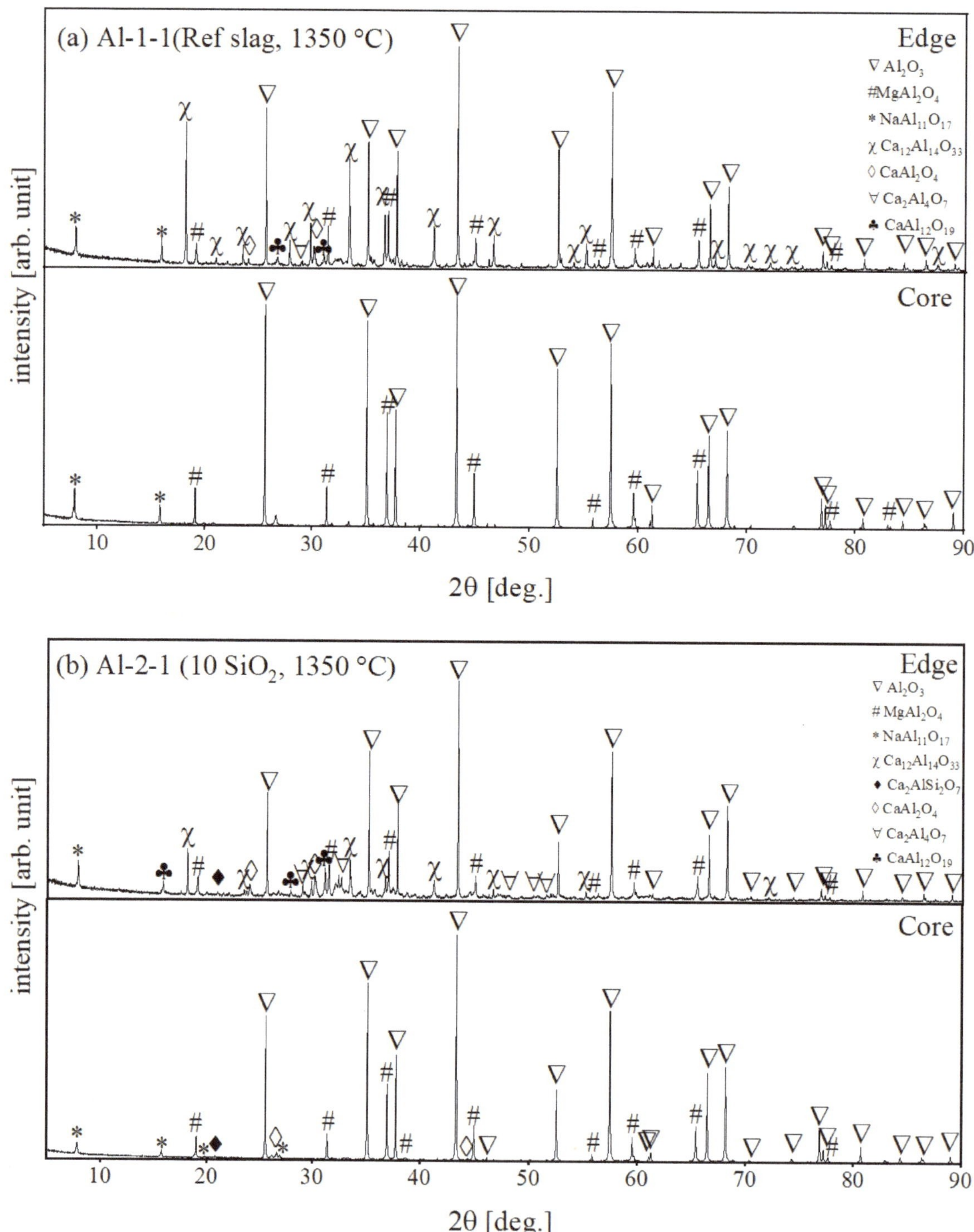

Figure 10. XRD patterns of (**a**) Al-1-1 (slag R) and (**b**) Al-2-1 samples (slag S), for 1350 °C.

In the edge parts of both samples, the original components of refractory (αAl_2O_3, spinel, βAl_2O_3) and slag ($C_{12}Al_{14}O_{33}$) were detected. Additionally, new aluminates appeared in low amounts in the edge parts of both samples; namely: $CaAl_2O_4$ and $CaAl_{12}O_{19}$. Mayenite from slag was not detected in the core parts of either sample.

The XRD patterns of Al-1-2 and Al-2-2 samples after corrosion at a higher temperature of 1450 °C are depicted in Figure 11a,b. After corrosion at 1450 °C, the core and edge parts of the sample were equally corroded by visual assessment; thus, only one averaged sample was subjected to XRD analysis. As observed from XRD patterns, there is an evident difference in phase composition between samples Al-1-2 (R slag) and Al-2-2 (S slag). Specifically, different types of new aluminate products appeared as a result of corrosion.

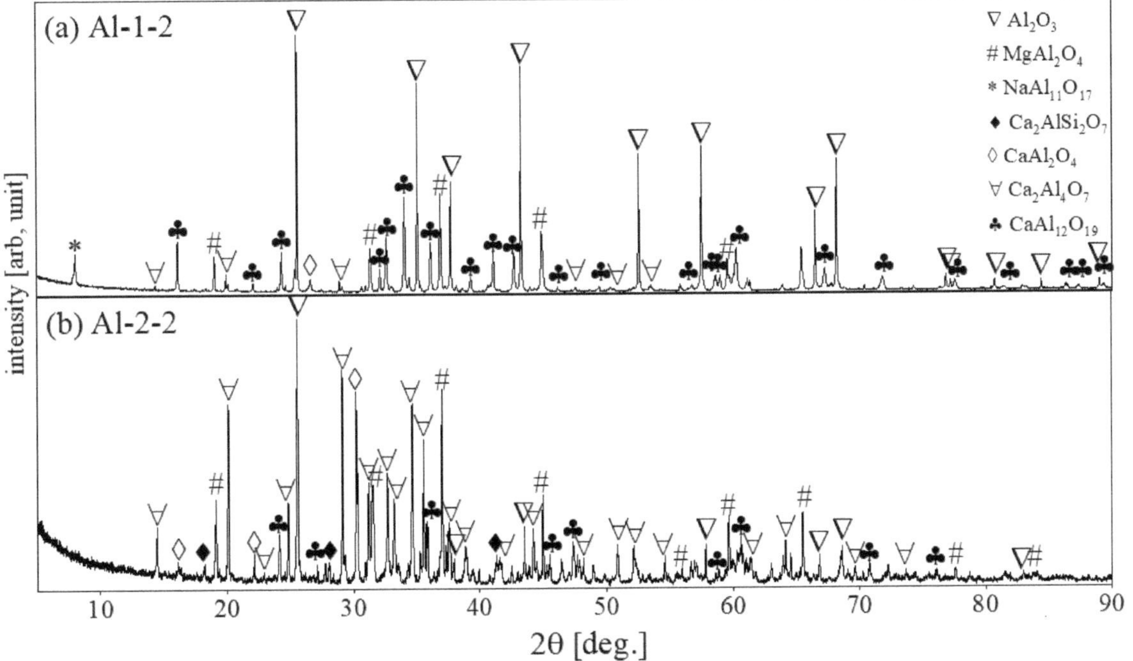

Figure 11. XRD patterns of (**a**) Al-1-2 (slag R) and (**b**) Al-2-2 samples (slag S), for 1450 °C.

In both samples, calcium monoaluminate (CA), calcium dialuminate (CA_2), and calcium hexaluminate (CA_6) were formed. However, peaks of gehlenite (C_2AS) were observed in the XRD pattern of only Al-2-2, which results from the increased silica content (10 wt.% of SiO_2) in slag S.

Another difference in the XRD patterns between samples Al-1-2 and Al-2-2 is the intensity of the peaks for CA, CA_2, and CA_6 phases. The XRD pattern of sample Al-1-2 (less SiO_2 content) contains higher peaks for CA and CA_6 phases, while the pattern of Al-2-2 is characterized by more peaks of CA_2. As can be seen by the predominant reflexes, a significant amount of CA_6 was formed in Al-1-2 (R), while in Al-2-2 the main corrosion product was CA_2. CA was identified in both samples in low amounts. From this, it can be seen that the greater formation of new aluminates at 1450 °C, compared to 1350 °C, is observed due to enhanced diffusion of Ca^{2+} towards components of Al_2O_3-$MgAl_2O_4$ refractory. The appearance of new phases was followed by decreasing the intensity of reflexes of the original refractory phases, due to their corrosion. No reflexes in after-corrosion were found for mayenite, the main phase of corrosive slag, confirming its total reaction.

In contrast to composition in the edge of samples A-1-1 and Al-2-1, in the XRD pattern of the samples, in Al-1-2 and Al-2-2 the peaks of calcium aluminate (CA) were considerably reduced while the peaks of calcium dialuminate (CA_2) and calcium hexaluminate (CA_6) raised up, which indicates progressing diffusion at 1450 °C of calcium ions (Ca^{2+}) to alumina (Al_2O_3), forming Al_2O_3-richer calcium aluminates. In addition, the formation of calcium hexaluminate (CA_6) reveals that the corrosion proceeds in the direction of the equilibrium conditions, as the excess of alumina is still present in the system. This behavior was correlated with decreasing numbers as well as the lower intensity of reflexes for αAl_2O_3, which is related to the dissolution of alumina grains in the molten slag during the corrosion process.

3.4.2. SEM/EDS Observations

Figure 12a–d show the SEM images of sample Al-1-1 corroded at 1350 °C. The phases indicated in the images were determined based on the chemical composition measured by EDS at a specific micro-area shown in Table 6. In this approach, we discovered secondary phases in the reaction zone produced due to corrosion of refractory by metallurgical slag.

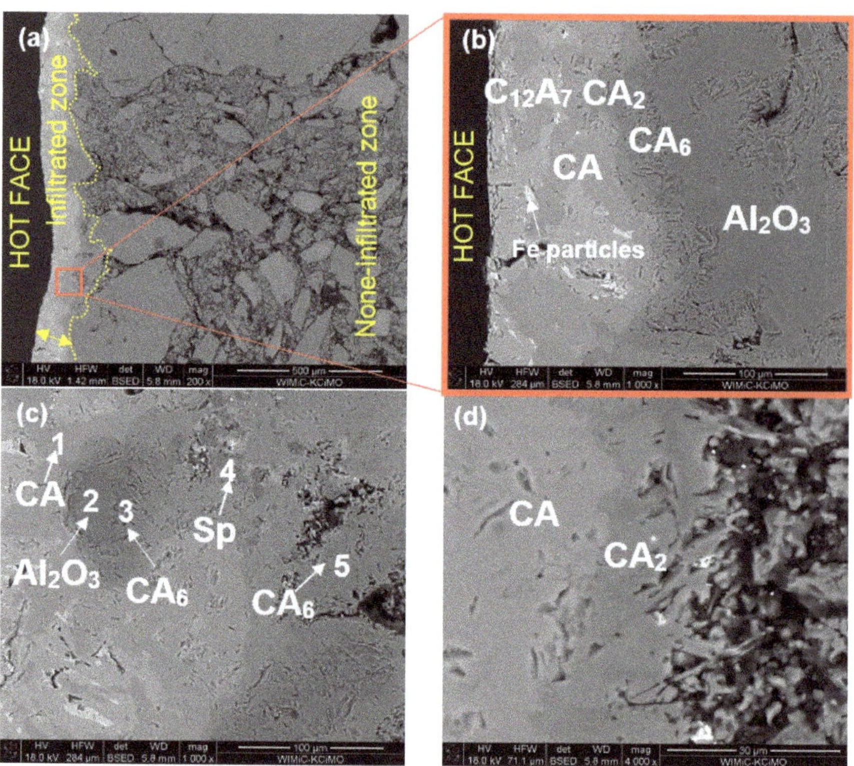

Figure 12. SEM images of sample Al-1-1 corroded by slag R (1350 °C, 10 h); (**a**,**b**): hot-face (slag/refractory interface); (**c**,**d**) higher magnification of interface zone.

Table 6. EDS chemical composition in selected micro-areas indicated in Figures 12–15.

Point	Elemental Composition (at.%) *					Assigned Phase
	Mg	Al	Si	Ca	Ti	
1	0.27	38.36	-	23.11	-	CA
2	-	52	-	-	-	Al_2O_3
3	1.05	50.04	-	4.38	-	CA_6
4	17.11	42.05	-	0.37	-	Sp
5	2.38	51.38	-	5.96	-	CA_6
6	16.27	41.65	-	0.29	-	Sp
7	0.93	43.99	45	13.17	-	CA_2
8	1.03	46.45	0.56	4.37	-	CA_6
9	1.43	15.24	5.95	13.30	3.03	C_2AS
10	0.70	34.48	0.44	14.40	-	CA
11	-	37.18	-	8.23	-	CA_2
12	-	41.83	-	2.99	-	CA_6
13	15.69	40.88	-	-	-	MA
14	-	41.18	-	13.87	-	CA_2
15	-	49.27	-	5.09	-	CA_6

* The rest to 100% is oxygen.

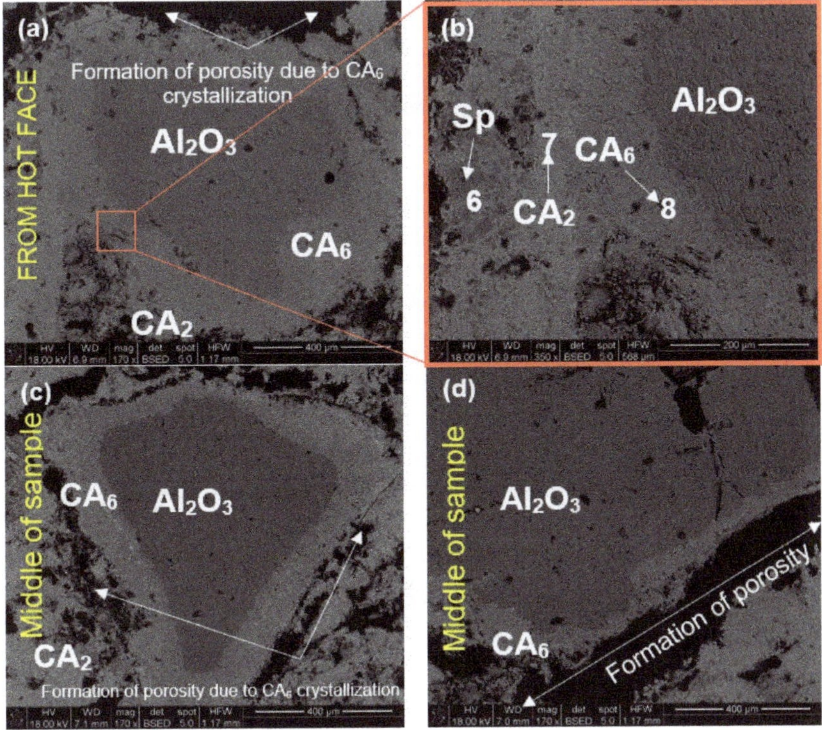

Figure 13. SEM images of sample Al-1-2 corroded by R slag (1450 °C, 10 h); (**a**) hot-face zone, (**b**) higher magnification of hot-face zone, (**c,d**) middle of the sample (core zone).

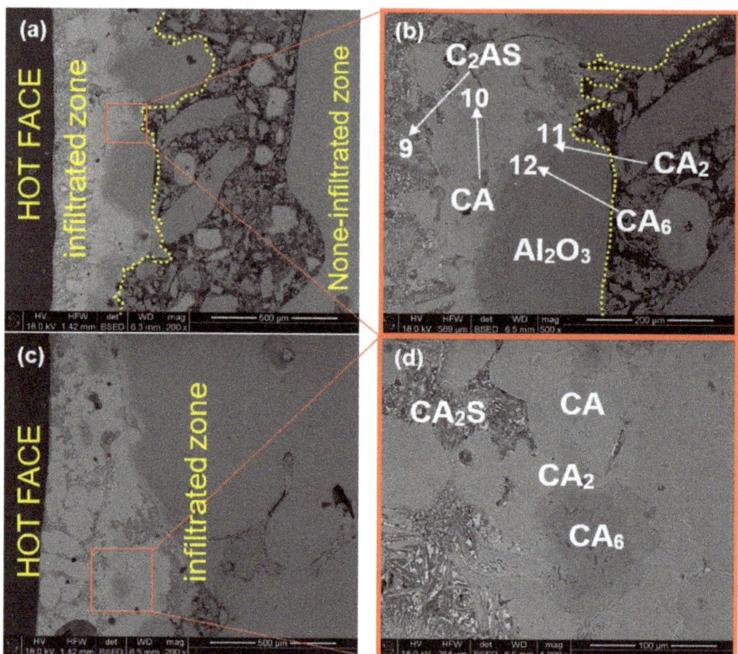

Figure 14. SEM images of sample Al-2-1 corroded by slag S (1350 °C, 10 h); (**a**,**c**) hot-face zone of the impregnated area, (**b**,**d**) high magnification of ho-face zone (slag/refractory interface).

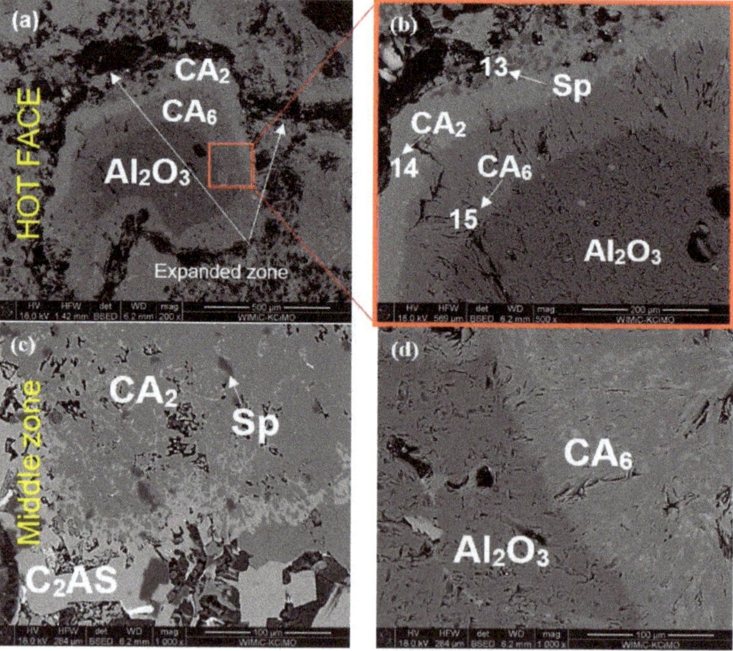

Figure 15. SEM images of sample Al-2-2 corroded by slag S (1450 °C, 10 h); (**a**) hot-face zone of the Al-2-2 sample, (**b**) higher-magnification of ho-face zone, (**c**) middle of the Al-2-2 sample, (**d**) higher magnification of middle zone.

As can be observed from the image of sample Al-1-1 corroded by R slag at 1350 °C (Figure 12a–d), slag infiltration affected the matrix compactness. The contact layer—mostly infiltrated by the slag—constituted about 150 μm. The magnified image (Figure 12b) clearly shows that this layer is composed of various types of calcium aluminates, located in the following direction from the contact zone: $C_{12}A_7$ (original slag component), CA, CA_2, CA_6. This confirms the progressive reaction of calcium aluminate from slag with alumina from refractory (phases in phase diagram Al_2O_3-CaO in the order of appearance: A-C_3A-$C_{12}A_7$-CA-CA_2-CA_6-C). In the depth above 150 μm from the contact zone, only Al_2O_3 was observed (Figure 12b). CA_6 crystallized in the form of flake crystals (Figure 12b,c). Such platelet morphology of CA_6 has previously been observed by numerous researchers [12,13,18,19]. The number of new phases was not significant in the volume of material.

The same sample but corroded at a higher temperature of 1450 °C (Al-1-2) was much more porous when compared to samples corroded at 1350 °C, as can be seen in Figure 13. This porosity is caused by the massive formation of platelet CA_6 crystals, which formed especially around Al_2O_3 grains, resulting in their separation from other components of the microstructure (Figure 13a,c,d). Spinel grains were dispersed among CA_2 polycrystals (Figure 13b). The progressive diffusion of Ca^{2+} from slag toward refractory, and Al^{3+} from refractory toward slag, resulted in the formation of calcium aluminates with increased Al content, which are evident from the generated layered microstructure zone $CA_2/CA_6/Al_2O_3$ in Figure 13b.

Figure 14a–d illustrate the SEM images of the Al-2-1 sample corroded by slag S at 1350 °C for 10 h. In this case, the infiltrated zone was much broader, from about 270 to over 700 μm. Gehlenite, C_2AS, was additionally detected in the after-corrosion refractory, which occurred prior to calcium aluminate layers CA, CA_2, CA_6 (Figure 14b,d). C_2AS, not observed for R slag, formed from the side of liquid slag S. This confirms the XRD results, which showed a small amount of gehlenite (Figure 10b). The calcium monoaluminate, CA, was observed in a gap between gehlenite, C_2AS, and calcium dialuminate, CA_2, in Figure 14b. Calcium hexaluminate, CA_6, and calcium dialuminate, CA_2, surrounded tightly the alumina aggregate as a result of intensive Ca^{2+} diffusion towards refractory grains. All the results were confirmed by the EDS analysis shown in Table 6.

After corrosion testing at 1450 °C (Al-2-2, Figure 15a–d), more progressive corrosion is clearly visible in the SEM images. A greater rate of diffusion is confirmed by the clearly distinguishable rings of secondary aluminates around alumina grains (Figure 15b). Intensive crystallization of platelet morphology CA_6 left great amounts of porosity. C_2AS was still present at 1450 °C (Figure 15c).

4. Discussion

4.1. Slag Viscosity

For evaluating the slag viscosity, different methods and models have been used in the literature. One of the most common models is the slag basicity, defined as a weight ratio of basic to acid oxides, according to Equation (2) [8].

$$\%B = \frac{\text{wt.\%CaO} + \text{wt.\%MgO} + \text{wt.\%MnO} + \text{wt.\%FeO (Basic oxides)}}{\text{wt.\%SiO}_2 + \text{wt.\%Al}_2O_3 + \text{wt.\%Fe}_2O_3 + \text{wt.\%P}_2O_5 \text{ (Acid oxides)}} \tag{2}$$

The XRF results presented in Table 1 indicate that the basicity of R and S slags are 0.99 and 0.88, respectively. Hence, slag S should possess higher viscosity and lower infiltration tendency compared to slag R [8]. Therefore, the increased viscosity of silica-enriched slag S was the reason for its higher flow point when compared to R slag in high-temperature microscopy testing (S-1345 °C and R-1320 °C).

According to the theory of Herasymenko [20], the slag structure can be assumed to be an ionic structure in the liquid state with cations including atoms of Si, P, Fe, Ca, Fe, Mn, Mg, Na, and non-metallic anions of O, S, and F. The small ions, with high charge, can attract oxygen to form complex structures, e.g., SiO_4, PO_4, AlO_3, FeO_2, and Fe_2O_5 in stable tetrahedral arrangements. These acid structural groups share the oxygen (so-called oxygen bridge) and are considered network formers; thus, they are responsible for the increased slag viscosity. Conversely, basic oxides such as CaO, MgO, FeO, and MnO release O^{2-} to the slag, thus breaking the network.

SiO_2 is a common component of numerous metallurgical slags. The structure of the liquid slag can be explained with respect to the silica addition. SiO_2 is an acid oxide and network former in the structure of slag. Furthermore, each Si^{4+} is surrounded tetrahedrally by O^{2-} ions, and each O^{2-} ion is connected to the other two oxygen ions. Because of the possibility of bonding in four tetrahedral directions, the slag structure exists as a three-dimensional polymerized network. Based on that, the viscosity of S slag is considered higher than R slag. Figure 16 shows the Si-O tetrahedral network in crystalline and molten slag.

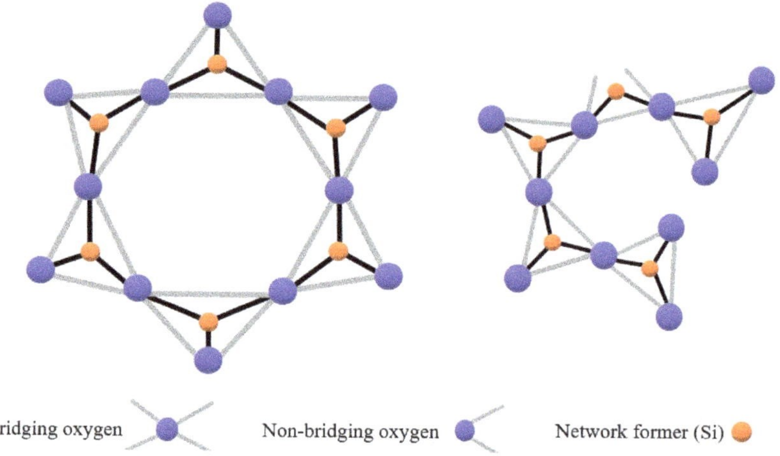

Figure 16. Network of Si-O tetrahedral in (**a**) crystalline and (**b**) molten slag [21].

The addition of basic oxides such as CaO (network breakers) causes revealing of the free oxygen O^{2-} and Ca^{2+} cations to a slag, which breaks Si-O bonds. Non-bridging oxygens de-polymerize the liquid slag and, consequently, make a charge balance for the CaO-SiO_2 system. If the slag becomes totally de-polymerized, all tetrahedral connections break down. Then, the slag is composed of free SiO_4^{4-}, O^{2-}, and Ca^{2+}. Figure 17 illustrates the structure of CaO-SiO_2-based slag, with tetrahedral connections broken in the system.

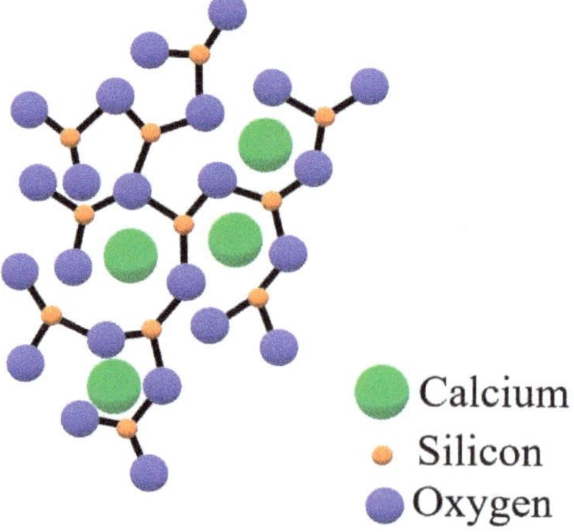

Figure 17. Structure of CaO-SiO_2 liquid slag [21].

4.2. Slag-Refractory Interactions

Based on the SEM images of the corroded samples Al-1-1 and Al-1-2 (Figures 12 and 13), calcium aluminate layers are distributed around the tabular alumina grains, which indicates the indirect dissolution mechanism of alumina-spinel refractory by metallurgical slag. The indirect dissolution of refractories has been studied by numerous authors [18,22–24]. According to the ternary phase diagram for CaO-Al_2O_3-SiO_2 (Figure 18) and XRD results, alumina aggregates in this work were indirectly dissolved forming $CaAl_2O_4$ (CA), $CaAl_4O_7$ (CA_2), and $CaAl_{12}O_{19}$ (CA_6). CA_6/CA_2 interfaces were clearly identified at the edges of tabular alumina grains by SEM/EDS. Previous studies on the corrosion of high-alumina spinel refractories [15,22] indicated the formation of dense layers around the coarse alumina grains. These layers consisted of CA, CA_2, and CA_6, and were suggested to impede corrosion during the wear process.

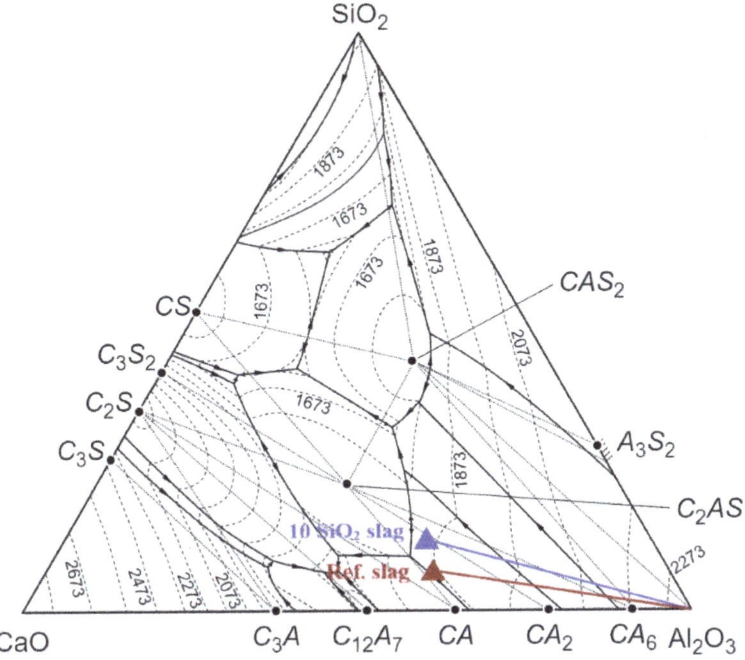

Figure 18. Ternary phase diagram of CaO-Al_2O_3-SiO_2 [25]. The starting compositions of the slag R (Ref. slag) and slag S ($10SiO_2$) are indicated in the diagram.

According to Khajornboon et al. [17], CA (sintered at 800–1000 °C) and CA_2 (sintered at 1000–1300 °C) phases are incompatible at 1450 °C, and from a thermodynamic point of view, they react with Al_2O_3 aggregate present in the microstructure forming CA_6. Moreover, due to phenomena occurring at interface CA_6/CA_2—diffusion of Ca^{2+} from CA_6 and the dissolution of Al_2O_3—CA_6 thickness grows. In addition, because of the diffusion of Ca^{2+} from slag to the interface slag/CA_2, the CA_2 layer grows. This phenomenon of indirect corrosion of alumina-spinel refractory is presented by model in Figure 19.

The results obtained in this work show that CA_6 previously formed around Al_2O_3 grains acts as an obstacle for further reaction. As can be observed in the SEM micrographs, the alumina aggregates are progressively attached by the calcium hexaluminate layer, and because of CA_6 formation, the erosion of alumina grains reduces, thus improving their corrosion resistance.

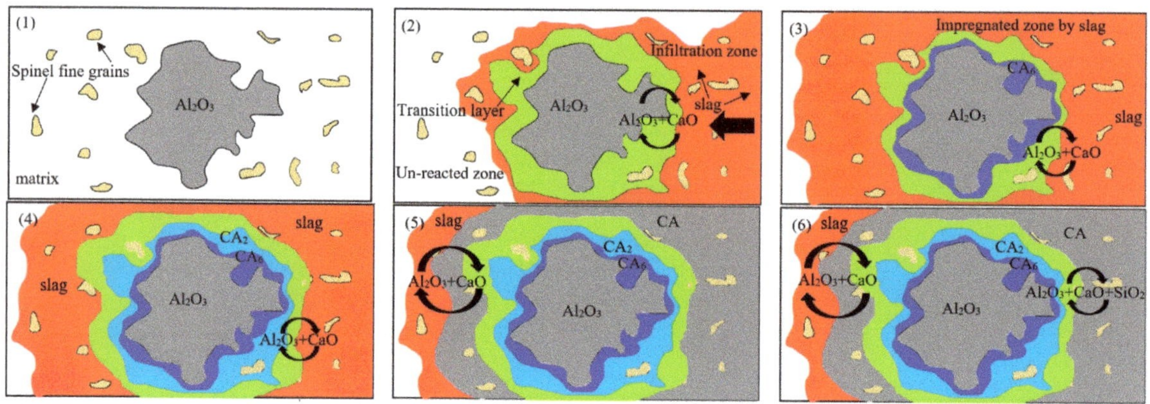

Figure 19. Model of corrosion mechanism of alumina-spinel refractory by metallurgical slag (green—transition layer in each corrosion stage, red—slag); (1) un-infiltrated Al-Sp microstructure, (2) first contact between CaO-Al$_2$O$_3$-SiO$_2$ (CAS) slag and Al-Sp refractory, (3) formation of CA$_6$ layer around alumina aggregates, (4) formation of CA$_6$, CA$_2$ layers (5) formation of CA$_6$, CA$_2$, and CA layers according to precipitation-dissolution mechanism.

According to the results shown in this study for Al$_2$O$_3$-MgAl$_2$O$_4$ refractory, and confirmed by the literature [8,23,24], CA$_6$ and CA$_2$ layer formation can prevent further corrosion of refractory aggregate. Nevertheless, it should be noted that there is about a 3.1% difference in density between CA$_6$ (3.78 g/cm^3) and alumina (3.95 g/cm^3), which contributes to the volumetric expansion during corrosion [6,24,26]. Consequently, CA$_6$ can cause the initiation of cracks, which intensifies the slag infiltration for longer corrosion periods. Therefore, despite the physical properties of alumina-spinel brick (porosity and density), a higher quantity of CA$_6$—formed during corrosion—can facilitate the slag penetration followed by deterioration of the functional properties of the refractory [16].

According to De Bilbao et al. [22,23], CA$_2$ phase is not a stable phase at elevated temperatures. Based on the dissolution/precipitation mechanism, the melted slag forms CA$_6$ and CA$_2$ phases in the initial stage of corrosion. In the first stage of corrosion, solid/solid reactions proceed, and since the slag contains Ca^{2+} and O^{2-} ions, which diffuse and react at interfaces Al$_2$O$_3$/CA$_6$ and CA$_6$/CA$_2$, the alumina layer is gradually dissolved, as shown in Figure 19. Moreover, the external CA$_2$ layer dissolves in the slag according to Equation (3).

$$CA_2 \rightarrow C_{(l)} + 2A_{(l)} \tag{3}$$

The liquid alumina Al$_{(l)}$, formed according to Equation (3), is therefore indirectly dissolved and revealed to slag. Then, Ca^{2+} cations are transported from the CA$_2$/slag interface toward the alumina grains and enter the CA$_2$/CA$_6$ interface, which is followed by reactions according to Equations (4) and (5):

$$3CA_2 \rightarrow 2C + CA_6 \tag{4}$$

$$CA_6 + 2C \rightarrow 3CA_2 \tag{5}$$

Reaction proceeding according to Equation (4) shifts the interface toward refractory alumina grain, while reaction proceeding according to Equation (5) follows in the opposite direction. Finally, the alumina aggregate shrinks, which facilitates cation diffusion through the CA$_6$ layer and enhances the reaction at the internal interface Al$_2$O$_3$/CA$_6$, as shown by Equation (6).

$$6A + C \rightarrow CA_6 \tag{6}$$

The crystal structures of CA_2 and CA_6, forming around alumina aggregates, are monoclinic and hexagonal, respectively. Therefore, the highest density of oxygen packing belongs to CA_6 and thanks to the crystallized close-packed system its cation diffusivity is lower compared to CA_2. Consequently, the content of cations that can leave the CA_2/CA_6 interface is lower than the cations that arrive in the CA_2/CA_6 interface. Therefore, the CA_2/CA_6 interface shifts toward the alumina aggregate, which is related to the increasing thickness of the CA_2 mono-layer. Then, the CA_2 mono-layer dissolves in the slag and its growth is limited as well.

The corrosion mechanism is shown in Figure 19. The green layer indicates the transitional layer in each step of corrosion. According to image (1), before the slag impregnation the alumina coarse grain (grey) is attached by the fine alumina-spinel grains (yellow) forming the matrix. At the initial corrosion stage (2), when the slag (red) infiltration begins, Al_2O_3 (from refractory) reacts with CaO (from slag) and, as a result, a transition layer (green) starts to form around the alumina grain. Subsequently, the impregnated zone extends and the calcium hexaluminate layer (CA_6, purple) forms around alumina aggregate as shown by (3). By prolonging the time of corrosion, calcium dialuminate (CA_2, blue) and calcium monoaluminate (CA, light grey) layers form around the A aggregate and, through this, the alumina aggregate is preserved from further corrosion. Finally, via extending the transitional zone, the chemical reactions continue until reaching thermodynamic equilibrium conditions.

The Rietveld method was employed to determine the quantity of each calcium aluminate phase in all samples (Al-1-1, Al-1-2 and Al-2-1, Al-2-2); the results are shown in Table 7. The amounts of gehlenite, C_2AS, in samples corroded by slag S (Al-1-2 and Al-2-2) were at a comparable levels of 11.5% and 10.6% for 1350 °C and 1450 °C, respectively. In contrast, about two times lower amounts of gehlenite formed in refractory corroded by slag R, which results from two times lower SiO_2 content in the original slag (5% SiO_2).

Table 7. Rietveld analysis of the test slag.

Slag	1350 °C			1450 °C		
	CA_6	CA_2	C_2AS	CA_6	CA_2	C_2AS
R slag	23.6	40.6	6.2	36.5	40.9	6.1
S slag	24.6	38.5	11.5	49.5	25.3	10.6

Gehlenite was formed as a corrosion product of a reaction between SiO_2 and CaO from slag with Al_2O_3 from refractory. C_2AS is a phase located in the ternary phase diagram SiO_2-CaO-Al_2O_3 (Figure 18), which melts at 1593 °C, thus increasing the content of the liquid in the matrix of material operating at high temperatures in steel devices.

The raised temperature of heat treatment causes the greater formation of CA_6, for both R slag and S slag. For 1350 °C—regardless of the type of slag—the corroded refractory contained similar content of CA_6 of about 24%. However, for corrosion at 1450 °C, the amount of CA_6 significantly increased (to 37% for R and to 50% for S) due to a greater rate of diffusion at higher temperatures. This will influence the compactness of the refractory during its operation.

5. Conclusions

This study showed the corrosion mechanism of alumina-spinel refractory by secondary metallurgical slag (B = 0.99) and slag with 10 wt.% addition of SiO_2 content (B = 0.88). For this purpose, an alumina-spinel disc sample was coated with a shell of slag to directly contact the external faces of the refractory sample. Subsequently, the coated samples were subjected to corrosion at 1350 and 1450 °C in an oxidizing atmosphere. The after-corrosion materials were analyzed in view of the alterations in their structure and microstructure. The following conclusions can be drawn from the results:

- The corrosion mechanism of alumina-spinel material tested against acid metallurgical slags is based on the indirect dissolution of alumina grains constituting the principal component of the refractory.
- The indirect dissolution of alumina covers the formation of new calcium aluminate layers around alumina grains, including CA_2 and CA_6, for both corrosive slags and heating at both 1350 °C and 1450 °C. By increasing the temperature to 1450 °C, the infiltration of refractory by slag increases; specifically, Ca^{2+} diffuses more intensively towards refractory followed by the formation of CA_2 and CA_6.
- CA_6 forms in platelet morphology and by expansion causes loosening of the microstructure, especially at increased temperatures when its amount is doubled.
- The low melting phase of gehlenite, $Ca_2Al_2SiO_7$, forms in phase composition of alumina-spinel samples corroded by both slags, and its amount is doubled for silica enriched S slag containing 10% SiO_2.

Author Contributions: Writing—original draft, S.D.; methodology and investigations, C.R., M.L.; supervision, formal analysis, R.P.; visualization and conceptualization, I.J.; and writing-reviewing and editing J.S. Corrosion of alumina-spinel refractory by secondary metallurgical slag using coating corrosion test. All authors have read and agreed to the published version of the manuscript.

Funding: The research was funded by statutory funds of the Faculty of Material Science and Ceramics, AGH University of Science and Technology, Krakow, Poland, Agreement no. 16.16.160.557, and the Marie Skłodowska-Curie Actions Innovative Training Networks in the frame of the project ATHOR—Advanced THermomechanical multiscale mOdelling of Refractory linings 764987 Grant.

Institutional Review Board Statement: Not applicable.

Informed Consent Statement: Not applicable.

Data Availability Statement: Not applicable.

Conflicts of Interest: The funders had no role in the design of the study; in the collection, analyses, or interpretation of data; in the writing of the manuscript, or in the decision to publish the results.

References

1. Kakroudi, M.G.; Vafa, N.P.; Asl, M.S.; Shokouhimehr, M. Effects of SiC content on thermal shock behavior and elastic modulus of cordierite—Mullite composites. *Ceram. Int.* **2020**, *46*, 23780–23784. [CrossRef]
2. Darban, S.; Kakroudi, M.G.; Vandchali, M.B.; Vafa, N.P.; Rezaei, F.; Charkhesht, V. Characterization of Ni-doped pyrolyzed phenolic resin and its addition to the Al_2O_3–C refractories. *Ceram. Int.* **2020**, *46*, 20954–20962. [CrossRef]
3. Schacht, C. *Refractories Handbook*; CRC Press: Boca Raton, FL, USA, 2004.
4. Ganesh, I.; Bhattacharjee, S.; Saha, B.P.; Johnson, R.; Rajeshwari, K.; Sengupta, R.; Rao, M.R.; Mahajan, Y.R. An efficient $MgAl_2O_4$ spinel additive for improved slag erosion and penetration resistance of high-Al_2O_3 and MgO–C refractories. *Ceram. Int.* **2002**, *28*, 245–253. [CrossRef]
5. Wang, Y.; Li, X.; Chen, P.; Zhu, B. Matrix microstructure optimization of alumina-spinel castables and its effect on high temperature properties. *Ceram. Int.* **2018**, *44*, 857–868. [CrossRef]
6. Braulio, M.A.L.; Rigaud, M.; Buhr, A.; Parr, C.; Pandolfelli, V.C. Spinel-containing alumina-based refractory castables. *Ceram. Int.* **2011**, *37*, 1705–1724. [CrossRef]
7. Reynaert, C.; Śnieżek, E.; Szczerba, J. Corrosion Tests for Refractory Materials Intended for the Steel Industry–A Review. *Ceram.–Silikáty* **2020**, *64*, 278–288. [CrossRef]
8. Poirier, J.; Smith, J.D.; Jung, I.H.; Kang, Y.B.; Eustathopoulos, N.; Blond, É.; Rigaud, M. *Corrosion of Refractories: The Fundamentals*; Göller Verlag: Baden-Baden, Germany, 2017.
9. Ludwig, M.; Wiśniewska, K.; Śnieżek, E.; Jastrzębska, I.; Prorok, R.; Szczerba, J. Effect of the chemical composition of slag on the corrosion of calcium zirconate material. *Mater. Chem. Phys.* **2021**, *258*, 123844. [CrossRef]
10. Lee, W.E.; Zhang, S. Melt corrosion of oxide and oxide-carbon refractories. *Int. Mater. Rev.* **1999**, *44*, 77–104. [CrossRef]
11. Song, J.; Liu, Y.; Lv, X.; You, Z. Corrosion Behavior of Al_2O_3 Substrate by SiO_2–MgO–FeO–CaO–Al_2O_3 Slag. *J. Mater. Res. Technol.* **2020**, *9*, 314–321. [CrossRef]
12. Braulio, M.A.L.; Martinez, A.T.; Luz, A.P.; Liebske, C.; Pandolfelli, V.C. Basic slag attack of spinel-containing refractory castables. *Ceram. Int.* **2011**, *37*, 1935–1945. [CrossRef]
13. Martinez, A.T.; Luz, A.P.; Braulio, M.A.L.; Pandolfelli, V.C. CA6 impact on the corrosion behavior of cement-bonded spinel-containing refractory castables: An analysis based on thermodynamic simulations. *Ceram. Int.* **2015**, *41*, 4714–4725. [CrossRef]

14. Madej, D.; Szczerba, J. Detailed studies on microstructural evolution during the high temperature corrosion of SiC-containing andalusite refractories in the cement kiln preheater. *Ceram. Int.* **2017**, *43*, 1988–1996. [CrossRef]
15. Jing, S.-Y.; Lin, L.-B.; Huang, N.-K.; Zhang, J.; Lu, Y. Investigation on lattice constants of Mg-Al spinels. *J. Mater. Sci. Lett.* **2000**, *19*, 225–227. [CrossRef]
16. Poirier, J.; Bouchetou, M.L. Corrosion of refractories, measurements and thermodynamic modeling. In Proceedings of the 10th International Conference of European Ceramic Society, Berlin, Germany, 17–21 June 2007; pp. 7–21.
17. Khajornboon, J.; Ota, K.; Washijima, K.; Shiono, T. Control of hexagonal plate-like microstructure of in-situ calcium hexaluminate in monolithic refractories. *J. Asian Ceram. Soc.* **2018**, *6*, 196–204. [CrossRef]
18. Sarpoolaky, H.; Zhang, S.; Argent, B.B.; Lee, W.E. Influence of grain phase on slag corrosion of low-cement castable refractories. *J. Am. Ceram. Soc.* **2001**, *84*, 426–434. [CrossRef]
19. Luz, A.P.; Martinez, A.T.; Braulio, M.A.L.; Pandolfelli, V.C. Thermodynamic evaluation of spinel containing refractory castables corrosion by secondary metallurgy slag. *Ceram. Int.* **2011**, *37*, 1191–1201. [CrossRef]
20. Herasymenko, P. Electrochemical theory of slag-metal equilibria. Part I.—Reactions of manganese and silicon in acid open-heart furnace. *Trans. Faraday Soc.* **1938**, *34*, 1245–1254. [CrossRef]
21. Çelikbilek, M.; Ersundu, A.E.; Aydın, S. Crystallization kinetics of amorphous materials. *Adv. Cryst. Process.* **2012**, *35347*, 127–162. Available online: https://books.google.co.jp/books?hl=zh-TW&lr=&id=4NqgDwAAQBAJ&oi=fnd&pg=PA127&dq=Crystallization+kinetics+of+amorphous+materials&ots=EiOAFn5HDL&sig=cXwDHgQxj1PfmhsSLn4LHR-1uGE&redir_esc=y#v=onepage&q=Crystallization%20kinetics%20of%20amorphous%20materials&f=false (accessed on 4 May 2022).
22. De Bilbao, E.; Dombrowski, M.; Pilliere, H.; Poirier, J. Time-resolved high-temperature X-ray diffraction for studying the kinetics of corrosion of high-alumina refractory by molten oxides. *Corros. Sci.* **2018**, *139*, 346–354. [CrossRef]
23. De Bilbao, E.; Dombrowski, M.; Traon, N.; Tonnesen, T.; Poirier, J.; Blond, E. Study of Reactive Impregnation and Phase Transformations During the Corrosion of High Alumina Refractories by Al_2O_3-CaO Slag. In *Advances in Science and Technology*; Trans Tech Publications Ltd.: Bäch, Switzerland, 2014; Volume 92, pp. 264–271.
24. Sako, E.Y.; Braulio, M.A.L.; Milanez, D.H.; Brant, P.O.; Pandolfelli, V.C. Microsilica role in the CA6 formation in cement-bonded spinel refractory castables. *J. Mater. Process. Technol.* **2009**, *209*, 5552–5557. [CrossRef]
25. Hasegawa, M. Thermodynamic Basis for Phase Diagrams. In *Treatise on Process Metallurgy*; Elsevier: Amsterdam, The Netherlands, 2014; pp. 527–556.
26. Zhang, S.; Rezaie, H.R.; Sarpoolaky, H.; Lee, W.E. Alumina dissolution into silicate slag. *J. Am. Ceram. Soc.* **2000**, *83*, 897–903. [CrossRef]

materials

Article

Corrosion Resistance of MgO and Cr_2O_3-Based Refractory Raw Materials to PbO-Rich Cu Slag Determined by Hot-Stage Microscopy and Pellet Corrosion Test

Maciej Ludwig [1,2,*], Edyta Śnieżek [1], Ilona Jastrzębska [1], Ryszard Prorok [1], Yawei Li [3,4], Ning Liao [3,4], Mithun Nath [3,4], Jozef Vlček [5] and Jacek Szczerba [1]

[1] Faculty of Materials Science and Ceramics, AGH University of Science and Technology, Al. Mickiewicza 30, 30-059 Kraków, Poland; e.sniezek@wp.pl (E.Ś.); ijastrz@agh.edu.pl (I.J.); rprorok@agh.edu.pl (R.P.); jszczerb@agh.edu.pl (J.S.)
[2] Forglass Sp. z.o.o., Wadowicka 8a, 30-415 Kraków, Poland
[3] The State Key Laboratory of Refractories and Metallurgy, Wuhan University of Science and Technology, Wuhan 430081, China; liyawei@wust.edu.cn (Y.L.); liaoning@wust.edu.cn (N.L.); mithunnath@wust.edu.cn (M.N.)
[4] National-Provincial Joint Engineering Research Center of High Temperature Materials and Lining Technology, Wuhan 430081, China
[5] Department of Thermal Engineering, Faculty of Materials Science and Technology VŠB, Technical University of Ostrava, 708 00 Ostrava, Czech Republic; jozef.vlcek@vsb.cz
* Correspondence: ludwig@agh.edu.pl

Citation: Ludwig, M.; Śnieżek, E.; Jastrzębska, I.; Prorok, R.; Li, Y.; Liao, N.; Nath, M.; Vlček, J.; Szczerba, J. Corrosion Resistance of MgO and Cr_2O_3-Based Refractory Raw Materials to PbO-Rich Cu Slag Determined by Hot-Stage Microscopy and Pellet Corrosion Test. *Materials* **2022**, *15*, 725. https://doi.org/10.3390/ma15030725

Academic Editor: Jose M. Bastidas

Received: 14 December 2021
Accepted: 10 January 2022
Published: 18 January 2022

Abstract: Chemical resistance of commercial refractory raw materials against Cu slag is critical to consider them as candidates for the production of refractories used in Cu metallurgy. In this study, we show the comparative results for the corrosion resistance of four commercial refractory raw materials—magnesia chromite co-clinkers FMC 45 and FMC 57, PAK, and fused spinel SP AM 70—against aggressive, low-melting PbO-rich Cu slag (Z1) determined by hot-stage microscopy (up to 1450 °C) and pellet test (1100 and 1400 °C). Samples were characterized after the pellet test by XRD, SEM/EDS, and examination of their physicochemical properties to explore the corrosion reactions and then assess comparatively their chemical resistance. Since many works have focused on corrosion resistance of refractory products, the individual refractory raw materials have not been investigated so far. In this work, we show that magnesia chromite co-clinker FMC 45 exhibits the most beneficial properties considering its application in the production of refractories for the Cu industry. Forsterite (Mg_2SiO_4) and güggenite (Cu_2MgO_3) solid solutions constitute corrosion products in FMC 45, and its mixture with slag shows moderate dimensional stability at high temperatures. On the other hand, the fused spinel SP AM 70 is the least resistant to PbO-rich Cu slag (Z1); it starts to sinter at 970 °C, followed by a fast 8%-shrinkage caused by the formation of güggenite solid solution in significant amounts.

Keywords: corrosion; MgO; Cr_2O_3; refractory; raw materials; Cu; slag; XRD; SEM

1. Introduction

The lifetime of refractory lining is one of the most critical factors during high-temperature processes. The repairs of the lining are both time-consuming and expensive, so high-quality refractories are desired and consistently developed. Refractory materials used in the Cu industry are not exposed to extremely high temperatures. However, they have to resist major thermal shocks. Moreover, they are subjected to chemical interactions with the surrounding environment, mechanical wear due to the movement of the stove charge, induced mechanical stresses in the linings, and hot erosion [1]. Refractory linings should withstand the influence of Cu, Cu matte, Cu slag, and SO_2. Melted Cu does not wet the ceramics; thus, it slightly affects the lifetime of the refractory. However, Cu slag is one of the most aggressive corrosive mediums towards refractories [2]. All factors—mechanical, thermal, and chemical stresses—exist simultaneously during pyrometallurgical Cu production [3–5], which intensifies the degradation of the refractory. Cu slag, composed of numerous oxides (Table 1), is generated during the smelting of Cu matte in shaft kilns and converters.

Table 1. Melting points of oxide components of slags from Cu production [6,7].

Oxide	As_2O_3	PbO	CuO	Fe_2O_3	SnO_2	SiO_2	ZnO	Al_2O_3	CaO	MgO
Melting point [°C]	312	897	1085	1565	1630	1723	1975	2020	2625	2825

During high-temperature process, the oxide components of the slag form low melting phases as a result of chemical reactions, e.g., fayalite $Fe_2[SiO_4]$ (T_M = 1205 °C) [8,9], cuprite Cu_2O (T_M = 1215 °C) [10], magnetite Fe_3O_4 (T_M = 1539 °C) [11], and lead silicate Pb_2SiO_4 (T_M = 747 °C) [7]. Therefore, since the 1950s, the most suitable refractories for Cu production have been magnesia–chromite refractories produced from chromite ore and sintered or fused magnesia. The magnesia–chromite refractories are fabricated at high temperatures from 1550 °C (silica-bonded) up to 2500 °C (directly bonded; produced by fusion of chromite grains and magnesia clinker) [1,5,12]. The former type of refractory shows excellent resistance in contact with Cu slag [13]; however, it contains Cr^{3+}, which tends to oxidize to carcinogenic Cr^{6+}. This phenomenon is accelerated when Cr-containing refractories are exposed to temperatures above 800 °C and contact with alkali or alkali earth oxides, especially CaO. Due to this fact, magnesia–chromite was completely withdrawn from the cement, lime, and glass industry [12]. Recycling of spent magnesia–chromite products containing hazardous Cr^{6+} is technologically complex, expensive, and economically not viable [1,14]. Moreover, their landfilling makes it challenging to properly protect and ensure both human and natural environment safety.

In recent years, many researchers have focused on the examination and enhancement of magnesia–chromite (MgO-Cr) refractories [15–17]. Chen et al. [2] compared the behavior of a direct-bonded MgO-Cr refractory and fused-grains-based refractories in contact with $Cu-Cu_xO-PbO$-based slag. The infiltration of the examined corrosive medium was found to be greater for the fused refractory product. The wetting behavior of chromite grains was found to be the most important factor in the corrosion protection of both types of refractories. The larger fused grains were a fair degree more resistant to corrosion. Other authors revealed the behavior of magnesia–chromite refractories against fayalite slag [18–23], Cu matte [21,24], fayalite-based slag with increased ZnO content [25–27], or calcium silicate slags [28]. In [23,29], authors presented the reduction of chromium oxide amount in refractories for the Cu industry. Currently, numerous works focus on the development of Cr-free refractories for Cu metallurgy. Jiang et al. [30] investigated the corrosion behavior of $MgO-MgAl_2O_4$-based refractories (M-MA refractories), produced at 1580 °C, against Cu, Cu_2O, and Cu matte. Cu did not infiltrate into $MgO-MgAl_2O_4$ refractories, and it only gathered on the surface of spinel grains, which confirms the non-wetting behavior of Cu towards the refractory Cu. In addition, MA showed relatively good resistance in contact with Cu matte. A slight reaction rate was observed for the system $MgAl_2O_4$-Cu matte, for which two layers were formed, but, beneficially, Al_2O_3 existed in greater amounts in observed layers. The worst corrosion resistance was observed for the interaction between M-MA refractory and Cu_2O. Cuprite penetrated deeply into the refractory and formed Cu-containing non-stoichiometric compound $2(Mg,Cu_2)O\cdot3Al_2O_3$, which was reported for the first time, and easily peeled off from the refractory body. The resistance of MA to liquid CuO_x was examined in [31] and to PbO-rich slag in [32]. Petkov et al. [14] compared the chemical behavior of six different Cr-free refractories with six different MgO-Cr refractories. Anode slag was the corrosive medium, which consisted of 50 wt.% CuO_x, 30–35 wt.% FeO, and 7–8 wt.% SiO_2. Two of these refractories (MgO-based brick with the addition of $ZrSiO_4$ and fused Al_2O_3-based brick) revealed extremely high rates of corrosion. Moreover, the addition of spinel (MA) did not improve the corrosion resistance of MgO-based bricks. Consistently, magnesia–chromite refractories exhibited the best performance. Cr-free solutions for the Cu industry are continuously under investigation, e.g., refractories from the system $MgO-Al_2O_3-SnO_2$ [33,34], Si_3N_4-SiC [35], or MgO doped with ZrO_2 nanoparticles [36]. The resistance of MgO to liquid CuO_x was examined in [37].

The influence of PbO-rich non-ferrous slag [38–41] and PbO-rich ferrous slag [38,41], or $PbO-CuO_x$-Cu mixture [2], on the corrosion resistance of MgO-Cr refractories is well recognized in the literature. In our recent work, we concluded that the higher the concentration of PbO in the lead-rich copper slags, the higher the degradation level of magnesia–chromite refractory [42]. However, the chemical interactions between crucial refractory raw materials with slag enriched in PbO and CuO_x are poorly investigated so far. Therefore, this work aimed to study the chemical interactions between commercial Cr-containing (magnesia–chromite co-clinkers FMC 45 and FMC57 as well as chromium ore PAK) and Cr-free refractory raw material (fused magnesia–aluminate spinel SP AM 70) against commercial PbO-rich Cu slag (Z1) and identify the most corrosion-resistant one.

2. Materials and Methods

The experiment consisted of three main stages: the chemical and structural characterization of four refractory raw materials and Cu slag, preparation of the samples for corrosion tests (refractory raw materials–Cu slag mixtures), and corrosion tests of mixtures by hot-stage microscopy (HSM) and pellet test (PT).

2.1. Characterization of Refractory Raw Materials and Cu Slag

Refractory raw materials used in the experiment were the following:

- Magnesia–chromite co-clinkers, FMC 45 and FMC 57

A semi-product for the production of magnesia–chromite co-clinkers is magnesia clinker, which is fabricated by burning magnesia carbonates at 1500–2000 °C and typically contains 95.0–99.8 wt.% MgO. Then, by burning the mixture of MgO clinker and chromite ore in the rotary kiln, or by fusion in an electric arc furnace, the magnesia–chromite co-clinkers are produced [6,43]. In addition to MgO and Cr_2O_3, the secondary components in magnesia–chromite co-clinkers are CaO, SiO_2, and Fe_2O_3 in a small percentage of contents. Details are provided in Table 2.

Table 2. Chemical composition (wt.%) of refractory raw materials by XRF.

Oxide	Magnesia–Chromite Co-Clinker FMC 45	Magnesia–Chromite Co-Clinker FMC 57	Pakistani Chromite Ore PAK	Fused Spinel SP AM 70
MgO	51.76	61.89	17.18	27.86
Al_2O_3	6.18	5.14	11.69	70.59
Cr_2O_3	24.87	19.40	49.76	-
Fe_2O_3	15.04	10.48	16.32	0.40
SiO_2	0.92	1.37	3.91	0.17
CaO	0.76	1.33	0.26	0.69
V_2O_5	0.06	0.07	0.09	-
Others	0.41	0.32	0.79	0.29

- Pakistani chromite ore, PAK

The phase composition of chromite ores is represented by a complex spinel solid solution $(FeO,MgO)·(Cr_2O_3,Al_2O_3,Fe_2O_3)$, containing 32–50 wt.% Cr_2O_3 [43].

Listed raw materials are commonly used during the fabrication of magnesia–chromite refractories applied as linings in slag zones of heating devices for the production of copper.

- Fused spinel, SP AM 70

Magnesia–aluminate spinel is a synthetic raw material produced by arc melting of magnesia and alumina. $MgAl_2O_4$ spinel is characterized by high refractoriness (T_m = 2122 °C [44]), a relatively high density of 3.58 g/cm^3, and a low thermal expansion coefficient of 8.83 × 10^{-6} 1/K (in the range 20–1200 °C) [45]. This material is applied in cement [46–48] and the steel industry [49–52].

The corrosive agent used in the experiment was industrial PbO-rich Cu slag, obtained after the first stage of the converting process. The examined raw materials and slag were characterized in terms of their chemical and phase composition on samples with grain size <63 μm, applying X-ray diffraction (XRD) and X-ray fluorescence (XRF) techniques. The diffraction patterns were collected by applying Philips PANalytical X'Pert-Pro diffractometer at room temperature, in the 2θ range of 5–90°, using a goniometer of 240 mm in diameter and Cu-Kα radiation. Chemical composition was measured using a PANalytical WDXRF Axios mAX spectrometer.

Samples of raw materials and slag used for chemical analysis (by XRF) and phase analysis (by XRD) as well as for both corrosion tests applied in the present study were powdered to below 63 μm. In the first step, bigger pieces of raw materials and slag (above 3 cm) were crushed using a laboratory jaw crusher. Then, they were subjected to a milling process using a centrifugal ball mill with ceramic grinders for 4 h. After this step, each component was sieved using a sieve with a mesh size of 63 μm, and oversized grains were manually milled using agate mortar. Eventually, the powders were mixed for 1 h.

The high-temperature behavior of the slag was tested by hot-stage microscopy (HSM) test to determine its sintering, melting, and flow temperatures and in situ dimensional changes, from room temperature to 1000 °C with a heating rate of 10 °C/min applying Carl Zeiss MH02 microscope. Sintering temperature was determined as corresponding to 2% shrinkage of linear sample dimensions. HSM method was described in detail in works [53,54].

2.2. Corrosion Tests of Raw Material–Slag Mixtures

2.2.1. Hot-Stage Microscopy

The powdered mixtures of raw materials and slag in mass ratio 1:1 were homogenized for 1 h and shaped into 3 mm cuboidal samples. Measurements were conducted using Carl Zeiss MH02 microscope with a heating rate of 10 °C/min to 1500 °C. Characteristic temperatures—sintering, melting, and flow—were determined based on in situ changes of the sample cross-section in microscopic view recorded by the camera during heating.

2.2.2. Pellet Test

The pellet test was conducted to identify and characterize the quasi-equilibrium reactions between the powdered mixture of refractory raw materials and Cu slag. Each refractory raw material was separately mixed and homogenized with Cu slag, ground below 63 μm, in the ball mill for 2 h in the mass ratio of 3:1 raw material to slag. The mixtures were uniaxially pressed under 65 MPa in cylindrical steel molds of 20 mm in diameter. The shaped samples were heat-treated in an air atmosphere at 1100 and 1400 °C in the laboratory electric furnace with a heating rate of 5 °C, soaked 3 h at maximum temperature, and finally cooled freely with the furnace. The temperature of 1100 °C was selected to evaluate the changes in phase composition of the samples before reaching a working temperature of copper converter (1300 °C), which can affect the performance of samples' components.

After the pellet test, the samples were subjected to examination of their bulk density and open porosity using a vacuum impregnator, based on the Archimedes' principle method. The XRD measurements for identification of the phase composition changes in corroded samples were conducted using Philips PANalytical X'Pert-Pro diffractometer at room temperature with 240 mm goniometer diameter applying Cu-Kα radiation, in the 2θ range of 5–90°. The microstructural analysis of the corroded samples was performed by a scanning electron microscope Nova Nano SEM 200 equipped with an energy-dispersive spectrometer (EDS) to measure the chemical composition in microareas. The samples for microscopic observations were prepared by embedding the material in the epoxy resin, polishing by standard ceramic technique, and coating the surface with a carbon layer.

3. Results

3.1. Characterization of Raw Materials

3.1.1. Chemical Composition of Raw Materials by XRF

According to Table 2, magnesia–chromite co-clinkers—FMC 45 and FMC 57—consisted mainly of MgO, Cr_2O_3, Fe_2O_3, and Al_2O_3. In FMC 45, MgO reached about 52%, while Cr_2O_3 content was under half that amount at 25%; other oxides were present in lower amounts, such as Fe_2O_3 at 15% and Al_2O_3 at 6%, while the range of the other oxides was below 1%. FMC 57 contained a higher quantity of MgO, equaling 62%. Simultaneously, FMC 57 was characterized by a lower content of Cr_2O_3, Fe_2O_3, and Al_2O_3 at 19%, 10%, and 5%, respectively. PAK contained about 50% Cr_2O_3, while the content of oxides MgO, Fe_2O_3 Al_2O_3, and SiO_2 reached 17%, 16%, 12%, and 4%, respectively. Fused spinel, SP AM 70, was characterized by contents of MgO and Al_2O_3 close to their amounts in the composition of stoichiometric $MgAl_2O_4$ [43]. The impurities in this Cr-free raw material existed mainly as CaO, SiO_2, and V_2O_5, equaling 0.9% in total.

The low content of SiO_2 and CaO in both magnesia–chromite co-clinkers <1.5% indicates their high suitability for refractory applications. On the other hand, in PAK the total content of Cr_2O_3 and Al_2O_3 was over 60% [55], and SiO_2 was slightly higher than recommended for high-temperature applications (<3.5 wt.%) [13]. In addition, SP AM 70 was characterized by a low amount of impurities. Thus, all test commercial raw materials were initially proposed as suitable for application in the Cu industry [43].

3.1.2. The Phase Composition of Raw Materials by XRD

XRD patterns of the test raw materials are presented in Figure 1. The main phase in both magnesia–chromite co-clinkers FMC 45 and FMC 57 as well as in chromite ore PAK (the only phase) was complex spinel solid solution, (Mg,Fe)[Fe,Cr,Al]$_2O_4$, where () and [] indicate tetrahedral and octahedral sites, respectively. MgO was the second phase in FMC 45, FMC 57, and fused spinel SP AM 70. No impurities were detected in all tested samples. In the PAK sample, the increased background is ascribed to the raised iron content among all tested samples (Table 2). This phenomenon was also observed elsewhere [56,57].

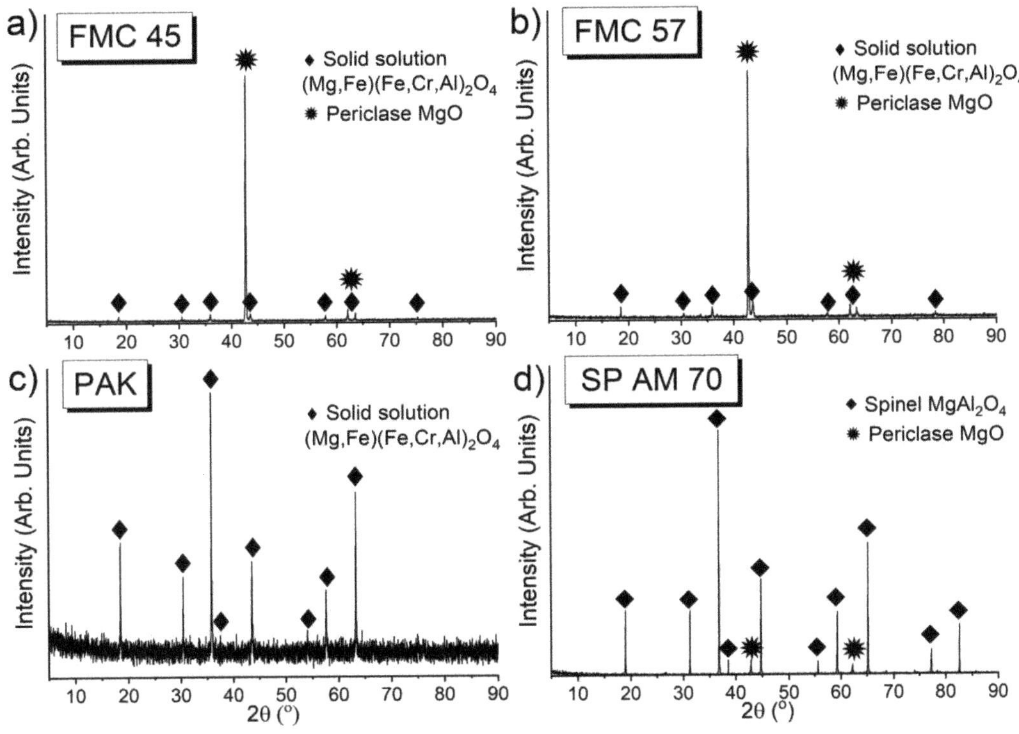

Figure 1. XRD patterns of (**a**) FMC 45, (**b**) FMC 57, (**c**) PAK, (**d**) SP AM 70.

3.2. Characterization of Cu Slag

3.2.1. Chemical Composition of Cu Slag by XRF

XRF chemical composition of Cu slag is shown in Table 3. PbO was the main component of the test slag, reaching about 40%. Due to the low melting point of PbO (897 °C) [7], it is highly corrosive towards refractory components and, additionally, reduces the viscosity of liquid slag [2]. Other oxides in the test slag were Fe_2O_3, CuO, and SiO_2, which existed in relatively high and comparable amounts of 18%, 16%, and 14%, respectively. The slag was contaminated by CaO and As_2O_3 in comparable amounts (3%), as well as by Al_2O_3 and ZnO (1.5%). In addition, low amounts of other oxides were detected such as Co_3O_4 (1.18%), Na_2O (0.97%), K_2O (0.29%), NiO (0.31%), SnO_2 (0.12%), and Cr_2O_3 (0.27%). Typically, Cu slags contain high levels of SiO_2 and Fe_xO_y [58,59]; hence, the properties of the selected slag in this work were predicted to be distinct. Compared to already investigated PbO-rich slags [38,41], the present slag is distinguished by a high level of aggressive Cu oxide.

Table 3. Chemical composition of test slag [42].

Oxide	PbO	Fe_2O_3	CuO	SiO_2	As_2O_3	CaO	Al_2O_3	ZnO	Others
wt. [%]	39.10	18.20	15.60	14.00	3.35	3.34	1.58	1.18	3.65

3.2.2. The Phase Composition of Cu Slag by XRD

Figure 2 depicts the phase composition of the test PbO-rich copper slag. Pb_2SiO_4 was the main phase of this slag, resulting from the high content of PbO and SiO_2. Other phases present in the slag were magnetite, Fe_3O_4, fayalite, Fe_2SiO_4, and cuprite, Cu_2O. Moreover, the increased background at the XRD pattern indicates that the slag contained a glassy phase, resulting from the fast cooling of the test slag [59].

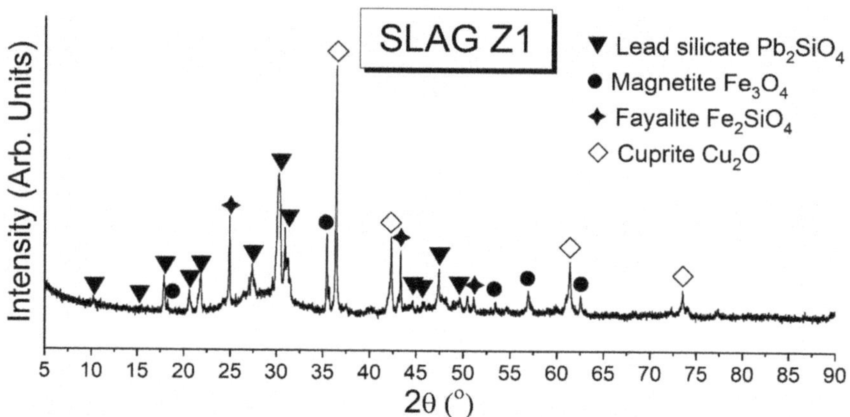

Figure 2. XRD pattern of the test slag [42].

3.2.3. Hot-Stage Microscopy Test of Cu Slag

Figure 3 presents the relative change in the linear dimension of the Cu slag cuboid sample as heated during the hot-stage microscopy test. During heating up in the range of 350–650 °C, the sample slightly expanded, as seen by the rising curve. Then, after the sharp descending stage, a second expansion in the narrow temperature range of 800–850 °C was observed. Finally, the sample started to sinter at 888 °C, followed by melting at 918 °C and full flow at 969 °C. The slag melted at a significantly lower temperature compared to typical fayalite-based slags (1420 °C in air [60] and above 1420 °C [61]) due to high Pb and Cu content.

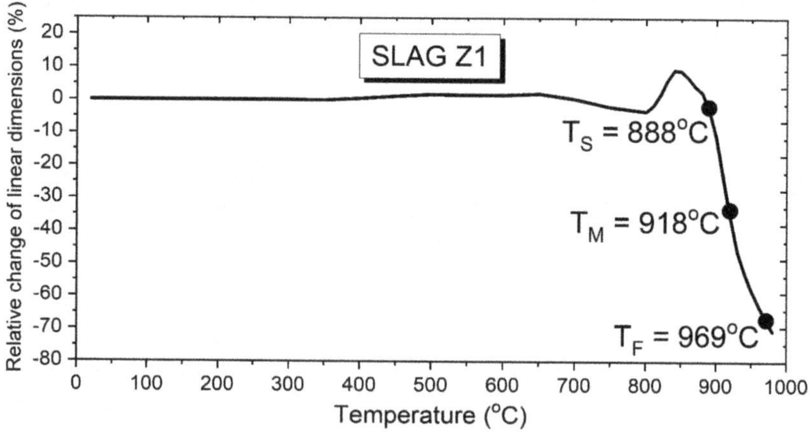

Figure 3. Results of hot-stage microscopy test of test slag; T_S-sintering temperature, T_M-melting temperature, T_F-flow temperature [42].

3.3. Corrosion of Raw Material–Cu Slag Mixtures

3.3.1. Hot-Stage Microscopy Test

Figure 4 depicts the hot-stage microscopy test results of the refractory raw materials FMC 45, FMC 57, PAK, and SP AM 70 mixed with PbO-rich Cu slag. The pictures of in situ changes in samples cross-sections are presented in Table 4. The highest sintering temperatures were observed for the raw materials containing high levels of Cr_2O_3—both magnesia–chromite co-clinkers and PAK, with the highest sintering point registered for FMC 45 of 1440 °C containing 25% Cr_2O_3. Such a high sintering temperature can be an effect of optimal content ratio Cr_2O_3/MgO (Table 2). Pakistani chromite ore PAK and magnesia–chromite co-clinker FMC 57 showed comparable sintering points of 1285 °C and 1260 °C, respectively. PAK is characterized by the highest Cr_2O_3 content of about 50% and, simultaneously, by the lowest MgO amount of 17% among all tested raw materials.

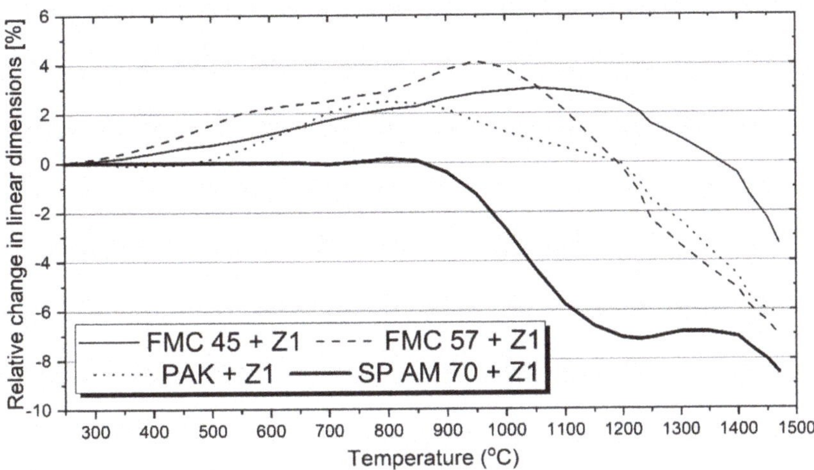

Figure 4. Results of hot-stage microscopy test for refractory raw material–Cu slag mixture. The sintering temperature was set as 2% linear shrinkage of the test sample.

Table 4. Microscopic view of the test raw material–Cu slag mixtures during the hot-stage microscopy test.

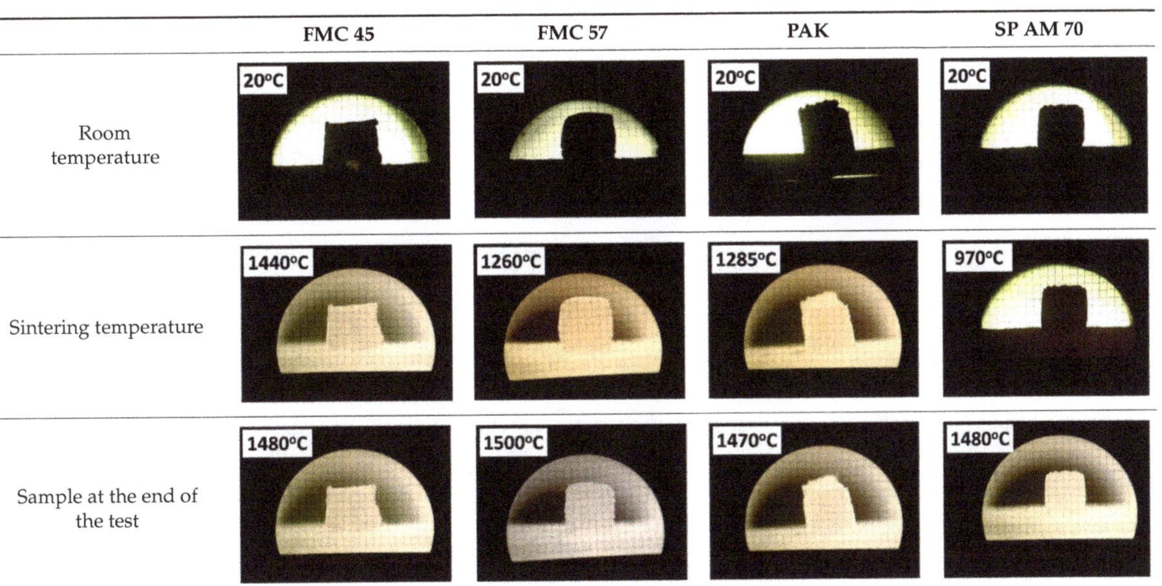

	FMC 45	FMC 57	PAK	SP AM 70
Room temperature	20°C	20°C	20°C	20°C
Sintering temperature	1440°C	1260°C	1285°C	970°C
Sample at the end of the test	1480°C	1500°C	1470°C	1480°C

The presented curves (Figure 4) show the expansion for all test raw materials. The most significant linear expansion of 4% was observed for the mixture of FMC 57 + slag at about 950 °C. This resulted from the highest level of MgO, which possesses a high thermal expansion coefficient ($\alpha_{MgO} = 15.60 \times 10^{-6}$ 1/K [62]). FMC 45 + slag reached a maximum expansion of about 3% at 1100 °C; for PAK, the maximum expansion was registered at 800 °C as 2.5%.

Conversely, no-chrome raw material—fused spinel SP AM 70—showed steady dimensions up to 850 °C, followed by very sharp shrinkage from 850 °C up to 1250 °C, reaching 7%. The dimensions were maintained up to 1400 °C, above which the sample started to shrink again. Additionally, SP AM 70 revealed the lowest sintering point of 970 °C and, through this test, was confirmed to be the least resistant to PbO-rich Cu slag. For this material, only shrinkage was observed. In contrast to the fused spinel (SP AM 70), PAK expanded much more during heating, which is typical for Cr-containing spinels. Notably, $FeCr_2O_4$ is characterized by a high thermal expansion coefficient of 12.38×10^{-6} 1/K (close to the thermal expansion coefficient of MgO) when compared to the thermal expansion coefficient of spinel *sensu stricto* of 8.83×10^{-6} 1/K (all coefficients were given for temperature range 20–1200 °C) [62]. None of the test mixtures started to flow when reaching the limit temperature of the test of 1500 °C.

3.3.2. Pellet Test

Phase Composition of the Raw Material–Slag Mixtures after the Pellet Test

Figure 5 shows that both magnesia–chromite co-clinkers—FMC 45 and FMC 57—showed the same phase composition after corrosion by pellet test at 1100 and 1400 °C. The new phases, which appeared at a lower temperature of 1100 °C, were forsterite, Mg_2SiO_4, and spinel solid solution, identified as a separate spinel phase probably due to different stoichiometry compared to the original spinel, as its reflexes are located just next to the original spinel ones. This phase disappeared at 1400 °C.

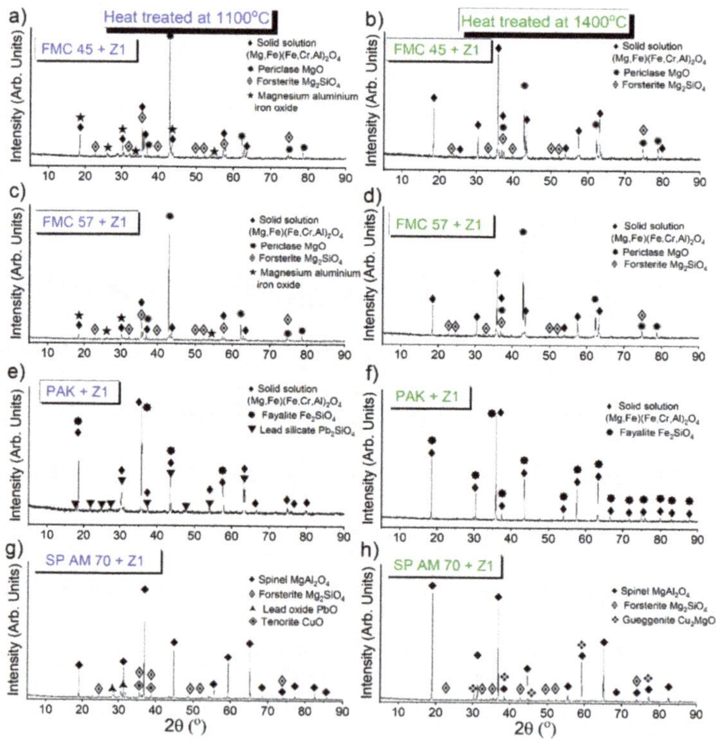

Figure 5. XRD patterns of raw material–slag mixtures after heating at 1100 °C—(**a**) FMC 45 + Z1, (**c**) FMC 57 + Z1, (**e**) PAK + Z1, (**g**) SP AM 70 + Z1 and 1400 °C—(**b**) FMC 45 + Z1, (**d**) FMC 57 + Z1, (**f**) PAK + Z1, (**h**) SP AM 70 + Z1.

In PAK, for corrosion at 1100 °C, two new phases formed, fayalite (Fe_2SiO_4) and lead silicate (Pb_2SiO_4), while the former disappeared when heating the mixture at 1400 °C. The clear phase identification of solution phases by XRD is very limited, as the method does not identify the compounds directly but only their crystalline structure by the position of the peaks. The replacement of one element by another in a solid solution may shift the lattice parameters and consequently the position of the peaks.

Corrosion of fused spinel SP AM–slag mixture at 1100 °C showed a more significant number of corrosion products: forsterite (Mg_2SiO_4), PbO, and CuO. Copper (II) oxide—tenorite—was detected (instead of Cu_2O) due to oxidizing atmosphere, as the investigation was done after cooling. According to the CuO-Cu_2O-MgO phase diagram [63], below 1021 °C, the phase transformation $Cu_2O \rightarrow CuO$ occurs. However, heating at a higher temperature of 1400 °C resulted in the CuO transformation into güggenite (Cu_2MgO_3) due to a direct reaction of tenorite with magnesia. Moreover, PbO disappeared during heating at 1400 °C as a result of its dissolution in other phases at increased temperatures.

In summary, all of the samples underwent active corrosion without the formation of any corrosion-protecting interfaces. At 1400 °C, forsterite (Mg_2SiO_4) was identified in FMC 45, FMC 57, and SP AM 70 (which possess above 20% of MgO), while fayalite (Fe_2SiO_4) was detected in PAK, which contains a significant iron content (16% of Fe_2O_3). Nevertheless, those two compounds create a continuous solid solution, called olivine, as per the diagram Mg_2SiO_4-Fe_2SiO_4 [64]. Endmembers detected by XRD are in fact solid solutions where Mg and Fa can replace each other more or less arbitrarily or be substituted by other elements of similar geometry such as Ca, Zn, or Ni. There are works [65] showing that even large-size Pb can substitute for Mg in oxide structure despite the great difference in cation radii (r_{Pb2+} = 119 pm, r_{Mg2+} = 72 pm); however, very scarce information on the system MgO-PbO exists in literature (and a low number of cards are available in crystallographic databases). In the discussed system Mg_2SiO_4-Fe_2SiO_4, the increased amount of Fe leads to a continuously descending melting point of the solid solution from 1890 °C for forsterite to 1205 °C for fayalite. Table 5 shows summarized phases detected using XRD method.

Table 5. Phases present in raw material–slag mixtures, after pellet test at 1100 °C and 1400 °C.

Temperature	FMC 45 + Z1	FMC 57 + Z1	PAK + Z1	SP AM 70 + Z1
1100 °C	Solid solution (Mg,Fe)(Fe,Cr,Al)$_2$O$_4$ Periclase MgO Forsterite Mg$_2$SiO$_4$ Magnesium aluminum oxide		Fayalite Fe$_2$SiO$_4$ Lead silicate Pb$_2$SiO$_4$ -	Spinel MgAl$_2$O$_4$ Lead Oxide PbO Forsterite Mg$_2$SiO$_4$ Tenorite CuO
1400 °C	Solid solution (Mg,Fe)(Fe,Cr,Al)$_2$O$_4$ Periclase MgO Forsterite Mg$_2$SiO$_4$		Fayalite Fe$_2$SiO$_4$ -	Spinel MgAl$_2$O$_4$ Forsterite Mg$_2$SiO$_4$ Güggenite Cu$_2$MgO$_3$

Bulk Density and Open Porosity of the Raw Material–Slag Mixtures after Pellet Test

Results of bulk density and open porosity for corrosive raw material–slag mixtures after pellet test are presented in Figure 6. The greatest density was detected for chromite ore PAK, being 3.3 g/cm^3 and 4.5 g/cm^3 for 1100 and 1400 °C, respectively. This behavior agreed with the low porosity of 25.6% for 1100 °C and 18.0% for 1400 °C. Out of the mixtures with magnesia raw materials, the purer FMC 57 showed a lower density at 1100 °C of 2.3 g/cm^3 than FMC 45. However, after heating at 1400 °C, it increased about 70% to 4.0 g/cm^3, which significantly exceeded the density of FMC 45 of 3.4 g/cm^3. Nevertheless, FMC 45 + slag reached the lowest porosity of 11.1% at 1400 °C, while for FMC 57 + slag, it equaled 20.9%.

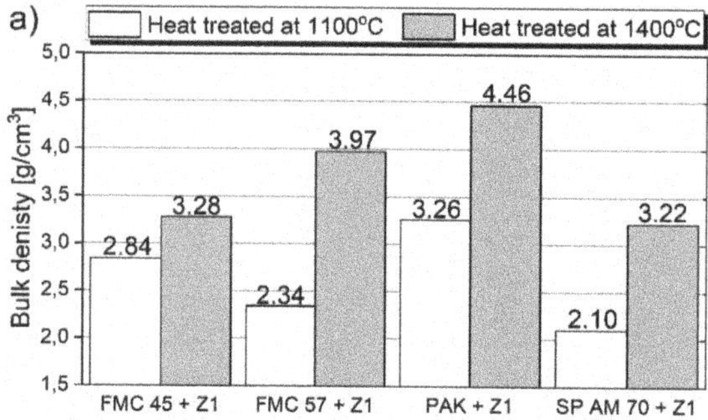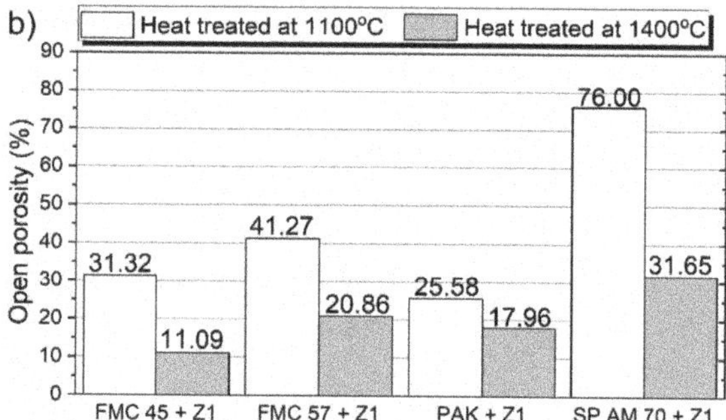

Figure 6. Bulk density (**a**) and open porosity (**b**) of samples after pellet corrosion test at 1100 °C and 1400 °C.

The slag mixture with spinel SP AM 70 exhibited the lowest density at 1100 and 1400 °C of 2.1 g/cm^3 and 3.2 g/cm^3, respectively. This behavior was accompanied by the extremely high porosity of 76% at lower temperatures and 31.7% at higher test temperatures. Such porosity can be associated with a high MgO diffusion rate from the spinel into slag at 1100 °C, leaving pores, which preceded güggenite formation, confirmed by XRD at 1400 °C (Figure 5h).

The results show the positive influence of Cr_2O_3 content on the high densification of the material linked with higher compactness of the sample represented by low open porosity. Moreover, the impurities in magnesia also lead to greater densification of the sample. This trend may influence lower penetration of materials by liquid slags present at operation conditions.

Microstructure Analysis of the Raw Material–Slag Mixtures after the Pellet Test

SEM microphotographs of FMC 45 corroded by Cu slag at 1100 and 1400 °C, together with EDS chemical analysis, are presented in Figure 7 and Table 6. Chemical analysis verified XRD results revealing magnesia and complex spinel solid solution in the samples after corrosion (Figure 7b, p. 1 and 3, respectively). For samples treated at 1400 °C, güggenite solid solution formed between magnesia and spinel grains (Figure 7b, p. 2), as a result of a reaction between magnesia from refractory and copper oxide from the slag. MgO grains at 1400 °C (Figure 7b) contained other dissolved elements such as Si (0.2%), Pb (0.8%), Cu (6.1%), and Fe (8.4%). Points 2 and 3 in Figure 7 represent average chemical compositions of magnesia grains containing PbO inclusions at 1100 °C, represented by dispersed small lightest-color microareas due to high molar weight. Scarce information on the system MgO-PbO exists in literature; thus, PbO solubility in MgO needs further study. Chen et al. [66] examined the $MgO-PbO-SiO_2$ system in the temperature range 700–1400 °C and found no phases containing only PbO and MgO, but they found three stable phases in the system $PbO-SiO_2$ ($PbSiO_3$, Pb_2SiO_4, and Pb_4SiO_6) and one ternary compound ($Pb_8Mg(Si_2O_7)_3$). Najem et al. [65] recently doped MgO nanoparticles with Pb^{2+}, but the maximum lead amount doped was low, from 0.03 to 0.05 at.%. X-ray diffractometry confirmed that only the cubic phase of space group Fm3m was generated, characteristic of MgO. The increased Pb^{2+} fraction doped in MgO caused the most intensive reflex (200) shift towards the lower 2θ values, and the increase in lattice parameter and crystallinity was observed. Moreover, the intensity of the most substantial peak decreased, attributed to the difference in the ionic radius between Pb^{2+} (r_{Pb2+} 119 pm) and Mg^{2+} (r_{Mg2+} = 72 pm).

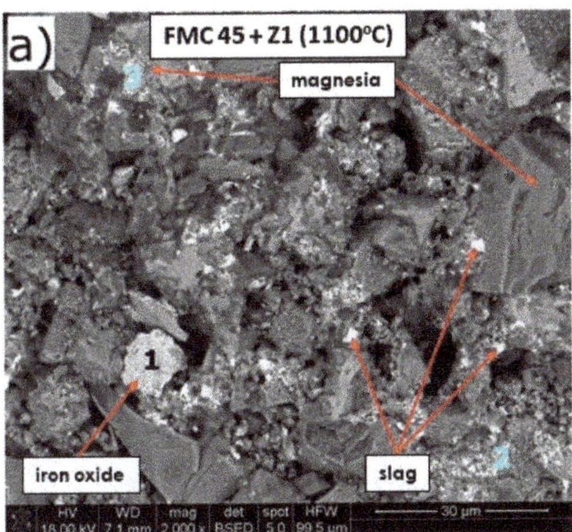

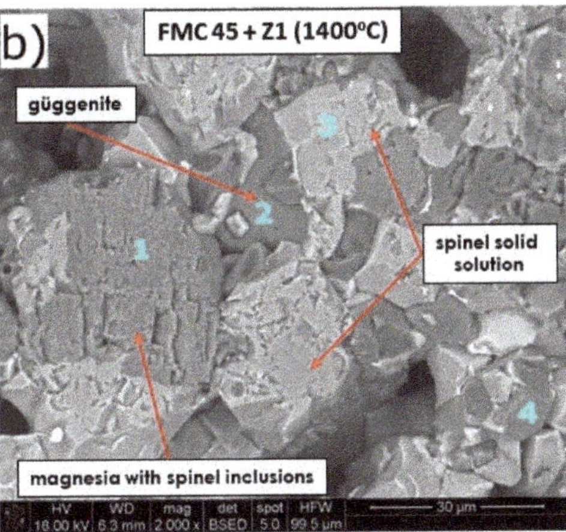

Figure 7. SEM images of magnesia–chromite co-clinker FMC 45 after pellet test at (**a**) 1100 °C, (**b**) 1400 °C.

Table 6. EDS analysis in microareas of mixture FMC 45 + slag after corrosion, in the points marked in Figure 7, together with corresponding phases.

Figure No.	Point	Phase	Chemical Composition, wt./mol% *								
			Pb	Cu	Mg	Cr	Al	Si	Fe	As	O
Figure 7a	1	Fe-Cr-O solid solution	1.1/0.2	-	1.2/2.3	12.2/10.8	0.9/1.5	2.7/4.2	70.0/57.9	-	4.3/12.5
	2	-	16.4/2.4	1.0/0.5	21.6/26.6	6.1/3.5	2.9/3.2	3.3/3.5	18.4/9.9	-	23.8/44.5
	3	-	19.0/2.6	2.3/1.0	25.1/28.9	3.8/2.0	1.9/2.0	5.7/5.7	10.1/5.1	-	28.3/49.7
Figure 7b	1	MgO with (Mg,Fe)(Fe,Cr,Al) spinel inclusions	0.8/0.1	6.1/2.3	44.9/44.4	4.6/2.1	1.0/0.9	0.2/0.1	8.4/3.6	1.7/0.6	29.5/44.3
	2	Güggenite s.s.	0.9/0.2	33.1/19.9	18.3/28.8	8.2/6.0	-	1.2/1.6	10.0/6.8	-	10.5/25.0
	3	Mg-rich spinel s.s. Mg-Fe-Cr-Al-O	1.3/0.2	1.3/0.5	15.9/16.7	11.4/5.6	7.4/7.1	1.1/1.0	20.6/9.5	-	34.9/55.9
	4	Cr-rich spinel solid solution Mg-Fe-Cr-Al-O	1.1/0.2	0.9/0.4	16.0/17.5	17.4/8.9	7.2/7.1	0.3/0.3	16.4/7.8	3.2/1.1	32.5/54.1

* The remaining elements are in minor amounts: Na, P, S, Ti, Ca, and Mn. s.s.—solid solution.

Spinel solid solutions were found to accept various stoichiometries and numerous ions into their structure (XRD showed secondary spinel as a separate phase). At 1400 °C, grains of spinel solid solution, as shown by point 3 in Figure 7b, were enriched in Mg (16%) and Fe (21%) with a relatively low content of Cr (11%), compared to other locations presented by point 4 where they were enriched with Cr (17%) and impoverished concerning Fe (16%). In addition, it can be seen that spinel solid solution contained low substitutions of Pb at the level of 1%. Similar to the binary system MgO-PbO, the solubility of Pb^{2+} in spinel solid solution in chromite ore was not found in the literature. Forsterite was not observed in SEM images, although it was identified by XRD, probably due to its low quantity in the sample.

Arsenic ions were detected only at 1400 °C in MgO (2%: point 1, Figure 7b) and in spinel solid solution grain (3%: point 4, Figure 7b).

The sample FMC 45 after corrosion at 1400 °C was visibly more densified compared to 1100 °C, with almost no slag phase in between the grains, which confirms the results showing its greater density and lower open porosity (Figure 6).

In FMC 57 for both 1100 and 1400 °C (Figure 8 and Table 7), magnesia and complex spinel solid solution were detected, which confirms XRD results of the corroded mixtures (Figure 5c,d). Pb from slag diffused and dissolved into the structure of complex spinel solid solution in low amounts at 1100 °C of 1% in p. 2 (Figure 8a) and 0.4% in p. 2 (Figure 8b) at 1400 °C.

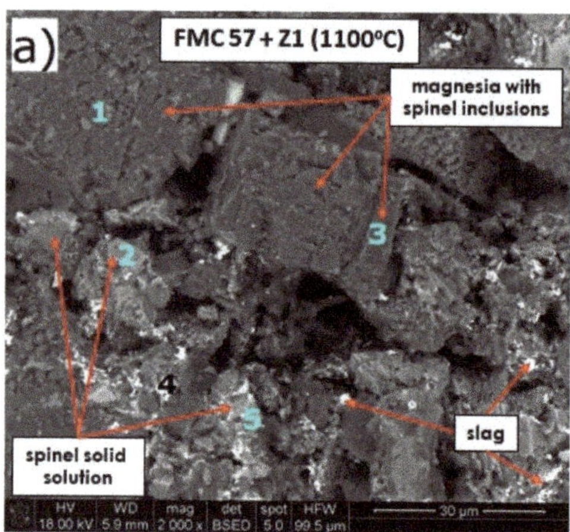
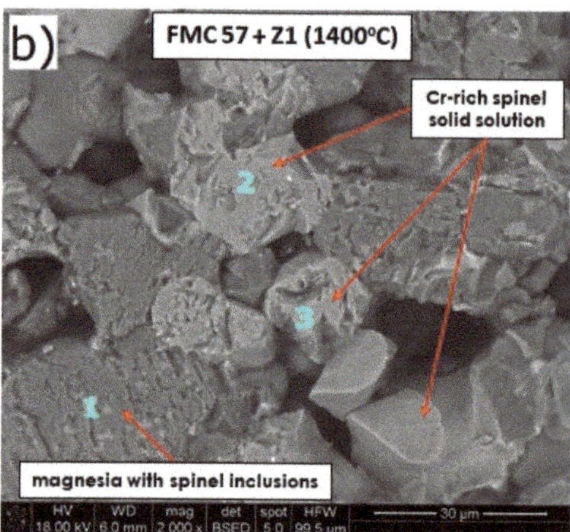

Figure 8. SEM images of magnesia–chromite co-clinker FMC 57 after pellet test at (**a**) 1100 °C and (**b**) 1400 °C.

Table 7. EDS analysis in microareas of mixture FMC 57 + slag after corrosion, in the points marked in Figure 8, together with corresponding phases.

Figure No.	Point	Phase	Chemical Composition, wt./mol% *								
			Pb	Cu	Mg	Cr	Al	Si	Fe	As	O
Figure 8a	1	MgO with (Mg,Fe)(Fe,Cr,Al) spinel inclusions	1.1/0.1	5.5/2.0	51.7/50.2	2.7/1.2	0.8/0.7	0.6/0.5	6.4/2.7	1.8/0.6	27.5/40.6
	2	Fe-rich spinel s.s.Mg-Fe-Cr-Al-O	1.0/0.1	2.7/1.2	16.5/18.9	13.5/7.2	5.5/5.6	0.1/0.1	30.2/15.0	-	29.1/50.6
	3	MgO s.s.	-	3.4/1.1	56.4/49.4	0.9/0.4	-	0.3/0.2	2.5/1.0	-	35.4/47.2
	4	-	33.2/6.0	2.8/1.6	17.4/26.7	6.4/4.6	0.8/1.1	5.9/7.8	4.3/2.9	4.1/2.0	15.8/37.0
	5	Fe-rich spinel s.s.Mg-Fe-Cr-Al-O	6.1/0.8	1.5/0.6	12.2/13.6	9.7/5.1	4.2/4.2	-	23.4/11.3	-	32.4/54.8
Figure 8b	1	MgO with spinel inclusions	0.4/0.1	0.7/0.3	30.7/32.4	15.2/7.5	4.6/4.4	0.6/0.5	16.0/7.4	-	28.6/45.8
	2	Cr-rich spinel s.s.Mg-Fe-Cr-Al-O	0.4/0.1	0.7/0.3	15.6/17.2	32.5/16.7	7.0/6.9	0.6/0.6	5.8/2.8	1.9/0.7	31.6/52.6
	3	Spinel s.s. Mg-Fe-Cr-Al-O	-	1.4/0.6	17.2/19.7	23.6/12.7	7.2/7.4	0.2/0.2	20.0/10.0	-	27.1/47.2

* The remaining elements are in minor amounts: Na, P, S, Ti, Ca, and Mn. s.s.—solid solution.

Among main slag components, MgO accepted Cu more than Pb—greater amounts of Cu dissolved at 1100 °C (5.5%: p. 1; 3.4%: p. 3; Figure 8a) than at 1400 °C (0.7%, p. 4, Figure 8a). Slightly higher amounts of Cu dissolved in spinel solid solution (2.7%—1100 °C, 0.7%—1400 °C).

Spinel solid solution grains detected in the material contained different amounts of constituent ions, e.g., at 1400 °C spinel grains mainly differed in the content of Cr and Fe (points 2,3 in Figure 8b). After treatment at 1400 °C, the slag phase was no longer visible in SEM images, compared to 1100 °C, at which point slag was distributed at grain boundaries.

Arsenic ions were detected—similarly to FMC 45 + slag—in MgO grains (about 2% in p. 1 and about 1% in p. 4, Figure 8a) and in spinel grains (2% in p. 2, Figure 8b).

The main phase detected in PAK by SEM/EDS (Figure 9 and Table 8) was Cr-rich complex spinel solid solution, the only phase component in the original PAK (XRD-Figure 1c). This phase was found to accept about 5% of Cu for every measured microarea (points 1,3 in Figure 9a; points 3,4 in Figure 9b). Large grains of chromite spinel solid solution of above 50 μm showed high corrosion resistance against Pb, as lead was not detected in them at 1100 °C (point 1, Figure 9a) and only in slight amounts of 1.7% at 1400 °C (point 4, Figure 9b).

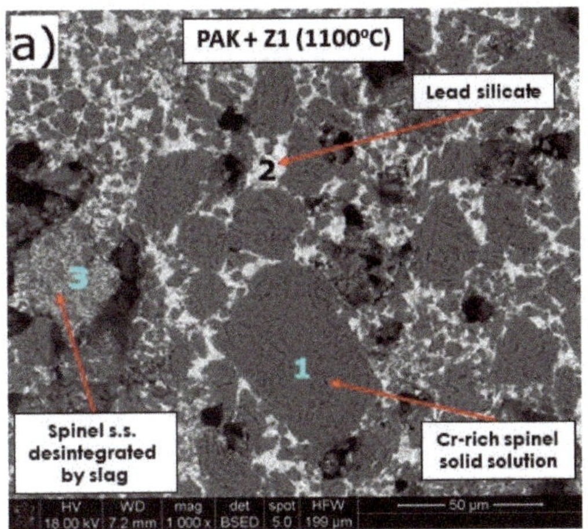
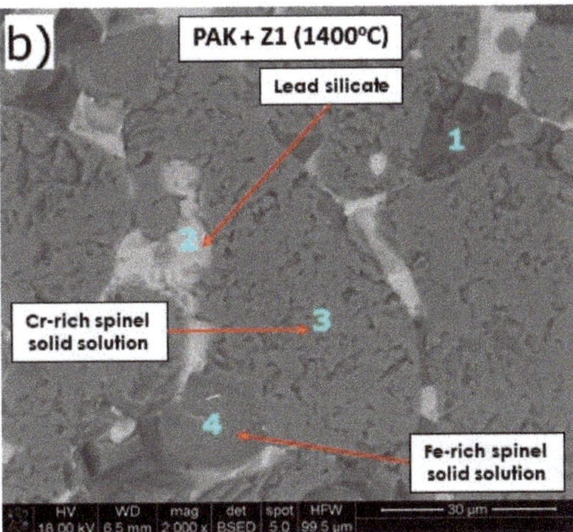

Figure 9. SEM images of PAK after pellet test at 1100 °C (**a**) and 1400 °C (**b**).

Table 8. EDS analysis in microareas of mixture PAK + slag after corrosion, in the points marked in Figure 9, together with corresponding phases.

Figure No.	Point	Phase	Chemical Composition, *wt./mol%* *								
			Pb	Cu	Mg	Cr	Al	Si	Fe	As	O
Figure 9a	1	Cr-rich spinel solid solution Mg-Fe-Cr-Al-O	-	5.8/2.7	10.7/13.2	40.1/23.0	8.1/8.9	0.4/0.4	3.6/1.9	3.5/1.4	25.1/46.7
	2	Lead silicate Pb_2SiO_4 s.s.	51.8/11.7	3.5/2.6	1.9/3.7	2.1/1.9	3.4/5.9	13.7/22.8	2.7/2.3	-	13.5/39.5
	3	-	19.5/3.3	4.6/2.6	7.1/10.3	20.4/13.8	4.0/5.2	5.3/6.6	13.9/8.8	-	20.7/45.6
Figure 9b	1	Olivine $(Ca,Mg,Fe)_2SiO_4$ **	0.4/0.4	0.5/0.2	9.6/9.8	1.5/0.7	3.9/3.6	25.6/22.6	6.9/3.1	-	30.5/47.1
	2	Lead silicate Pb_2SiO_4 s.s.	35.9/6.3	2.7/1.5	1.7/2.5	2.9/2.0	6.0/8.1	19.5/25.3	3.1/2.0	-	19.4/44.2
	3	Cr-rich spinel s.s. Mg-Fe-Cr-Al-O	-	4.9/2.5	9.9/13.0	44.3/27.2	4.0/4.7	-	9.4/5.4	2.1/0.9	22.1/44.1
	4	Fe,Cr-rich spinel s.s. Mg-Fe-Cr-Al-O	1.7/0.4	7.1/4.8	6.1/10.6	32.9/27.0	2.4/3.7	0.9/1.3	34.6/26.4	2.7/1.5	7.8/20.9

* The remaining elements are in minor amounts: Na, P, S, Ti, Ca, and Mn. ** Additionally, 18.6 wt.%/8.3 mol.% of Ca was detected, which substitutes Mg or Fe in olivine. S.s.—solid solution.

The new phases—corrosion products in the mixture PAK + slag—confirm XRD results: lead silicate Pb_2SiO_4 (light grey areas; 1100 °C, 1400 °C: point 2 in Figure 9a,b) and olivine $[Mg,Fe]_2SiO_4$ (dark grey area; 1400 °C: point 1, Figure 9b). The amount of secondary Pb_2SiO_4 was significant in SEM images. Pb_2SiO_4 dissolved comparable, low amounts of Mg^{2+} and Cr^{3+} of about 2–3%. 2.5% of arsenic ions were registered only in Cr–spinel solid solution (point 1 in Figure 9a; point 3,4 in Figure 9b).

SEM/EDS analysis of the test mixture of fused spinel SP AM 70 + slag (Figure 10, Table 9) revealed güggenite as corrosion product at 1400 °C (also identified by XRD—Figure 5h). Güggenite is a non-stoichiometric phase that is stable over a wide range of compositions, as shown by phase equilibria of the MgO-Cu_2O-CuO system in [63]. Güggenite did not accept any Pb. It dissolved very slight amounts of Cr of 0.2% and Al of 0.4% and an increased amount of Fe of 4.2%. This phase appeared in SEM images as light-grey microareas. It shall be mentioned that there is scarce information in the literature about the properties of this phase. The slag phase, visible as light-grey microareas (p. 2, Figure 10a), was enriched with Mg (23%), which comes from the dissolution of spinel from refractory by aggressive slag. Si was not detected to greatly infiltrate the spinel phase, as maximum Si content in spinel grains was 0.3% at 1100 °C, and it increased to 5.1% at 1400 °C.

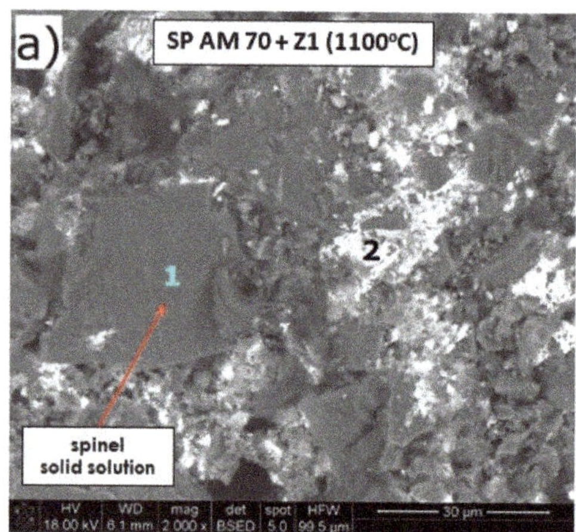

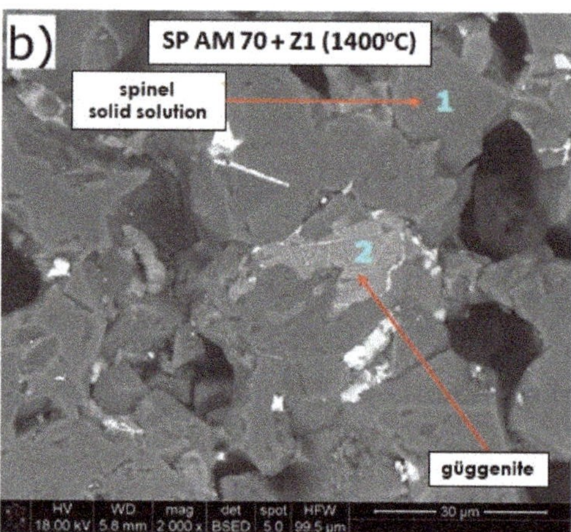

Figure 10. SEM images of fused spinel SP AM 70 after pellet test at 1100 °C (**a**) and 1400 °C (**b**).

Table 9. EDS analysis in microareas of mixture PAK + slag after corrosion, in the points marked in Figure 10, together with corresponding phases.

Figure No.	Point	Phase	Chemical Composition, wt./mol% *								
			Pb	Cu	Mg	Cr	Al	Si	Fe	As	O
Figure 10a	1	Spinel $MgAl_2O_4$ s.s.	0.7/0.1	0.6/0.2	18.1/16.4	0.5/0.2	43.8/35.8	0.3/0.2	0.7/0.3	-	33.0/45.5
	2	-	5.3/0.7	15.3/6.2	23.0/24.2	0.3/0.2	2.0/1.9	14.5/13.2	2.0/0.9	-	30.5/48.9
Figure 10b	1	Spinel $MgAl_2O_4$ s.s.	-	1.3/0.5	16.7/15.5	0.5/0.2	40.5/33.7	0.2/0.2	5.1/2.0	-	33.2/46.6
	2	Güggenite s.s.	-	25.0/10.4	41.5/45.3	0.2/0.1	0.4/0.4	-	4.2/2.0	-	23.3/38.6

* The remaining elements are in minor amounts: Na, P, S, Ti, Ca, and Mn. s.s.—solid solution.

Spinel phase $MgAl_2O_4$ showed relatively high resistance to Pb-rich copper slag at high temperatures, as only about 0.7% of Pb and 0.6% of Cu were detected in the spinel grains at 1100 °C, while at 1400 °C only Cu was observed at the level of 1.3%, and no Pb was detected. Spinel accepted above 5% of Fe at 1400 °C (and 0.7% at 1100 °C), and this behavior is characteristic for spinel structure, which tolerates ions with radii 40–80 pm (ferrous ions: $r_{Fe2+,IV} = 61.5$ pm, $r_{Fe2+,VI} = 74$ pm; ferric ions: $r_{Fe3+,IV} = 49$ pm, $r_{Fe3+,VI} = 64.5$ pm) [67–71]. Arsenic ions were not registered in any analyzed region.

4. Discussion

The study presents comparative results of chemical resistance of four commercial raw materials—two kinds of magnesia–chromite co-clinkers (FMC 45 and FMC 57), chromite ore PAK, and fused spinel SP AM 70—commonly used for the production of refractories. So far outside this study, only refractory products have been investigated in terms of their chemical resistance to copper slags. Table 10 summarizes and compares the results obtained in this work with results obtained for refractory products in other works.

The slag used in the experiment was PbO-rich slag containing about 40% PbO, 18% Fe_2O_3, 16% CuO, and 14% SiO_2, derived from the Cu converting process. The oxides existed in the slag in the form of four main phases—Pb_2SiO_4, Fe_3O_4, Fe_2SiO_4, and Cu_2O—which are aggressive towards refractory material as they possess low melting points: Pb_2SiO_4 at 747 °C [7] Fe_2SiO_4 at 1220 °C [72], and Cu_2O at 1230 °C [73].

The presence of these low-melting-point components affected the high-temperature behavior of the raw materials tested by hot-stage microscopy (HSM), especially fused spinel SP AM 70, which started to sinter at 970 °C followed by drastic shrinkage reaching 7% at 1400 °C. This test showed the low resistance of this kind of raw material to PbO-rich Cu slag and confirmed that it should not be used as the main refractory component for non-ferrous metallurgy where similar slags occur.

The opposite behavior was found for Pakistani chromite ore PAK, which showed high resistance to PbO-rich slag and the best dimensional stability up to 1200 °C (the most crucial temperature range for the copper industry). A similar high-temperature behavior during the HSM test was observed for FMC 45 and 57, although they expanded more. The highest sintering point of 1440 °C, determined as 2% linear shrinkage of the test sample, was registered for the mixture FMC 45 + slag.

Comparing both HSM and pellet corrosion tests, FMC 45 is the most corrosion-resistant against PbO-rich slag among all test raw materials. The new phases that appeared after corrosion in FMC 45 were forsterite (Mg_2SiO_4) due to reaction between MgO from refractory and SiO_2 from slag (acc. to Reaction (1)) and spinel solid solution having different stoichiometry than the original, which disappeared at 1400 °C. Here, FMC 45 revealed extremely low open porosity (11%). Forsterite formed in relatively low amounts in FMC 45 and FMC 57, as it was not observed during SEM/EDS analysis (Figure 7a,b and Figure 8a,b), and this was confirmed by low intensity of reflexes in XRD patterns (Figure 5a,b).

$$MgO_{ref} + SiO_{2slag} \rightarrow Mg_2SiO_{4solid\ solution} \tag{1}$$

Although forsterite is a high-temperature phase, melting congruently at 1890 °C [6], thus improving thermal resistance of the material, and possessing good chemical stability and excellent insulation properties, it is simultaneously characterized by a relatively low density (3.21 g/cm^3). This is lower compared to the rest of the predominant phases— $\rho_{MgO} = 3.59\ g/cm^3$, $\rho_{(Fe,Mg)Cr2O4} = 4.70\ g/cm^3$—resulting in the major loosening of the microstructure and causing, if it occurs in excess, so-called "forsterite bursting", detrimental for the performance of refractories. The microstructure disintegration was previously reported for massive formation of forsterite as corrosion product [2,14,22,39,40,67,74,75].

Aggressive components of the slag such as Cu and Pb dissolved in the MgO grains of FMC 45 and FMC 57. For instance, 2% of Cu dissolved in MgO grains of FMC 45 at 1100 °C, increasing to 6.1% at 1400 °C. For FMC 57, 5.5% of Cu was detected in MgO at 1100 °C. As per the MgO-CuO_x phase diagram [63], MgO can dissolve up to 21% of Cu ions at 700 °C (2), maintained until 1048 °C, then decreases to 10% and 5% at 1200 °C and 1400 °C, respectively, which proves the observed results.

$$MgO_{refractory} + CuO_{x,slag} \rightarrow (Cu,Mg)O_{solid\ solution} \tag{2}$$

Together with Cu, Fe diffused into periclase grains in significant amounts of about 8% (point 1, Figure 7b). This phenomenon is commonly observed [2–4,14,25] due to the similar geometry of ions in both oxides (r_{Mg2+} = 72 pm, r_{Fe2+} = 77 pm) [76]. In general, Pb diffused more willingly into Cr–spinel solid solution grains than MgO grains, which was previously observed in our recent work [42] for interactions between MgO-Cr refractory and PbO-rich slags and in [2] for interactions between MgO-Cr refractory and Cu-Cu_xO-PbO melt. The study on the MgO-PbO system, as well as solubility of Pb^{2+} in Cr–spinel solid solution, requires future study due to a gap in the literature, despite its significance in the industry. Si tends to diffuse mainly to MgO grains, with its greatest level of 5.7% at 1100 °C, but it remained unreacted (dissolved in magnesia). In contrast, Mg diffused into slag, which was observed in FMC57 (26% Mg, p. 4, Figure 8a) or SP AM 70 (24% Mg, p. 2, Figure 10a).

FMC 57 and PAK, although possessing various phase compositions, showed similar high-temperature behavior in the HSM test, with sintering points for both raw materials above 1250 °C, followed by a 6% shrinkage up to the limit temperature of measurement of 1500 °C. Large grains of Cr–spinel solid solution in PAK, especially those enriched with Cr and Fe, showed excellent resistance to Pb infiltration. In contrast, the small ones were surrounded by the corrosion product—low melting phase of non-stoichiometric (as confirmed by SEM/EDS) lead silicate, Pb_2SiO_4 (T_m = 747 °C [7]), appeared as light-grey microareas. Pb_2SiO_4 constituted the main phase component of the original slag identified by XRD; it was generated in slag as a result of a reaction between PbO (slag component) and SiO_2 (flux) during slag formation in Cu converting (3). This phase existed in equilibrium with the main component of PAK—complex Cr–spinel solid solution—in the conditions of the pellet corrosion test both at 1100 °C and 1400 °C.

$$2PbO_{slag} + SiO_{2,flux\ or\ concentrate} \rightarrow Pb_2SiO_{4slag} \tag{3}$$

Cr–spinel solid solution accepted more Cu than Pb, as the average amount of Cu in spinel grains was 5%, while the content of Pb was about 1.7%. Overall, the amount of Cu accepted by the phases of PAK was doubled when compared to FMC 45. The second component of the slag—fayalite Fe_2SiO_4—accepted Mg from PAK and formed olivine solid solution $[Mg,Fe]_2SiO_4$, which is commonly found as a corrosion product in MgO-containing refractories when fayalite-based Cu slags accompany the process [14,74]. With the increased content of MgO in the slag or refractory, part of Fe^{2+} (r_{Fe2+} = 74 pm) in Fe_2SiO_4 is replaced by Mg^{2+} of smaller ionic radius (r_{Mg2+} = 66 pm), which leads to a decrease in lattice parameter reflected by the overall right shift of XRD reflexes [77]. As long as the MgO is more than FeO in olivine, the melting temperature of solid solution is high [64], and the material is safe for refractory application up to 1350 °C, typical for Cu production. PAK showed relatively high density after corrosion at 1400 °C of 4.70 g/cm^3, which is the effect of high Cr_2O_3 content (ρ_{Cr2O3} = 5.22 g/cm^3). In contrast to the shrinkage of SP AM 70 during heating, PAK—chromium-containing raw material—expands much during heating, which may result from changing Fe^{2+}/Fe^{3+} ratio during heating in the air—a partial reduction of iron oxides in $(Mg,Fe)(Fe,Cr,Al)_2O_4$ up to 1200 °C, followed by their re-oxidation [68].

Fused spinel SP AM 70 showed the lowest corrosion resistance to PbO-rich slag among all tested refractory raw materials. This study showed that the spinel phase of low Cr_2O_3 content is poorly resistant to Cu slag. SP AM 70 mixed with slag started to sinter as at 970 °C, which is especially unfavorable due to the fast and significant shrinkage of the material afterward. Such behavior resulted from the massive formation of a new corrosion product—güggenite. Güggenite formed in significant amounts in SP AM 70, as reflected by the high intensity of reflexes in XRD (Figure 5h), and this new phase was also observed by SEM/EDS (Figure 10b). As shown in phase equilibria of the $MgO-Cu_2O-CuO$ system in [63], güggenite is a non-stoichiometric phase that is stable over a wide range of compositions from $CuMgO_2$ to Cu_3MgO_4. Güggenite was found to exist in three crystallographic forms, namely, Güggenite A (orthorhombic, Pmmn (59)), güggenite B (orthorhombic, space group I(0)), and güggenite X (monoclinic). Güggenite A (67% CuO, 33% MgO) was obtained [63] in air at 1000 °C, and it was found to transform into güggenite X above 1050 °C. Güggenite B (75% CuO, 25% MgO) was obtained above 1050 °C. Güggenite can exist in equilibria with CuO (tenorite) and Cu_2O (cuprite); with tenorite (monoclinic) at temperatures below 1000 °C; and with cuprite (cubic) above 1021 °C. This phase was lately shown in MgO-based material corroded by CuO from matte [77]. In our work, the CuO_x-MgO phase that appeared is characterized by the lower Cu/Mg ratio of 0.7 for FMC45 (p. 2, Figure 7b) and 0.2 for PAK (p. 2, Figure 10b), while theoretically Cu/Mg in güggenite can vary from 1 (for $CuMgO_2$) to 3 (for Cu_3MgO_4). However, in [63] where authors investigated compositions with the general formula $Mg_{1-x}Cu_xO$ (x = 0 ÷ 1.0), güggenite A appeared from x = 0.17 (950 °C), which corresponds to Cu/Mg = 0.2. This confirms the phase observed in the present work of low Cu/Mg ratio is in fact güggenite. Thus, güggenite can occur in a wider composition range, depending on temperature.

As can be seen in the magnified image (Figure 10b), the güggenite phase is likely to separate CuO of needle morphology. The needles-like small crystals were previously found in tenorite by several authors [77]. In fact, the presence of CuO needles in the neighborhood of güggenite can result from the fact that güggenite underwent decomposition during free cooling, as it decomposes at 1062 °C. Similarly, very scarce information is available on the system PbO-MgO, which requires further study.

$$MgO_{refractory} + 2CuO_{slag} \rightarrow Cu_2MgO_{3solid\ solution} \tag{4}$$

Cu_2MgO_3 was previously found [67] during corrosion of MgO-Cr refractory by slag in the air atmosphere. Güggenite formed during cooling of the sample below 1100 °C; however, its physicochemical properties are poorly recognized and require future studies. According to the ternary system $CuO-Cu_2O-MgO$ [63], Cu_2MgO_3 is stable below 1062 °C. This means that the presence of güggenite is unfavorable in ab refractory, which operates in temperature range 900–1100 °C, as its decomposition above 1062 °C causes spalling of the refractory structure, e.g., during refractory reheating [67]. Moreover, forsterite formed in SP AM 70, but its amount was relatively low, as confirmed by the slight intensity of XRD reflexes characteristic for this phase. Despite corrosion resistance of spinel solid solution grains (Figure 1, Table 9; [31]), discussed phenomena of formation of new phases (forsterite, güggenite) contributed to the overall poor resistance of SP AM 70 to Cu slag. In addition, in [19] authors observed the liquid Cu-slag penetrated throughout the interfaces in a $MgO-MgAl_2O_4$-based refractory.

In the present work, low amounts of As were detected (~2%), which dissolved mostly in MgO and spinel grains. As assumed by Reinharter et al. [78], As_2O_3 tend to react mainly with basic components of refractories.

Simplified corrosion resistance mechanism was given in Figure 11.

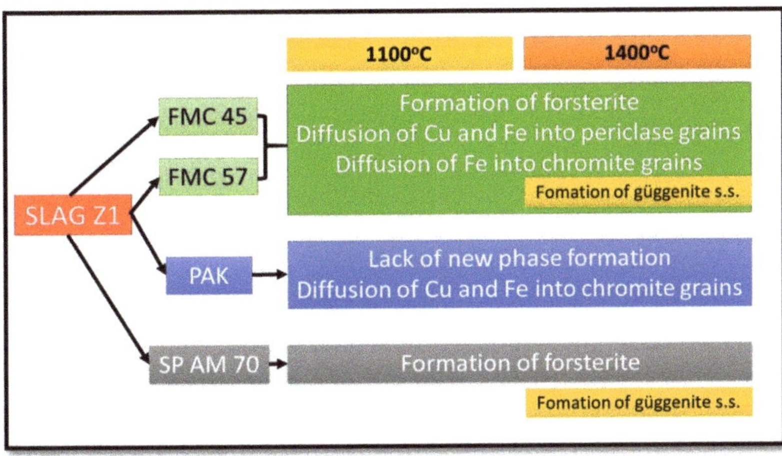

Figure 11. The main mechanisms of the corrosion interaction between refractory raw materials and slag Z1 at 1100 °C and 1400 °C.

Table 10. Comparison of results obtained in the present study with other works. MgO-Cr—magnesia–chromite product, Al$_2$O$_3$-Cr—alumina–chromite product.

Research Work	Testing Materials and Conditions	Massive Formation of Forsterite	Dissolution of Mg into Melt	Diffusion of Fe and Cu into MgO Grains	Diffusion of Fe into Chromite Grains	Formation of Güggenite in Contact with Cr-Containing Material	Formation of Güggenite in Contact with Cr-Free Material	Formation of Corrosion Protective/Interface Layer
Present work	Raw materials: FMC 45, FMC 57, PAK, SP AM 70 Corrosive agent: PbO-rich copper slag Temperature: 1100 °C and 1400 °C	+	+	+	+	+	+	
[42]	Product: MgO-Cr Corrosive agent: PbO-rich copper slags Temperature: 1300 °C	+	+	+	+	+		
[40]	Product: MgO-Cr Corrosive agent: PbO-based slag Temperature: 1300 °C	+	+		+			+
[39]	Product: MgO-Cr Corrosive agent: PbO-SiO$_2$-MgO slag Temperature: 1200 °C	+	+		+			
[2]	Product: MgO-Cr Corrosive agent: Cu-Cu$_x$O-PbO Temperature: 1200 °C		+	+	+			
[25]	Product: MgO-Cr Corrosive agent: Fayalite slag with increased ZnO content Temperature: 1200 °C		+		+			+
[74]	Product: MgO-Cr Corrosive agent: Copper smelting slag Temperature: 1250 °C	+	+	+	+	+		
[29]	Product: Al$_2$O$_3$-Cr Corrosive agent: Fayalite slag with increased ZnO content Temperature: 1200 °C		Dissolution of Al in slag		+			+
[30]	Product: MgAl$_2$O$_4$ Corrosive agent: Cu$_2$OT emperature: 1300 °C							+
[36]	Product: MgO doped with ZrO$_2$ nanoparticles Corrosive agent: Fayalite slag Temperature: 1450 °C	+	+					

5. Conclusions

- Four commercial refractory raw materials commonly used in refractories for Cu metallurgy—two magnesia–chromite co-clinkers (FMC 45 and FMC 57), a chromite ore (PAK), and a fused spinel (SP AM 70)—were comparatively investigated against PbO-rich Cu slag by hot-stage microscopy and pellet corrosion test.
- Test slag, containing the high levels of PbO of 39% and CuO$_x$ of 16%, was characterized by a low melting point of 969 °C, determined by hot-stage microscopy test.
- From the results of both pellet and hot-stage microscopy corrosion tests, the most beneficial behavior was determined for FMC 45. It exhibited relatively stable dimensions during heating with 2% shrinkage, which corresponded to sample sintering at 1440 °C, being the highest sintering point among all tested raw materials. The only corrosion product was forsterite, which formed in slight amounts.

- Fused spinel (SP AM 70) was the least resistant to PbO-rich slag, as it started to sinter as first at 970 °C, followed by a fast and high 8% shrinkage of the material afterward. Güggenite solid solution formed due to the reaction between CuO_x from slag and MgO from refractory. This phase is potentially detrimental, as it decomposes above 1062 °C, leading to the spalling of the material during reheating. Moreover, forsterite (Mg_2SiO_4) formed during corrosion can be harmful in larger amounts due to its significant volume.
- Despite the different phase composition, PAK and FMC 57 showed comparable corrosion resistance and high-temperature behavior; thus, they constitute the most promising prospective raw materials for refractories dedicated to non-ferrous metallurgy where aggressive Pb-Cu-O slags occur.

Author Contributions: Writing—original draft preparation, M.L.; methodology, E.Ś., M.N., J.V., R.P.; supervision, I.J., J.S., R.P.; writing—review and editing, Y.L., N.L., M.N., J.V., formal analysis, Y.L., N.L.; investigation; M.L., J.S., I.J., R.P.; visualization, E.Ś., M.L.; conceptualization, J.S., J.V., Y.L., E.Ś., N.L., M.N. All authors have read and agreed to the published version of the manuscript.

Funding: This research was supported by the funds of The National Centre for Research and Development, grant no. LIDER/14/0086/L-12/20/NCBR/2021, and by the statutory funds of the Faculty of Material Science and Ceramics AGH University of Science and Technology, Poland Agreement no. 16.16.160.557.

Data Availability Statement: Data available on request due to restrictions eg. privacy or ethical The data presented in this study are available on request from the corresponding author. The data are not publicly available due to technical or time limitations.

Conflicts of Interest: The authors declare no conflict of interest.

References

1. Malfliet, A.; Lotfian, S.; Scheunis, L.; Petkov, V.; Pandelaers, L.; Jones, P.T.; Blanpain, B. Degradation mechanisms and use of refractory linings in copper production processes: A critical review. *J. Eur. Ceram. Soc.* **2014**, *34*, 849–876. [CrossRef]
2. Chen, L.; Li, S.; Jones, P.T.; Guo, M.; Blanpain, B.; Malfliet, A. Identification of magnesia-chromite refractory degradation mechanisms of secondary copper smelter linings. *J. Eur. Ceram. Soc.* **2016**, *36*, 2119–2132. [CrossRef]
3. Petkov, V.; Jones, P.T.; Blanpain, B. Optimisation of an anode furnace refractory lining using distinct magnesia-chromite refractory types. *World Metall.—ERZMETALL* **2007**, *60*, 208–217.
4. Gregurek, D.; Schmidl, J.; Reinharter, K.; Reiter, V.; Spanring, A. Copper Anode Furnace: Chemical, Mineralogical and Thermo-Chemical Considerations of Refractory Wear Mechanisms. *JOM* **2018**, *70*, 2428–2434. [CrossRef]
5. Schlesinger, M.E. Refractories for copper production. *Miner. Process. Extr. Metall. Rev.* **1996**, *16*, 125–140. [CrossRef]
6. Nadachowski, F. *Zarys Technologii Materiałów Ogniotrwałych*; Śląskie Wydawnictwa Techniczne: Katowice, Poland, 1995.
7. Pashkeev, Y.; Vlasov, V.N. To the problem of phase equilibrium in PbO-SiO_2 system. *Izvestiia-vysshie Uchebnye Zavedeniia Tsvetnaia Metallurgiia* **1999**, *4*, 3–6.
8. Wu, M.P.; Eriksson, G.; Pelton, A.D. Calculated phase diagram of the system FeO-SiO_2 in equilibrium with metallic Fe with total iron expressed as "FeO". *ISIJ Int.* **1993**, *33*, 26–35. [CrossRef]
9. Gabasiane, T.S.; Danha, G.; Mamvura, T.A.; Mashifana, T.; Dzinomwa, G. Characterization of copper slag for beneficiation of iron and copper. *Heliyon* **2021**, *7*, e06757. [CrossRef] [PubMed]
10. Takeda, Y. Phase Diagram of CaO-FeO-Cu_2O slag under copper saturation. *Acta Crystallogr.* **1956**, *9*, 211–225. [CrossRef]
11. Bechta, S.V.; Krushinov, E.V.; Al'myashev, V.I.; Vitol, S.A.; Mezentseva, L.P.; Petrov, Y.B.; Lopukh, D.B.; Khabenskii, V.B.; Barrachin, M.; Hellmann, S. T-X phase diagram for the join Fe-Fe_2O_3. *Russ. J. Inorg. Chem.* **2006**, *51*, 325–331.
12. Mcewan, N.; Courtney, T.; Parry, R.A.; Knupfer, P. Chromite—A cost-effective refractory raw material for refractories in various metallurgical applications. *S. Afr. Pyrometallurgy* **2011**, 359–373.
13. Seetharaman, S. Industrial Processes. In *Treatise on Process Metallurgy*; Elsevier: Amsterdam, The Netherlands, 2013; Volume 3.
14. Petkov, V.; Jones, P.T.; Boydens, E.; Blanpain, B.; Wollants, P. Chemical corrosion mechanisms of magnesia-chromite and chrome-free refractory bricks by copper metal and anode slag. *J. Eur. Ceram. Soc.* **2007**, *27*, 2433–2444. [CrossRef]
15. Golestani Fard, F.; Talimian, A. Improving Corrosion Behaviour of Magnesia-chrome Refractories by Addition of Nanoparticles. *Refract. Worldforum* **2014**, *6*, 93–98.
16. Lotfian, N.; Nourbakhsh, A.A.; Mirsattari, S.N.; Mackenzie, K.J.D. Functionalization of nano-$MgCr_2O_4$ additives by silanol groups: A new approach to the development of magnesia-chrome refractories. *Ceram. Int.* **2021**, *47*, 31724–31731. [CrossRef]

17. Lotfian, N.; Nourbakhsh, A.A.; Mirsattari, S.N.; Saberi, A.; Mackenzie, K.J.D. A comparison of the effect of nanostructured $MgCr_2O_4$ and $FeCr_2O_4$ additions on the microstructure and mechanical properties of direct-bonded magnesia-chrome refractories. *Ceram. Int.* **2020**, *46*, 747–754. [CrossRef]

18. Pérez, I.; Moreno-Ventas, I.; Parra, R.; Ríos, G. Comparative analysis of refractory wear in the copper-making process by a novel (industrial) dynamic test. *Ceram. Int.* **2019**, *45*, 1535–1544. [CrossRef]

19. Rigaud, M.; Palco, S.; Paransky, E. New refractory materials for the copper industry. In Proceedings of the Tehran International Conference on Refractories, Tehran, Iran, 4–6 May 2004.

20. Pérez, I.; Moreno-Ventas, I.; Ríos, G. Post-mortem study of magnesia-chromite refractory used in Peirce-Smith Converter for copper-making process, supported by thermochemical calculations. *Ceram. Int.* **2018**, *44*, 13476–13486. [CrossRef]

21. Pérez, I.; Moreno-Ventas, I.; Ríos, G. Chemical degradation of magnesia-chromite refractory used in the conversion step of the pyrometallurgical copper-making process: A thermochemical approach. *Ceram. Int.* **2018**, *44*, 18363–18375. [CrossRef]

22. Gregurek, D.; Prietl, T.; Breyner, S.B.; Ressler, A.; Berghofer, N.M. Innovative Magnesia-Chrome Fused Grain Material For Non-Ferrous Metals Refractory Applications. *S. Afr. Inst. Min. Metall.* **2012**, 251–260.

23. Xu, L.; Chen, M.; Wang, N.; Gao, S. Chemical wear mechanism of magnesia-chromite refractory for an oxygen bottom-blown copper-smelting furnace: A post-mortem analysis. *Ceram. Int.* **2021**, *47*, 2908–2915. [CrossRef]

24. Lange, M.; Garbers-Craig, A.M.; Cromarty, R. Wear of magnesia-chrome refractory bricks as a function of matte temperature. *J. S. Afr. Inst. Min. Metall.* **2014**, *114*, 4.

25. Chen, L.; Guo, M.; Shi, H.; Huang, S.; Jones, P.T.; Blanpain, B.; Malfliet, A. Effect of ZnO level in secondary copper smelting slags on slag/magnesia-chromite refractory interactions. *J. Eur. Ceram. Soc.* **2016**, *36*, 1821–1828. [CrossRef]

26. Chen, L.; Guo, M.; Shi, H.; Scheunis, L.; Jones, P.T.; Blanpain, B.; Malfliet, A. The influence of ZnO in fayalite slag on the degradation of magnesia-chromite refractories during secondary Cu smelting. *J. Eur. Ceram. Soc.* **2015**, *35*, 2641–2650. [CrossRef]

27. Xu, L.; Chen, M.; Wang, N.; Gao, S.; Wu, Y. Degradation mechanisms of magnesia-chromite refractory bricks used in oxygen side-blown reducing furnace. *Ceram. Int.* **2020**, *46*, 17315–17324. [CrossRef]

28. Hu, S.; Huang, A.; Jia, Q.; Zhang, S.; Chen, L. Degradation of magnesia-chromite refractory in ZnO-containing ferrous calcium silicate slags. *Ceram. Int.* **2021**, *47*, 11276–11284. [CrossRef]

29. Chen, L.; Malfliet, A.; Vleugels, J.; Blanpain, B.; Guo, M. Degradation mechanisms of alumina-chromia refractories for secondary copper smelter linings. *Corros. Sci.* **2018**, *136*, 409–417. [CrossRef]

30. Jiang, Y.; Chen, M.; Chen, J.; Zhao, B. Interactions of $MgO \cdot Al_2O_3$ spinel with Cu, Cu_2O and copper matte at high temperature. *Ceram. Int.* **2018**, *44*, 14108–14112. [CrossRef]

31. Hellstén, N.; Taskinen, P. Experimental phase relations between MgO-saturated magnesium-aluminate spinel ($MgAl_2O_4$) and CuO_x-rich liquid. *Ceram. Int.* **2017**, *43*, 11116–11122. [CrossRef]

32. De Wilde, E.; Bellemans, I.; Campforts, M.; Guo, M.; Blanpain, B.; Moelans, N.; Verbeken, K. Sessile drop evaluation of high temperature copper/spinel and slag/spinel interactions. *Trans. Nonferrous Met. Soc. China* **2016**, *26*, 2770–2783. [CrossRef]

33. Jedynak, L.; Wojsa, J.; Podwórny, J.; Wala, T. Refractories from $MgO-Al_2O_3-SnO_2$ system for metallurgical applications. In Proceedings of the 11th International Conference and Exhibition of the European Ceramic Society, Krakow, Poland, 21–25 June 2009; Volume 2009, pp. 812–817.

34. Jedynak, L.; Wojsa, J. Bezchromowe, niewypalane materiały ogniotrwałe. *Pr. Inst. Szkła Ceram. Mater. Ogniotrwałych i Bud.* **2010**, *3*, 69–79.

35. Yurkov, A.; Malakho, A.; Avdeev, V. Corrosion behavior of silicon nitride bonded silicon carbide refractory material by molten copper and copper slag. *Ceram. Int.* **2017**, *43*, 4241–4245. [CrossRef]

36. Gómez-Rodríguez, C.; Antonio-Zárate, Y.; Revuelta-Acosta, J.; Verdej, L.F.; Fernández-González, D.; López-Perales, J.F.; Rodríguz-Castellanos, E.A.; García-Quiñonez, L.V.; Castillo-Rodríguez, G.A. Research and development of novel refractory of MgO doped with ZrO_2 nanoparticles for copper slag resistance. *Materials* **2021**, *14*, 2277. [CrossRef]

37. Hellstén, N.; Hamuyuni, J.; Taskinen, P. High-temperature phase equilibria of Cu-O-MgO system in air. *Thermochim. Acta* **2016**, *623*, 107–111. [CrossRef]

38. Scheunis, L.; Campforts, M.; Jones, P.T.; Blanpain, B.; Malfliet, A. Spinel saturation of a PbO based slag as a method to mitigate the chemical degradation of magnesia-chromite bricks. *J. Eur. Ceram. Soc.* **2016**, *36*, 4291–4299. [CrossRef]

39. Scheunis, L.; Fallah-Mehrjardi, A.; Campforts, M.; Jones, P.T.; Blanpain, B.; Malfliet, A.; Jak, E. The effect of a temperature gradient on the phase formation inside a magnesia-chromite refractory in contact with a non-ferrous $PbO-SiO_2-MgO$ slag. *J. Eur. Ceram. Soc.* **2015**, *35*, 2933–2942. [CrossRef]

40. Scheunis, L.; Fallah Mehrjardi, A.; Campforts, M.; Jones, P.T.; Blanpain, B.; Jak, E. The effect of phase formation during use on the chemical corrosion of magnesia-chromite refractories in contact with a non-ferrous $PbO-SiO_2$ based slag. *J. Eur. Ceram. Soc.* **2014**, *34*, 1599–1610. [CrossRef]

41. Scheunis, L.; Campforts, M.; Jones, P.T.; Blanpain, B.; Malfliet, A. The influence of slag compositional changes on the chemical degradation of magnesia-chromite refractories exposed to PbO-based non-ferrous slag saturated in spinel. *J. Eur. Ceram. Soc.* **2015**, *35*, 347–355. [CrossRef]

42. Ludwig, M.; Śnieżek, E.; Jastrzębska, I.; Piwowarczyk, A.; Wojteczko, A.; Li, Y.; Szczerba, J. Corrosion of magnesia-chromite refractory by PbO-rich copper slags. *Corros. Sci.* **2022**, *195*, 109949. [CrossRef]

43. Galos, K. *Surowce krajowego przemysłu materiałów ogniotrwałych w świetle przemian gospodarczych*; Polska Akademia Nauk: Kraków, Poland, 1999.

44. Jung, I.; Decterov, S.A.; Pelton, A.D. Critical Thermodynamic Evaluation and Optimization of the MgO-Al_2O_3, CaO-MgO-Al_2O_3, and MgO-Al_2O_3-SiO_2 systems. *J. Phase Equilibria Diffus.* **2004**, *25*, 329–345. [CrossRef]

45. Beals, R.J.; Cook, R.L. Directional Dilatation of Crystal Lattices at Elevated Temperatures. *J. Am. Ceram. Soc.* **1957**, *40*, 279–284. [CrossRef]

46. Wu, G.; Yan, W.; Schafföner, S.; Lin, X.; Ma, S.; Zhai, Y.; Liu, X.; Xu, L. Effect of magnesium aluminate spinel content of porous aggregates on cement clinker corrosion and adherence properties of lightweight periclase-spinel refractories. *Constr. Build. Mater.* **2018**, *185*, 102–109. [CrossRef]

47. Hartenstein, J.; Gueguen, E.; Moulin, J.; Schepers, A. Imroving properties of magnesia-aluminate spinel bricks for cement rotary kilns by using micro-fine MgO particles. In Proceedings of the UNITECR Congress, Santiago, Chile, 26–29 September 2017.

48. Ohno, M.; Yoshikawa, S.; Toda, H.; Fujii, M.; Chiba, H.; Ozeki, F. Magnesia-Spinel Brick With Good Coating Adhesion and High Resistance To Corrosion and Spalling for Cement Rotary Kilns. In Proceedings of the UNITECR Congress, Santiago, Chile, 26–29 September 2017.

49. Braulio, M.A.L.; Martinez, A.G.T.; Luz, A.P.; Liebske, C.; Pandolfelli, V.C. Basic slag attack of spinel-containing refractory castables. *Ceram. Int.* **2011**, *37*, 1935–1945. [CrossRef]

50. Sako, E.Y.; Braulio, M.A.L.; Zinngrebe, E.; Van Der Laan, S.R.; Pandolfelli, V.C. Fundamentals and applications on in situ spinel formation mechanisms in Al_2O_3-MgO refractory castables. *Ceram. Int.* **2012**, *38*, 2243–2251. [CrossRef]

51. Luz, A.P.; Braulio, M.A.L.; Tomba Martinez, A.G.; Pandolfelli, V.C. Slag attack evaluation of in situ spinel-containing refractory castables via experimental tests and thermodynamic simulations. *Ceram. Int.* **2012**, *38*, 1497–1505. [CrossRef]

52. Sako, E.Y.; Braulio, M.A.L.; Pandolfelli, V.C. The corrosion resistance of microsilica-containing Al_2O_3-MgO and Al_2O_3-spinel castables. *Ceram. Int.* **2012**, *38*, 4783–4789. [CrossRef]

53. Panna, W.; Wyszomirski, P.; Kohut, P. Application of hot-stage microscopy to evaluating sample morphology changes on heating. *J. Therm. Anal. Calorim.* **2016**, *125*, 1053–1059. [CrossRef]

54. Ludwig, M.; Wiśniewska, K.; Śnieżek, E.; Jastrzębska, I.; Prorok, R.; Szczerba, J. Effect of the chemical composition of slag on the corrosion of calcium zirconate material. *Mater. Chem. Phys.* **2021**, *258*, 123844. [CrossRef]

55. Serry, M.A.; Othman, A.G.M.; Girgis, L.G.; Telle, R. Assessment of coclinkered shaped magnesite-chromite refractories processed from Egyptian materials. *Am. Ceram. Soc. Bull.* **2007**, *86*, 9101–9106.

56. Tripathy, S.K.; Murthy, Y.R.; Singh, V.; Suresh, N. Processing of Ferruginous Chromite Ore by Dry High-Intensity Magnetic Separation. *Miner. Process. Extr. Metall. Rev.* **2016**, *37*, 196–210. [CrossRef]

57. Motasim, M.; Al-Tigani, H.; Mohamed, A.A.; Abdullah, A.; Seifelnasr, S. Mineralogical and Chemical Characterization of Disseminated Low-Grade Sudanese Chromite Ore in Gedarif State at Umm Saqata-Qala Elnahal. *J. Environ. Anal. Chem.* **2019**, *6*, 1000261. [CrossRef]

58. Gorai, B.; Jana, R.K. Premchand, Characteristics and utilisation of copper slag—A review. *Resour. Conserv. Recycl.* **2003**, *39*, 299–313. [CrossRef]

59. Wang, X.; Geysen, D.; Padilla, S.V.; D'Hoker, N.; Van Gerven, T.; Blanpain, B. Characterization of Copper Slag. In *REWAS 2013 Enabling Materials Resource Sustainability*; Springer: Cham, Switzerland, 2013; pp. 54–68. [CrossRef]

60. Gregurek, D.; Majcenovic, C.; Budna, K.; Schmidl, J.; Spanring, A. *Wear Phenomena in Non-Ferrous Metal Furnaces*; Springer International Publishing: Cham, Switzerland, 2018.

61. Gregurek, D.; Wenzl, C.; Reiter, V.; Studnicka, H.L.; Spanring, A. Slag Characterization: A Necessary Tool for Modeling and Simulating Refractory Corrosion on a Pilot Scale. *JOM* **2014**, *66*, 1677–1686. [CrossRef]

62. Madhusudhan Rao, A.S.; Narender, K. Studies on thermophysical properties of CaO and MgO by γ-ray attenuation. *J. Thermodyn.* **2014**, *2014*, 123478. [CrossRef]

63. Paranthaman, M.; David, K.A.; Lindemer, T.B. Phase equilibria of the MgO-Cu_2O-CuO system. *Mater. Res. Bull.* **1997**, *32*, 165–173. [CrossRef]

64. Beattie, P. Olivine-melt and orthopyroxene-melt equilibria. *Contrib. Mineral. Petrol.* **1993**, *115*, 103–111. [CrossRef]

65. Najem, I.A.; Edrees, S.J.; Rasin, F.A. Structural and Magnetic Characterisations of Pb-Doped MgO Nanoparticles by a Modified Pechini Method. *IOP Conf. Ser. Mater. Sci. Eng.* **2020**, *987*, 012027. [CrossRef]

66. Chen, S.; Zhao, B.; Jak, E.; Hayes, P.C. Experimental study of phase equilibria in the PbO-MgO-SiO_2 system. *Metall. Mater. Trans. B* **2001**, *32*, 11–16. [CrossRef]

67. O'Neill, H.S.C.; Navrotsky, A. Simple spinels; crystallographic parameters, cation radii, lattice energies, and cation distribution. *Am. Mineral.* **1983**, *68*, 181–194.

68. Jastrzębska, I.; Szczerba, J.; Błachowski, A.; Stoch, P. Structure and microstructure evolution of hercynite spinel ($Fe^{2+}Al_2O_4$) after annealing treatment. *Eur. J. Mineral.* **2017**, *29*, 62–71. [CrossRef]

69. Jastrzębska, I.; Szczerba, J.; Stoch, P. Structural and microstructural study on the Arc-Plasma Synthesized (APS) $FeAl_2O_4$-$MgAl_2O_4$ transitional refractory compound. *High Temp. Mater. Process.* **2017**, *36*, 299–303. [CrossRef]

70. Jastrzebska, I.; Bodnar, W.; Witte, K.; Burkel, E.; Stoch, P.; Szczerba, J. Structural properties of Mn-substituted hercynite. *Nukleonika* **2017**, *62*, 95–100. [CrossRef]

71. Chen, M.; Chen, J.; Zhao, B. Corrosion resistances of Cr-free refractories to copper smelting slags. In *Advances in Molten Slags, Fluxes, Salts, Proceedings of the 10th International Conference on Molten Slags, Fluxes Salts, Seattle, DC, USA, 22–25 May 2016*; Springer: Cham, Switzerland, 2016; pp. 1101–1108. [CrossRef]

72. Kirgintsev, A.N. To the thermodynamics of the FeO-SiO$_2$ phase diagram. *Geochem. Int.* **1998**, *36*, 874–876.

73. Schramm, L.; Behr, G.; Löser, W.; Wetzig, K. Thermodynamic reassessment of the Cu-O phase diagram. *J. Phase Equilib. Diff.* **2005**, *26*, 605–612. [CrossRef]

74. Chen, M.; Jiang, Y.; Cui, Z.; Wei, C.; Zhao, B. Chemical Degradation Mechanisms of Magnesia–Chromite Refractories in the Copper Smelting Furnace. *JOM* **2018**, *70*, 2443–2448. [CrossRef]

75. Liu, G.; Li, Y.; Zhu, T.; Xu, Y.; Liu, J.; Sang, S.; Li, Q.; Li, Y. Influence of the atmosphere on the mechanical properties and slag resistance of magnesia-chrome bricks. *Ceram. Int.* **2020**, *46*, 11225–11231. [CrossRef]

76. Shannon, R.D.; Prewitt, C.T. Effective ionic radii in oxides and fluorides. *Acta Crystallogr. Sect. B Struct. Crystallogr. Cryst. Chem.* **1969**, *25*, 925–946. [CrossRef]

77. Available online: https://www.sciencedirect.com/topics/chemistry/copper-oxide (accessed on 10 November 2021).

78. Reinharter, K.; Gregurek, D.; Majcenovic, C.; Schmidl, J.; Spanring, A. Influence of arsenic on the chemical wear of magnesia-chromite refractories in copper smelting furnaces. In *Extraction*; The Minerals, Metals & Materials Series; Springer: Cham, Switzerland, 2018.

Article

High-Temperature Chemical Stability of Cr(III) Oxide Refractories in the Presence of Calcium Aluminate Cement

Tengteng Xu [1,2], Yibiao Xu [1,2,*], Ning Liao [1,2], Yawei Li [1,2,*] and Mithun Nath [1,2,*]

[1] The State Key Laboratory of Refractories and Metallurgy, Wuhan University of Science and Technology, Wuhan 430081, China; xutengteng1992@163.com (T.X.); liaoning@wust.edu.cn (N.L.)

[2] National-Provincial Joint Engineering Research Center of High Temperature Materials and Lining Technology, Wuhan University of Science and Technology, Wuhan 430081, China

* Correspondence: xuyibiao@wust.edu.cn (Y.X.); liyawei@wust.edu.cn (Y.L.); mithunnath@wust.edu.cn (M.N.)

Abstract: Al_2O_3-CaO-Cr_2O_3 castables are used in various furnaces due to excellent corrosion resistance and sufficient early strength, but toxic Cr(VI) generation during service remains a concern. Here, we investigated the relative reactivity of analogous Cr(III) phases such as Cr_2O_3, $(Al_{1-x}Cr_x)_2O_3$ and in situ Cr(III) solid solution with the calcium aluminate cement under an oxidizing atmosphere at various temperatures. The aim is to comprehend the relative Cr(VI) generation in the low-cement castables (Al_2O_3-CaO-Cr_2O_3-O_2 system) and achieve an environment-friendly application. The solid-state reactions and Cr(VI) formation were investigated using powder XRD, SEM, and leaching tests. Compared to Cr_2O_3, the stability of $(Al_{1-x}Cr_x)_2O_3$ against CAC was much higher, which improved gradually with the concentration of Al_2O_3 in $(Al_{1-x}Cr_x)_2O_3$. The substitution of Cr_2O_3 with $(Al_{1-x}Cr_x)_2O_3$ in the Al_2O_3-CaO-Cr_2O_3 castables could completely inhibit the formation of Cr(VI) compound $CaCrO_4$ at 500–1100 °C and could drastically suppress $Ca_4Al_6CrO_{16}$ generation at 900 to 1300 °C. The Cr(VI) reduction amounting up to 98.1% could be achieved by replacing Cr_2O_3 with $(Al_{1-x}Cr_x)_2O_3$ solid solution. However, in situ stabilized Cr(III) phases as a mixture of $(Al_{1-x}Cr_x)_2O_3$ and $Ca(Al_{12-x}Cr_x)O_{19}$ solid solution hardly reveal any reoxidation. Moreover, the CA_6 was much more stable than CA and CA_2, and it did not participate in any chemical reaction with $(Al_{1-x}Cr_x)_2O_3$ solid solution.

Keywords: Al_2O_3-CaO-Cr_2O_3-O_2 system; $(Al_{1-x}Cr_x)_2O_3$; $Ca(Al_{12-x}Cr_x)O_{19}$; Cr(VI) compounds; leaching test

Citation: Xu, T.; Xu, Y.; Liao, N.; Li, Y.; Nath, M. High-Temperature Chemical Stability of Cr(III) Oxide Refractories in the Presence of Calcium Aluminate Cement. *Materials* **2021**, *14*, 6590. https://doi.org/10.3390/ma14216590

Academic Editors: Jacek Szczerba and Ilona Jastrzębska

Received: 30 August 2021
Accepted: 26 October 2021
Published: 2 November 2021

Publisher's Note: MDPI stays neutral with regard to jurisdictional claims in published maps and institutional affiliations.

1. Introduction

Cr_2O_3-containing refractories possess remarkable corrosion resistance due to their extremely low solubility and high chemical stability against molten slag. Therefore, they are widely used as lining materials in incinerators, gasifiers, glass furnaces, non-ferrous smelting, etc. [1–6]. In addition, refractories as castables have become a popular choice in recent decades because of the energy-saving manufacturing process, convenient for installation and repair works, where binders' chemistry plays a crucial role [7–9]. Calcium alumina cement (CAC) binders are the most widely used since they exhibit fast setting and strength development, stable thermo-mechanical behaviour, and resistance to slag attack [10]. However, Cr_2O_3 can oxidize into toxic Cr(VI) products at high temperatures upon reaction with alkali or alkaline earth metals/oxides/compounds under an oxidizing atmosphere [11–13]. The Cr(VI) compounds pose a severe threat to humans and the environment since they are toxic, carcinogenic, highly soluble in water, and quickly enter the food cycle [14]. Therefore, it is of significant environmental and practical significance to inhibit the generation of Cr(VI) when applying Al_2O_3-CaO-Cr_2O_3 castables as lining materials.

The Al_2O_3-CaO-Cr_2O_3 system was not investigated in detail earlier though numerous Cr(VI) reduction techniques were described for other applications [15–17]. Generally, Cr(VI) formation was closely related to the atmosphere and basicity of other components in the

Cr$_2$O$_3$-containing materials [18,19]. For the Cr$_2$O$_3$-containing refractory linings, since the operating conditions and service atmosphere in a given furnace can hardly be changed in practical production, most related work has focused on Cr(VI) minimization using some additives at high temperatures. The acidic components such as SiO$_2$, TiO$_2$, Fe$_2$O$_3$, and P$_2$O$_5$ can effectively suppress the Cr(III) oxidation during thermal treatment of Cr$_2$O$_3$-containing refractories [19–22]. However, these oxide additives usually result in low melting point phases in the matrix, deteriorating either the thermo-mechanical properties or the slag corrosion resistance [23,24]. Previous research indicated that incorporating chromium into solid solution phases can reduce the risk of Cr(VI) formation in the Cr$_2$O$_3$-containing materials [25–27]. For example, the investigation of the Al$_2$O$_3$-Cr$_2$O$_3$-CaO-MgO pure system confirmed that composite spinel Mg(Al,Cr)$_2$O$_4$ could co-exist with CA$_2$, where chromium existed in +3 state [25,27]. Again, Wu et al. [28] studied the effect of temperature on Cr(VI) formation in a pure (Al,Cr)$_2$O$_3$ system with CAC in air atmosphere, where (Al$_{1-x}$Cr$_x$)$_2$O$_3$ was found to be chemically stable against CAC up to 1100 °C, beyond which Ca$_4$Al$_6$CrO$_{16}$ (hauyne) phase predominantly start from.

Nath et al. investigated the phase evolution of the Al$_2$O$_3$-CaO-Cr$_2$O$_3$ refractories castables after treatment at various temperatures, where CaO from cement facilitated the conversion of Cr(III) into Cr(VI) [29]. The main phase of CAC (CA and CA$_2$) react with Cr$_2$O$_3$ in the air to produce CaCrO$_4$ and Ca$_4$Al$_6$CrO$_{16}$ at mid-temperature (700–1100 °C). At the same time, nearly all the Cr$_2$O$_3$ would convert into (Al$_{1-x}$Cr$_x$)$_2$O$_3$ ($0 < x < 1$) and Ca(Al,Cr)$_{12}$O$_{19}$ solid solution at 1500 °C, which leads to a significant decrease of Cr(VI) compounds amounts [20,29]. Thus, (Al$_{1-x}$Cr$_x$)$_2$O$_3$ solid solutions having high refractoriness and good chemical stability could be better performing materials with better mechanical properties and slag corrosion resistance [30–33]. Based on the above research, it can be inferred that substituting Cr$_2$O$_3$ with (Al$_{1-x}$Cr$_x$)$_2$O$_3$ solid solution as a starting component in the Al$_2$O$_3$-CaO-Cr$_2$O$_3$ castables could be a feasible way to inhibit the formation of Cr(VI) at various temperatures, especially at mid-temperature (700–1100 °C), without compromising other properties. However, systematic work relating to the effect of (Al$_{1-x}$Cr$_x$)$_2$O$_3$ solid solution on Cr(VI) formation in Al$_2$O$_3$-CaO-Cr$_2$O$_3$ refractory castables is rare.

The present work aims to inhibit the formation of Cr(VI) compounds in Al$_2$O$_3$-CaO-Cr$_2$O$_3$ castables by substituting Cr$_2$O$_3$ with (Al$_{1-x}$Cr$_x$)$_2$O$_3$ solid solution as starting chromium-containing constituent. Firstly, (Al$_{1-x}$Cr$_x$)$_2$O$_3$ solid solutions were pre-synthesized at 1300 to 1650 °C. Secondly, Al$_2$O$_3$-CaO-Cr$_2$O$_3$ castables with the pre-synthesized (Al$_{1-x}$Cr$_x$)$_2$O$_3$ solid solution were fabricated and treated at the temperature range of 300–1500 °C in the air since castables would be put to use without firing and a temperature gradient occurs in any furnace linings in actual practice. The phase evolution and Cr(VI) generation of the Al$_2$O$_3$-CaO-Cr$_2$O$_3$ castables with temperature and the corresponding mechanism were studied using XRD and related software, SEM, and leaching tests. Furthermore, since the (Al$_{1-x}$Cr$_x$)$_2$O$_3$ and Ca(Al,Cr)$_{12}$O$_{19}$ could be formed in the Al$_2$O$_3$-CaO-Cr$_2$O$_3$ castables at high temperature [20], castables with Cr$_2$O$_3$ were pre-heated at 1500 °C to produce the in situ formed (Al$_{1-x}$Cr$_x$)$_2$O$_3$, whose effect on the Cr(VI) formation for the castables at various temperature was also evaluated.

2. Materials and Methods

Tabular alumina (Al$_2$O$_3$) of various size fractions, 5–3 mm, 3–1 mm, 1–0 mm, and $\leq$0.045 mm, were procured from Zhejiang Zili Alumina Materials Technology Co., Ltd., Shangyu, China. Reactive α-alumina fines of size fraction $\leq$ 0.005 mm were purchased from Kaifeng Tenai Co., Ltd., Kaifeng, China. Industrial-grade fused chromium oxide (Cr$_2$O$_3$) (size $\leq$ 0.074 mm) was obtained from Luoyang Yuda Refractories Co., Ltd., Luoyang, China. The hydraulic calcium aluminate cement binder, Secar 71 (CA and CA$_2$ phases), was procured from Imerys Aluminates, Tianjin, China. An organic defloculant, FS 65 (Wuhan Sanndar Chemical Co., Ltd., Wuhan, China), was used as the dispersant. The detailed chemical composition of raw materials is shown in Table 1.

Table 1. The chemical composition of raw materials (wt%).

Raw Materials	SiO_2	Al_2O_3	CaO	Fe_2O_3	MgO	Na_2O	K_2O	Cr_2O_3
Tabular alumina	0.08	99.30	-	0.02	-	0.28	-	-
Reactive α-alumina	0.28	98.87	0.07	0.13	0.12	0.10	0.005	-
Fused chromium oxide	0.82	0.59	0.38	0.73	0.27	0.14	0.01	94.02
Calcium aluminate cement	0.40	70.6	28.4	0.20	-	-	-	-

Cr_2O_3 and Al_2O_3 fine powders with a mass ratio of 8:17 were dry-mixed, pressed into pellets, and then treated at 1300, 1600, and 1650 °C for 3 h in the air to obtain the mixture of Al_2O_3, Cr_2O_3 and $(Al_{1-x}Cr_x)_2O_3$ solid solution and pre-synthesized $(Al_{1-x}Cr_x)_2O_3$ solid solution. Thus, obtained pellets were then pulverized to 200-mesh powders before adding them into the castables. The specimen with Cr_2O_3 and Al_2O_3 powders as initial raw materials was labelled as R, while specimens with $(Al_{1-x}Cr_x)_2O_3$ solid solution pre-synthesized at 1300 °C, 1600 °C, and 1650 °C were designated as S13, S16, and S165, respectively. Specimen R was pre-heated at 1500 °C for 3h (labelled as F15) to produce the in situ formed $(Al_{1-x}Cr_x)_2O_3$, whose effect on the Cr(VI) formation in the castables at various temperatures was also evaluated then. The castables were formulated based on the Andreasen distribution coefficient (q) value of 0.31, and the specific formulation is shown in Table 2. Each batch was dry-mixed for 3 min in a Hobart mixer followed by wet-mixing (4.0 wt% water, 25 °C) for further 3 min, and then castables were moulded in a vibrating table (1 min) into bars of size 160 mm × 40 mm × 40 mm at room temperature. All specimens were cured at 25 °C and 75% ± 5% relative humidity for 24 h in a standard cement maintainer and dried at 110 °C for 24 h in an electric air oven after demoulding. Dried specimens R, S13, S16, S165, and specimen F15 were finally heated in the temperature range of 300–1500 °C for 3h at peak temperature in air.

Table 2. The formulation of fine powders undergoing reaction within castables (30wt% of total).

Code	Aggregates Al_2O_3 (wt%)	Fine Powders (wt%)				Pre-Treatment Temperature (°C)
		Al_2O_3	Cr_2O_3	CAC	$(Al_{1-x}Cr_x)_2O_3$	
R	70	17	8	5	-	-
F15	70	17	8	5	-	In situ treated at 1500
S13	70	-	-	5	25	$(Al_{1-x}Cr_x)_2O_3$ made at 1300
S16	70	-	-	5	25	$(Al_{1-x}Cr_x)_2O_3$ made at 1600
S165	70	-	-	5	25	$(Al_{1-x}Cr_x)_2O_3$ made at 1650

Note: 0.1 wt% additional organic dispersant was added to each formulation to make the castables. CAC designates calcium aluminate cement (Here, a commercial Secar 71 cement was used). Each batch contains 8 wt% of Cr_2O_3.

To figure out relative oxidation, the mechanisms of the Cr(VI) generation in the castables and the corresponding chemical reactions, fine powders of CAC and CA_6 were mixed with Cr_2O_3 and pre-synthesized $(Al_{1-x}Cr_x)_2O_3$ (Table 3). Then, the mixed powders were pressed into Φ20 mm × 20 mm cylindrical specimens under a pressure of 50 MPa. After being treated at 900 °C and 1300 °C for 3 h in the air atmosphere, the phase compositions and microstructures of the specimens were analyzed by XRD and SEM, respectively.

Table 3. The formulation of cylindrical specimens (wt%).

Specimens	CAC	CA_6	Cr_2O_3	$(Al_{1-x}Cr_x)_2O_3$
C-C	50		50	
C-S	50			50
CH-C		50	50	
CH-S		50		50

Note: CAC (Secar 71) contains mixture of $CaAl_2O_4$ (63%) and $Ca_2Al_4O_7$ (35%), $CA_6 = CaAl_{12}O_{19}$.

The crystalline phase compositions were identified by X-ray diffraction (XRD) patterns using a X'Pert Pro diffractometer (PANalytical, Almelo, Netherlands) (Copper Kα radiation (λ = 1.5418 Å) at 40 kV/40 mA, step size 0.02 over a 2θ range of 5–90°) and analyzed by the software of X'pert Pro High Score (Philips, Almelo, Netherlands). Lattice parameters were calculated using X'pert Pro High Score (Philips, Almelo, Netherlands) and Celref 2.0 software. Microstructure morphology was analyzed by scanning electron microscopy (SEM, Nova 400 Nano-SEM, FEI Company, Hillsboro, OR, USA) equipped with energy dispersive spectroscopy (EDS, Oxford, High Wycombe, UK).

Cr(VI) leachability was evaluated using the leaching test according to TRGS 613 standard procedure, which is suitable for determining water-soluble Cr(VI) compounds in cement and products containing cement. Leaching specimens were prepared by crushing and grinding thoroughly before passing through a 200-mesh sieve ($\leq$74 µm). Fine samples were stirred with deionized water as a leaching solution using a magnetic stirrer at a speed of 300 rpm for 15 min (at room temperature) with a solid–liquid ratio of 1:20. Then, leachates were obtained through a 0.45-µm membrane filter with a glass fibre by vacuum. The Cr(VI) concentration in the leachates was determined using a colorimetric method. The Cr(VI) can react in acid condition with the 1,5-diphenylcarbazide (DPC) to form 1,5-diphenylcarbazone, a red complex (0.02–0.2 mg/L chrome). Then, the absorbance of the leachates after the DPC method was recorded at 540 nm, using a 722 Vis spectrophotometer (Jinghua Instruments, Shanghai, China).

3. Results and Discussion

3.1. Pre-Synthesis of $(Al_{1-x}Cr_x)_2O_3$ Powders

The pre-synthesized powders of the $(Al_{1-x}Cr_x)_2O_3$ solid solution at different temperatures are observed by XRD (Figure 1). It could be found that both corundum and eskolaite existed as separate phases after dry mixing at 25 °C. After treated at 1300 °C, the eskolaite disappeared with a noticeable reduction of the peak intensity of corundum, while a new phase identified as $(Al_{1-x}Cr_x)_2O_3$ solid solution was generated. With the increase in the heat treatment temperature, the peak intensity of corundum reduced gradually until disappearance at 1650 °C, while the peak intensity of $(Al_{1-x}Cr_x)_2O_3$ solid solution increased steadily. After treated at temperatures up to 1650 °C, only the $(Al_{1-x}Cr_x)_2O_3$ solid solution could be detected in the specimens. So, it could be inferred that we have added the $(Al_{1-x}Cr_x)_2O_3$ solid solution with the remnant of corundum and eskolaite (sample S13 and S16), while that of S165 is a complete $(Al_{1-x}Cr_x)_2O_3$ solid solution. In addition, the lattice parameters of the $(Al_{1-x}Cr_x)_2O_3$ solid solution were calculated in comparison with Al_2O_3 (reference code: JCPDS 01-081-2266, $a = b = 4.7569$ Å and $c = 12.9830$ Å) and Cr_2O_3 (reference code: JCPDS 00-038-1479, $a = b = 4.9540$ Å and $c = 13.5842$ Å). Since Al_2O_3 has smaller lattice parameters than Cr_2O_3, the $(Al_{1-x}Cr_x)_2O_3$ solid solution reveals smaller lattice parameters than Cr_2O_3. With the increasing temperature, the $(Al_{1-x}Cr_x)_2O_3$ solid solution showed decreasing lattice parameters as more Al_2O_3 dissolution is expected at higher temperatures. For example, the lattice parameter $a = b = 4.8607$ Å at 1300 °C (for sample S13) decreased to $a = b = 4.8291$ Å at 1600 °C (for sample S16).

3.2. Cr(VI) Leachability

The Cr(VI) concentration in Al_2O_3-CaO-Cr_2O_3 castables treated at various temperatures was evaluated by leaching test according to the TRGS 613 standard procedure (Figure 2). The details of Cr(VI) reduction compared to the reference specimen R is presented in Table 4. With the addition of the pre-synthesized $(Al_{1-x}Cr_x)_2O_3$ solid solution, a noticeable decrease in the Cr(VI) concentration was observed. The specimens with $(Al_{1-x},Cr_x)_2O_3$ pre-synthesized at higher temperature exhibited relatively lower Cr(VI) concentrations at the same heat treatment temperature (exception for specimen S165 at 1300 and 1500 °C). For example, at 700 °C, the total amount of Cr(VI) reduced drastically from 1233.2 mg/kg in specimen R (without $(Al_{1-x}Cr_x)_2O_3$) to 223.7 mg/kg in specimen S13 (a reduction of 81.9%), and reduced further to 24.0 mg/kg in specimen S165 (a decrease of 98.1%). However, at 1300 °C, specimen S165 exhibited an even higher Cr(VI) concentration than the reference specimen. Moreover, the temperature corresponding to the maximum Cr(VI) concentration shifted from 900 °C for R to 1100 °C for the pre-synthesized $(Al_{1-x}Cr_x)_2O_3$. The specimen F15, pre-heated at 1500 °C, exhibited extremely low Cr(VI) concentration at all heat treatment temperatures studied. It was concluded that the chromium would present as Cr(III) together within the solid solution of $(Al_{1-x}Cr_x)_2O_3$ and $Ca(Al,Cr)_{12}O_{19}$ after the pre-heating treatment at 1500 °C [20]. Therefore, it is plausible that the reoxidation of these stable solid solution phases did not occur. Although the mid-temperature (700–1100 °C) was favourable for Cr(VI) formation, the total amount of Cr(VI) in F15 was still only 13.0-17.3 mg/kg (a decrease of ~98.9–99.1% compared to specimen R). These values are below the allowable Cr(VI) limit of the Environmental Protection Agency (EPA), United States (5 mg/L is equivalent to 100 mg/kg) [34].

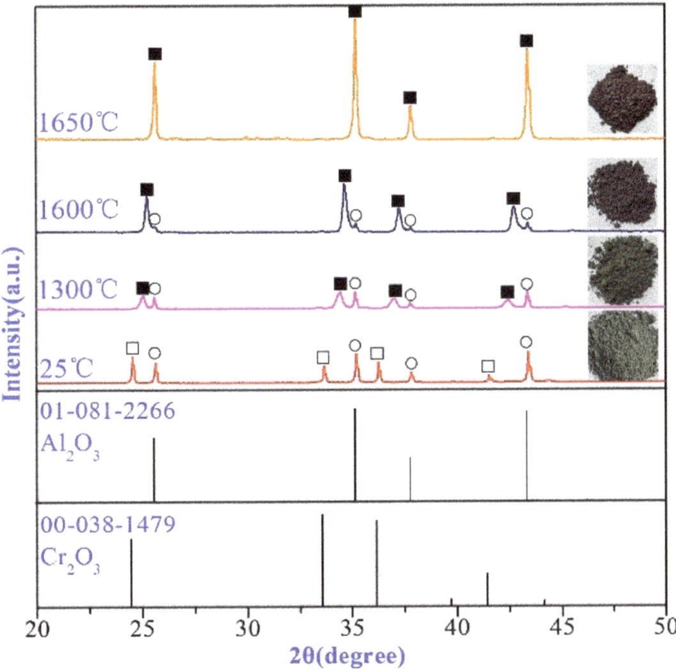

Figure 1. XRD pattern of $(Al_{1-x}Cr_x)_2O_3$ solid solution pre-synthesized at different temperatures. ○-Corundum (Al_2O_3), ■-$(Al_{1-x},Cr_x)_2O_3$ solid solution, □-Eskolaite (Cr_2O_3).

Table 4. Relative Cr(VI) reduction (%) of specimens compared to R at different temperatures.

Specimens	Temperature (°C)							
	110	300	500	700	900	1100	1300	1500
S13	43.7	−19.5	61.7	81.9	16.1	21.2	10.5	12.6
S16	47.4	48.6	93.4	95.0	57.2	24.0	28.0	38.7
S165	38.5	58.0	87.4	98.1	67.6	35.8	−91.4	−202.4
F15	-	-	-	98.9	99.1	99.0	93.5	−30.8

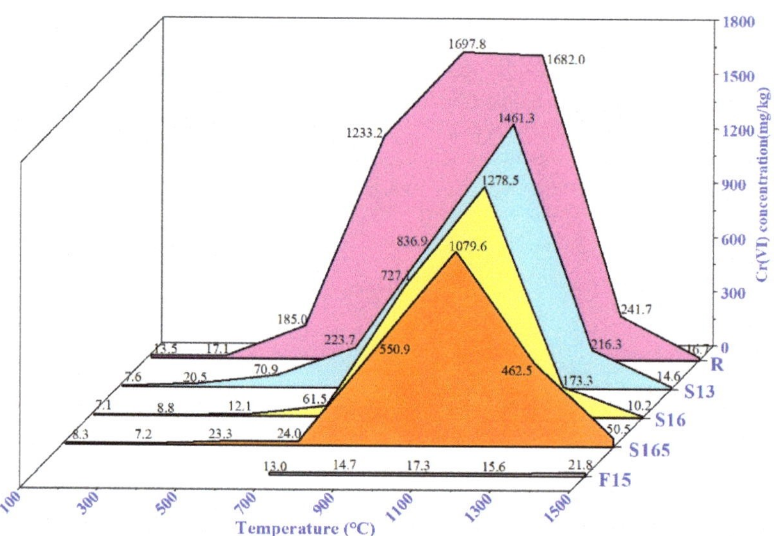

Figure 2. Cr(VI) concentration as a function of temperature in the Al$_2$O$_3$-CaO-Cr$_2$O$_3$ castables.

3.3. Phase Evolution of the Castables

To study the effect of the pre-synthesized (Al$_{1-x}$Cr$_x$)$_2$O$_3$ solid solution on the phase evolution of the castables, phase compositions of the specimens treated at 110–1500 °C were analyzed (Figure 3). In all samples, the main phase corundum and the NaAl$_{11}$O$_{17}$ impurity could be detected at all temperatures, and hydrate phase C$_3$AH$_6$ was generated at 110 °C but then disappeared at 300 °C due to dehydration. For specimen R, the CaCrO$_4$ phase could be detected at 300 °C, whose peak intensity increased with the increase in temperature from 300 °C to 900 °C but then decreased with further increasing temperature until disappearance at 1300 °C. The Ca$_4$Al$_6$CrO$_{16}$ was generated at 900 °C, whose peak intensity reached a maximum at 1100 °C but dropped down with further increasing temperature until disappearance at 1500 °C. Moreover, eskolaite existing in the range of 110 °C to 1100 °C reduced in peak intensity with temperature and disappeared at 1300 °C, while the (Al$_{1-x}$Cr$_x$)$_2$O$_3$ solid solution and CaAl$_{12}$O$_{19}$ increased in peak intensity after generating at 1100 °C and 1300 °C, respectively. However, for specimens S13, S16, and S165, no CaCrO$_4$ phase was detected at 300–1100 °C, indicating chromium that in the (Al$_{1-x}$Cr$_x$)$_2$O$_3$, the CAC in this temperature range would not oxidize the solid solution. At 900–1300 °C, although the Ca$_4$Al$_6$CrO$_{16}$ phase was still formed in these specimens with pre-synthesized (Al$_{1-x}$Cr$_x$)$_2$O$_3$, the peak intensity of Cr(VI) compound was much lower compared with sample R. The peak intensity of the Ca$_4$Al$_6$CrO$_{16}$ phase reached the maximum at 1100 °C in Al$_2$O$_3$-CaO-Cr$_2$O$_3$ castables, and therefore, the highest Cr(VI) concentration for the specimens with pre-synthesized (Al$_{1-x}$Cr$_x$)$_2$O$_3$ were detected at 1100 °C (Figure 3b). In general, the substitution of Cr$_2$O$_3$ with (Al$_{1-x}$Cr$_x$)$_2$O$_3$ in the Al$_2$O$_3$-CaO-Cr$_2$O$_3$ castables can almost completely restrict the formation of CaCrO$_4$ compounds at 300–1100 °C and effectively lower the Cr(VI) compound Ca$_4$Al$_6$CrO$_{16}$ formation at 900–1300 °C, which was following the results of Cr(VI) leachability shown in Figure 2. After being treated at 1500 °C, only the corundum (with NaAl$_{11}$O$_{17}$ impurity), the (Al$_{1-x}$Cr$_x$)$_2$O$_3$ solid solution, and the CA$_6$ phases were found in specimens R, S13, S16, and S165. The enlarged XRD patterns of the castables (Figure 3c) indicated that samples with (Al$_{1-x}$Cr$_x$)$_2$O$_3$ pre-synthesized at higher temperature exhibited relative lower peak intensity of the CA$_6$ phase after being treated at 1300 °C. In addition, specimen F15, which had the same phase compositions as the other four specimens treated at 1500 °C, showed hardly any phase changes with the subsequent heat treatment temperature.

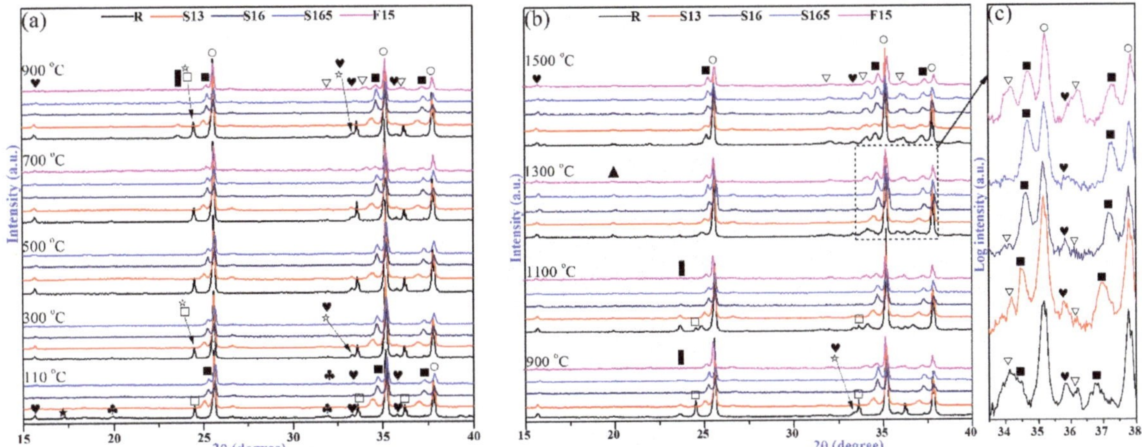

Figure 3. XRD patterns of Al$_2$O$_3$-CaO-Cr$_2$O$_3$ castables, (**a**) 110–900 °C, (**b**) 900–1500 °C, (**c**) 1300 °C. ○—Corundum (Al$_2$O$_3$), ■—(Al$_{1-x}$,Cr$_x$)$_2$O$_3$ solid solution, □—Eskolaite (Cr$_2$O$_3$), ☆—CaCrO$_4$, ▮—Hauyne (Ca$_4$Al$_6$CrO$_{16}$), ▽—CA$_6$ (CaAl$_{12}$O$_{19}$), ♥—NaAl$_{11}$O$_{17}$, ♣—C$_3$AH$_6$ (Ca$_3$Al$_2$(OH)$_{12}$).

3.4. Reaction Mechanism

The above results demonstrated that in the Al$_2$O$_3$-CaO-Cr$_2$O$_3$ castables, chromium and calcium would exist in the state of Cr$_2$O$_3$/(Al$_{1-x}$Cr$_x$)$_2$O$_3$, and CAC/CA$_6$, respectively, which affects the formation and concentration of Cr(VI) compounds CaCrO$_4$ and Ca$_4$Al$_6$CrO$_{16}$ at mid-temperature (700–1100 °C). Fine powders of CAC/CA$_6$ were mixed with Cr$_2$O$_3$/(Al$_{1-x}$Cr$_x$)$_2$O$_3$ (pre-synthetized at 1650 °C) to figure out the mechanisms of the Cr(VI) generation in the castables and the corresponding chemical reactions. Then, the mixed powders were treated at 900 and 1300 °C for 3 h in the air; the XRD patterns and SEM microstructure are summarized in Figures 4 and 5, respectively. The plausible chemical reaction equations discussed below in various samples heated at 900 °C and 1300 °C are listed in Table 5. In addition, the qualitative EDS spot analysis (atomic%) was shown in Table 6, corresponding to the fractured surface in Figure 5. Needless to mention that the uneven surface of the specimen would only reveal the non-stoichiometric composition to identify the different phases associated with different morphology.

Table 5. Chemical reaction equations in cylindrical specimens.

Specimens	900 °C	1300 °C
C-C	(1)	(3) (4) (5)
C-S	(2)	(2)
CH-C	-	(5) (6)
CH-S	-	-

$$4CaAl_2O_4 + 2Cr_2O_3 + 3O_2 \rightarrow 4CaCrO_4 + 4Al_2O_3 \qquad (1)$$
$$16CaAl_2O_4 + y(Al_{1-x},Cr_x)_2O_3 + 3O_2 \rightarrow 4Ca_4Al_6CrO_{16} + (y + 2)Al_2O_3 \qquad (2)$$
$$16CaAl_2O_4 + 2Cr_2O_3 + 3O_2 \rightarrow 4Ca_4Al_6CrO_{16} + 4Al_2O_3 \qquad (3)$$
$$16CaAl_4O_7 + 2Cr_2O_3 + 3O_2 \rightarrow 4Ca_4Al_6CrO_{16} + 20Al_2O_3 \qquad (4)$$
$$(1 - x)Al_2O_3 + xCr_2O_3 \rightarrow (Al_{1-x},Cr_x)_2O_3 \qquad (5)$$
$$16CaAl_{12}O_{19} + 2Cr_2O_3 + 3O_2 \rightarrow 4Ca_4Al_6CrO_{16} + 84Al_2O_3 \qquad (6)$$

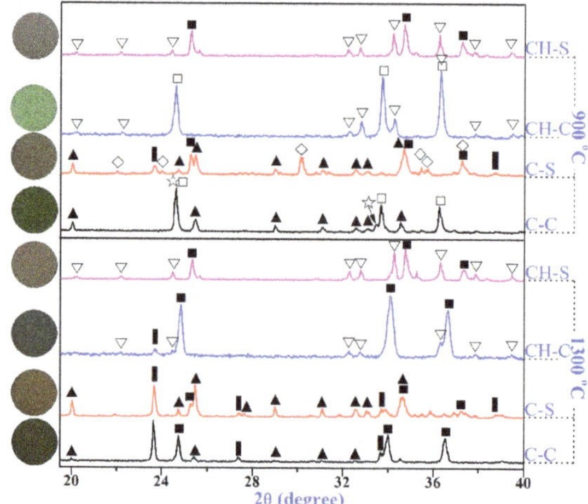

Figure 4. XRD pattern and corresponding images of cylindrical specimens heated at 900 °C and 1300 °C. ■—$(Al_{1-x},Cr_x)_2O_3$ solid solution. □—Eskolaite (Cr_2O_3), ☆—$CaCrO_4$, ▮—Hauyne $(Ca_4Al_6CrO_{16})$, ▲—CA_2 $(CaAl_4O_7)$, ◇—CA $(CaAl_2O_4)$, ▽—CA_6 $(CaAl_{12}O_{19})$.

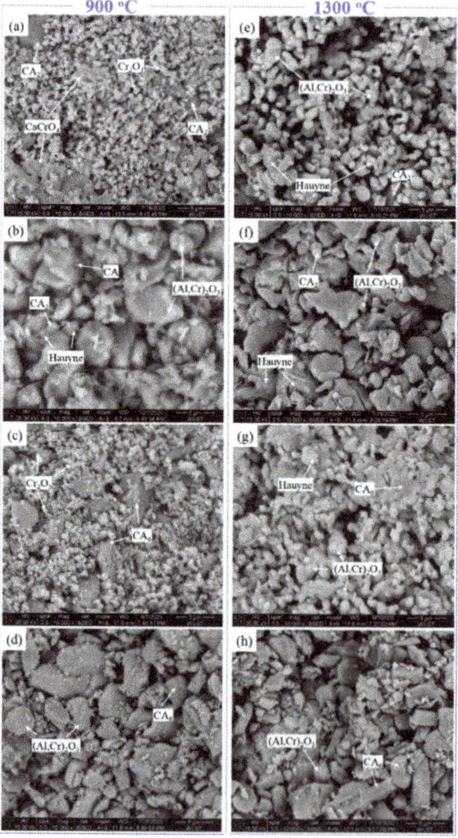

Figure 5. SEM images of cylindrical specimens heated at 900 °C and 1300 °C, (**a,e**) C-C; (**b,f**) C-S; (**c,g**) CH-C; (**d,h**) CH-S.

Table 6. Examples of qualitative EDS spot analysis of the samples (atomic%) for identifying various phases in Figure 5.

Phase	Al	Ca	Cr	O
$CaCrO_4$	-	28.36	41.27	30.37
$Ca_6Al_4CrO_{16}$	34.27	18.58	6.34	40.82
$CaAl_2O_4$	40.39	16.94	-	42.68
$CaAl_4O_7$	49.87	4.13	-	46.00
$(Al,Cr)_2O_3$	48.68	-	15.27	36.05

After being treated at 900 °C, the CA phase disappeared in specimen C-C with the formation of many granular $CaCrO_4$ grains (Figure 5a) via reaction 1. However, the sample C-S was still composed of the initial main phases (CA, CA_2, and $(Al_{1-x}Cr_x)_2O_3$) (Figure 5b) in addition to forming minute amounts of $Ca_4Al_6CrO_{16}$ (reaction 2). As the heat treatment temperature increased to 1300 °C, plenty of chrome-hauyne and $(Al_{1-x}Cr_x)_2O_3$ solid solution (Figure 5e) were generated in specimen C-C (via reactions 3–5), accompanied by the disappearance of CA and significant reduction in CA_2 phase, while specimen C-S possessed relative lower peak intensity of $Ca_4Al_6CrO_{16}$ although it had similar phases as C-C. Combining the observations of Cr(VI) in Figure 2, with the phase evolution results (Figures 3 and 4), it can be deduced that compared with Cr_2O_3, the $(Al_{1-x}Cr_x)_2O_3$ solid solution was more stable that would not form $CaCrO_4$ and could effectively hinder the $Ca_4Al_6CrO_{16}$ formation when contacted with CAC. Therefore, the substitution of Cr_2O_3 with $(Al_{1-x}Cr_x)_2O_3$ can effectively lower the Cr(VI) concentration of the castables after being treated at various temperatures (Figure 2). Furthermore, the castables with $(Al_{1-x}Cr_x)_2O_3$ pre-synthesized at higher temperature exhibited lower Cr(VI) concentration, implying that the stability of the $(Al_{1-x}Cr_x)_2O_3$ improved gradually with the Al_2O_3 proportion in the solid solution. In addition, in comparison with the CA_2 phase, CA was more likely to react with $Cr_2O_3/(Al_{1-x}Cr_x)_2O_3$ resulting in the formation of Cr(VI) compounds.

For specimens CH-C, no new phases occurred after heat treatment at 900 °C, and only a minuscule amount of chrome-hauyne was generated at 1300 °C (Figure 5g) via Eqs. 6, which also produced Al_2O_3 that subsequently interacted with Cr_2O_3 to develop the $(Al_{1-x}Cr_x)_2O_3$ solid solution via Eqs. 5. It is worth mentioning that no changes in the phase compositions were detected in specimen CH-S after heat treatment at both 900 °C and 1300 °C (Figure 4). These observations demonstrated that calcium in CA_6 was much more stable than in CA and CA_2, which only caused slight oxidation of Cr_2O_3 and would not take chemical reaction with $(Al_{1-x}Cr_x)_2O_3$ solid solution. Therefore, specimen F15, in which chromium and calcium existed in $(Al_{1-x}Cr_x)_2O_3$ and CA_6, respectively, showed no changes in phase composition and extremely low Cr(VI) concentration at various heat treatment temperatures. In the Al_2O_3-CaO-Cr_2O_3 castables, CA_6 could be generated from the reaction between CAC and Al_2O_3 powders in the matrix at 1300 °C (Figure 4). However, for specimen S165, since no Al_2O_3 existed in the $(Al_{1-x}Cr_x)_2O_3$ powder pre-synthesized at 1650 °C, the calcium would still exist as CA and CA_2 rather than CA_6 at 1300 °C. As a result, specimen S165 possessed an even higher Cr(VI) concentration than the reference specimen R at 1300 °C, suggesting that CA and CA_2 can more easily react with $(Al_{1-x}Cr_x)_2O_3$ to produce $Ca_4Al_6CrO_{16}$ compared with CA_6.

4. Conclusions

In the present work, $(Al_{1-x}Cr_x)_2O_3$ solid solution was pre-synthesized at a different temperatures for the inhibition of the formation of Cr(VI) in Al_2O_3-CaO-Cr_2O_3 castables was systematically investigated. The summarized conclusions are as follows:

(1) Compared with Cr_2O_3, the stability of the $(Al_{1-x}Cr_x)_2O_3$ solid solution in contact with CAC was much higher. Furthermore, the substitution of Cr_2O_3 with $(Al_{1-x}Cr_x)_2O_3$ in the Al_2O_3-CaO-Cr_2O_3 castables can completely inhibit the mid-temperature (300–1100 °C) formation of Cr(VI) compound $CaCrO_4$ and relatively higher temperature Cr(VI) phase $Ca_4Al_6CrO_{16}$ (hauyne) drastically reduced at 900

to 1300 °C. Therefore, replacing Cr_2O_3 with $(Al_{1-x}Cr_x)_2O_3$ can effectively lower the Cr(VI) concentration of the castables after being treated at various temperatures, and a reduction in Cr(VI) amounts up to 98.1% with $(Al_{1-x}Cr_x)_2O_3$ addition could be achieved. Most importantly, Cr(III) present within the in situ $(Al_{1-x}Cr_x)_2O_3$ and $Ca(Al,Cr)_{12}O_{19}$ solid solution phases showed maximum reoxidation resistance and thus need further investigation.

(2) In comparison with the CA_2 phase, CA was more likely to react with $Cr_2O_3/(Al_{1-x}Cr_x)_2O_3$, resulting in Cr(VI) compound formation. Simultaneously, calcium in CA_6 was much more stable than in CA and CA_2, which only caused slight oxidation of Cr_2O_3 and would not undergo a chemical reaction with $(Al_{1-x}Cr_x)_2O_3$ solid solution. Thus, incorporating some Al_2O_3 powders in the matrix of the Al_2O_3-CaO-Cr_2O_3 castables to form CA_6 at a temperature above 1300 °C was also essential for inhibiting Cr(VI) formation when using $(Al_{1-x}Cr_x)_2O_3$ solid solution as a substitute for Cr_2O_3.

Author Contributions: Conceptualization and validation, T.X., M.N. and Y.X.; methodology and visualization, T.X. and M.N.; formal analysis, data curation, and investigation, T.X.; software, T.X. and Y.X.; writing—original draft preparation, T.X.; writing—review and editing, M.N.; resources and supervision, Y.L.; project administration, N.L.; funding acquisition, M.N. All authors have read and agreed to the published version of the manuscript.

Funding: This work was funded by the National Natural Science Foundation of China (NSFC) (Nos. 51950410587, 51872211, and U1908227).

Institutional Review Board Statement: Not applicable.

Informed Consent Statement: Not applicable.

Data Availability Statement: Data sharing not applicable.

Acknowledgments: The authors would like to acknowledge all the administrative and technical divisions of WUST for providing respective facilities.

Conflicts of Interest: The authors declare no conflict of interest.

References

1. Weinberg, A.V.; Varona, C.; Chaucherie, X.; Goeuriot, D.; Poirier, J. Extending refractory lifetime in rotary kilns for hazardous waste incineration. *Ceram. Int.* **2016**, *42*, 17626–17634. [CrossRef]
2. Kaneko, T.K.; Zhu, J.; Howell, N.; Rozelle, P.; Sridhar, S. The effects of gasification feedstock chemistries on the infiltration of slag into the porous high chromia refractory and their reaction products. *Fuel* **2014**, *115*, 248–263. [CrossRef]
3. Nath, M.; Song, S.; Liu, H.; Li, Y. $CaAl_2Cr_2O_7$: Formation, Synthesis, and Characterization of a New Cr(III) Compound under Air Atmosphere in the Al_2O_3-CaO-Cr_2O_3 System. *Ceram. Int.* **2019**, *45*, 16476–16481. [CrossRef]
4. Nath, M.; Song, S.; Liao, N.; Xu, T.; Liu, H.; Tripathi, H.S.; Li, Y. Co-existence of a Cr^{3+} phase ($CaAl_2Cr_2O_7$) with hydraulic calcium aluminate phases at high-temperature in the Al_2O_3-CaO-Cr_2O_3 system. *Ceram. Int.* **2021**, *47*, 2624–2630. [CrossRef]
5. Perez, I.; Moreno-Ventas, I.; Rios, G. Chemical degradation of magnesia-chromite refractory used in the conversion step of the pyrometallurgical copper-making process: A thermochemical approach. *Ceram. Int.* **2018**, *44*, 18363–18375. [CrossRef]
6. Wu, Y.; Song, S.; Xue, Z.; Nath, M. Formation mechanisms and leachability of hexavalent chromium in Cr_2O_3-containing refractory castables of Electric Arc Furnace cover. *Mater. Metallur. Trans. B* **2019**, *50*, 808–815. [CrossRef]
7. da Luz, A.P.; Braulio, M.A.L.; Pandolfelli, V.C. *Refractory Castable Engineering*; Goller Verlag GmbH: Baden, Germany, 2015.
8. Tomšů, F.; Palčo, S. Refractory monolithics versus shaped refractory products. *Interceram. Int. Ceram. Rev.* **2017**, *66*, 20–23. [CrossRef]
9. Nath, M.; Song, S.; Li, Y.; Xu, Y. Effect of Cr_2O_3 addition on corrosion mechanism of refractory castables for waste melting furnaces and concurrent formation of hexavalent chromium. *Ceram. Int.* **2018**, *44*, 2383–2389. [CrossRef]
10. Heikal, M.; Radwan, M.M.; Al-Duaij, O.K. Physico-mechanical characteristics and durability of calcium aluminate blended cement subject to different aggressive media. *Constr. Build. Mater.* **2015**, *78*, 379–385. [CrossRef]
11. Verbinnen, B.; Billen, P.; Coninckxloo, M.V.; Vandecasteele, C. Heating temperature dependence of Cr(III) oxidation in the presence of alkali and alkaline earth salts and subsequent Cr(VI) leaching behavior. *Environ. Sci. Technol.* **2013**, *47*, 5858–5863. [CrossRef]
12. Garcia-Ramos, E.; Romero-Serrano, A.; Zeifert, B.; Flores-Sanchez, P.; Hallen-Lopez, M.; Palacios, E.G. Immobilization of chromium in slags using MgO and Al_2O_3. *Steel. Res. Int.* **2008**, *79*, 332–339. [CrossRef]

13. Mao, L.; Gao, B.; Deng, N.; Liu, L.; Cui, H. Oxidation behavior of Cr(III) during thermal treatment of chromium hydroxide in the presence of alkali and alkaline earth metal chlorides. *Chemosphere* **2016**, *145*, 1–9. [CrossRef] [PubMed]
14. Saha, R.; Nandi, R.; Saha, B. Sources and toxicity of hexavalent chromium. *J. Coord. Chem.* **2011**, *64*, 1782–1806. [CrossRef]
15. Bishop, M.E.; Glasser, P.; Dong, H.; Arey, B.; Kovarik, L. Reduction and immobilization of hexavalent chromium by microbially reduced Fe-bearing clay minerals. *Geochim. Et Cosmochim. Acta* **2014**, *133*, 186–203. [CrossRef]
16. Lee, Y.; Nassaralla, C.L. Minimization of hexavalent chromium in magnesite-chrome refractory. *Met. Mater. Trans. A* **1997**, *28*, 855–859. [CrossRef]
17. Chen, J.; Jiao, F.; Zhang, L.; Yao, H.; Ninomiya, Y. Elucidating the mechanism of Cr(VI) formation upon the interaction with metal oxides during coal oxy-fuel combustion. *J. Hazar. Mater.* **2013**, *261*, 260–268. [CrossRef] [PubMed]
18. Xu, T.; Nath, M.; Xu, Y.; Li, Y.; Liao, N.; Sang, S. Thermal evolution of Al_2O_3–CaO–Cr_2O_3 castables in different atmospheres. *Ceram. Int.* **2021**, *47*, 11043–11051. [CrossRef]
19. Mao, L.; Deng, N.; Liu, L.; Cui, H.; Zhang, W. Effects of Al_2O_3, Fe_2O_3, SiO_2 on Cr(VI) formation during heating of solid waste containing Cr(III). *Chem. Eng. J.* **2016**, *304*, 216–222. [CrossRef]
20. Nath, M.; Song, S.; Xu, T.; Wu, Y.; Li, Y. Effective inhibition of Cr(VI) in the Al_2O_3-CaO-Cr_2O_3 refractory castables system through silica gel assisted in-situ secondary phase tuning. *J. Clean Prod.* **2019**, *233*, 1038–1046. [CrossRef]
21. Mao, L.; Deng, N.; Liu, L.; Cui, H.; Zhang, W. Inhibition of Cr(III) oxidation during thermal treatment of simulated tannery sludge: The role of phosphate. *Chem. Eng. J.* **2016**, *294*, 1–8. [CrossRef]
22. Lin, S.H.; Chen, C.N.; Juang, R.S. Structure and thermal stability of toxic chromium(VI) species doped onto TiO_2 powders through heat treatment. *J. Environ. Manag.* **2009**, *90*, 1950–1955. [CrossRef] [PubMed]
23. Martinez, A.G.T.; Luz, A.P.; Braulio, M.A.L.; Pandolfelli, V.C. Creep behavior modeling of silica fume containing Al_2O_3-MgO refractory castables. *Ceram. Int.* **2012**, *38*, 327–332. [CrossRef]
24. Bie, C.; Sang, S.; Li, Y.; Xu, Y.; Qiao, J.; Fang, M. Effects of firing and operating atmospheres on microstructure and properties of phosphate bonded Cr_2O_3-Al_2O_3-ZrO_2 bricks. *China's Refract.* **2015**, *49*, 168–174.
25. Wu, Y.; Song, S.; Garbers-Craig, A.M.; Xue, Z. Formation and leachability of hexavalent chromium in the Al_2O_3-CaO-MgO-Cr_2O_3 system. *J. Eur. Ceram. Soc.* **2018**, *38*, 2649–2661. [CrossRef]
26. Liu, H.; Song, S.; Garbers-Craig, A.M.; Xue, Z. High-temperature stability of $Mg(Al,Cr)_2O_4$ spinel co-existing with calcium aluminates in air. *Ceram. Int.* **2019**, *45*, 16166–16172. [CrossRef]
27. Klyucharov, Y.V.; Eger, V.G. On the reaction between magnesiochromite and calcium oxide. *Refractories* **1963**, *4*, 137–144. [CrossRef]
28. Wu, Y.; Song, S.; Xue, Z.; Nath, M. Effect of temperature on hexavalent chromium formation in $(Al,Cr)_2O_3$ with calcium aluminate cement in air. *ISIJ Int.* **2019**, *59*, 1178–1183. [CrossRef]
29. Nath, M.; Song, S.; Garbers-Craig, A.M.; Li, Y. Phase evolution with temperature in chromium-containing refractory castables used for waste melting furnaces and Cr(VI) leachability. *Ceram. Int.* **2018**, *44*, 20391–20398. [CrossRef]
30. Nath, M.; Liao, N.; Xu, T.; Xu, Y.; Wu, Y.; Song, S.; Li, Y. Recent research on Cr-containing refractory castables. *Refract. Worldforum* **2021**, *13*, 48–57.
31. Nath, M.; Ghosh, A.; Tripathi, H.S. Hot corrosion behaviour of Al_2O_3–Cr_2O_3 refractory by molten glass at 1200 °C under static condition. *Corros. Sci.* **2016**, *102*, 153–160. [CrossRef]
32. Nath, M.; Kumar, P.; Song, S.; Li, Y.; Tripathi, H.S. Thermo-mechanical stability of bulk $(Al_{1-x}Cr_x)_2O_3$ solid solution. *Ceram. Int.* **2019**, *45*, 12411–12416. [CrossRef]
33. Nath, M.; Dana, K.; Gupta, S.; Tripathi, H.S. Hot corrosion behavior of slip-cast alumina-chrome refractory crucible against molten glass. *Mater. Corros.* **2014**, *65*, 742–747. [CrossRef]
34. Toxicological Review of Hexavalent Chromium. In *CAS No. 18540-29-9*; EPA: Washington, DC, USA, 1998.

MDPI

St. Alban-Anlage 66

4052 Basel

Switzerland

www.mdpi.com

Materials Editorial Office

E-mail: materials@mdpi.com

www.mdpi.com/journal/materials